Lecture Notes in Computer Science   11564

Commenced Publication in 1973
Founding and Former Series Editors:
Gerhard Goos, Juris Hartmanis, and Jan van Leeuwen

More information about this series at http://www.springer.com/series/7412

Bernhard Burgeth · Andreas Kleefeld ·
Benoît Naegel · Nicolas Passat ·
Benjamin Perret (Eds.)

Mathematical Morphology and Its Applications to Signal and Image Processing

14th International Symposium, ISMM 2019
Saarbrücken, Germany, July 8–10, 2019
Proceedings

 Springer

Editors
Bernhard Burgeth (iD)
Universität des Saarlandes
Saarbrücken, Germany

Benoît Naegel (iD)
Université de Strasbourg
Illkirch Cedex, France

Benjamin Perret (iD)
Université Paris-Est
Noisy le Grand Cedex, France

Andreas Kleefeld (iD)
Forschungszentrum Jülich GmbH
Jülich, Germany

Nicolas Passat (iD)
Université de Reims Champagne-Ardenne
Reims Cedex 2, France

ISSN 0302-9743   ISSN 1611-3349 (electronic)
Lecture Notes in Computer Science
ISBN 978-3-030-20866-0   ISBN 978-3-030-20867-7 (eBook)
https://doi.org/10.1007/978-3-030-20867-7

LNCS Sublibrary: SL6 – Image Processing, Computer Vision, Pattern Recognition, and Graphics

This Springer imprint is published by the registered company Springer Nature Switzerland AG
The registered company address is: Gewerbestrasse 11, 6330 Cham, Switzerland

Preface

This volume contains the articles accepted for presentation at the 14th International Symposium on Mathematical Morphology (ISMM 2019), held in Saarbrücken, Germany, during July 8–10, 2019. The 13 previous editions of this conference were very successful, and the ISMM has established itself as the main scientific event in the field.

We received 54 high-quality papers, each of which was sent to at least three Program Committee members for review. Based on 191 detailed reviews, we accepted 41 papers (including 40 regular papers, plus one related to an invited plenary talk). The authors of these 41 articles are from 11 different countries: Austria, Brazil, France, Germany, Greece, India, Italy, The Netherlands, Sweden, UK, and USA.

The ISMM 2019 papers highlight the current trends and advances in mathematical morphology, be they purely theoretical contributions, algorithmic developments, or novel applications, where real-life imaging and computer vision problems are tackled with morphological tools.

We wish to thank the members of the Program Committee for their efforts in reviewing all submissions on time and giving extensive feedback. We would like to take this opportunity to thank all the other people involved in this conference: firstly, the Steering Committee for giving us the chance to organize ISMM 2019; secondly, the three invited speakers, David Coeurjolly, Fred A. Hamprecht, and Boguslaw Obara, for accepting to share their recognized expertise; and finally, the most important component of any scientific conference, the authors for producing the high-quality and original contributions. We acknowledge the EasyChair conference management system that was invaluable in handling the paper submission, the review process, and putting this volume together. We also acknowledge Springer for making possible the publication of these proceedings in the LNCS series.

April 2019

Bernhard Burgeth
Andreas Kleefeld
Benoît Naegel
Nicolas Passat
Benjamin Perret

.

Organization

ISMM 2019 was organized by Universität des Saarlandes, Germany, with the support of Forschungszentrum Jülich GmbH, ESIEE Paris, Université de Strasbourg, and Université de Reims Champagne-Ardenne.

Organizing Committee

Burgeth, Bernhard (General Chair, Program Co-chair)	Universität des Saarlandes, Germany
Kleefeld, Andreas (Program Co-chair)	Forschungszentrum Jülich GmbH, Germany
Naegel, Benoît (Program Co-chair)	Université de Strasbourg, France
Passat, Nicolas (Program Co-chair)	Université de Reims Champagne-Ardenne, France
Perret, Benjamin (Program Co-chair)	ESIEE Paris, France

Steering Committee

Angulo, Jesús	Mines ParisTech, France
Barrera, Junior	University of Sao Paulo, Brazil
Benediktsson, Jón Atli	University of Iceland, Iceland
Bloch, Isabelle	Télécom ParisTech, France
Borgefors, Gunilla	Uppsala University, Sweden
Chanussot, Jocelyn	Grenoble INP, France
Keshet, Renato	Hewlett Packard Laboratories, Israel
Kimmel, Ron	Technion, Israel
Luengo Hendriks, Cris	Flagship Biosciences Inc., Colorado, USA
Maragos, Petros	National Technical University of Athens, Greece
Najman, Laurent	ESIEE Paris, France
Ronse, Christian	Université de Strasbourg, France
Salembier, Philippe	Universitat Politècnica de Catalunya, Spain
Schonfeld, Dan	University of Illinois at Chicago, USA
Soille, Pierre	European Commission, Joint Research Centre, Italy
Talbot, Hugues	CentraleSupélec, France
Wilkinson, Michael H. F.	University of Groningen, The Netherlands

Program Committee

Angulo, Jesús	Mines ParisTech, France
Aptoula, Erchan	Gebze Technical University, Turkey
Asplund, Teo	Uppsala University, Sweden
Balázs, Péter	University of Szeged, Hungary
Bertrand, Gilles	ESIEE Paris, France
Baldacci, Fabien	Université de Bordeaux, France
Bilodeau, Michel	Mines ParisTech, France
Bloch, Isabelle	Télécom ParisTech, France
Blusseau, Samy	Mines ParisTech, France
Borgefors, Gunilla	Uppsala University, Sweden
Bosilj, Petra	University of Lincoln, UK
Boutry, Nicolas	EPITA, France
Breuß, Michael	Brandenburg University of Technology Cottbus - Senftenberg, Germany
Brunetti, Sara	Università degli Studi di Siena, Italy
Burgeth, Bernhard	Universität des Saarlandes, Germany
Caissard, Thomas	Université Lyon 1, France
Carlinet, Edwin	EPITA, France
Cavallaro, Gabriele	Forschungszentrum Jülich GmbH, Germany
Chanda, Bhabatosh	Indian Statistical Institute, India
Chanussot, Jocelyn	Grenoble INP, France
Chassery, Jean-Marc	CNRS, France
Čomić, Lidija	University of Novi Sad, Serbia
Comon, Pierre	CNRS, France
Couprie, Michel	ESIEE Paris, France
Cousty, Jean	ESIEE Paris, France
Curic, Vladimir	Klarna, Sweden
Dalla Mura, Mauro	Grenoble INP, France
Damiand, Guillaume	CNRS, France
Debayle, Johan	École Nationale Supérieure des Mines de Saint-Étienne, France
Decencière, Étienne	Mines ParisTech, France
Dela Haije, Tom	Datalogisk Institut, Denmark
Didas, Stephan	Umwelt-Campus Birkenfeld, Germany
Dokládal, Petr	Mines ParisTech, France
Evans, Adrian	University of Bath, UK
Ferri, Massimo	Università di Bologna, Italy
Feschet, Fabien	Université Clermont Auvergne, France
Frouin, Frédérique	Inserm, France
Genctav, Asli	Middle East Technical University, Turkey
Géraud, Thierry	EPITA, France
Grélard, Florent	Université Lyon 2, France
Grossiord, Éloïse	IUCT - Oncopôle, France
González-Díaz, Rocío	Universidad de Sevilla, Spain

Gonzalez-Lorenzo, Aldo	Aix-Marseille Université, France
Guimarães, Silvio	PUC Minas, Brazil
Jalba, Andrei	Technische Universiteit Eindhoven, The Netherlands
Jeulin, Dominique	Mines ParisTech, France
Jiménez, María José	Universidad de Sevilla, Spain
Kiran, Bangalore Ravi	NAVYA Group, France
Kleefeld, Andreas	Forschungszentrum Jülich GmbH, Germany
Krähenbühl, Adrien	Université de Strasbourg, France
Kropatsch, Walter	Vienna University of Technology, Austria
Kumaki, Takeshi	Ritsumeikan University, Japan
Kurtz, Camille	Université Paris Descartes, France
Lagorre, Corinne	Université Paris-Est Créteil, France
Lefèvre, Sébastien	Université Bretagne Sud, France
Lindblad, Joakim	Uppsala University, Sweden
Loménie, Nicolas	Université Paris Descartes, France
Lozes, François	Université de Caen Normandie, France
Machairas, Vaïa	Mines ParisTech, France
Malmberg, Filip	Uppsala University, Sweden
Maragos, Petros	National Technical University of Athens, Greece
Marcotegui, Beatriz	Mines ParisTech, France
Mari, Jean-Luc	Aix-Marseille Université, France
Mazo, Loïc	Université de Strasbourg, France
Meinke, Jan	Forschungszentrum Jülich GmbH, Germany
Merveille, Odyssée	Université de Strasbourg, France
Meyer, Fernand	Mines ParisTech, France
Modersitzki, Jan	University of Lübeck, Germany
Molina-Abril, Helena	Universidad de Sevilla, Spain
Monasse, Pascal	École des Ponts ParisTech, France
Morard, Vincent	GE Healthcare, France
Naegel, Benoît	Université de Strasbourg, France
Najman, Laurent	ESIEE Paris, France
Normand, Nicolas	Université de Nantes, France
Noyel, Guillaume	International Prevention Research Institute, France
Obara, Boguslaw	Durham University, UK
Ouzounis, Georgios	Arlo Technologies, Inc., USA
Passat, Nicolas	Université de Reims Champagne-Ardenne, France
Pinoli, Jean-Charles	École Nationale Supérieure des Mines de Saint-Étienne, France
Pluta, Kacper	Technion, Israel
Puybareau, Élodie	EPITA, France
Randrianasoa, Jimmy F.	Télécom Saint-Étienne, France
Real, Pedro	Universidad de Sevilla, Spain
Robic, Julie	Clarins Laboratories, France
Ronse, Christian	Université de Strasbourg, France
Rousseau, François	IMT Atlantique, France
Salembier, Philippe	Universitat Politècnica de Catalunya, Spain

Contents

Multivariate Morphology

Computational Morphology

Machine Learning

Segmentation

Applications in Engineering

Applications in (Bio)medical Imaging

Theory

Tropical Geometry, Mathematical Morphology and Weighted Lattices

Petros Maragos[✉]

National Technical University of Athens, Athens, Greece
maragos@cs.ntua.gr

Abstract. Mathematical Morphology and Tropical Geometry share the same max/min-plus scalar arithmetic and matrix algebra. In this paper we summarize their common ideas and algebraic structure, generalize and extend both of them using weighted lattices and a max-⋆ algebra with an arbitrary binary operation ⋆ that distributes over max, and outline applications to geometry, image analysis, and optimization. Further, we outline the optimal solution of max-⋆ equations using weighted lattice adjunctions, and apply it to optimal regression for fitting max-⋆ tropical curves on arbitrary data.

Keywords: Tropical Geometry · Morphology · Weighted lattices

1 Introduction

Max-plus convolutions have appeared in morphological image analysis, convex analysis and optimization [3,13,17,19,23,25], and nonlinear dynamical systems [2,21]. Max-plus or its dual min-plus arithmetic and corresponding matrix algebra have been used in operations research and scheduling [10]; discrete event control systems, max-plus control and optimization [1,2,6,8]; idempotent mathematics [16]. Max-plus arithmetic is an idempotent semiring; as such it is covered by the theory of dioids [12]. The dual min-plus has been called 'tropical semiring' and has been used in finite automata [15] and tropical geometry [18].

Max and min operations (or more generally supremum and infimum) form the algebra of lattices, which has been used to generalize Euclidean morphology based on Minkowski set operations and their extensions to functions via level sets to morphology on complete lattices [13,14,24]. The scalar arithmetic of morphology on functions has been mainly flat; a few exceptions include max-plus convolutions in [25] and related operations of the image algebra in [22]. Such non-flat morphological operations and their generalizations to a max-⋆ algebra have been systematized using the theory of weighted lattices [20,21]. This connects morphology with max-plus algebra and tropical geometry.

Mathematical Morphology and Tropical Geometry share the same max/min-plus scalar arithmetic and max/min-plus matrix algebra. In this paper we summarize their common ideas and algebraic structure, extend both of them using

© Springer Nature Switzerland AG 2019
B. Burgeth et al. (Eds.): ISMM 2019, LNCS 11564, pp. 3–15, 2019.
https://doi.org/10.1007/978-3-030-20867-7_1

weighted lattices, and outline applications to geometry, image analysis and optimization. We begin with some elementary concepts from morphological operators and tropical geometry. Then, we extend the underlying max-plus algebra to a max-$\star$ algebra where matrix operations and signal convolutions are generalized using a (max,$\star$) arithmetic with an arbitrary binary operation $\star$ that distributes over max. This theory is based on complete weighted lattices and allows for both finite- and infinite-dimensional spaces. Finally, we outline the optimal solution of systems of max-$\star$ equations using weighted lattice adjunctions and projections, and apply it to optimal regression for fitting max-$\star$ tropical curves on data.

2   Background: Morphology on Flat Lattices

We view images, signals and vectors as elements of complete lattices $(\mathcal{L}, \vee, \wedge)$, like the set $\mathrm{Fun}(E, \overline{\mathbb{R}})$ of functions with domain E and values in $\overline{\mathbb{R}} = \mathbb{R} \cup \{-\infty, +\infty\}$, and consider operators on $\mathcal{L}$, i.e., mappings from $\mathcal{L}$ to itself.

Monotone Operators: A lattice operator ψ is called *increasing* (resp. *decreasing*) if it is order preserving (resp. inverting). Examples of increasing operators are the lattice homomorphisms which preserve suprema and infima. If a lattice homomorphism is also a bijection, then it becomes an automorphism. Four fundamental types of increasing operators are: *dilations* δ and *erosions* ε that satisfy respectively $\delta(\bigvee_i X_i) = \bigvee_i \delta(X_i)$ and $\varepsilon(\bigwedge_i X_i) = \bigwedge_i \varepsilon(X_i)$ over arbitrary (possibly infinite) collections; *openings* α that are increasing, idempotent and antiextensive; *closings* β that are increasing, idempotent and extensive. Openings and closings are *lattice projections*. Examples of decreasing operators are the dual homomorphisms, which interchange suprema with infima. A lattice dual automorphism is a bijection that interchanges suprema with infima. A *negation* ν is a dual automorphism that is also involutive.

Residuation and Adjunctions: An increasing operator ψ on a complete lattice $\mathcal{L}$ is called **residuated** [5] if there exists an increasing operator $\psi^\sharp$ such that

$$\psi\psi^\sharp \leq \mathbf{id} \leq \psi^\sharp\psi \tag{1}$$

$\psi^\sharp$ is called the **residual** of ψ and is the closest to being an inverse of ψ. Specifically, the residuation pair $(\psi, \psi^\sharp)$ can solve inverse problems of the type $\psi(X) = Y$ either exactly since $\hat{X} = \psi^\sharp(Y)$ is the greatest solution of $\psi(X) = Y$ if a solution exists, or approximately since $\hat{X}$ is the greatest *subsolution* in the sense that $\hat{X} = \bigvee\{X : \psi(X) \leq Y\}$. On complete lattices an increasing operator ψ is residuated (resp. a residual $\psi^\sharp$) if and only if it is a dilation (resp. erosion). The residuation theory has been used for solving inverse problems in matrix algebra [2,9,10] over the max-plus or other idempotent semirings.

Dilations and erosions on a complete lattice $\mathcal{L}$ come in pairs (δ, ε) of operators; such a pair is called **adjunction** on $\mathcal{L}$ if

$$\delta(X) \leq Y \iff X \leq \varepsilon(Y) \quad \forall X, Y \in \mathcal{L} \tag{2}$$

The double inequality (2) is equivalent to the inequality (1) satisfied by a residuation pair of increasing operators if we identify the residuated map ψ with δ and its residual $\psi^{\sharp}$ with ε. There is a one-to-one correspondence between the two operators of an adjunction; e.g., given a dilation δ, there is a unique erosion $\varepsilon(Y) = \bigvee \{X \in \mathcal{L} : \delta(X) \leq Y\}$ such that (δ, ε) is adjunction. An adjunction (δ, ε) automatically yields two lattice projections, an opening $\alpha = \delta\varepsilon$ and a closing $\beta = \varepsilon\delta$, such that $\delta\varepsilon \leq \mathbf{id} \leq \varepsilon\delta$. There are also other types of lattice projections studied in [9].

3   Weighted Lattices

Lattice-Ordered Monoids and Clodum: A lattice $(\mathcal{K}, \vee, \wedge)$ is often endowed with a third binary operation, called symbolically the 'multiplication' $\star$, under which $(\mathcal{K}, \star)$ is a group, or a monoid, or just a semigroup [4]. Consider now an algebra $(\mathcal{K}, \vee, \wedge, \star, \star')$ with four binary operations, which we call a *lattice-ordered double monoid*, where $(\mathcal{K}, \vee, \wedge)$ is a lattice, $(\mathcal{K}, \star)$ is a monoid whose 'multiplication' $\star$ distributes over $\vee$, and $(\mathcal{K}, \star')$ is a monoid whose 'multiplication' $\star'$ distributes over $\wedge$. These distributivities imply that both $\star$ and $\star'$ are increasing. To the above definitions we add the word *complete* if $\mathcal{K}$ is a complete lattice and the distributivities involved are infinite. We call the resulting algebra a *complete lattice-ordered double monoid*, in short *clodum* [19–21]. Previous works on minimax or max-plus algebra have used alternative names for structures similar to the above definitions which emphasize semigroups and semirings instead of lattices [2,10,12]; see [21] for similarities and differences. We precisely define an algebraic structure $(\mathcal{K}, \vee, \wedge, \star, \star')$ to be a **clodum** if:

(C1) $(\mathcal{K}, \vee, \wedge)$ is a complete distributive lattice. Thus, it contains its least $\bot := \bigwedge \mathcal{K}$ and greatest element $\top := \bigvee \mathcal{K}$. The supremum $\vee$ (resp. infimum $\wedge$) plays the role of a generalized *'addition'* (resp. *'dual addition'*).

(C2) $(\mathcal{K}, \star)$ is a monoid whose operation $\star$ plays the role of a generalized *'multiplication'* with identity ('unit') element e and is a dilation.

(C3) $(\mathcal{K}, \star')$ is a monoid with identity e' whose operation $\star'$ plays the role of a generalized *'dual multiplication'* and is an erosion.

Remarks: (i) As a lattice, $\mathcal{K}$ is not necessarily infinitely distributive, although in this paper all our examples will be such.
(ii) The clodum 'multiplications' $\star$ and $\star'$ do not have to be commutative.
(iii) The least (greatest) element $\bot$ ($\top$) of $\mathcal{K}$ is both the 'zero' element for the 'addition' $\vee$ ($\wedge$) and an absorbing null for the 'multiplication' $\star$ ($\star'$).
(iv) We avoid degenerate cases by assuming that $\vee \neq \star$ and $\wedge \neq \star'$. However, $\star$ may be the same as $\star'$, in which case we have a self-dual 'multiplication'.
(v) A clodum is called *self-conjugate* if it has a lattice negation $a \mapsto a^*$.

If $\star = \star'$ over $G = \mathcal{K} \setminus \{\bot, \top\}$ where $(G, \star)$ is a group and $(G, \vee, \wedge)$ a conditionally complete lattice, then the clodum $\mathcal{K}$ becomes a richer structure which we call a *complete lattice-ordered group*, in short **clog**. In any clog the distributivity between $\vee$ and $\wedge$ is of the infinite type and the 'multiplications' $\star$ and $\star'$ are commutative. Then, for each $a \in G$ there exists its 'multiplicative inverse' a^{-1} such that $a \star a^{-1} = e$. Further, the 'multiplication' $\star$ and its self-dual $\star'$ (which coincide over G) can be extended over the whole $\mathcal{K}$ by involving the null elements. A clog becomes self-conjugate by setting $a^* = a^{-1}$ if $\bot < a < \top$, $\top^* = \bot$, and $\bot^* = \top$. In a clog $\mathcal{K}$ the $\star$ and $\star'$ coincide in all cases with only one exception: the combination of the least and greatest elements; thus, we can denote the clog algebra as $(\mathcal{K}, \vee, \wedge, \star)$.

Example 1. (a) *Max-plus* clog $(\overline{\mathbb{R}}, \vee, \wedge, +, +')$: $\vee / \wedge$ denote the standard sup/inf on $\overline{\mathbb{R}}$, $+$ is the standard addition on $\overline{\mathbb{R}}$ playing the role of a 'multiplication' $\star$ with $+'$ being the 'dual multiplication' $\star'$; the operations $+$ and $+'$ are identical for finite reals, but $a + (-\infty) = -\infty$ and $a +' (+\infty) = +\infty$ for all $a \in \overline{\mathbb{R}}$. The identities are $e = e' = 0$, the nulls are $\bot = -\infty$ and $\top = +\infty$, and the conjugation mapping is $a^* = -a$.
(b) *Max-times* clog $([0, +\infty], \vee, \wedge, \times, \times')$: The identities are $e = e' = 1$, the nulls are $\bot = 0$ and $\top = +\infty$, and the conjugation mapping is $a^* = 1/a$.
(c) *Max-min* clodum $([0, 1], \vee, \wedge, \min, \max)$: As 'multiplications' we have $\star = \min$ and $\star' = \max$. The identities and nulls are $e' = \bot = 0$, $e = \top = 1$. A possible conjugation mapping is $a^* = 1 - a$. Additional clodums that are not clogs are discussed in [19, 21] using more general fuzzy intersections and unions.
(d) *Matrix* max-$\star$ clodum: $(\mathcal{K}^{n \times n}, \vee, \wedge, \boxtimes, \boxtimes')$ where $\mathcal{K}^{n \times n}$ is the set of $n \times n$ matrices with entries from a clodum $\mathcal{K}$, $\vee / \wedge$ denote here elementwise matrix sup/inf, and $\boxtimes$, $\boxtimes'$ denote max-$\star$ and min-$\star'$ matrix 'multiplications':

$$\mathbf{C} = \mathbf{A} \boxtimes \mathbf{B} = [c_{ij}], \ c_{ij} = \bigvee_{k=1}^{n} a_{ik} \star b_{kj} \ , \ \mathbf{D} = \mathbf{A} \boxtimes' \mathbf{B} = [d_{ij}], \ d_{ij} = \bigwedge_{k=1}^{n} a_{ik} \star' b_{kj}$$

This is a clodum with non-commutative 'multiplications'.

Complete Weighted Lattices: Consider a nonempty collection $\mathcal{W}$ of mathematical objects, which will be our space; examples of such objects include the vectors in $\overline{\mathbb{R}}^n$ or signals in $\mathrm{Fun}(E, \overline{\mathbb{R}})$. Also, consider a clodum $(\mathcal{K}, \vee, \wedge, \star, \star')$ of scalars with *commutative* operations $\star, \star'$ and $\mathcal{K} \subseteq \overline{\mathbb{R}}$. We define *two internal operations* among vectors/signals X, Y in $\mathcal{W}$: their supremum $X \vee Y : \mathcal{W}^2 \to \mathcal{W}$ and their infimum $X \wedge Y : \mathcal{W}^2 \to \mathcal{W}$, which we denote using the same supremum symbol ($\vee$) and infimum symbol ($\wedge$) as in the clodum, hoping that the differences will be clear to the reader from the context. Further, we define *two external operations* among any vector/signal X in $\mathcal{W}$ and any scalar c in $\mathcal{K}$: a 'scalar multiplication' $c \star X : (\mathcal{K}, \mathcal{W}) \to \mathcal{W}$ and a 'scalar dual multiplication' $c \star' X : (\mathcal{K}, \mathcal{W}) \to \mathcal{W}$, again by using the same symbols as in the clodum. Now, we define $\mathcal{W}$ to be a **weighted lattice** space over the clodum $\mathcal{K}$ if for all $X, Y, Z \in \mathcal{W}$ and $a, b \in \mathcal{K}$ all the axioms of Table 3 in [21] hold. These axioms bear a striking similarity with those of a linear space. One difference is that the vector/signal addition ($+$) of linear spaces is now replaced by two dual superpositions, the lattice supremum ($\vee$) and infimum ($\wedge$); further, the scalar multiplication ($\times$) of linear spaces is now replaced by two operations $\star$ and $\star'$ which are dual to each other. Only one major property of the linear spaces is missing from the weighted lattices: the existence of 'additive inverses'. We define the weighted lattice $\mathcal{W}$ to be a **complete weighted lattice (CWL)** space if (i) $\mathcal{W}$ is closed under any (possibly infinite) suprema and infima, and (ii) the distributivity laws between the scalar operations $\star$ ($\star'$) and the supremum (infimum) are of the infinite type. Note that, a commutative clodum is a complete weighted lattice over itself.

4 Vector and Signal Operators on Weighted Lattices

We focus on CWLs whose underlying set is a *space* $\mathcal{W} = \mathrm{Fun}(E, \mathcal{K})$ of *functions* $f : E \rightarrow \mathcal{K}$ with values from a clodum $(\mathcal{K}, \vee, \wedge, \star, \star')$ of scalars as in Examples 1(a), (b), (c). Such functions include n-dimensional vectors if $E = \{1, 2, ..., n\}$ or d-dimensional signals of continuous $(E = \mathbb{R}^d)$ or discrete domain $(E = \mathbb{Z}^d)$. Then, we extend *pointwise* the supremum, infimum and scalar multiplications of $\mathcal{K}$ to the functions: e.g., for $F, G \in \mathcal{W}$, $a \in \mathcal{K}$ and $x \in E$, we define $(F \vee G)(x) := F(x) \vee G(x)$ and $(a \star F)(x) := a \star F(x)$. Further, the scalar operations $\star$ and $\star'$, extended pointwise to functions, distribute over any suprema and infima, respectively. If the clodum $\mathcal{K}$ is self-conjugate, then we can extend the conjugation $(\cdot)^*$ to functions F pointwise: $F^*(x) \triangleq (F(x))^*$.

Elementary increasing operators on $\mathcal{W}$ are those that act as **vertical translations** (in short V-translations) of functions. Specifically, pointwise 'multiplications' of functions $F \in \mathcal{W}$ by scalars $a \in \mathcal{K}$ yield the *V-translations* τ_a and *dual V-translations* τ'_a, defined by $\tau_a(F)](x) := a \star F(x)$ and $\tau'_a(F)](x) := a \star' F(x)$. A function operator ψ on $\mathcal{W}$ is called **V-translation invariant** if it commutes with any V-translation τ, i.e., $\psi \tau = \tau \psi$. Similarly for dual translations.

Every function $F(x)$ admits a representation as a supremum of V-translated impulses placed at all points or as infimum of dual V-translated impulses:

$$F(x) = \bigvee_{y \in E} F(y) \star q_y(x) = \bigwedge_{y \in E} F(y) \star' q'_y(x) \qquad (3)$$

where $q_y(x) = e$ at $x = y$ and $\perp$ else, whereas $q'_y(x) = e'$ at $x = y$ and $\top$ else. By using the V-translations and the representation of functions with impulses, we can build more complex increasing operators. We define operators δ as **dilation V-translation invariant (DVI)** and operators ε as **erosion V-translation invariant (EVI)** iff for any $c_i \in \mathcal{K}$, $F_i \in \mathcal{W}$

$$\mathrm{DVI} : \delta(\bigvee_i c_i \star F_i) = \bigvee_i c_i \star \delta(F_i), \quad \mathrm{EVI} : \varepsilon(\bigwedge_i c_i \star' F_i) = \bigwedge_i c_i \star' \varepsilon(F_i) \quad (4)$$

The structure of a DVI or EVI operator's output is simplified if we express it via the operator's impulse responses. Given a dilation δ on $\mathcal{W}$, its **impulse response map** is the map $H : E \rightarrow \mathrm{Fun}(E, \mathcal{K})$ defined at each $y \in E$ as the output function $H(x, y)$ from δ when the input is the impulse $q_y(x)$. Dually, for an erosion operator ε we define its *dual impulse response map* H' via its outputs when excited by dual impulses: for $x, y \in E$

$$H(x, y) \triangleq \delta(q_y)(x), \quad H'(x, y) \triangleq \varepsilon(q'_y)(x) \tag{5}$$

Applying a DVI operator δ or an EVI operator ε to (3) and using the definitions in (5) proves the following unified representation.

Theorem 1. *(a) An operator δ on $\mathcal{W}$ is DVI iff its output can be expressed as*

$$\delta(F)(x) = \bigvee_{y \in E} H(x, y) \star F(y) \tag{6}$$

(b) An operator ε on $\mathcal{W}$ is EVI iff its output can be expressed as

$$\varepsilon(F)(x) = \bigwedge_{y \in E} H'(x, y) \star' F(y) \tag{7}$$

On signal spaces, the operations (6) and (7) are *shift-varying nonlinear convolutions*.

Weighted Lattice of Vectors: Consider now the nonlinear vector space $\mathcal{W} = \mathcal{K}^n$, equipped with the pointwise partial ordering $\mathbf{x} \leq \mathbf{y}$, supremum $\mathbf{x} \vee \mathbf{y} = [x_i \vee y_i]$ and infimum $\mathbf{x} \wedge \mathbf{y} = [x_i \wedge y_i]$ between any vectors $\mathbf{x}, \mathbf{y} \in \mathcal{W}$. Then, $(\mathcal{W}, \vee, \wedge, \star, \star')$ is a complete weighted lattice. Elementary increasing operators are the *vector V-translations* $\tau_a(\mathbf{x}) = a \star \mathbf{x} = [a \star x_i]$ and their duals $\tau'_a(\mathbf{x}) = a \star' \mathbf{x}$, which 'multiply' a scalar a with a vector $\mathbf{x}$ elementwise. A vector transformation on $\mathcal{W}$ is called (dual) V-translation invariant if it commutes with any vector (dual) V-translation. By defining as 'impulses' the impulse vectors $\mathbf{q}_j = [q_j(i)]$ and their duals $\mathbf{q}'_j = [q'_j(i)]$, where the index j signifies the position of the identity, each vector $\mathbf{x} = [x_1, ..., x_n]^T$ has a representation as a max of V-translated impulse vectors or as a min of V-translated dual impulse vectors. More complex examples of increasing operators on this vector space are the max-$\star$ and the min-$\star'$ 'multiplications' of a matrix $\mathbf{A}$ with an input vector $\mathbf{x}$,

$$\delta_{\mathbf{A}}(\mathbf{x}) \triangleq \mathbf{A} \boxplus \mathbf{x}, \quad \varepsilon_{\mathbf{A}}(\mathbf{x}) \triangleq \mathbf{A} \boxplus' \mathbf{x} \tag{8}$$

which are the prototypes of any vector transformation that obeys a sup-$\star$ or an inf-$\star'$ superposition.

Theorem 2. *(a) Any vector transformation on the complete weighted lattice $\mathcal{W} = \mathcal{K}^n$ is DVI iff it can be represented as a matrix-vector max-$\star$ product $\delta_{\mathbf{A}}(\mathbf{x}) = \mathbf{A} \boxplus \mathbf{x}$ where $\mathbf{A} = [a_{ij}]$ with $a_{ij} = \{\delta(\mathbf{q}_j)\}_i$.*
(b) Any vector transformation on $\mathcal{K}^n$ is EVI iff it can be represented as a matrix-vector min-$\star'$ product $\varepsilon_{\mathbf{A}}(\mathbf{x}) = \mathbf{A} \boxplus' \mathbf{x}$ where $\mathbf{A} = [a_{ij}]$ with $a_{ij} = \{\varepsilon(\mathbf{q}'_j)\}_i$.

Given a vector dilation $\delta(\mathbf{x}) = \mathbf{A} \boxplus \mathbf{x}$, there corresponds a unique erosion ε so that (δ, ε) is a *vector adjunction* on $\mathcal{W}$, i.e. $\delta(\mathbf{x}) \leq \mathbf{y} \iff \mathbf{x} \leq \varepsilon(\mathbf{y})$. We can find the adjoint vector erosion by decomposing both vector operators based on scalar operators (η, ζ) that form a *scalar adjunction* on $\mathcal{K}$:

$$\eta(a, v) \leq w \Longleftrightarrow v \leq \zeta(a, w) \tag{9}$$

If we use as scalar 'multiplication' a commutative binary operation $\eta(a, v) = a \star v$ that is a dilation on $\mathcal{K}$, its scalar adjoint erosion becomes

$$\zeta(a, w) = \sup\{v \in \mathcal{K} : a \star v \leq w\} \tag{10}$$

which is a (possibly non-commutative) binary operation on $\mathcal{K}$. Then, the original vector dilation $\delta(\mathbf{x}) = \mathbf{A} \boxtimes \mathbf{x}$ is decomposed as

$$\{\delta(\mathbf{x})\}_i = \bigvee_j \eta(a_{ij}, x_j) = \bigvee_j a_{ij} \star x_j, \quad i = 1, ..., n \tag{11}$$

whereas its adjoint vector erosion is decomposed as

$$\{\varepsilon(\mathbf{y})\}_j = \bigwedge_i \zeta(a_{ij}, y_i), \quad j = 1, ..., n \tag{12}$$

The latter can be written as a min-ζ matrix-vector multiplication

$$\varepsilon(\mathbf{y}) = \mathbf{A}^T \square'_\zeta \mathbf{y} \tag{13}$$

Further, if $\mathcal{K}$ is a *clog*, then $\zeta(a, w) = a^* \star' w$ and hence

$$\varepsilon(\mathbf{y}) = \mathbf{A}^* \boxtimes' \mathbf{y}, \tag{14}$$

where $\mathbf{A}^* = [a_{ji}{}^*]$ is the *adjoint* (i.e. conjugate transpose) of $\mathbf{A} = [a_{ij}]$.

Weighted Lattice of Signals: Consider the set $\mathcal{W} = \text{Fun}(E, \mathcal{K})$ of all signals $f : E \to \mathcal{K}$ with values from $\mathcal{K}$. The signal translations are the operators $\tau_{k,v}(f)(t) = f(t - k) \star v$ and their duals. A signal operator on $\mathcal{W}$ is called *(dual) translation invariant* iff it commutes with any such (dual) translation. This translation-invariance contains both a vertical translation and a horizontal translation (shift). Consider now operators Δ on $\mathcal{W}$ that are dilations and translation-invariant. Then Δ is both DVI in the sense of (4) and shift-invariant. We call such operators **dilation translation-invariant (DTI)** systems. Applying Δ to an input signal f decomposed as supremum of translated impulses yields its output as the sup-$\star$ convolution $\circledast$ of the input with the system's impulse response $h = \Delta(q)$, where $q(x) = e$ if $x = 0$ and $\perp$ else:

$$\Delta(f)(x) = (f \circledast h)(x) = \bigvee_{y \in E} f(y) \star h(x - y) \tag{15}$$

Conversely, every sup-$\star$ convolution is a DTI system. As done for the vector operators, we can also build signal operator pairs $(\Delta, \mathcal{E})$ that form adjunctions.

Given Δ we can find its adjoint $\mathcal{E}$ from scalar adjunctions (η, ζ). Thus, by (9) and (10), if $\eta(h, f) = h \star f$, the adjoint signal erosion becomes

$$\mathcal{E}(g)(y) = \bigwedge_{x \in E} \zeta[h(x - y), g(x)] \tag{16}$$

Further, if $\mathcal{K}$ is a clog, then

$$\mathcal{E}(g)(y) = \bigwedge_{x \in E} g(x) \star' h^*(x - y) \tag{17}$$

5   Tropical Geometry and CWL Generalizations

Tropical geometry [18] is an extension of analytic Euclidean geometry where the traditional arithmetic of the real field $(\mathbb{R}, +, \times)$ involved in the analytic expressions of geometric objects is replaced by the arithmetic of the min-plus tropical semiring $(\mathbb{R}_{\min}, \wedge, +)$; some authors use its max-plus dual semiring $(\mathbb{R}_{\max}, \vee, +)$. We use both semirings as part of the weighted lattice - clog $(\overline{\mathbb{R}}, \vee, \wedge, +)$. For example, the analytic expressions for the Euclidean line $y_{\text{e-line}} = ax + b$ and parabola $y_{\text{e-parab}} = ax^2 + bx + c$ become the tropical curves shown in Fig. 1 and described by the max-plus polynomials

$$y_{\text{t-line}} = \max(a + x, b), \quad y_{\text{t-parab}} = \max(a + 2x, b + x, c) \tag{18}$$

The above examples generalize to multiple dimensions or higher degrees and show us the way to tropicalize any classic n-variable polynomial (linear combination of power monomials) $\sum_i a_i z_1^{u_1^i} \cdots z_n^{u_n^i}$ defined over $\mathbb{R}^n$ where $\mathbf{u}^i = (u_1^i, ..., u_n^i)$ is some nonnegative integer vector: replace the sum with max and log the individual terms so that the multiplicative coefficients become additive and the powers become integer multiples of the indefinite log variables. Thus, a general max-polynomial $p : \mathbb{R}^n \to \mathbb{R}$ has the expression:

$$p(\mathbf{x}) = \bigvee_{i=1}^{k} b_i + \mathbf{c}_i^T \mathbf{x}, \quad \mathbf{x} = (x_1, ... x_n) \tag{19}$$

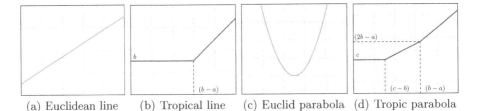

(a) Euclidean line   (b) Tropical line   (c) Euclid parabola   (d) Tropic parabola

Fig. 1. Euclidean curves and their tropical versions.

where k is the rank of p. Further, we can assume as in [6] real coefficient vectors $\mathbf{c}_i \in \mathbb{R}^n$. An interesting geometric object related to a max-polynomial p is its *Newton polytope* (conv$(\cdot)$ denotes convex hull)

$$\text{New}(p) \triangleq \text{conv}\{\mathbf{c}_i : i = 1, ..., \text{rank}(p)\} \tag{20}$$

This satisfies several important properties [7] (see Fig. 2 for an example):

$$\text{New}(p_1 \vee p_2) = \text{conv}(\text{New}(p_1) \cup \text{New}(p_2)) \tag{21}$$
$$\text{New}(p_1 + p_2) = \text{New}(p_1) \oplus \text{New}(p_2) \tag{22}$$

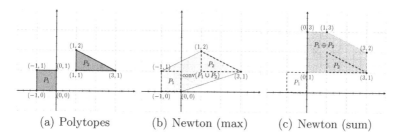

(a) Polytopes   (b) Newton (max)   (c) Newton (sum)

Fig. 2. (a) Newton polytopes of two max-polynomials $p_1(x, y) = \max(x+y, 3x+y, x+2y)$ and $p_2(x, y) = \max(0, -x, y, y-x)$, (b) their max $p_1 \vee p_2$, and (c) their sum $p_1 + p_2$.

In pattern analysis problems on Euclidean spaces $\mathbb{R}^{n+1}$ we often use halfspaces $\mathcal{H}(\mathbf{a}, b) := \{\mathbf{x} \in \mathbb{R}^n : \mathbf{a}^T \mathbf{x} \leq b\}$, polyhedra (finite intersections of halfspaces) and polytopes (compact polyhedra formed as the convex hull of a finite set of points). Replacing linear inner products $\mathbf{a}^T\mathbf{x}$ with max-plus versions yields *tropical halfspaces* [11] with parameters $\mathbf{a} = [a_i], \mathbf{b} = [b_i] \in \mathbb{R}^{n+1}$:

$$\mathcal{T}(\mathbf{a}, \mathbf{b}) \triangleq \{\mathbf{x} \in \mathbb{R}^n : \max(a_{n+1}, \bigvee_{i=1}^{n} a_i + x_i) \leq \max(b_{n+1}, \bigvee_{i=1}^{n} b_i + x_i)\} \tag{23}$$

where $\min(a_i, b_i) = -\infty \; \forall i$. Examples of regions formed by such tropical halfspaces are shown in Fig. 3. Obviously, their separating boundaries are tropical lines. Such regions were used in [7] as morphological perceptrons.

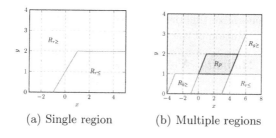

(a) Single region   (b) Multiple regions

Fig. 3. Regions formed by tropical halfspaces in $\mathbb{R}^2$.

In the same way that weighted lattices generalize max-plus morphology and extend it to other types of clodum arithmetic, we can extend the above objects of max-plus tropical geometry to other max-$\star$ geometric objects. For example, over a clodum $(\mathcal{K}, \vee, \wedge, \star, \star')$, we can generalize tropical lines as $y = \max(a \star x, b)$ and tropical planes as $z = \max(a \star x, b \star y, c)$. Figure 4 shows some generalized tropical lines where the $\star$ operation is sum, product and min. Further, we can generalize max-plus halfspaces (23) to max-$\star$ tropical halfspaces:

$$\mathcal{T}(\mathbf{a}, \mathbf{b}) \triangleq \left\{ \mathbf{x} \in \mathcal{K}^n : \mathbf{a}^T \boxtimes \begin{pmatrix} \mathbf{x} \\ e \end{pmatrix} \leq \mathbf{b}^T \boxtimes \begin{pmatrix} \mathbf{x} \\ e \end{pmatrix} \right\} \tag{24}$$

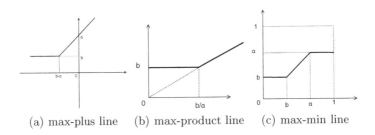

(a) max-plus line   (b) max-product line   (c) max-min line

Fig. 4. Max-$\star$ tropical lines $y = \max(a \star x, b)$: (a) $\star = +$, (b) $\star = \times$, (c) $\star = \wedge$.

6   Applications to Optimization and Machine Learning

6.1   Solving Max-$\star$ Equations

Consider a scalar clodum $(\mathcal{K}, \vee, \wedge, \star, \star')$, a matrix $\mathbf{A} \in \mathcal{K}^{m \times n}$ and a vector $\mathbf{b} \in \mathcal{K}^m$. The set of solutions of the max-$\star$ equation

$$\mathbf{A} \boxtimes \mathbf{x} = \mathbf{b} \tag{25}$$

over $\mathcal{K}$ is either empty or forms a sup-semillatice. A related problem in applications of max-plus algebra to scheduling is when a vector $\mathbf{x}$ represents start times, a vector $\mathbf{b}$ represents finish times and the matrix $\mathbf{A}$ represents processing delays. Then, if $\mathbf{A} \boxtimes \mathbf{x} = \mathbf{b}$ does not have an exact solution, it is possible to find the optimum $\mathbf{x}$ such that we minimize a norm of the earliness subject to zero lateness. We generalize this problem from max-plus to max-$\star$ algebra. The optimum will be the solution of the following constrained minimization problem:

$$\text{Minimize } \|\mathbf{A} \boxtimes \mathbf{x} - \mathbf{b}\| \text{ s.t. } \mathbf{A} \boxtimes \mathbf{x} \leq \mathbf{b} \tag{26}$$

where the norm $\|\cdot\|$ is either the ℓ_∞ or the ℓ_1 norm. While the two above problems have been solved in [10] for the max-plus case, we provide next a more general result using adjunctions for the general case when $\mathcal{K}$ is just a clodum or a general clog.

Theorem 3 (*[21]*).   *Consider a vector dilation* $\delta(\mathbf{x}) = \mathbf{A} \boxtimes \mathbf{x}$ *over a clodum* $\mathcal{K}$ *and let* ε *be its adjoint vector erosion. (a) If Eq. (25) has a solution, then*

$$\hat{\mathbf{x}} = \varepsilon(\mathbf{b}) = \mathbf{A}^T \square_\zeta' \mathbf{b} = [\bigwedge_i \zeta(a_{ij}, b_i)] \tag{27}$$

is its greatest solution, where ζ *is the scalar adjoint erosion of* $\star$ *as in (10).*
(b) If $\mathcal{K}$ *is a clog, the solution (27) becomes*

$$\hat{\mathbf{x}} = \mathbf{A}^* \boxtimes' \mathbf{b} \tag{28}$$

(c) The solution to problem (26) is generally (27), or (28) in the case of a clog.

A main idea for solving (26) is to consider vectors $\mathbf{x}$ that are *subsolutions* in the sense that $\delta(\mathbf{x}) = A \boxtimes \mathbf{x} \leq \mathbf{b}$ and find the greatest such subsolution $\hat{\mathbf{x}} = \varepsilon(\mathbf{b})$, which yields either the greatest exact solution of (25) or an optimum approximate solution in the sense of (26). This creates a lattice projection onto the max-$\star$ span of the columns of $\mathbf{A}$ via the opening $\delta(\varepsilon(\mathbf{b})) \leq \mathbf{b}$ that best approximates $\mathbf{b}$ from below.

6.2   Regression for Optimal Fitting Tropical Lines to Data

We examine a classic problem in machine learning, fitting a line to data by minimizing an error norm, in the light of tropical geometry. Given data $(x_i, y_i) \in \mathbb{R}^2$, $i = 1, ..., n$, if we wish to fit a Euclidean line $y = ax + b$ by minimizing the ℓ_2 error norm, the optimal (least-squares) solution for the parameters a, b is

$$\hat{a}_{\mathrm{LS}} = \frac{n \sum_i x_i y_i - (\sum_i x_i)(\sum_i y_i)}{n \sum_i (x_i)^2 - (\sum_i x_i)^2}, \quad \hat{b}_{\mathrm{LS}} = \frac{1}{n} \sum_i (y_i - \hat{a}_{\mathrm{LS}} x_i) \tag{29}$$

Suppose now we wish to fit a general tropical line $y = \max(a \star x, b)$ by minimizing the ℓ_1 error norm. The equations to solve become:

$$\underbrace{\begin{bmatrix} x_1 & e \\ \vdots & \vdots \\ x_n & e \end{bmatrix}}_{\mathbf{X}} \boxtimes \begin{bmatrix} a \\ b \end{bmatrix} = \underbrace{\begin{bmatrix} y_1 \\ \vdots \\ y_n \end{bmatrix}}_{\mathbf{y}} \tag{30}$$

By Theorem 3, the optimal (min ℓ_1 error) solution for any clodum arithmetic is

$$\begin{bmatrix} \hat{a} \\ \hat{b} \end{bmatrix} = \mathbf{X}^T \square_\zeta' \mathbf{y} = \begin{bmatrix} \bigwedge_i \zeta(x_i, y_i) \\ \bigwedge_i \zeta(e, y_i) \end{bmatrix} \tag{31}$$

where ζ is the scalar adjoint erosion (10) of $\star$. If $\mathcal{K}$ is a clog like in the max-plus and max-product case, then $\zeta(a, w) = a^* \star' w$. Next we write in detail the solution for the tropical line for the three special cases where the scalar arithmetic is based on the max-plus clog, max-product clog and the max-min clodum (the shapes of the corresponding lines are shown in Fig. 4):

$$(\hat{a}, \hat{b}) = \begin{cases} \bigwedge_i y_i - x_i, \ \bigwedge_i y_i), & \text{max-plus} \ (\star = +) \\ \bigwedge_i y_i/x_i, \ \bigwedge_i y_i), & \text{max-times} \ (\star = \times) \\ \bigwedge_i \max([y_i \geq x_i], y_i), \ \bigwedge_i y_i), & \text{max-min} \ (\star = \wedge) \end{cases} \quad (32)$$

where $[\cdot]$ denotes Iverson's bracket in the max-min case. Thus, the above approach allows to optimally fit tropical lines to arbitrary data. Figure 5 shows an example. It can also be generalized to higher-degree curves and to high-dimensional data.

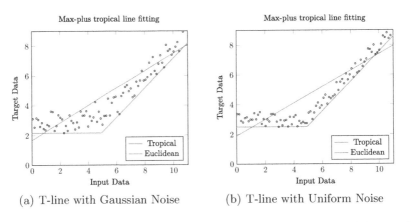

(a) T-line with Gaussian Noise   (b) T-line with Uniform Noise

Fig. 5. (a) Red curve: Optimal fitting via (32) of a max-plus tropical line to data $y = \max(x - 2, 3)$ corrupted by additive i.i.d. Gaussian noise $N(0, 0.25)$. Blue line: Euclidean line fitting via least squares. (b) Same experiment as in (a) but with uniform noise $U(-0.5, 0.5)$. (Color figure online)

Acknowledgements. I wish to thank V. Charisopoulos and E. Theodosis for insightful discussions and help with the figures. This work was co-financed by the European Regional Development Fund of the EU and Greek national funds through the Operational Program Competitiveness, Entrepreneurship and Innovation, under the call 'Research – Create – Innovate' (T1EDK-01248, "i-Walk").

References

1. Akian, M., Gaubert, S., Guterman, A.: Tropical polyhedra are equivalent to mean payoff games. Int. J. Algebra Comput. **22**(1), 125001 (2012)
2. Baccelli, F., Cohen, G., Olsder, G.J., Quadrat, J.P.: Synchronization and Linearity: An Algebra for Discrete Event Systems. Wiley, Chichester (1992). web edn. 2001
3. Bellman, R., Karush, W.: On a new functional transform in analysis: the maximum transform. Bull. Am. Math. Soc. **67**, 501–503 (1961)
4. Birkhoff, G.: Lattice Theory. American Mathematical Society, Providence (1967)

5. Blyth, T.S., Janowitz, M.F.: Residuation Theory. Pergamon Press, Oxford (1972)
6. Butkovič, P.: Max-linear Systems: Theory and Algorithms. Springer, London (2010). https://doi.org/10.1007/978-1-84996-299-5
7. Charisopoulos, V., Maragos, P.: Morphological perceptrons: geometry and training algorithms. In: Angulo, J., Velasco-Forero, S., Meyer, F. (eds.) ISMM 2017. LNCS, vol. 10225, pp. 3–15. Springer, Cham (2017). https://doi.org/10.1007/978-3-319-57240-6_1
8. Cohen, G., Dubois, D., Quadrat, J., Viot, M.: A linear system theoretic view of discrete event processes and its use for performance evaluation in manufacturing. IEEE Trans. Autom. Control **30**, 210–220 (1985)
9. Cohen, G., Gaubert, S., Quadrat, J.: Duality and separation theorems in idempotent semimodules. Linear Alegbra Appl. **379**, 395–422 (2004)
10. Cuninghame-Green, R.: Minimax Algebra. Springer, Heidelberg (1979). https://doi.org/10.1007/978-3-642-48708-8
11. Gaubert, S., Katz, R.D.: Minimal half-spaces and external representation of tropical polyhedra. J. Algebr. Comb. **33**, 325–348 (2011)
12. Gondran, M., Minoux, M.: Graphs, Dioids and Semirings: New Models and Algorithms. Springer, Boston (2008). https://doi.org/10.1007/978-0-387-75450-5
13. Heijmans, H.: Morphological Image Operators. Academic Press, Boston (1994)
14. Heijmans, H., Ronse, C.: The algebraic basis of mathematical morphology I. Dilations and erosions. Comput. Vis. Graph. Image Process. **50**, 245–295 (1990)
15. Perin, J.-E.: Tropical semirings. In: Gunawardena, J. (ed.) Idempotency, pp. 50–69. Cambridge University Press, Cambridge (1998). https://doi.org/10.1017/CBO9780511662508.004
16. Litvinov, G.L., Maslov, V.P., Shpiz, G.B.: Idempotent functional analysis: an algebraic approach. Math. Notes **69**(5), 696–729 (2001)
17. Lucet, Y.: What shape is your conjugate? A survey of computational convex analysis and its applications. SIAM Rev. **52**(3), 505–542 (2010)
18. Maclagan, D., Sturmfels, B.: Introduction to Tropical Geometry. American Mathematical Society, Providence (2015)
19. Maragos, P.: Lattice image processing: a unification of morphological and fuzzy algebraic systems. J. Math. Imaging Vis. **22**, 333–353 (2005)
20. Maragos, P.: Representations for morphological image operators and analogies with linear operators. In: Hawkes, P. (ed.) Advances in Imaging and Electron Physics, vol. 177, pp. 45–187. Academic Press/Elsevier Inc. (2013)
21. Maragos, P.: Dynamical systems on weighted lattices: general theory. Math. Control Signals Syst. **29**(21) (2017). https://doi.org/10.1007/s00498-017-0207-8
22. Ritter, G.X., Wilson, J.N.: Handbook of Computer Vision Algorithms in Image Algebra, 2nd edn. CRC Press, Boca Raton (2001)
23. Serra, J.: Image Analysis and Mathematical Morphology. Academic Press, London (1982)
24. Serra, J. (ed.): Image Analysis and Mathematical Morphology. Theoretical Advances, vol. 2. Academic Press, London (1988)
25. Sternberg, S.R.: Grayscale morphology. Comput. Vis. Graph. Image Process. **35**, 333–355 (1986)

From Structuring Elements
to Structuring Neighborhood Systems

Alexandre Goy[1(✉)], Marc Aiguier[1], and Isabelle Bloch[2]

[1] MICS, CentraleSupélec, Université Paris-Saclay, Gif-sur-Yvette, France
{alexandre.goy,marc.aiguier}@centralesupelec.fr
[2] LTCI, Télécom ParisTech, Université Paris-Saclay, Paris, France
isabelle.bloch@telecom-paristech.fr

Abstract. In the context of mathematical morphology based on structuring elements to define erosion and dilation, this paper generalizes the notion of a structuring element to a new setting called structuring neighborhood systems. While a structuring element is often defined as a subset of the space, a structuring neighborhood is a subset of the subsets of the space. This yields an extended definition of erosion; dilation can be obtained as well by a duality principle. With respect to the classical framework, this extension is sound in many ways. It is also strictly more expressive, for any structuring element can be represented as a structuring neighborhood but the converse is not true. A direct application of this framework is to generalize modal morpho-logic to a topological setting.

Keywords: Structuring element · Neighborhood · Filter · Morpho-logic · Topology

1 Introduction

The motivation of this paper is to apply mathematical morphology in logic (*morpho-logic*), in particular for spatial reasoning. Morpho-logic was initially introduced for propositional logic [5], and proved useful to model knowledge, beliefs or preferences, and to model classical reasoning methods such as revision, fusion or abduction [6,7]. Extensions to modal logic [8] and first-order logic [10] were then proposed. The framework of satisfaction systems and stratified institutions was then proposed as a more general setting encompassing many logics [1–3]. In modal logic, morphological operators can be seen as modalities, with generally strong properties. However, the modalities of *topological* modal logic cannot be obtained when considering the usual definition of structuring element as a set or a binary relation; furthermore, the properties of these modalities are not those of erosion and dilation but closer to those of opening and closing, because of the double quantification ∀/∃ in their definition. The starting point of this paper is to try to see these modalities as weaker forms of erosion and dilation, derived from a lax notion of structuring element.

© Springer Nature Switzerland AG 2019
B. Burgeth et al. (Eds.): ISMM 2019, LNCS 11564, pp. 16–28, 2019.
https://doi.org/10.1007/978-3-030-20867-7_2

Let X denote a set and $\mathcal{P}(X)$ its powerset. The context we consider is the one of deterministic mathematical morphology, with dilations and erosions defined on $\mathcal{P}(X)$ from structuring elements. The main idea of this paper is to consider a structuring element not as an element of $\mathcal{P}(X)$, or a function from X into $\mathcal{P}(X)$, but as a function taking values in $\mathcal{P}(\mathcal{P}(X))$. We call it structuring neighborhood system, in accordance with the topological flavor of the considered setting. The aim of this paper is then to study the structure of the set of such structuring neighborhoods, and to establish their properties. In particular we show that many properties are satisfied, but some, mostly related to adjunctions, may be lost.

The paper is organized as follows. In Sect. 2, we recall some results on dilations and erosions defined using classical structuring elements. In Sect. 3 we introduce the proposed definition of structuring neighborhood systems. Their set can be endowed with a lattice structure and with a monoid structure, as in the classical setting [11, 15]. Properties are then derived. We also introduce a weaker form of erosion, which is proved to occur exactly when the structuring neighborhood system is a filter (in the sense of logic). Finally in Sect. 4 we show that the proposed framework leads to good results on topological modal logic, thus achieving our initial aim.

2   Structure on Structuring Elements

2.1   Mathematical Morphology Based on Structuring Elements

In the context of deterministic mathematical morphology, a class of basic operators is often defined based on the notion of structuring element (see e.g. [4, 12, 16]). A general definition of a structuring element is the following.

Definition 1 (Structuring element). *A structuring element is a function* $b : X \to \mathcal{P}(X)$.

Example 1 (Translations in $\mathbb{R}^n$*).* Let B be a subset of $\mathbb{R}^n$ and $b : \mathbb{R}^n \to \mathcal{P}(\mathbb{R}^n)$ be defined by $b(x) = B_x$ (the translated of B at position x). This is a well-known structuring element in translation invariant morphological image processing.

The two following examples come from the morphological study of logic, also called morpho-logic. See [8] for the modal case and [3] for a more general account of morpho-logic. A wider class of logical systems will be presented in Sect. 4.

Example 2 (Modal morpho-logic). Let $\langle W, R \rangle$ be a Kripke frame, i.e. W is a set of worlds, and $R \subseteq W \times W$ is a binary relation often called accessibility relation. One can define a structuring element $b : W \to \mathcal{P}(W)$ by $b(w) = \{w' \in W \mid (w, w') \in R\}$, i.e. the set of worlds accessible from w.

Example 3 (First-order morpho-logic). Let $\mathsf{Var} = \{x, y, z, ...\}$ be a set of variables and M be a set. A function $f : \mathsf{Var} \to M$ is called a *variable affectation*. For any $x \in \mathsf{Var}$, one can define the x-structuring element $b_x : M^{\mathsf{Var}} \to \mathcal{P}(M^{\mathsf{Var}})$ by $b(f) = \{g \in M^{\mathsf{Var}} \mid \forall y \neq x, g(y) = f(y)\}$.

The two following simple examples will be useful in what follows.

Example 4 (Singleton). The singleton structuring element $\mathsf{sgt} : X \to \mathcal{P}(X)$ is defined by $\mathsf{sgt}(x) = \{x\}$.

Example 5 (Symmetric). Let $b : X \to \mathcal{P}(X)$ be a structuring element. Its symmetric structuring element $b^{\dagger}$ is defined by $b^{\dagger}(x) = \{y \in X \mid x \in b(y)\}$.

Let $b : X \to \mathcal{P}(X)$ be a structuring element. Erosion $\varepsilon[b]$ and dilation $\delta[b]$ are two operators $\mathcal{P}(X) \to \mathcal{P}(X)$ defined for all $U \in \mathcal{P}(X)$ by

$$\varepsilon[b](U) = \{x \in X \mid b(x) \subseteq U\} \tag{1}$$

$$\delta[b](U) = \{x \in X \mid b(x) \cap U \neq \emptyset\} \tag{2}$$

Remark 1. Usually, dilation is rather defined using the symmetric structuring element by $\delta[b](U) = \{x \in X \mid b^{\dagger}(x) \cap U \neq \emptyset\}$. This yields good properties, especially adjunction [11]. Report to Remark 2 to understand the choice of not using $b^{\dagger}$.

Let $\mathsf{StEl}(X) = \mathcal{P}(X)^X$ denote the set of all structuring elements on X, and let $\mathsf{Op}(X) = \mathcal{P}(X)^{\mathcal{P}(X)}$ be the set of all operators on $\mathcal{P}(X)$, which contains all erosions and dilations.

2.2   Lattice Structure

One can define a partial order on $\mathsf{StEl}(X)$ using pointwise inclusion: $b \leq c$ iff for all $x \in X$, $b(x) \subseteq c(x)$. This order endows $\mathsf{StEl}(X)$ with a complete lattice structure, where for all $(b_i)_{i \in I} \in \mathsf{StEl}(X)^I$,

$$\left(\bigwedge_{i \in I} b_i \right)(x) = \bigcap_{i \in I} b_i(x) \tag{3}$$

$$\left(\bigvee_{i \in I} b_i \right)(x) = \bigcup_{i \in I} b_i(x) \tag{4}$$

The greatest element is the full structuring element, defined by $\mathsf{ful}(x) = X$, and the least element is the empty structuring element, defined by $\mathsf{emp}(x) = \emptyset$. There is also a similar complete lattice structure on $\mathsf{Op}(X)$.

2.3   Monoid Structure

One can define an internal composition law $\star$ on $\mathsf{StEl}(X)$. Let $b, c : X \to \mathcal{P}(X)$ be structuring elements. Let $(b \star c)(x) = \{z \in X \mid \exists y \in b(x), z \in c(y)\}$. The operation $\star$ is associative, with neutral element sgt. This turns $(\mathsf{StEl}(X), \star, \mathsf{sgt})$ into a monoid. There is also a monoid structure $(\mathsf{Op}(X), \circ, \mathsf{id})$ where $\circ$ is the usual composition of functions and $\mathsf{id}(U) = U$ for all $U \in \mathcal{P}(X)$. Note that $\varepsilon[\mathsf{sgt}] = \delta[\mathsf{sgt}] = \mathsf{id}$.

We will need the following result in the next section.

Proposition 1. *For any $b, c \in \mathsf{StEl}(X)$, one has, for all $x \in X, U \in \mathcal{P}(X)$, $(b \star c)(x) \subseteq U$ iff $b(x) \subseteq \varepsilon[c](U)$.*

Proof. First note by unfolding the definitions that $(b \star c)(x) = \delta[c^\dagger](b(x))$. The wanted property then becomes $\delta[c^\dagger](b(x)) \subseteq U \iff b(x) \subseteq \varepsilon[c](U)$. This is true, by the standard result of adjunction between $\varepsilon[c]$ and $\delta[c^\dagger]$.   □

2.4 Usual Properties

Table 1 contains a review of classical properties of erosion and dilation that will be generalized in the following section. We also give new interpretations of some of these properties with respect to the lattice and monoid structures of $\mathsf{StEl}(X)$ and $\mathsf{Op}(X)$.

Table 1. Properties of classical erosion and dilation

Usual name	Statement	Interpretation
Duality	$\varepsilon[b](X \setminus U) = X \setminus \delta[b](U)$	/
Monotonicity	$U \subseteq V \Rightarrow \varepsilon[b](U) \subseteq \varepsilon[b](V)$	$\varepsilon[b]$ isotone
	$U \subseteq V \Rightarrow \delta[b](U) \subseteq \delta[b](V)$	$\delta[b]$ isotone
	$b \leq c \Rightarrow \varepsilon[c](U) \subseteq \varepsilon[b](U)$	$\varepsilon[-]$ antitone
	$b \leq c \Rightarrow \delta[b](U) \subseteq \delta[c](U)$	$\delta[-]$ isotone
Anti-extensivity	$\mathsf{sgt} \leq b \Leftrightarrow \varepsilon[b](U) \subseteq U$	$\varepsilon[-]$ antitone
Extensivity	$\mathsf{sgt} \leq b \Leftrightarrow U \subseteq \delta[b](U)$	$\delta[-]$ isotone
Preservation of X	$\varepsilon[b](X) = X$	$\varepsilon[b]$ erosion
Preservation of $\emptyset$	$\delta[b](\emptyset) = \emptyset$	$\delta[b]$ dilation
Commutation with inf	$\varepsilon[b](\bigcap_{i \in I} U_i) = \bigcap_{i \in I} \varepsilon[b](U_i)$	$\varepsilon[b]$ erosion
	$\varepsilon[b](U \cap V) = \varepsilon[b](U) \cap \varepsilon[b](V)$	/
Commutation with sup	$\delta[b](\bigcup_{i \in I} U_i) = \bigcup_{i \in I} \delta[b](U_i)$	$\delta[b]$ dilation
	$\delta[b](U \cup V) = \delta[b](U) \cup \delta[b](V)$	/
Associativity of dilation	$\varepsilon[b \star c] = \varepsilon[b] \circ \varepsilon[c]$	$\varepsilon[-]$ monoid homomorphism
	$\delta[b \star c] = \delta[b] \circ \delta[c]$	$\delta[-]$ monoid homomorphism
Adjunction	$U \subseteq \varepsilon[b](V) \Leftrightarrow \delta[b^\dagger](U) \subseteq V$	Galois connection

3 Structuring Neighborhoods

3.1 Introducing Neighborhoods Systems

In this section, we will generalize all notions of Sect. 2 using a lax notion of structuring element, called structuring neighborhood system, or structuring neighborhood for short.[1]

---

[1] While $b(x)$ is often considered as a neighborhood of x according to a given topology on X, here our notion of neighborhood system refers to a set of subsets of X.

Definition 2 (Structuring neighborhood). *A structuring neighborhood is a function $b : X \to \mathcal{P}(\mathcal{P}(X))$.*

Example 6 (Topology). Assume that X is endowed with a topology τ. The associated structuring neighborhood is $b(x) = \{U \in \mathcal{P}(X) \mid \exists O \in \tau, x \in O$ and $O \subseteq U\}$ (the set of topological neighborhoods of x).

Definition 3 (From element to neighborhood). *Let $b : X \to \mathcal{P}(X)$ be a structuring element in the sense of the previous section. Then one can define a structuring neighborhood $\bar{b} : X \to \mathcal{P}(\mathcal{P}(X))$ by $\bar{b}(x) = \{U \in \mathcal{P}(X) \mid b(x) \subseteq U\}$ (the set of all supersets of $b(x)$).*

Example 7 (Singleton). The structuring element sgt becomes the structuring neighborhood $\overline{\mathsf{sgt}}$, which is given explicitly by $\overline{\mathsf{sgt}}(x) = \{U \in \mathcal{P}(X) \mid x \in U\}$.

Let $b : X \to \mathcal{P}(\mathcal{P}(X))$ be a structuring neighborhood.

Definition 4. *Define the operator $\bar{\varepsilon}[b] : \mathcal{P}(X) \to \mathcal{P}(X)$ by*

$$\bar{\varepsilon}[b](U) = \{x \in X \mid U \in b(x)\} \tag{5}$$

Note that for the moment, nothing guarantees that the operator $\bar{\varepsilon}[b]$ is an erosion. One can also define another operator $\bar{\delta}[b]$ using a duality principle, namely $\bar{\delta}[b](U) = X \setminus \bar{\varepsilon}[b](X \setminus U)$. This leads to

$$\bar{\delta}[b](U) = \{x \in X \mid X \setminus U \notin b(x)\} \tag{6}$$

Remark 2. When it comes to define $\bar{\delta}[b]$, we could not have used a *symmetric* structuring neighborhood. Indeed, there is no obvious definition of such an object. This is also why, for consistency purposes, no symmetric was involved in our definition of dilation in Sect. 2. As the symmetric structuring element is a crucial component to get the adjunction-related properties, structuring neighborhood will lack these results.

Let $\mathsf{StNb}(X) = \mathcal{P}(\mathcal{P}(X))^X$ denote the set of all structuring neighborhoods on X. Definition 3 induces a map $\overline{} : \mathsf{StEl}(X) \to \mathsf{StNb}(X)$ which will be called the *plunge map* because it is injective. The following proposition shows that this extension of erosion and dilation is sound with respect to the classical case:

Proposition 2. *For any $b \in \mathsf{StEl}(X)$, $\bar{\varepsilon}[\bar{b}] = \varepsilon[b]$ and $\bar{\delta}[\bar{b}] = \delta[b]$.*

Proof. By definition, $x \in \bar{\varepsilon}[\bar{b}](U) \Leftrightarrow U \in \bar{b}(x) \Leftrightarrow b(x) \subseteq U \Leftrightarrow x \in \varepsilon[b](U)$. For the other one, use duality: $\bar{\delta}[\bar{b}](U) = X \setminus \bar{\varepsilon}[\bar{b}](X \setminus U) = X \setminus \varepsilon[b](X \setminus U) = \delta[b](U)$. □

Remark 3. This yields immediately $\bar{\varepsilon}[\overline{\mathsf{sgt}}] = \varepsilon[\mathsf{sgt}] = \mathrm{id}$ and $\bar{\delta}[\overline{\mathsf{sgt}}] = \delta[\mathsf{sgt}] = \mathrm{id}$.

3.2   Lattice Structure

One can define a partial order on $\mathsf{StNb}(X)$ using pointwise *reversed* inclusion. More precisely, for any $b, c \in \mathsf{StNb}(X)$, define $b \preceq c$ if and only if for all $x \in X$, $b(x) \supseteq c(x)$. The reason for which the inclusion is reversed in the definition of $\preceq$ lies in the following result.

Proposition 3. *For any* $b, c \in \mathsf{StEl}(X)$, $b \leq c$ *if and only if* $\bar{b} \preceq \bar{c}$.

Proof. Assume $b \leq c$, take $x \in X$, $U \in \mathcal{P}(X)$, and assume that $U \in \bar{c}(x)$ i.e. $c(x) \subseteq U$. As $b \leq c$, we have $b(x) \subseteq c(x)$, hence $b(x) \subseteq U$ i.e. $U \in \bar{b}(x)$. This completes the proof that $\bar{b} \preceq \bar{c}$. For the other implication, assume $\bar{b} \preceq \bar{c}$. Let $x \in X$ and show that $b(x) \subseteq c(x)$. As $c(x) \in \bar{c}(x)$ and $\bar{c}(x) \subseteq \bar{b}(x)$, one has $c(x) \in \bar{b}(x)$, so by definition of $\bar{b}$ this yields $b(x) \subseteq c(x)$. Hence, $b \leq c$.   □

Then, the set $\mathsf{StNb}(X)$ has a complete lattice structure, with, for any family $(b_i)_{i \in I}$ in $\mathsf{StNb}(X)$:

$$\left(\bigwedge_{i \in I} b_i \right)(x) = \bigcup_{i \in I} b_i(x) \tag{7}$$

$$\left(\bigvee_{i \in I} b_i \right)(x) = \bigcap_{i \in I} b_i(x) \tag{8}$$

The empty structuring element emp gives rise to a structuring neighborhood $\overline{\mathsf{emp}}$, defined by $\overline{\mathsf{emp}}(x) = \mathcal{P}(X)$, which turns out to be least element of this lattice. The greatest structuring neighborhood is called the *void* and is given by $\mathsf{void}(x) = \emptyset$. It is not the image of the full structuring element by the plunge function; indeed, $\overline{\mathsf{ful}}(x) = \{X\}$. Actually, void is the image of no structuring element by the plunge map, because for any structuring element b, $b(x) \in \bar{b}(x)$; henceforth, the plunge map is not surjective.

3.3   Monoid Structure

One can define an internal composition law in $\mathsf{StNb}(X)$. Given $b, c \in \mathsf{StNb}(X)$, let $(b \star c)(x) = \{U \in \mathcal{P}(X) \mid \bar{\varepsilon}[c](U) \in b(x)\}$. The symbol $\star$ is used for both the composition in $\mathsf{StEl}(X)$ and the one in $\mathsf{StNb}(X)$ because its meaning can be determined unambiguously from the type of the maps b and c.

Proposition 4. *Let* $b, c \in \mathsf{StNb}(X)$ *be structuring neighborhoods. Let* $\bar{\varepsilon}$ *and* $\bar{\delta}$ *be the two operators introduced in Definition 4. Then, one has* $\bar{\varepsilon}[b \star c] = \bar{\varepsilon}[b] \circ \bar{\varepsilon}[c]$ *and* $\bar{\delta}[b \star c] = \bar{\delta}[b] \circ \bar{\delta}[c]$.

Proof. Concerning the first operator, $x \in \bar{\varepsilon}[b \star c](U) \iff U \in (b \star c)(x) \iff \bar{\varepsilon}[c](U) \in b(x) \iff x \in \bar{\varepsilon}[b](\bar{\varepsilon}[c](U))$. The case of the other operator is obtained by duality.   □

Proposition 5. *The operation* $\star$ *is associative and* $\overline{\mathsf{sgt}}$ *is its neutral element.*

Proof. The element $\overline{\mathsf{sgt}}$ is neutral because $U \in (b \star \overline{\mathsf{sgt}})(x) \iff \overline{\varepsilon}[\overline{\mathsf{sgt}}](U) \in b(x) \iff U \in b(x)$, and $U \in (\overline{\mathsf{sgt}} \star b)(x) \iff \overline{\varepsilon}[b](U) \in \overline{\mathsf{sgt}}(x) \iff x \in \overline{\varepsilon}[b](U) \iff U \in b(x)$. For associativity, use Proposition 4:

$$U \in (b \star (c \star d))(x) \iff \overline{\varepsilon}[c \star d](U) \in b(x) \tag{9}$$
$$\iff (\overline{\varepsilon}[c] \circ \overline{\varepsilon}[d])(U) \in b(x) \tag{10}$$
$$\iff \overline{\varepsilon}[d](U) \in (b \star c)(x) \tag{11}$$
$$\iff U \in ((b \star c) \star d)(x) \tag{12}$$

$\square$

Proposition 5 yields that $(\mathsf{StNb}(X), \star, \overline{\mathsf{sgt}})$ is a monoid. According to Proposition 4 and Remark 3, the functions $\overline{\varepsilon}[-], \overline{\delta}[-] : \mathsf{StNb}(X) \to \mathsf{Op}(X)$ are then monoid homomorphisms. This is analogous to the structuring element case, where $\varepsilon[-], \delta[-] : \mathsf{StEl}(X) \to \mathsf{Op}(X)$ are monoid homomorphisms. We have the following additional result.

Proposition 6. *The plunge map* $\overline{=} : \mathsf{StEl}(X) \to \mathsf{StNb}(X)$ *is a monoid homomorphism.*

Proof. It does obviously preserve the neutral element, for sgt is neutral in $\mathsf{StEl}(X)$ and $\overline{\mathsf{sgt}}$ is neutral in $\mathsf{StNb}(X)$. Let $b, c \in \mathsf{StEl}(X)$. What remains to show is that $\overline{b} \star \overline{c} = \overline{b \star c}$. Computations result in

$$U \in (\overline{b} \star \overline{c})(x) \iff \overline{\varepsilon}[\overline{c}](U) \in \overline{b}(x) \tag{13}$$
$$\iff \varepsilon[c](U) \in \overline{b}(x) \tag{14}$$
$$\iff b(x) \subseteq \varepsilon[c](U) \tag{15}$$
$$\iff (b \star c)(x) \subseteq U \tag{16}$$
$$\iff U \in (\overline{b \star c})(x) \tag{17}$$

where we used Proposition 1 for the equivalence between Eqs. 15 and 16. $\square$

Proposition 2 and all monoid homomorphisms that have been discussed so far can be summed up in the commutative diagram of Fig. 1.

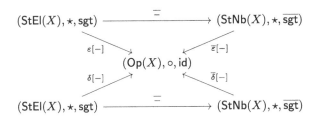

Fig. 1. Overview of the monoid-preserving structure of morphological operators

3.4 Usual Properties: Towards Filters

Many properties of classical erosion and dilation can be found back at the price of some necessary and sufficient conditions on the structuring neighborhood. Table 2 sums them up.

Definition 5 (Upper family). *Let $b \in \mathsf{StNb}(X)$. It is an* upper family *if for all x in X, all V in $\mathcal{P}(X)$ and all U in $b(x)$, $V \supseteq U$ implies $V \in b(x)$.*

Definition 6 (Preservation). *Let $b \in \mathsf{StNb}(X)$. It* preserves intersections *if for all $x \in X$ and all $(U_i)_{i \in I} \in \mathcal{P}(X)$, $\bigcap_{i \in I} U_i \in b(x) \iff \forall i \in I, U_i \in b(x)$. It preserves finite intersections if for all $x \in X$ and all $U, V \in \mathcal{P}(X)$, $U \cap V \in b(x) \iff U, V \in b(x)$.*

Table 2. Properties of $\bar{\varepsilon}[b]$ and $\bar{\delta}[b]$ with respect to b

Usual name	Statement	Equivalent condition on b
Duality	$\bar{\varepsilon}[b](X \setminus U) = X \setminus \bar{\delta}[b](U)$	True
Monotonicity	$U \subseteq V \Rightarrow \bar{\varepsilon}[b](U) \subseteq \bar{\varepsilon}[b](V)$	b upper family
	$U \subseteq V \Rightarrow \bar{\delta}[b](U) \subseteq \bar{\delta}[b](V)$	b upper family
	$b \preceq c \Rightarrow \bar{\varepsilon}[c](U) \subseteq \bar{\varepsilon}[b](U)$	True
	$b \preceq c \Rightarrow \bar{\delta}[b](U) \subseteq \bar{\delta}[c](U)$	True
Anti-extensivity	$\overline{\mathsf{sgt}} \preceq b \Leftrightarrow \bar{\varepsilon}[b](U) \subseteq U$	True
Extensivity	$\overline{\mathsf{sgt}} \preceq b \Leftrightarrow U \subseteq \bar{\delta}[b](U)$	True
Preservation of X	$\bar{\varepsilon}[b](X) = X$	$b \preceq \overline{\mathsf{ful}}$
Preservation of $\emptyset$	$\bar{\delta}[b](\emptyset) = \emptyset$	$b \preceq \overline{\mathsf{ful}}$
Commutation	$\bar{\varepsilon}[b](\bigcap_{i \in I} U_i) = \bigcap_{i \in I} \bar{\varepsilon}[b](U_i)$	b preserves $\cap$
with inf	$\bar{\varepsilon}[b](U \cap V) = \bar{\varepsilon}[b](U) \cap \bar{\varepsilon}[b](V)$	b preserves finite $\cap$
Commutation	$\bar{\delta}[b](\bigcup_{i \in I} U_i) = \bigcup_{i \in I} \bar{\delta}[b](U_i)$	b preserves $\cap$
with sup	$\bar{\delta}[b](U \cup V) = \bar{\delta}[b](U) \cup \bar{\delta}[b](V)$	b preserves finite $\cap$
Associativity	$\bar{\varepsilon}[b \star c] = \bar{\varepsilon}[b] \circ \bar{\varepsilon}[c]$	True
of dilation	$\bar{\delta}[b \star c] = \bar{\delta}[b] \circ \bar{\delta}[c]$	True

Preservation of intersections does confer $\bar{\varepsilon}[b]$ the name of erosion, but this is a very strong requirement. Following the path of Example 6 (topological neighborhoods are preserved by finite intersections, but not by all intersections), it would be more reasonable to only ask b to preserve *finite* intersections. This is why we define hereafter a notion that is weaker than erosion.

Definition 7 (Weak erosion). *Let $E \in \mathsf{Op}(X)$. It is called a* weak erosion *if $E(X) = X$ and for all $U, V \in \mathcal{P}(X)$, $E(U \cap V) = E(U) \cap E(V)$.*

Proposition 7. *If b preserves finite intersections, then b is an upper family.*

Proof. Assume that b preserves finite intersections. Let $U, V \in \mathcal{P}(X)$ such that $V \supseteq U$ and $U \in b(x)$. Then $U \cap V = U \in b(x)$. By hypothesis, this yields $V \in b(x)$, so that b is an upper family. □

We are naturally led to a mathematical construction that is very important in topology and logic: filters.

Definition 8 (Filter). *A filter on X is an element $\mathfrak{f}$ of $\mathcal{P}(\mathcal{P}(X))$ such that*

1. $X \in \mathfrak{f}$,
2. $U, V \in \mathfrak{f} \Rightarrow U \cap V \in \mathfrak{f}$,
3. $V \supseteq U, U \in \mathfrak{f} \Rightarrow V \in \mathfrak{f}$.

Proposition 8. *The operator $\bar{\varepsilon}[b]$ is a weak erosion if, and only if, the set $b(x)$ is a filter for every $x \in X$.*

Proof. Assume that $\bar{\varepsilon}[b]$ is a weak erosion. As $\bar{\varepsilon}[b](X) = X$, one has $b \preceq \overline{\mathsf{ful}}$ so $X \in b(x)$ for all $x \in X$, and then Condition *(1)* for a filter is true. As $\bar{\varepsilon}[b]$ distributes over finite intersections, b preserves finite intersections: this gives both Condition *(2)* and (according to Proposition 7) the upper family property, i.e. Condition *(3)* is true. Conversely, assume that for every $x \in X$, $b(x)$ is a filter. Then $X \in b(x)$ so $\bar{\varepsilon}[b](X) = X$. Furthermore, Condition *(2)* of filters implies that $\bar{\varepsilon}[b](U \cap V) \supseteq \bar{\varepsilon}[b](U) \cap \bar{\varepsilon}[b](V)$; the other inclusion comes directly from Condition *(3)*. □

Remark 4. A variant of Proposition 8 is that $\bar{\varepsilon}[b]$ is an erosion if and only if the set $b(x)$ is *an Alexandrov filter* for every $x \in X$, i.e. $b(x)$ is a filter that is stable under arbitrary intersections. It turns out that in this case, defining $\cap b(x) = \bigcap_{U \in b(x)} U$ yields $\bar{\varepsilon}[b] = \varepsilon[\cap b]$. In other words, any erosion obtained via a structuring neighborhood can already be obtained via a well-chosen structuring element. Asking only for weak erosions allows us to strictly enhance the expressiveness of this framework (see Example 8).

Example 8 (Interior in $\mathbb{R}$). Take $X = \mathbb{R}$ and τ be its usual topology. As it does not commute with infinite intersections, the interior operator is not an erosion. However, it is a weak erosion and can be modeled in our framework using the structuring element given in Example 6.

4 Application: Morpho-Logic

The fact that erosions commute with *infinite* infima is particularly important in mathematical morphology. However, some interesting applications are still within reach of weak erosions, which only commute with *finite* infima. The most noticeable of them is maybe logic. Indeed, in logic, the infimum of two formulas is given by their conjunction. As formulas are built through an inductive process in a finite number of steps, no infinite conjunctions will ever arise. Weak erosions obtained from structuring neighborhoods turn out to be useful to model logical phenomena that were previously beyond the scope of mathematical morphology. In this section we will discuss how our framework handles neighborhood logic, which is based on structuring neighborhoods. For the sake of succintness, sometimes only erosion will be addressed, but all dual expected results also hold for dilation.

4.1   Neighborhood Modal Logic

The logical study of neighborhood structures has ancient origins (see e.g. [9]). A pretty exhaustive account is given in the recent book of Pacuit [13]. Neighborhood models are a generalization of Kripke models (see Example 2) towards a second-order semantics of modalities.

Syntax. Let P be a countable set whose elements will be called *propositional variables* and denoted by letters $p, q, r \ldots$ The set $\mathcal{F}$ of formulas of modal logic is defined by the following grammar:

$$\varphi, \psi ::= p \mid \neg\varphi \mid \varphi \wedge \psi \mid \Box_i \varphi \tag{18}$$

where i runs through some index set I and p runs through P. There are thereby $|I|$ different modal operators $\Box_i$. Other connectives like $\vee$, $\rightarrow$, $\Diamond_i$ are all expressible from those above: $\varphi \vee \psi = \neg(\neg\varphi \wedge \neg\psi)$, $\varphi \rightarrow \psi = \neg(\varphi \wedge \neg\psi)$ and $\Diamond_i \varphi = \neg\Box_i \neg\varphi$. Given an arbitrary $p \in P$, define also the logical constants $\top = p \vee \neg p$ and $\bot = p \wedge \neg p$.

Semantics. A *model* is a triple $\langle W, (N_i)_{i \in I}, V \rangle$ where W is a set whose elements are called *worlds* or *states*, $N_i \in \mathsf{StNb}(W)$ is a structuring neighborhood and $V : P \rightarrow \mathcal{P}(W)$ is a function called the *valuation*. The semantics of formulas with respect to a model $\langle W, N, V \rangle$ consists of a relation $\models$ included in $W \times \mathcal{F}$. The assertion $w \models \varphi$ intuitively means that the formula φ is true at state w. Its negation is denoted $w \not\models \varphi$, i.e. the formulat φ is not true at w. The relation $\models$ is defined by structural induction on formulas as follows:

- $w \models p$ if and only if $w \in V(p)$,
- $w \models \neg\varphi$ if and only if $w \not\models \varphi$,
- $w \models \varphi \wedge \psi$ if and only if $w \models \varphi$ and $w \models \psi$,
- $w \models \Box_i \varphi$ if and only if $\{w' \in W \mid w' \models \varphi\} \in N_i(w)$.

The set $\{w' \in W \mid w' \models \varphi\}$ will also be denoted by $[\![\varphi]\!]$. Note that the definition of the satisfaction relation amounts to:

- $[\![p]\!] = V(p)$,
- $[\![\neg\varphi]\!] = W \setminus [\![\varphi]\!]$,
- $[\![\varphi \wedge \psi]\!] = [\![\varphi]\!] \cap [\![\psi]\!]$,
- $[\![\Box_i \varphi]\!] = \bar{\varepsilon}[N_i]([\![\varphi]\!])$.

Define an equivalence relation $\equiv$ on $\mathcal{F}$ by $\varphi \equiv \psi$ iff $[\![\varphi]\!] = [\![\psi]\!]$. Table 2 then gives rise to the logical rules listed in Table 3, where a model $\langle W, (N, N', N''), V \rangle$ is fixed, with $N'' = N \star N'$. The modalities of N, N', N'' are denoted respectively by $\square, \square', \square''$, and also $\lozenge, \lozenge', \lozenge''$. Note that $\varphi \equiv \top$ means that φ is true at every state; hence $\varphi \equiv \top$ will be simply abbreviated to φ in the left column of Table 2.

Table 3. Properties of logical systems with respect to N

Logical rule	Equivalent condition on N
$\square \neg \varphi \equiv \neg \lozenge \varphi$	True
$(\varphi \rightarrow \psi) \rightarrow (\square \varphi \rightarrow \square \psi)$	N upper family
$(\varphi \rightarrow \psi) \rightarrow (\lozenge \varphi \rightarrow \lozenge \psi)$	N upper family
$N \preceq N' \Rightarrow \square' \varphi \rightarrow \square \varphi$	True
$N \preceq N' \Rightarrow \lozenge \varphi \rightarrow \lozenge' \varphi$	True
$\overline{\mathsf{sgt}} \preceq N \Leftrightarrow \square \varphi \rightarrow \varphi$	True
$\overline{\mathsf{sgt}} \preceq N \Leftrightarrow \varphi \rightarrow \lozenge \varphi$	True
$\square \top \equiv \top$	$N \preceq \overline{\mathsf{ful}}$
$\lozenge \bot \equiv \bot$	$N \preceq \overline{\mathsf{ful}}$
$\square(\varphi \wedge \psi) \equiv \square \varphi \wedge \square \psi$	N preserves finite $\cap$
$\lozenge(\varphi \vee \psi) \equiv \lozenge \varphi \vee \lozenge \psi$	N preserves finite $\cap$
$\square'' \varphi \equiv \square \square' \varphi$	True
$\lozenge'' \varphi \equiv \lozenge \lozenge' \varphi$	True

The case of classical modal logic was studied by Bloch [8]. It corresponds to the case $N = \bar{b}$ where b is given in Example 2; also all properties on the left of the above table are satisfied because $\bar{\varepsilon}[N] = \bar{\varepsilon}[\bar{b}] = \varepsilon[b]$ is then an erosion.

4.2 Topological Modal Logic

A more specific case is the one of topological modal logic, where the space W is endowed with a topology τ. The set $[\![\square \varphi]\!]$ (resp. $[\![\lozenge \varphi]\!]$) is defined to be the topological interior (resp. closure) of $[\![\varphi]\!]$ with respect to τ. The question of whether these modalities can be represented as a pair erosion/dilation was raised in [3]. By extending the notion of a structuring element, the framework developed in this paper brings a positive answer to this issue. Indeed, the topological $\square$ operator can be obtained as $\bar{\varepsilon}[N]$ using the structuring neighborhood of Example 6:

$$N(w) = \{U \in \mathcal{P}(W) \mid \exists O \in \tau, w \in O, O \subseteq U\} \tag{19}$$

It is a standard fact that the interior operator distributes over binary intersections and that the interior of the whole set is itself, making this operator a weak erosion. Consistently with Proposition 8, this reflects the other standard fact that the set $N(w)$ of topological neighborhoods is a filter for every $w \in W$.

5   Conclusion

In this paper, we have shown that moving from $\mathcal{P}(X)$ to $\mathcal{P}(\mathcal{P}(X))$ to define structuring neighborhoods leads to strong algebraic structures (lattice and monoid), as well as to a large set of properties of dilations and erosions, at the price of losing the adjunction property. The restriction to finite intersections resulted in a weaker definition of erosion, equivalently asking the structuring neighborhood to be a filter. This setting was then applied to morpho-logic, e.g. topological modal logic. The logic inherits the properties of the morphological operators. A potential application could then be spatial logics based on spatial relations, and their use for image understanding.

For further extensions, for instance fuzzy systems, a coalgebraic treatment might be relevant. Specifically, the theory of coalgebras has identified the concept underlying modal operators semantics as *predicate liftings* [14]. General erosion could be defined by $\varepsilon[b](U) = b^{-1}(\lambda(U))$ where the predicate lifting λ shapes the type of the system; modal properties can then be studied at the level of λ.

Other perspectives include replacing $\mathcal{P}(X)$ or $\mathcal{P}(\mathcal{P}(X))$ by any complete lattice, or further moving to a categorical setting by generalizing structuring elements as coalgebras, dilation and erosion as adjoint functors, and so on.

References

1. Aiguier, M., Atif, J., Bloch, I., Hudelot, C.: Belief revision, minimal change and relaxation: a general framework based on satisfaction systems, and applications to description logics. Artif. Intell. **256**, 160–180 (2018)
2. Aiguier, M., Atif, J., Bloch, I., Pino Pérez, R.: Explanatory relations in arbitrary logics based on satisfaction systems, cutting and retraction. Int. J. Approx. Reason. **102**, 1–20 (2018)
3. Aiguier, M., Bloch, I.: Dual logic concepts based on mathematical morphology in stratified institutions: applications to spatial reasoning. CoRR abs/1710.05661 (2017). http://arxiv.org/abs/1710.05661
4. Bloch, I., Heijmans, H., Ronse, C.: Mathematical morphology (Chap. 13). In: Aiello, M., Pratt-Hartman, I., van Benthem, J. (eds.) Handbook of Spatial Logics, pp. 857–947. Springer, Dordrecht (2007). https://doi.org/10.1007/978-1-4020-5587-4_14
5. Bloch, I., Lang, J.: Towards mathematical morpho-logics. In: 8th International Conference on Information Processing and Management of Uncertainty in Knowledge based Systems IPMU 2000, Madrid, Spain, vol. III, pp. 1405–1412 (2000)
6. Bloch, I., Lang, J., Pino Pérez, R., Uzcátegui, C.: Morphologic for knowledge dynamics: revision, fusion, abduction. Technical report arXiv:1802.05142, arXiv cs.AI, February 2018
7. Bloch, I., Pérez, R.P., Uzcategui, C.: A unified treatment of knowledge dynamics. In: International Conference on the Principles of Knowledge Representation and Reasoning, KR2004, Canada, pp. 329–337 (2004)
8. Bloch, I.: Modal logics based on mathematical morphology for qualitative spatial reasoning. J. Appl. Non-Classical Log. **12**(3–4), 399–423 (2002)
9. Chellas, B.F.: Modal Logic: An Introduction. Cambridge University Press, Cambridge (1980)

10. Gorogiannis, N., Hunter, A.: Merging first-order knowledge using dilation operators. In: Hartmann, S., Kern-Isberner, G. (eds.) FoIKS 2008. LNCS, vol. 4932, pp. 132–150. Springer, Heidelberg (2008). https://doi.org/10.1007/978-3-540-77684-0_11

11. Heijmans, H.J.A.M., Ronse, C.: The algebraic basis of mathematical morphology - part I: dilations and erosions. Comput. Vis. Graph. Image Process. **50**, 245–295 (1990)

12. Najman, L., Talbot, H. (eds.): Mathematical Morphology: From Theory to Applications. ISTE-Wiley, Hoboken (2010)

13. Pacuit, E.: Nijssen: Neighborhood Semantics for Modal Logic. Springer, Heidelberg (2017). https://doi.org/10.1007/978-3-319-67149-9

14. Pattinson, D.: Coalgebraic modal logic: soundness, completeness and decidability of local consequence. Theor. Comput. Sci. **309**(1), 177–193 (2003)

15. Ronse, C.: Lattice-theoretical fixpoint theorems in morphological image filtering. J. Math. Imaging Vis. **4**(1), 19–41 (1994)

16. Serra, J.: Image Analysis and Mathematical Morphology. Academic Press, London (1982)

Hierarchical Laplacian and Its Spectrum in Ultrametric Image Processing

Jesús Angulo[✉]

CMM-Centre de Morphologie Mathématique, MINES ParisTech,
PSL-Research University, Paris, France
`jesus.angulo@mines-paristech.fr`

Abstract. The Laplacian of an image is one of the simplest and useful image processing tools which highlights regions of rapid intensity change and therefore it is applied for edge detection and contrast enhancement. This paper deals with the definition of the Laplacian operator on ultrametric spaces as well as its spectral representation in terms of the corresponding eigenfunctions and eigenvalues. The theory reviewed here provides the computational framework to process images or signals defined on a hierarchical representation associated to an ultrametric space. In particular, image regularization by ultrametric heat kernel filtering and image enhancement by hierarchical Laplacian are illustrated.

Keywords: Ultrametric space · Semigroups on ultrametric spaces · Hierarchical Laplacian · Ultrametric image processing

1 Introduction

The Laplace operator (or Laplacian) of an image is one of the simplest and useful image processing tools, since it highlights regions of rapid intensity change and therefore it is applied for edge detection (zero crossing edge detector [9]) and contrast enhancement by subtraction from the original image.

The Laplace operator is defined as the divergence of the gradient. When applied to a function f on $\mathbb{R}^n$, the Laplacian produces a scalar function given by $L(f) = \nabla \circ \nabla f = \Delta f = \sum_{i=1}^{n} \frac{\partial^2 f}{\partial x_i^2}$. The second order differentiation becomes in the discrete counterpart a second order difference; i.e., in 1D, $f'(x) = (f(x + h/2) - f(x - h/2))/h$ and $f''(x) = (f(x + h) + f(x - h) - 2f(x))/h^2$. Hence, the discrete Laplace operator of an image f is given by $L(f)(x) = \sum_{z \in \tilde{N}(x)} f(z) - |\tilde{N}| f(x)$, where $\tilde{N}(x)$ is the local neighbourhood of pixel x, excluding itself. This transformation can be calculated using a convolution filter with the corresponding kernels. The Laplace operator is also naturally defined on Riemanian manifolds, the so-called Laplace–Beltrami operator, which has been also widely used in image and shape processing [11]. In the literature of mathematical morphology, the notion of morphological (flat) Laplacian is defined as a second-order operator given by the difference between the gradient by dilation

© Springer Nature Switzerland AG 2019
B. Burgeth et al. (Eds.): ISMM 2019, LNCS 11564, pp. 29–40, 2019.
https://doi.org/10.1007/978-3-030-20867-7_3

and the gradient by erosion [10], and can be interpreted just as the ∞-Laplace operator [8].

The goal of this paper is to consider the definition of the Laplacian operator on ultrametric spaces as well as its spectral representation in terms of the corresponding eigenfunctions and eigenvalues. Indeed, the theory of the heat semigroup and Laplacian operator in the case of ultrametric spaces has been developed in a series of papers by Bendikov and collaborators [2–5]. Our purpose here is just to review the main results of this theory and to bring it to the context of ultrametric image processing.

Let us precise that ultrametric image or data processing means in our context. The image domain can be considered as an ultrametric space, where the pixels/vertices of the image are hierarchically organized into clusters at different aggregation levels. Interaction between image pixels or vertices is associated to the ultrametric distance. This kind of representation is naturally used in image segmentation, in the case of morphological image segmentation, hierarchical representations are ubiquitous [6,12,13], and the algorithmic ingredients to construct them are typically minimum spanning trees and quasi-flat zones. Once an image has been endowed with a hierarchical structure (i.e., the image domain is an ultrametric space), the image can be not only segmented, but also filtered out, enhanced and so on, according to such representation. In this context, we have introduced the corresponding ultrametric morphological semigroups and scale-spaces [1]. This paper is a step forwards in the program of revisiting classical image/data processing on ultrametric representations.

The rest of the paper is organized as follows. Section 2 provides some background definitions on ultrametric spaces. The notion of heat semigroup on ultrametric spaces is briefly reminded in Sect. 3, which is required to review the notion of Laplacian on an ultrametric space in Sect. 4. Section 5 considers in particular the case of the Laplacian on a discrete ultrametric space, which is the relevant case for ultrametric image processing. Some illustrative examples of ultrametric image Laplacian are discussed in Sect. 6. Finally, Sect. 7 closes the paper with some conclusions and perspectives.

2   Preliminaries

Ultrametric Space. Let (X, d) be a metric space. The metric d is called an ultrametric if it satisfies the ultrametric inequality, i.e., $d(x,y) \leq \max\{d(x,z), d(z,y)\}$, that is obviously stronger than the usual triangle inequality. In this case (X, d) is called an ultrametric space.

Consider the metric balls $B_r(x) = \{y \in X : d(x,y) \leq r\}$. The ultrametric inequality implies that any two metric balls are either disjoint or one contains the other. In particular, every two balls of the same radius are disjoint or coincide. Thus, every point inside a ball is its center. As a consequence of these properties, the collection of all distinct balls of the same radius r forms a partition X; for increasing values of r, the balls are also increasing, hence we obtain a family of nested partitions of X which forms a hierarchy. This is a fundamental property of ultrametric spaces.

Throughout the rest of the paper, we consider the triplet (X, d, μ), where (X, d) is a compact separable ultrametric space; compactness involves that all balls are compact and separability that the set of all values of metric d is at most countable and all distinct balls of a given radius $r > 0$ form at most a countable partition of X. The measure μ is a Radom measure on X with full support, such that $0 < \mu(B_r(x)) < \infty$ for all $x \in X$ and $r > 0$.

Discrete Ultrametric Space. An ultrametric space (X, d) is called discrete if: (i) the set X is countable, (ii) all ultrametric balls $B_r(x)$ are finite, and (iii) the distance function d takes only integer numbers. Given a discrete ultrametric space (X, d), one can consider any measure μ on 2^X such that $0 < \mu(x) < \infty$ for any $x \in X$ and $\mu(X) = \infty$. For example, μ can be a counting measure.

Set of Ultrametric Balls $\mathcal{B}$. Denote by $\mathcal{B}$ be the family (countable set) of all balls C in X with positive radii, and thus positive measure $\mu(C) > 0$. Using the hierarchical interpretation of ultrametric balls, for any ball $C \in \mathcal{B}$ such that $C \neq X$ there is a unique parent (or predecessor) ball B which contains C. In this case, C is called the child (or successor) of B. For any ball B with $\operatorname{diam}(B) > 0$ the number $\deg(B)$ of its children satisfies $2 \leq \deg(B) < \infty$. Moreover, all the children of B are disjoint and their union is equal to B. Let us denote by $C \prec B$ that C is a child of B. The functions $\{1_C : C \prec B\}$ are linearly independent (i.e., orthogonal) and therefore we have

$$\langle 1_C, 1_{C'} \rangle = 0, \ C \prec B, C' \prec B, \quad \text{and} \quad \sum_{C \prec B} 1_C = 1_B. \tag{1}$$

Intrinsic Ultrametric d_*. Let us first note that different ultrametric distances on X can produce the same set $\mathcal{B}$. In the case of (X, d), for any $x, y \in X$, we denote $B(x \curlywedge y)$ the minimal ball containing both x and y. The intrinsic ultrametric is defined as: $d_*(x, y) = \operatorname{diam}(B(x \curlywedge y))$ when $x \neq y$ and $d_*(x, y) = 0$ when $x = y$. Note that the ultrametrics d and d_* generates the same set of balls, with a feasible change of their diameter function.

3   Heat Semigroup and Heat Kernel on Ultrametric Spaces

We review in this section some background material on ultrametric heat semigroups from [2,3]. The first operator, the averaging one, is the building block for the ultrametric heat kernel theory.

Averaging Operator Q_r. Let us define the family $\{Q_r\}_{r>0}$ of averaging operators acting on non-negative (or bounded) Borel functions $f : X \to \mathbb{R}$ by

$$Q_r f(x) = \frac{1}{\mu(B_r(x))} \int_{B_r(x)} f \, d\mu. \tag{2}$$

As we can see, that corresponds to computing the mean value at each class of the partition of radius r of the hierarchy associated to d_*. It is symmetric in x, y

because $B_r(x) = B_r(y)$ for any $y \in B_r(x)$. We set $Q_0 := \mathrm{id}$. If $\mu(X) = \infty$ then $Q_\infty = 0$, while in the case $\mu(X) < \infty$ we have $Q_\infty f = \frac{1}{\mu(X)} \int_X f d\mu$. We have $Q_r f \geq 0$ if $f \geq 0$ and $Q_r 1 = 1$.

Q_r is a Markov operator on the space of bounded Borel functions on $L^2(X, \mu)$, which satisfies the following (supremal) semigroup property [3]:

$$Q_r Q_s = Q_s Q_r = Q_{\max\{r,s\}}, \quad \text{with} \quad Q_r Q_r f = Q_r f. \tag{3}$$

Ultrametric Heat Semigroup $\{P_t f\}_{t>0}$. Let us choose a distance probability distribution function σ that satisfies the following assumptions: $\sigma : [0, \infty] \to [0, 1]$ is a strictly monotone increasing left-continuous function such that $\sigma(0+) = 0$ and $\sigma(\infty) = 1$. Typically, $\sigma(r) = \exp(-1/r)$. The operator P determined by the triple (d, μ, σ)

$$Pf = \int_0^\infty Q_r f \, d\sigma(r) \tag{4}$$

is an isotropic Markov operator which determines a discrete time Markov chain $\{X_n\}_{n \in \mathbb{N}}$ on X with the following transition rule: X_{n+1} is μ-uniformly distributed in $B_r(X_n)$ where the radius r is chosen at random according to the probability distribution σ.

The Markov operator P is non-negative definite, which allows us to define P_t for all $t \geq 0$ using the power t of distribution function σ. Then $\{P_t\}_{t \geq 0}$ is a symmetric strongly continuous Markov semigroup where P_t admits for $t > 0$ the following representation:

$$P_t f(x) = \int_0^\infty Q_r f(x) \, d\sigma^t(r), \tag{5}$$

which satisfies the following (additive) semigroup property

$$P_s P_t = P_{s+t}. \tag{6}$$

The family $\{P_t\}_{t \geq 0}$ is a strongly continuous symmetric Markov semigroup on $L^2(X, \mu)$.

Ultrametric Heat Kernel $p_t(x, y)$. For any $t > 0$, the operator P_t admits an integral kernel $p_t(x, y)$, i.e., for all $f \in \mathcal{B}_b$ and $f \in L^2$, one has

$$P_t f(x) = \int_X p_t(x, y) f(y) d\mu(y),$$

where $p_t(x, y)$ is given by

$$p_t(x, y) = \int_{[d(x,y), \infty)} \frac{d\sigma^t(r)}{\mu(B_r(x))}.$$

The function $p_t(x, y)$ is called the heat kernel of the semigroup $\{P^t\}$. For all $x, y \in X$ and $t > 0$, we have: (i) $p_t(x, y) > 0$; (ii) $p_t(x, y) = p_t(y, x)$; (iii) $p_t(x, y) \leq$

$\min\{p_t(x,x), p_t(y,y)\}$. The function $(x,y) \mapsto \frac{1}{p_t(x,y)}$ if $x \neq y$ and 0 if $x = y$ is an ultrametric.

Let us introduce the intrinsic ultrametric distance associated to (d, μ, σ):

$$d_*(x,y) = [-\log \sigma\,(d(x,y))]^{-1}.$$

For any $r \geq 0$, set

$$r \mapsto s = [-\log \sigma\,(r)]^{-1}.$$

Using the fact that σ is strictly monotone increasing, the following identity holds $\forall x \in X$: $B_s^*(x) = B_r(x)$. Consequently, the metrics d and d_* determine the same set of balls and the same topology.

Heat Kernel and Heat Semigroup in Ultrametric Discrete Space. For practical applications and in particular, in ultrametric image processing, we deal with discrete spaces. In such a framework, the computation of the heat semigroup becomes very efficient since the integral of ultrametric distance becomes a weighted sum of averaging operators. The latter only involves an operator at each level of the hierarchy.

Let (X, d, μ) be a discrete ultrametric space with a counting measure μ on 2^X. The idea now is to replace the ultrametric distance distribution function σ by a discrete distribution of weighting values in $(0, 1]$. Namely, choose a sequence $\{c_k\}_{k=0}^\infty$ of strictly positive reals such that $\sum_{k=0}^\infty c_k = 1$. Let us introduce $s_k = \sum_{i=0}^k c_i$ such that $0 < s_{k-1} < s_k < 1$, $k = 0, 1, \cdots$ and $s_k \to 1$ as $k \to \infty$; $s_{-1} = 0$.

Using the distribution of weights $\{s_k\}_{k=0}^\infty$, the ultrametric averaging operator, heat semigroup and heat kernel acting on function f are respectively given:

$$Q_k f(x) = \frac{1}{\mu\,(B_k(x))} \sum_{y \in B_k(x)} f(y); \tag{7}$$

$$P_t f(x) = \sum_{k=0}^\infty s_k^t\,(Q_k f(x) - Q_{k+1} f(x)) = \sum_{k=0}^\infty \left(s_k^t - s_{k-1}^t\right) Q_k f(x); \tag{8}$$

$$p_t(x,y) = \sum_{k=d(x,y)}^\infty \left(s_k^t - s_{k-1}^t\right) \frac{1}{\mu\,(B_k(x))}. \tag{9}$$

Note that the second equality in (8) is based on the Abel transformation (i.e., summation by parts of sequences) [2]. Since $\left\{s_k^t - s_{k-1}^t\right\}_{k=0}^\infty$, $t > 0$, is a stochastic sequence, the operator P_t is Markov. The semigroup indentity here $P_t P_s = P_{t+s}$ follows from the functional calculus.

The equivalence of the continuous semigroup (5) and the discrete counterpart (8) is obtained by taking density function: $\sigma_k = 1 - s_k = \sum_{l>k} c_l$, $k \geq 0$. Clearly $\{\sigma_k\}_{k=0}^\infty$ can be any sequence of positive real that satisfies the conditions: $\sigma_{k+1} < \sigma_k < 1$, $k = 0, 1, \cdots$, and $\sigma_k \to 0$ as $k \to \infty$.

4   Laplacian on an Ultrametric Space

Laplace Operator as a Spectral Decomposition on (d, μ, σ). Let the spectral resolution $\{E_\lambda\}$ be defined as:

$$E_\lambda = \begin{cases} Q_{1/\lambda}, & \lambda > 0, \\ 0, & \lambda \leq 0, \end{cases} \quad \text{for any } \lambda \in \mathbb{R}.$$

Note that $E_{0+} = Q_\infty$. Using the change of variable $s = 1/\lambda$, the spectral decomposition of $\mathcal{L}$ in the classical form becomes

$$\mathcal{L} = \int_{[0,+\infty)} \lambda dE_\lambda = -\int_{(0,\infty)} \frac{1}{s} dQ_s. \tag{10}$$

The L^2-spectrum of the Laplacian $\mathcal{L}$. For any ball $C \in \mathcal{B}$ define, on the one hand, the function ϕ_C on X as

$$\phi_C = \frac{1}{\mu(C)} \mathbf{1}_C - \frac{1}{\mu(B)} \mathbf{1}_B = \left[\frac{1}{\mu(C)} - \frac{1}{\mu(B)} \right] \mathbf{1}_C - \frac{1}{\mu(B)} \mathbf{1}_{B \setminus C}, \tag{11}$$

where B is the parent ball of C, i.e., $\mathbf{1}_B = \mathbf{1}_C + \mathbf{1}_{B \setminus C}$. On the other hand, for any $C \in \mathcal{B}$, set also

$$\lambda(C) = \frac{1}{\text{diam}(C)}. \tag{12}$$

If $C = X$, then set $\phi_C \equiv 1$ and $\lambda(C) = 0$. Using the linear independence of the indicators of the child balls (1), one has that, for any parent B, $\sum_{C \prec B} \mu(C) \phi_C = 0$.

The following result reveals the spectral nature of the pair $(\phi_C, \lambda(C))$.

Proposition 1. *We can see that ϕ_C is an eigenfunction of $\mathcal{L}$ with the eigenvalue $\lambda(C)$, i.e.,*

$$\mathcal{L}\phi_C = \lambda(C)\phi_C.$$

Proof. For $C \in \mathcal{B}$ of radius $r = \text{diam}(C)$, where B is the parent one of radius r'. Any ball of radius $s < r'$ either is disjoint with C or is contained in C, which implies that $\mathbf{1}_C$ is constant in any such ball. It follows that, for any $s < r'$, we have $Q_s \mathbf{1}_C = \mathbf{1}_C$ and, similarly $Q_s \mathbf{1}_B = \mathbf{1}_B$, whence

$$Q_s \phi_C = \phi_C.$$

For $s \geq r'$ any ball of radius s either contains both balls C, B or is disjoint from B. Since the averages of the two functions $\frac{1}{m(C)} \mathbf{1}_C$ and $\frac{1}{m(B)} \mathbf{1}_B$ over any ball containing C and B are equal, we obtain that in this case $Q_s \phi_C = 0$. It follows that

$$\mathcal{L}\phi_C = -\int_{(0,\infty)} \frac{1}{s} Q_s \phi_C \, ds = \frac{1}{r'} \phi_C = \lambda(C) \phi_C.$$

More generaly, we have the following result on the complete representation of Laplacian on an ultrametric space as a base of ball-based eigenfunctions (Fig. 1).

Theorem 1 ((Bendikov et al. 2014) [3]). *For any $C \in \mathcal{B}$ the function ϕ_C is an eigenfunction of $\mathcal{L}$ with the eigenvalue $\lambda(C)$. The family $\{\phi_C : C \in \mathcal{K}\}$ is complete in $L^2(X, \mu)$. Consequently, the operator $\mathcal{L}$ has a complete system of compactly supported eigenfunctions, called the Haar system associated to (X, d, μ).*

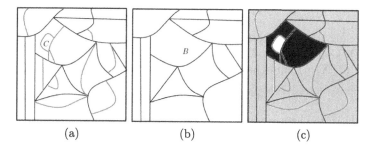

(a)   (b)   (c)

Fig. 1. Eigenfunction of an ultrametric ball: (a) and (b) depict two partitions of a discrete ultrametric space, where the ball B is the father of ball C; (c) eigenfunction of ball C, ϕ_C, in grey, the value of the function is 0, in white, it is a positive value and in black, a negative one.

Theorem 2 ((Benkikov et al. 2014) [3]). *The spectrum spec $\mathcal{L}$ of the Laplacian $\mathcal{L}$ is given by*

$$\operatorname{spec} \mathcal{L} = \overline{\left\{ \frac{1}{r} : r \in \Lambda \right\}} \cup \{0\}, \quad where \ \{0\} \ is \ for \ \lambda(X) = 0.$$

The Hierarchical Laplacian $\mathcal{L}$ on a Function. Given the ultrametric measurable set (X, d, μ), starting from (10), one defines (pointwise) the Laplacian on a function $f \in \mathcal{F}$ as [3]:

$$\mathcal{L} f(x) = \sum_{C \in \mathcal{B}:\, x \in B} \kappa(C) \left(f(x) - \frac{1}{\mu(C)} \int_C f \, d\mu \right), \tag{13}$$

with the positive scaling function given typically by $\kappa(C) = \operatorname{diam}(C)^{-1} - \operatorname{diam}(B)^{-1}$, where B is the parent ball of C.

The operator $(\mathcal{L}, \mathcal{F})$ acts in $L^2(X, \mu)$, is symmetric, self-adjoint and admits a complete system of eigenfunctions $\{\phi_C\}$ given by (11), where $C \subset B$ run over all nearest neighboring balls in $\mathcal{B}$ [3]. For any two distinct balls C and C', the functions ϕ_C and $\phi_{C'}$ are orthogonal in $L^2(X, \mu)$. The corresponding eigenvalue is given by $\lambda(C) = \operatorname{diam}(C)^{-1}$. Note that one does not need to specify the ultrametric d for the function $\kappa(C)$ and the corresponding eigenvalue $\lambda(C)$. One only requires the family of balls $\mathcal{B}$, or in other words, the intrinsic ultrametric distance $d_*(x, y) = \operatorname{diam}(C)$. However, the practical computation of the hierarchical Laplacian (13) requires a sum over all the balls of $\mathcal{B}$.

5   Laplace Operator on a Discrete Ultrametric Space

Let us now focus on a discrete ultrametric space (X, d). Using the following functional identity (see Introduction) $P = \exp(\mathcal{L})$, whence $\mathcal{L} = \log \frac{1}{P}$, and taking $t = 1$ in the discrete heat kernel expression (9), it holds [2]:

$$\mathcal{L} = \sum_{k=0}^{\infty} \left(\log \frac{1}{s_k} \right) (Q_k - Q_{k+1}) = \sum_{k=0}^{\infty} l_k \, (Q_k - Q_{k+1}),$$

with $l_k = -\log s_k$, where the series converges in the strong operator topology of $L^2(X, m)$. Consequently, $\mathcal{L}$ is a bounded, non-negative definite, self-adjoint operator in $L^2(X, m)$, and the discrete spectrum of $\mathcal{L}$ is given by

$$\mathrm{spec}_{L^2}\mathcal{L} = \{l_k\}_{k=0}^{\infty} \cup \{0\}.$$

Thus each l_k is an eigenvalue of $\mathcal{L}$. In particular, the point 0 of $\mathrm{spec}_{L^\infty}\mathcal{L}$ is an eigenvalue of multiplicity 1.

Applied to any $f \in L^2(X)$, we have now

$$\mathcal{L} f = \sum_{k=0}^{\infty} l_k \, (Q_k f - Q_{k+1} f) = l_0 f - \sum_{k=1}^{\infty} (l_{k-1} - l_k) \, Q_k f. \tag{14}$$

Here also the practical interest of this expression of Laplacian of f is obvious, since the basic ingredient is the set of averaging operators Q_k on f.

Considering the counting measure for Q_k, discrete ultrametric Laplacian (14) can be written as:

$$\mathcal{L} f(x) = \alpha \left(f(x) - \frac{1}{\alpha} \sum_{k=1}^{\infty} \frac{\beta_k}{|B_k(x)|} \sum_{z \in B_k(x)} f(z) \right), \tag{15}$$

with $\alpha = \log \frac{1}{s_0} > 0$ and $\beta_k = \log \frac{s_k}{s_{k-1}} = \log \frac{\sum_{j=0}^{k} c_j}{\sum_{j=0}^{k-1} c_j} > 0$. Up to the minus sign of $\mathcal{L} = -\Delta$, this form (15) can be compared with the discrete Laplacian on $\mathbb{Z}^n$ (see Introduction):

$$\Delta f(x) = -|\widetilde{N}| \left(f(x) - \frac{1}{|\widetilde{N}|} \sum_{z \in \widetilde{N}(x)} f(z) \right).$$

If f is a non-negative function on X such that $\mathcal{L} f \equiv 0$, then $f = \mathrm{const}$.

Deformation of Laplace Operator. Let $\mathcal{L}$ be the Laplace operator of the semigroup $\{P_t\}_{t \geq 0}$. Let $\xi : [0, \infty) \to [0, \infty)$ be a continuous, strictly monotone increasing function with $\xi(0) = 0$. Then, the operator $\xi(\mathcal{L})$ is the Laplace operator of the semigroup $\left\{ P_t^{\xi} \right\}_{t \geq 0}$, defined by [2]

$$P^{\xi} = \sum_{k=0}^{\infty} c_k^{\xi} Q_k,$$

where the stochastic sequence of c_k^ξ is given by $c_0^\xi = e^{-\xi(l_0)}$; $c_k^\xi = e^{-\xi(l_k)} - e^{-\xi(l_{k-1})}$; $k \geq 1$.

Example. Taking the deformation $\xi(t) = 1 - e^{-\alpha t}$, we have:

$$\mathcal{L}_\xi = \xi(\mathcal{L}) = \sum_{k=0}^{\infty} \xi(l_k)(Q_k f - Q_{k+1} f) \sum_{k=0}^{\infty}(1 - s_k^\alpha)(Q_k f - Q_{k+1} f) = \mathrm{id} - P_\alpha.$$

Thus, the residue between f and the heat operator at scale α, $P_\alpha f$, can be seen as a Laplacian, "deformation" of the standard one by $1 - e^{-\alpha \mathcal{L} f}$.

Ultrametric Infinity Laplace Operator. Let us first consider the following result.

Theorem 3 ((Benkikov et al. 2012) [2]). *For any $p \in [1, +\infty]$, the operator $\mathcal{L}$ can be extended as a bounded operator acting on $L^p = L^p(X, \mu)$. Moreover, we have*

$$spec_{L^p} \mathcal{L} = spec_{L^2} \mathcal{L}, \quad \text{for any } p \in [1, +\infty].$$

Thus, the spectrum of the ultrametric discrete L^p Laplacian is the same for any p, including the ultrametric infinity Laplacian. This is an important difference with respect to the Euclidean and metric-space cases, where the infinity Laplacian is a very particular operator [8]. Indeed, the infinity Laplacian is a 2nd-order partial differential operator, which in the discrete case of a graph with edges E is defined as [14]: $\Delta_\infty u(x) = \sup_{y\,:\,(x,y)\in E} u(y) + \inf_{y\,:\,(x,y)\in E} u(y) - 2u(x)$, which corresponds just to the definition of the morphological Laplacian [10] on the graph of edges E; i.e., dilation + erosion - 2 function. Inspired by this equivalence, we can propose a morphological inspired infinity Laplacian which mimics the discrete one on a graph by using the notion of (discrete) ultrametric dilation D_t and erosion E_t introduced in [1]:

$$\mathcal{L}_t^\infty f(x) = 2f(x) - (D_t f(x) + E_t f(x))$$

$$= 2f(x) - \left(\sup_{0\leq k\leq\infty}\left\{ \sup_{y\in B_k(x)} f(y) \wedge b_{k,t} \right\} + \inf_{0\leq k\leq\infty}\left\{ \inf_{y\in B_k(x)} f(y) \vee (M - b_{k,t}) \right\} \right),$$

where $b_{k,t} = M - t^{-1} c_k$ is the so-called discrete structuring function a scale t, with M just the maximum value of $f(x)$. Note that the change of sign between Δ_∞ and $\mathcal{L}_t^\infty$ is only a convention to be consistent with the ultrametric Laplacian $\mathcal{L}f$. This morphological operator can be compared to (15). Note also that the spectral representation of such ultrametric operators requires naturally the study of the eigenvalues and eigenfunctions of morphological operators.

6   Application to Ultrametric Image Processing

For the examples that we consider here, the ultrametric space (X, d) is built from a minimum spanning tree (MST). First, let G be an edge-weighted undirected neighbor graph with points $x \in X$ as vertices and all edge weights as nonnegative values. An MST of G is a spanning tree that connects all the vertices together with the minimal total weighting for its edges, and let $d(x, y)$ be the largest edge weight in the path of the MST between x and y. Then the vertices of the graph G, with distance measured by d form an ultrametric space. By thresholding the corresponding MST at k, $0 \leq k \leq K$, a set of partitions is obtained which produces all balls $B_k(x)$. In particular, for the case of the discrete images used in the examples, G is a 4-connected pixel neighbor graph and the edge weights are the grey-level difference.

In addition, the sequence used in the examples is just $c_k = k/20$, with $k = 1, \cdots, 20$ (ultrametric distances are quantified into $K = 20$ levels).

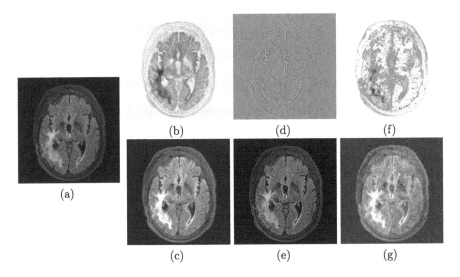

Fig. 2. Discrete ultrametric Laplace operator: (a) original image $f(x)$, (b) negative of Laplacian $-\mathcal{L}f(x)$, (c) enhanced image by adding the Laplacian $f(x) + \mathcal{L}f(x)$, (d) classical image $\Delta f(x)$, (e) enhanced image by $f(x) - \Delta f(x)$, (f) negative ultrametric morphological Laplacian $-\mathcal{L}_t^\infty f(x)$, with $t = 0.01$ (g) enhanced image by adding morphological Laplacian $f(x) + \mathcal{L}_{t=0.01}^\infty f(x)$.

Figure 2 provides an example of the application of discrete Laplace operator on an image endowed with its intrinsic ultrametric distance, which is compared with the standard image Laplacian and with the morphological ultrametric Laplacian, as well as the corresponding enhanced images obtained by adding the Laplacian. From this illustration it can be noted that the enhancement obtained by the hierarchical Laplace operator is more "regional" than the one by the classical Laplacian. This is expected by the fact that the classical Laplacian is just a second derivative in local neighborhoods. We have observed in similar results from other images that the ultrametric Laplacian improves image dynamics without introducing artifacts. The morphological inspired infinity Laplacian provides a stronger detection of contrasted bright/dark classes of the hierarchy

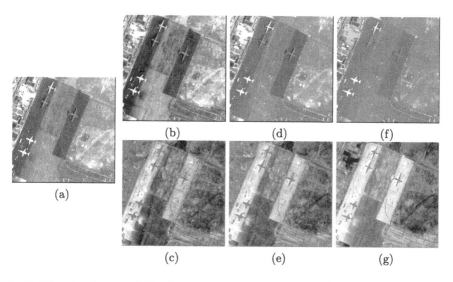

Fig. 3. Discrete ultrametric Laplace operator and heat kernel: (a) original image $f(x)$, (b) enhanced image by adding the ultrametric Laplacian $f(x) + \mathcal{L}f(x)$, (c) corresponding negative of Laplacian $-\mathcal{L}f(x)$, (d) and (f) ultrametric heat kernel $P_t f(x)$, with respectively $t = 0.01$ and $t = 0.1$, (e) and (g) deformation of Laplace operator corresponding to heat kernel $-\mathcal{L}_{\xi(t)} f(x)$, with respectively $t = 0.01$ and $t = 0.1$. In residues (e) and (g) the intensities has been stretched.

and consequently, the corresponding enhancement is stronger than the ultrametric Laplacian. Note that in the case of the morphological Laplacian, there is a scale parameter t, used for the ultrametric multiscale dilation and erosion.

Another comparison is depicted in Fig. 3. This time, besides the ultrametric Laplacian and corresponding enhanced image, two examples of the ultrametric heat kernel operator applied on the image are included, with different values of scale parameter t. This heat kernel operator provided a smoothing, or ultrametric regularization of the image, similar to a Gaussian filtering, but without any effect of blurring since the underlying ultrametric space contains the hierarchical organization of images zones. The associated deformation of Laplacian (see Sect. 5) for the heat kernel images are also given, where obviously, the parameter t allows a control of the zones to be detected.

7   Conclusions and Perspectives

The theory reviewed in this paper provides the computational framework to process images or signals defined on a hierarchical representation associated to an ultrametric distance space. In particular, image regularization by ultrametric heat kernel filtering and image enhancement by hierarchical Laplacian have been illustrated. The spectrum of the corresponding Laplacian is just related to the evolution of the size of nested classes along the hierarchy. Our ongoing work will study, on the one hand, the interest of the associated ultrametric pseudo-differential equations [7] in the context of ultrametric images/data processing and on the other hand, the applicative interest of the spectrum of the hierarchical Laplacian to describe shapes and point clouds endowed with an ultrametric structure.

References

1. Angulo, J., Velasco-Forero, S.: Morphological semigroups and scale-spaces on ultrametric spaces. In: Angulo, J., Velasco-Forero, S., Meyer, F. (eds.) ISMM 2017. LNCS, vol. 10225, pp. 28–39. Springer, Cham (2017). https://doi.org/10.1007/978-3-319-57240-6_3
2. Bendikov, A., Grigor'yan, A., Pittet, C.: On a class of Markov semigroups on discrete ultra-metric spaces. Potential Anal. **37**(2), 125–169 (2012)
3. Bendikov, A., Grigor'yan, A., Pittet, C., Woess, W.: Isotropic Markov semigroups on ultrametric spaces. Uspekhi Mat. Nauk **69**(4), 3–102 (2014)
4. Bendikov, A., Krupski, P.: On the spectrum of the hierarchical Laplacian. Potential Anal. **41**(4), 1247–1266 (2014)
5. Bendikov, A., Woess, W., Cygan, W.: Oscillating heat kernels on ultrametric spaces. J. Spectr. Theory, arXiv:1610.03292 (2017)
6. Cousty, J., Najman, L., Kenmochi, Y., Guimarães, S.: Hierarchical segmentations with graphs: quasi-flat zones, minimum spanning trees, and saliency maps. J. Math. Imaging Vis. **60**(4), 479–502 (2018)
7. Khrennikov, A., Kozyrev, S., Zúñiga-Galindo, W.: Ultrametric Pseudo Differential Equations and Applications. Encyclopedia of Mathematics and Its Applications. Cambridge University Press, Cambridge (2018)
8. Lindqvist, P.: Notes on the Infinity Laplace Equation. Springer, Cham (2016)
9. Marr, D., Hildreth, E.C.: Theory of edge detection. Proc. Roy. Soc. Lond. Seri. B **207**, 187–217 (1980)
10. van Vliet, L.J., Young, I.T., Beckers, G.L.: A nonlinear operator as edge detector in noisy images. Comput. Vis. Graph. Image Process. **45**, 167–195 (1989)
11. Wetzler, A., Aflalo, Y., Dubrovina, A., Kimmel, R.: The Laplace-Beltrami operator: a ubiquitous tool for image and shape processing. In: Hendriks, C.L.L., Borgefors, G., Strand, R. (eds.) ISMM 2013. LNCS, vol. 7883, pp. 302–316. Springer, Heidelberg (2013). https://doi.org/10.1007/978-3-642-38294-9_26
12. Meyer, F.: Hierarchies of partitions and morphological segmentation. In: Kerckhove, M. (ed.) Scale-Space 2001. LNCS, vol. 2106, pp. 161–182. Springer, Heidelberg (2001). https://doi.org/10.1007/3-540-47778-0_14
13. Meyer, F.: Watersheds on weighted graphs. Pattern Recog. Lett. **47**, 72–79 (2014)
14. Peres, Y., Schramm, O., Sheffield, S., Wilson, D.B.: Tug-of-war and the Infinity Laplacian. J. Am. Math. Soc. **22**(1), 167–210 (2009)

A Link Between the Multiplicative and Additive Functional Asplund's Metrics

Guillaume Noyel[1,2]([⊠]) [iD]

[1] University of Strathclyde, Institute of Global Public Health,
Dardilly, Lyon, France
[2] International Prevention Research Institute, Dardilly, Lyon, France
guillaume.noyel@i-pri.org

Abstract. Functional Asplund's metrics were recently introduced to perform pattern matching robust to lighting changes thanks to double-sided probing in the Logarithmic Image Processing (LIP) framework. Two metrics were defined, namely the LIP-multiplicative Asplund's metric which is robust to variations of object thickness (or opacity) and the LIP-additive Asplund's metric which is robust to variations of camera exposure-time (or light intensity). Maps of distances - i.e. maps of these metric values - were also computed between a reference template and an image. Recently, it was proven that the map of LIP-multiplicative Asplund's distances corresponds to mathematical morphology operations. In this paper, the link between both metrics and between their corresponding distance maps will be demonstrated. It will be shown that the map of LIP-additive Asplund's distances of an image can be computed from the map of the LIP-multiplicative Asplund's distance of a transform of this image and vice-versa. Both maps will be related by the LIP isomorphism which will allow to pass from the image space of the LIP-additive distance map to the positive real function space of the LIP-multiplicative distance map. Experiments will illustrate this relation and the robustness of the LIP-additive Asplund's metric to lighting changes.

Keywords: Pattern Recognition · Insensitivity to lighting changes ·
Logarithmic Image Processing · Mathematical Morphology ·
Asplund's metric · Double-sided probing ·
Relation between functional Asplund's metrics

1 Introduction

Functional Asplund's metrics defined in the Logarithmic Image Processing (LIP) framework [11] are useful tools to compare images. They possess the interesting property of *insensitivity to lighting changes*. Two functional metrics were introduced by Jourlin et al.: (i) firstly, the LIP-multiplicative Asplund's metric which is based on the LIP-multiplicative law is robust to a variation of object

© Springer Nature Switzerland AG 2019
B. Burgeth et al. (Eds.): ISMM 2019, LNCS 11564, pp. 41–53, 2019.
https://doi.org/10.1007/978-3-030-20867-7_4

opacity (or thickness) [9, 12] and (ii) secondly, the LIP-additive Asplund's metric which is defined with the LIP-additive law is robust to a variation of the light intensity (or the camera exposure-time) [11]. They both extend to grey level images the Asplund's metric for binary shapes [1, 6] which is insensitive to an homothety (or a magnification) of the shapes. In each functional metric, one image is selected as a probe and is compared to the other one on its two sides after a LIP-multiplication by a scalar (i.e. an homothety) or a LIP-addition of a constant (i.e. an intensity translation). Maps of Asplund's distances can also be computed between the neighbourhood of each pixel and a probe function defined on a sliding window [12]. Noyel and Jourlin have shown [23, 24] that the map of LIP-multiplicative of Asplund's distance is a combination of Mathematical Morphology (MM) operations [8, 16, 17, 27, 28].

In the literature, other approaches of double-sided probing were defined. E.g., in the hit-or-miss transform [27] and in its extension to grey level images [15], a unique structuring element was matched with the image from above and from below. In [2], Banon et al. translated a unique template two times along the grey level axis. An erosion and an anti-dilation were used to count the pixels whose values were in between the two translated templates. In an approach inspired by the computation of the Hausdorff distance, Odone et al. [25] checked if the grey values of an image were included in an "interval" around the other. The interval was obtained by a 3D dilation of the other image and was vertically translated for each point of the first image. If a sufficient number of the image points were in the "interval", then the template was considered as matched. Barat et al. [3] showed that the last three methods correspond to a neighbourhood of functions (i.e. a tolerance tube) defined by a specific metric for each method. Their topological approach constituted a unified framework named virtual double-sided image probing (VDIP). The metrics were defined in the equivalence class of functions with a grey level addition of a constant. Only the patterns which were in the tolerance tube were selected. In practice, the tolerance tube was computed as a difference between a grey-scale dilation and an erosion. However, even if the approach of Barat et al. was based on a double-sided probing, such a method was not insensitive to lighting variations. It simply removed the variations due to an addition of a constant which had no optical justification contrary to the LIP functional Asplund's metrics. The existence of lighting variations in numerous settings such as medical images [19], industrial control [21, 22], driving assistance [7], large databases [20] or security [5] gives a prime importance to the functional Asplund's metrics defined in the LIP framework.

The aim of this paper is to study the existence of a link between the two functional Asplund's metrics. The paper is organised as follows. Firstly, the LIP framework will be reviewed. The definition and the properties of the LIP-multiplicative and the LIP-additive Asplund's metrics will then be recalled. Secondly, a link between the two metrics will be demonstrated. Finally, the results will be illustrated before concluding.

2   Background

In this section, we will present the LIP model and the two functional Asplund's metrics.

2.1   Logarithmic Image Processing

The LIP model was introduced by Jourlin et al. [10,11,14]. It is based on a famous optical law, namely the Transmittance law, which gives it nice optical properties, especially for processing low-light images. The LIP model is also consistent with the *human visual system* as shown in [4]. This gives to that model the important property to process images acquired by reflection as a human eye would do. Let $\mathcal{I} = [0, M[^D$ be the set of grey level images defined on a domain $D \subset \mathbb{R}^n$ with values in $[0, M[\subset \mathbb{R}$. For 8-bit digitised images, M is equal to $2^8 = 256$. For an image $f \in \mathcal{I}$ acquired by transmission, the transmittance T_f of the semi-transparent object which generates f is equal to $T_f = 1 - f/M$. According to the transmittance law, the transmittance $T_{f \triangle g}$ of the superimposition of two objects which generate f and g is equal to the (point-wise) product "." of their transmittances T_f and T_g:

$$T_{f \triangle g} = T_f.T_g. \tag{1}$$

By replacing the transmittances by their expressions in Eq. 1, one can deduce the LIP-addition of two images $f \triangle g$:

$$f \triangle g = f + g - fg/M. \tag{2}$$

The LIP-multiplication $\triangle$ of an image f by a real number λ is deduced from Eq. 2 by considering that the addition $f \triangle f$ may be written as $2 \triangle f$:

$$\lambda \triangle f = M - M (1 - f/M)^{\lambda}. \tag{3}$$

A LIP-negative function $\triangle f$ can be defined by the equality $f \triangle (\triangle f) = 0$ which allows to write the LIP-difference $f \triangle g$ between two images f and g:

$$\triangle f = (-f)/(1 - f/M) \tag{4}$$

$$f \triangle g = (f - g)/(1 - g/M). \tag{5}$$

Remark 1. $\triangle f$ is not always an image, as it may have negative values. $f \triangle g$ is an image iff $f \geq g$. As $\triangle f$ may take values in the interval $] - \infty, M[$, it is called a function. The set of functions of which the values are less than M is denoted $\mathcal{F}_M$ and is equal to $] - \infty, M[^D$.

Remark 2. Contrary to the classical grey scale, the LIP-scale is inverted: 0 corresponds to the white extremity when no obstacle is placed between the source and the sensor, whereas M corresponds to the black extremity when no light passes through the object.

2.2   The LIP-multiplicative Asplund's Metric

The LIP-multiplicative Asplund's metric was defined in [9, 12]. Let $\mathcal{I}^* = \,]0, M[^D$ be the space of images with strictly positive values.

Definition 1 (LIP-multiplicative Asplund's metric). *Let f and $g \in \mathcal{I}^*$ be two grey level images. A probing image is selected, e.g. g, and two numbers are defined: $\lambda = \inf \{\alpha, f \leq \alpha \,\triangle\, g\}$ and $\mu = \sup \{\alpha, \alpha \,\triangle\, g \leq f\}$. The LIP-multiplicative Asplund's metric $d_{As}^{\triangle}$ is defined by:*

$$d_{As}^{\triangle}(f, g) = \ln (\lambda/\mu). \tag{6}$$

Property 1 ([12]). The LIP-multiplicative Asplund's metric is theoretically invariant under lighting changes caused by variations of the semi-transparent object opacity (or thickness) which are modelled by a LIP-multiplication: $\forall \alpha \in \mathbb{R}^{*+}$: $d_{As}^{\triangle}(f, g) = d_{As}^{\triangle}(\alpha \,\triangle\, f, g)$.

To be mathematically rigorous, $d_{As}^{\triangle}$ is a metric in the space of equivalence classes $\mathcal{I}^{\triangle}$ of the images $f^{\triangle}$ and $g^{\triangle}$, where $f^{\triangle} = \{h \in \mathcal{I}/\exists k > 0, k \,\triangle\, f = h\}$ [11, chap 3]. However, one can keep the notations we used because $\forall (f^{\triangle}, g^{\triangle}) \in (\mathcal{I}^{\triangle})^2, d_{As}^{\triangle}(f^{\triangle}, g^{\triangle}) = d_{As}^{\triangle}(f_1, g_1)$, where $d_{As}^{\triangle}(f_1, g_1)$ is the Asplund's distance between any elements f_1 and g_1 of the equivalence classes $f^{\triangle}$ and $g^{\triangle}$. The relation $(\exists k > 0, k \,\triangle\, f = h)$ is an equivalence relation which satisfies the three properties of reflexivity, symmetry and transitivity [11, 23]

Remark 3 (Terminology). When the Asplund's metric $d_{As}^{\triangle}$ is applied between two images f and g, the real value $d_{As}^{\triangle}(f, g)$ is called the Asplund's distance between f and g. The distance is the value of the metric between both images.

Let $b \in \,]0, M[^{D_b}$ be a probe function defined on a domain $D_b \subset D$. A map of Asplund's distances between an image $f \in \mathcal{I}^*$ and a probe b can be introduced as follows for each pixel $x \in D$ of the image f.

Definition 2 (Map of LIP-multiplicative Asplund's distances [12]**).** *The map of Asplund's distances $As_b^{\triangle} : \mathcal{I}^* \to (\mathbb{R}^+)^D$ is defined by:*

$$As_b^{\triangle} f(x) = d_{As}^{\triangle}(f_{|D_b(x)}, b). \tag{7}$$

$f_{|D_b(x)}$ is the restriction of f to the neighbourhood $D_b(x)$ centred on $x \in D$.

The map of the least upper bounds (mlub) λ_b and the map of the greatest lower bounds (mglb) μ_b can also be defined as follows. Let $\overline{\mathcal{I}} = [0, M]^D$ be the set of images with the value M included.

Definition 3 (LIP-multiplicative maps of the least upper and of the greatest lower bounds[23]**).** *Given $\overline{\mathbb{R}}^+ = [0, +\infty]$, let $f \in \overline{\mathcal{I}}$ be an image and $b \in \,]0, M[^{D_b}$ a probe. Their map of the least upper bounds (mlub)*

$\lambda_b : \overline{\mathcal{I}} \to (\overline{\mathbb{R}^+})^D$ and their map of the greatest lower bounds (mglb) $\mu_b : \overline{\mathcal{I}} \to (\overline{\mathbb{R}^+})^D$ are respectively defined by:

$$\lambda_b f(x) = \inf_{h \in D_b} \{\alpha, f(x+h) \le \alpha \triangle b(h)\}, \tag{8}$$

$$\mu_b f(x) = \sup_{h \in D_b} \{\alpha, \alpha \triangle b(h) \le f(x+h)\}. \tag{9}$$

The map of LIP-multiplicative Asplund's distance has also the property to be theoretically insensitive to lighting changes caused by variations of the object opacity (or thickness) which are modelled by a LIP-multiplication.

Let us define $\widetilde{f} = \ln(1 - f/M)$, where $f \in \overline{\mathcal{I}}$. The relation between the map of LIP-multiplicative Asplund's distances and MM has been demonstrated in [23] with the following Propositions 1 and 2.

Proposition 1. *Given $f \in \overline{\mathcal{I}}$, the mlub λ_b and mglb μ_b are equal to:*

$$\lambda_b f(x) = \vee\{\widetilde{f}(x+h)/\widetilde{b}(h), h \in D_b\}, \tag{10}$$
$$\mu_b f(x) = \wedge\{\widetilde{f}(x+h)/\widetilde{b}(h), h \in D_b\}. \tag{11}$$

If in addition $f > 0$, the map of Asplund's distances expression $As_b^{\triangle}$ becomes:

$$As_b^{\triangle} f = \ln(\lambda_b f/\mu_b f). \tag{12}$$

Proposition 2. *The mlub λ_b and the mglb μ_b are a dilation and an erosion, respectively, between the two complete lattices $(\overline{\mathcal{I}}, \le)$ and $((\overline{\mathbb{R}^+})^D, \le)$ [8,28].*

Indeed, $\forall f, g \in \overline{\mathcal{I}}$, the mlub distributes over supremum $\lambda_b(f \vee g) = \lambda_b(f) \vee \lambda_b(g)$ and the mglb distributes over infimum $\mu_b(f \wedge g) = \mu_b(f) \wedge \mu_b(g)$ [23].

Moreover, as shown in [24], the mlub, mglb and distance map will be expressed in Proposition 3 with the usual morphological operations for grey level functions, namely the dilation $(f \oplus b)(x) = \vee_{h \in D_b}\{f(x-h) + b(h)\}$ and the erosion $(f \ominus b)(x) = \wedge_{h \in D_b}\{f(x+h) - b(h)\}$. The symbols $\oplus$ and $\ominus$ represent the extension to functions of the Minkowski operators between sets.

Proposition 3. *The mlub λ_b, the mglb μ_b and the distance map $As_b^{\triangle}$ are equal to [24]:*

$$\lambda_b f = \exp(\widehat{f} \oplus (-\widehat{b})), \tag{13}$$
$$\mu_b f = \exp(\widehat{f} \ominus \widehat{b}), \tag{14}$$

$$As_b^{\triangle} f = \left[\widehat{f} \oplus (-\widehat{b})\right] - \left[\widehat{f} \ominus \widehat{b}\right]. \tag{15}$$

$\overline{b}$ is the reflected structuring function defined by $\forall x \in \overline{D}_b$, $\overline{b}(x) = b(-x)$ [29] and $\widehat{f}$ is the function defined by $\widehat{f} = \ln(-\widetilde{f}) = \ln(-\ln(1 - f/M))$.

As the operations of dilation $\oplus$ and erosion $\ominus$ exist in numerous image analysis software, Eq. 15 facilitates the programming of the map of LIP-multiplicative Asplund's distances $As_b^{\triangle} f$ of the image f.

2.3   The LIP-additive Asplund's Metric

Jourlin has proposed a definition for the LIP-additive Asplund's metric [11, chap. 3]. Let us present it in a more precise way.

Definition 4 (LIP-additive Asplund's metric). *Let f and $g \in \mathcal{F}_M$ be two functions, we select a probing function, e.g. g, and we define the two numbers: $c_1 = \inf \{c, f \leq c \mathbin{\triangle} g\}$ and $c_2 = \sup \{c, c \mathbin{\triangle} g \leq f\}$, where c lies in the interval $]-\infty, M[$. The LIP-additive Asplund's metric $d_{As}^{\triangle}$ is defined according to:*

$$d_{As}^{\triangle}(f,g) = c_1 \mathbin{\triangle} c_2. \tag{16}$$

Property 2 ([11, chap. 3]). The LIP-additive Asplund's metric is theoretically invariant under lighting changes caused by variations of the light intensity (or camera exposure-time) which are modelled by a LIP-addition of a constant: $\forall k \in\ ]-\infty, M[, d_{As}^{\triangle}(f,g) = d_{As}^{\triangle}(f \mathbin{\triangle} k, g)$ and $d_{As}^{\triangle}(f, f \mathbin{\triangle} k) = 0$.

Remark 4. As for the LIP-multiplicative Asplund's metric, it would be more rigorous to define the LIP-additive Asplund's metric on the equivalence classes $f^{\triangle}$ and $g^{\triangle}$, where $f^{\triangle} = \{h \in \mathcal{F}_M / \exists k \in\ ]-\infty, M[, f \mathbin{\triangle} k = h\}$.

A map of Asplund's distances between an image and a probe can also be defined for each image pixel.

Definition 5 (Map of LIP-additive Asplund's distances). *Let $f \in \mathcal{F}_M$ be a function and $b \in\ ]-\infty, M[^{D_b}$ a probe. The map of Asplund's distances is the mapping $As_b^{\triangle} : \mathcal{F}_M \to \mathcal{I}$ defined by:*

$$As_b^{\triangle} f(x) = d_{As}^{\triangle}(f_{|D_b(x)}, b). \tag{17}$$

The map of LIP-additive Asplund's distance has also the property to be theoretically insensitive to lighting changes caused by variations of light intensity (or camera exposure-time) which are modelled by a LIP-addition of a constant.

3   Linking the LIP-multiplicative and the LIP-additive Asplund's Metrics

In this section, first, the map of LIP-additive Asplund's distances will be expressed with neighbourhood operations. Then the link between both maps and the metrics will be studied. Finally, this link will be briefly discussed.

3.1   General Expression of the Map of LIP-additive Asplund's Distances

From Definition 4, the maps of the least upper bounds $c_{1_b} f$ and of the greatest lower bounds $c_{2_b} f$ can be defined by computing the constant c_1 and c_2 between a probe $b \in\ ]-\infty, M[^{D_b}$ and the function restriction $f_{|D_b(x)}$.

Definition 6 (LIP-additive maps of the least upper and of the greatest lower bounds). *Let $f \in \overline{\mathcal{F}}_M$ be a function and $b \in]-\infty, M]^{D_b}$ a probe. Their map of the least upper bounds (mlub) $c_{1_b} : \overline{\mathcal{F}}_M \to \overline{\mathcal{F}}_M$ and their map of the greatest lower bounds (mglb) $c_{2_b} : \overline{\mathcal{F}}_M \to \overline{\mathcal{F}}_M$ are defined by:*

$$c_{1_b} f(x) = \inf_{h \in D_b} \{c, f(x+h) \leq c \triangle b(h)\} \tag{18}$$

$$c_{2_b} f(x) = \sup_{h \in D_b} \{c, c \triangle b(h) \leq f(x+h)\}. \tag{19}$$

The mlub expression can be rewritten as follows, $\forall x \in D$:

$$
\begin{aligned}
c_{1_b} f(x) &= \inf\{c, c \geq f(x+h) \triangle b(h), h \in D_b\} \\
&= \vee\{f(x+h) \triangle b(h), h \in D_b\},
\end{aligned}
\tag{20}
$$

where the last equality is due to the complete lattice structure. In a similar way, the mglb c_{2_b} becomes:

$$c_{2_b} f(x) = \wedge\{f(x+h) \triangle b(h), h \in D_b\}. \tag{21}$$

The general expression of the map of LIP-additive Asplund's distances between f and b is therefore:

$$As_b^{\triangle} f = c_{1_b} f \triangle c_{2_b} f. \tag{22}$$

This last expression will be useful to establish the link with the map of LIP-additive Asplund's distances.

3.2 Link Between the Maps of Distances (and the Metrics)

First of all, an isomorphism is needed between the lattice $(\overline{\mathbb{R}^+})^D$ of the LIP-multiplicative mlub $\lambda_b f$, or mglb $\mu_b f$, and the lattice $[-\infty, M]^D$ of the LIP-additive mlub $c_{1_b} f$, or mglb $c_{2_b} f$. This isomorphism $\xi : [-\infty, M]^D \to \overline{\mathbb{R}}^D$ and its inverse ξ^{-1} were both defined in [13, 18] by:

$$\xi(f) = -M \ln (1 - f/M) \tag{23}$$

$$\xi^{-1}(f) = M(1 - \exp(-f/M)). \tag{24}$$

Remark 5. One can notice that $\xi(f) = -M\tilde{f}$. This relation will be useful in the proof hereinafter.

There exists the following relation between the distance maps.

Proposition 4. *Let $f \in \mathcal{I}^*$ be an image, $f^c = M - f$ its complement and $b \in]0, M]^{D_b}$ a structuring function. The map of LIP-additive Asplund's distances is related to the map of LIP-multiplicative distances by the following equation:*

$$As_b^{\triangle} f = \frac{1}{M}\xi\left(As_{(\xi(b))^c}^{\triangle}(\xi(f))^c\right), \tag{25}$$

where $(\xi(f))^c = M - \xi(f) \in]-\infty, M]^D$.

Corollary 1. *Using the following variable changes $f_1 = (\xi(f))^c$ and $b_1 = (\xi(b))^c$, with $f_1 \in \mathcal{F}_M$ and $b_1 \in\,]-\infty, M[^{D_b}$, an equivalent equation is obtained:*

$$As_{b_1}^{\triangle} f_1 = \xi^{-1}\left(M.As_{\xi^{-1}(b_1^c)}^{\triangle}\xi^{-1}(f_1^c)\right). \tag{26}$$

Remark 6. Equation 25 can also be written as:
$As_{(\xi(b))^c}^{\triangle}(\xi(f))^c = \xi^{-1}(M.As_b^{\triangle} f) = M(1 - \exp(-As_b^{\triangle} f))$. As the map of LIP-multiplicative Asplund's distances $As_b^{\triangle} f$ of f is an element of $[0, +\infty[^D$, the map of LIP-additive distances $As_{(\xi(b))^c}^{\triangle}(\xi(f))^c$ of $(\xi(f))^c$ is an element of $[0, M[^D = \mathcal{I}$ and is therefore an image.

Remark 7. The same relation exists between both functional Asplund's metrics:

$$d_{As}^{\triangle}(f, g) = \frac{1}{M}\xi(d_{As}^{\triangle}([\xi(f)]^c, [\xi(g)]^c)). \tag{27}$$

Proof. Let $f \in \overline{\mathcal{I}} = [0, M]^D$ be an image, there is:

$$(M - f) \triangle (M - b) = M\frac{(M - f) - (M - b)}{M - (M - b)} = M\frac{b - f}{b} = M\left(1 - \frac{f}{b}\right). \tag{28}$$

Equation 28 is set in Eqs. 20 and 21, which gives: $\forall x \in D$,

$$c_{1_{(\xi(b))^c}}^{\triangle}(\xi(f))^c(x) = \vee_{h \in D_b}\{(M - \xi(f)(x + h)) \triangle (M - \xi(b)(h))\}$$

$$= \vee_{h \in D_b}\left\{M\left(1 - \frac{\xi(f)(x + h)}{\xi(b)(h)}\right)\right\}$$

$$= M\left(1 - \wedge_{h \in D_b}\left\{\frac{\xi(f)(x + h)}{\xi(b)(h)}\right\}\right)$$

$$= M\left(1 - \wedge_{h \in D_b}\left\{\frac{-M\tilde{f}(x + h)}{-M\tilde{b}(h)}\right\}\right)$$

$$= M(1 - \wedge_{h \in D_b}\{\tilde{f}(x + h)/\tilde{b}(h)\})$$

$$= M(1 - \mu_b f(x)), \tag{29}$$

$$c_{2_{(\xi(b))^c}}^{\triangle}(\xi(f))^c(x) = M(1 - \vee_{h \in D_b}\{\tilde{f}(x + h)/\tilde{b}(h)\})$$

$$= M(1 - \lambda_b f(x)). \tag{30}$$

By combining Eqs. 29 and 30 with Eqs. 28, 12 and 24, one deduces the following equations:

$$As_{(\xi(b))^c}^{\triangle}(\xi(f))^c = c_{1_{(\xi(b))^c}}^{\triangle}(\xi(f))^c \triangle c_{2_{(\xi(b))^c}}^{\triangle}(\xi(f))^c = (M - M\mu_b f) \triangle (M - M\lambda_b f)$$

$$= M\left(1 - \frac{\mu_b f}{\lambda_b f}\right) = M(1 - \exp(-As_b^{\triangle} f))$$

$$= \xi^{-1}(M.As_b^{\triangle} f)$$

$$\Leftrightarrow As_b^{\triangle} f = \frac{1}{M}\xi\left(As_{(\xi(b))^c}^{\triangle}(\xi(f))^c\right).$$

Moreover, when f lies in $\overline{\mathcal{I}}$, the function $\xi(f)$ lies in $[0, \infty]^D$ and $(\xi(f))^c = M - \xi(f)$ lies in $[-\infty, M]^D$. $\square$

3.3 Discussion

Equations 26 and 25 show that the maps of LIP-multiplicative and LIP-additive Asplund's distances as well as their corresponding metrics are related by the isomorphism ξ. These relations allow to compute one distance map of an image by mean of the other distance map of a transform of this image. E.g. the LIP-additive map $As_{b_1}^{\triangle} f_1$ of the function f_1 can be computed by using the program of the LIP-multiplicative map $As_{\xi^{-1}(b_1^c)}^{\triangle} \xi^{-1}(f_1^c)$ of the transformed function $\xi^{-1}(f_1^c)$. However, both equations do not directly link both distance maps of the image f. E.g. the LIP-multiplicative map of the image f, $As_b^{\triangle} f$, is not directly related to the LIP-additive map of the image f, $As_b^{\triangle} f$.

The relation given in Eqs. 26 and 25 is not surprising. Indeed, the map of LIP-additive Asplund's distances $As_{(\xi(b))^c}^{\triangle} (\xi(f))^c$ is an image which lies in $[0, M]^D$, whereas the map of LIP-multiplicative Asplund's distances $As_b^{\triangle} f$ lies in $(\mathbb{R}^+)^D$. The isomorphism ξ allows to pass from the image space of the LIP-additive distance map $As_{(\xi(b))^c}^{\triangle} (\xi(f))^c$ to the real function space of the LIP-multiplicative distance map $As_b^{\triangle} f$.

4 Illustration

Figure 1 illustrates relation 25, where the map of LIP-additive Asplund's distances $As_b^{\triangle}$ is deduced from the map of LIP-multiplicative distances $As_{\xi^{-1}(b^c)}^{\triangle}$. Moreover, it shows the theoretical insensitivity of the map of LIP-additive Asplund's distances to a lighting change simulated by the LIP-addition of a constant. For this experiment, an image of a parrot [26] (Fig. 1a) was selected in the Yahoo Flickr Creative Commons 100 Million Dataset (YFCC100M) [30] and converted into a luminance image f in grey levels (Fig. 1b). A darkened image f^{dk} is obtained by LIP-adding a constant 200 to the luminance image f: $f^{dk} = f \mathbin{\triangle} 200$ (Fig. 1c). This operation simulates a decreasing of the camera exposure-time or a decreasing of the light intensity. In order to detect the parrot's eye, a probe b is designed. A white ring - with a height of 161 grey levels - surrounds a black disk whose grey value is equal to 4 (Fig. 1d). The LIP-additive distance map $As_b^{\triangle} f$ of the image f (Fig. 1f and g) is computed from the LIP-multiplicative distance map $As_{\xi^{-1}(b^c)}^{\triangle} \xi^{-1}(f^c)$ of the function $\xi^{-1}(f^c)$ (Fig. 1e) using Eq. 25. The centre of the parrot's eye corresponds to the minimum of this former map, $As_b^{\triangle} f$ (Fig. 1f and g), which can be easily extracted by a threshold. It is remarkable that the detection of an object in a low-light and complex image can be performed by a simple threshold of its map of Asplund's distances.

Moreover, the LIP-additive distance map $As_b^{\triangle} f^{dk}$ between the darkened image f^{dk} and the probe b - designed for the brightest image f - is also equal to the LIP-additive distance map $As_b^{\triangle} f$ between the brightest image f and the

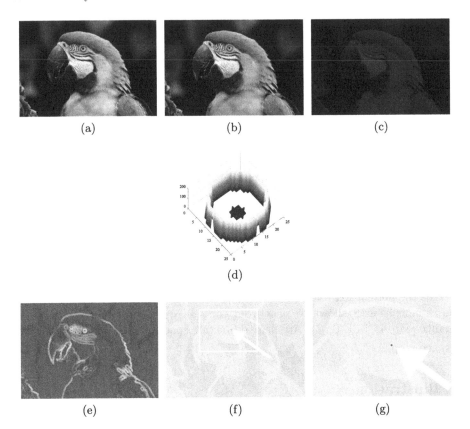

Fig. 1. (a) Colour image of a parrot [26] and (b) its luminance f. (c) Darkened image f^{dk} obtained by a LIP addition of a constant 200: $f = f^{dk} \triangle 200$. (d) Topographic surface of the probe b. (e) Map of LIP-multiplicative Asplund's distances $As^{\triangle}_{\xi^{-1}(b^c)} \xi^{-1}(f^c)$ of the function $\xi^{-1}(f^c)$ which is used to compute the (f) map of LIP-additive Asplund's distances $As^{\triangle}_b f$ of the image f. It is also equal to the distance map $As^{\triangle}_b f^{dk}$ between the darkened image f^{dk} and the same probe b. (g) Zoom in of the map (f). Both white arrows indicate the minimum of the map $As^{\triangle}_b f$ corresponding to the eye centre. (Color figure online)

probe b (Fig. 1f). This result shows the insensitivity of the map of LIP-additive Asplund's distance to a variation of camera exposure-time which is simulated by a LIP-addition of a constant.

In addition, the LIP-additive distance map was also computed directly with Eq. 17 and compared to the LIP-additive distance map obtained from Eq. 25 (Fig. 1f). Both maps of distances were equal with a numerical precision corresponding to the rounding error of the computer. Therefore, relation 25 is numerically verified.

5 Conclusion

A link between the maps of LIP-multiplicative and LIP-additive Asplund's distances has therefore been successfully demonstrated. The relation is based on the LIP-isomorphism which allows to pass from the image space of the LIP-additive distance map to the positive real function space of the LIP-multiplicative distance map. However, there does not exist a link between the maps of LIP-multiplicative and LIP-additive Asplund's distances of the same image. Nevertheless, the proven relation is not only interesting from a theoretical point of view but also from a practical point of view. Indeed, it allows to compute one map of an image from the other map of a transform of this image. Experiments have verified the relation from a numerical point of view. They have also illustrated the main interest of the map of LIP-additive Asplund's distances, i.e. its insensitivity to lighting changes modelled by a LIP-addition and corresponding to a variation of the camera exposure-time or of the light intensity. Such properties open the way to numerous applications where the lighting conditions are partially controlled.

References

1. Asplund, E.: Comparison between plane symmetric convex bodies and parallelograms. Math. Scand. **8**, 171–180 (1960). https://doi.org/10.7146/math.scand.a-10606
2. Banon, G., Faria, S.: Morphological approach for template matching. In: Proceedings of X Brazilian Symposium on Computer Graphics and Image Process, pp. 171–178. IEEE Computer Society, October 1997. https://doi.org/10.1109/SIGRA.1997.625169
3. Barat, C., Ducottet, C., Jourlin, M.: Virtual double-sided image probing: a unifying framework for non-linear grayscale pattern matching. Pattern Recogn. **43**(10), 3433–3447 (2010). https://doi.org/10.1016/j.patcog.2010.04.020
4. Brailean, J., Sullivan, B., Chen, C., Giger, M.: Evaluating the EM algorithm for image processing using a human visual fidelity criterion. In: International Conference on Acoustics, Speech, and Signal Process, ICASSP-1991, vol 4, pp. 2957–2960, April 1991. https://doi.org/10.1109/ICASSP.1991.151023
5. Foresti, G.L., Micheloni, C., Snidaro, L., Remagnino, P., Ellis, T.: Active video-based surveillance system: the low-level image and video processing techniques needed for implementation. IEEE Sig. Process. Mag. **22**(2), 25–37 (2005). https://doi.org/10.1109/MSP.2005.1406473
6. Grünbaum, B.: Measures of symmetry for convex sets. In: Proceedings of Symposium in Pure Mathematics, vol. 7, pp. 233–270. American Mathematical Society, Providence, RI (1963). https://doi.org/10.1090/pspum/007
7. Hautière, N., Aubert, D., Jourlin, M.: Measurement of local contrast in images, application to the measurement of visibility distance through use of an onboard camera. Traitement du Signal **23**(2), 145–158 (2006). http://hdl.handle.net/2042/5826
8. Heijmans, H.: Morphological image operators. In: Advances in Imaging and Electron Physics: Supplement, vol. 25. Academic Press, Boston (1994). https://books.google.fr/books?id=G-hRAAAAMAAJ

9. Jourlin, M., Couka, E., Abdallah, B., Corvo, J., Breugnot, J.: Asplünd's metric defined in the Logarithmic Image Processing (LIP) framework: a new way to perform double-sided image probing for non-linear grayscale pattern matching. Pattern Recognit. **47**(9), 2908–2924 (2014). https://doi.org/10.1016/j.patcog.2014.03.031

10. Jourlin, M., Pinoli, J.: Logarithmic image processing: the mathematical and physical framework for the representation and processing of transmitted images. In: Hawkes, P.W. (ed.) Advances in Imaging and Electron Physics, vol. 115, pp. 129–196. Elsevier, Amsterdam (2001). https://doi.org/10.1016/S1076-5670(01)80095-1

11. Jourlin, M.: Logarithmic image processing: theory and applications. In: Advances in Imaging and Electron Physics, vol. 195. Elsevier Science, Amsterdam (2016). https://doi.org/10.1016/S1076-5670(16)30078-7

12. Jourlin, M., Carré, M., Breugnot, J., Bouabdellah, M.: Chapter 7 - Logarithmic image processing: additive contrast, multiplicative contrast, and associated metrics. In: Hawkes, P.W. (ed.) Advances in Imaging and Electron Physics, vol. 171, pp. 357–406. Elsevier, Amsterdam (2012). https://doi.org/10.1016/B978-0-12-394297-5.00007-6

13. Jourlin, M., Pinoli, J.C.: Image dynamic range enhancement and stabilization in the context of the logarithmic image processing model. Sig. Process. **41**(2), 225–237 (1995). https://doi.org/10.1016/0165-1684(94)00102-6

14. Jourlin, M., Pinoli, J.: A model for logarithmic image-processing. J. Microsc. **149**(1), 21–35 (1988). https://doi.org/10.1111/j.1365-2818.1988.tb04559.x

15. Khosravi, M., Schafer, R.: Template matching based on a grayscale hit-or-miss transform. IEEE Trans. Image Process. **5**(6), 1060–1066 (1996). https://doi.org/10.1109/83.503921

16. Matheron, G.: Eléments pour une théorie des milieux poreux. Masson, Paris (1967)

17. Najman, L., Talbot, H.: Mathematical Morphology: From Theory to Applications, 1st edn. Wiley, Hoboken (2013). https://doi.org/10.1002/9781118600788

18. Navarro, L., Deng, G., Courbebaisse, G.: The symmetric logarithmic image processing model. Digit. Sig. Process. **23**(5), 1337–1343 (2013). https://doi.org/10.1016/j.dsp.2013.07.001

19. Noyel, G., Angulo, J., Jeulin, D., Balvay, D., Cuenod, C.A.: Multivariate mathematical morphology for DCE-MRI image analysis in angiogenesis studies. Image Anal. Stereol. **34**(1), 1–25 (2014)

20. Noyel, G., Thomas, R., Bhakta, G., Crowder, A., Owens, D., Boyle, P.: Superimposition of eye fundus images for longitudinal analysis from large public health databases. Biomed. Phys. Eng. Express **3**(4), 045015 (2017). https://doi.org/10.1088/2057-1976/aa7d16

21. Noyel, G.: Method of monitoring the appearance of the surface of a tire. International PCT patent WO2011131410 (A1), also published as: US9002093 (B2), FR2959046 (B1), JP5779232 (B2), EP2561479 (A1), CN102844791 (B), BR112012025402 (A2), October 2011. https://patentscope.wipo.int/search/en/WO2011131410

22. Noyel, G., Jeulin, D., Parra-Denis, E., Bilodeau, M.: Method of checking the appearance of the surface of a tyre. International PCT patent WO201304 5593 (A1), also published as US9189841 (B2), FR2980735 (B1), EP2761587 (A1), CN103843034 (A), April 2013. https://patentscope.wipo.int/search/en/WO20130 45593

23. Noyel, G., Jourlin, M.: Double-sided probing by map of Asplund's distances using logarithmic image processing in the framework of mathematical morphology. In: Angulo, J., Velasco-Forero, S., Meyer, F. (eds.) ISMM 2017. LNCS, vol. 10225, pp. 408–420. Springer, Cham (2017). https://doi.org/10.1007/978-3-319-57240-6_33

24. Noyel, G., Jourlin, M.: A simple expression for the map of Asplund's distances with the multiplicative Logarithmic Image Processing (LIP) law. In: 12th European Congress for Stereology and Image Analysis, Kaiserslautern, Germany, September 2017. https://arxiv.org/abs/1708.08992

25. Odone, F., Trucco, E., Verri, A.: General purpose matching of grey level arbitrary images. In: Arcelli, C., Cordella, L.P., di Baja, G.S. (eds.) IWVF 2001. LNCS, vol. 2059, pp. 573–582. Springer, Heidelberg (2001). https://doi.org/10.1007/3-540-45129-3_53

26. Parrot: Parrot image from the YFCC100M dataset (2008). Licence CC BY 2.0. https://www.flickr.com/photos/mdpettitt/2744081052

27. Serra, J.: Image Analysis and Mathematical Morphology: Theoretical Advances, vol. 2. Academic Press, Boston (1988). https://books.google.fr/books?id=BpdT AAAAYAAJ

28. Serra, J., Cressie, N.: Image Analysis and Mathematical Morphology, vol. 1. Academic Press, New York (1982). https://books.google.fr/books?id=RQIUAQ AAIAAJ

29. Soille, P.: Morphological Image Analysis: Principles and Applications, 2nd edn. Springer, Heidelberg (2004). https://doi.org/10.1007/978-3-662-05088-0

30. Thomee, B., et al.: YFCC100M: the new data in multimedia research. Commun. ACM **59**(2), 64–73 (2016). https://doi.org/10.1145/2812802

Discrete Topology and Tomography

An Equivalence Relation Between Morphological Dynamics and Persistent Homology in 1D

Nicolas Boutry[1]([⊠]), Thierry Géraud[1], and Laurent Najman[2]

[1] EPITA Research and Development Laboratory (LRDE),
Le Kremlin-Bicêtre, France
nicolas.boutry@lrde.epita.fr
[2] Université Paris-Est, LIGM, Équipe A3SI, ESIEE, Marne-la-Vallée, France

Abstract. We state in this paper a strong relation existing between Mathematical Morphology and Morse Theory when we work with 1D $\mathfrak{D}$-Morse functions. Specifically, in Mathematical Morphology, a classic way to extract robust markers for segmentation purposes, is to use the dynamics. On the other hand, in Morse Theory, a well-known tool to simplify the Morse-Smale complexes representing the topological information of a $\mathfrak{D}$-Morse function is the persistence. We show that pairing by persistence is equivalent to pairing by dynamics.

Keywords: Mathematical Morphology · Morse Theory · Dynamics · Persistence

1 Introduction

In *Mathematical Morphology* [16–18], *dynamics* [11,12,19], defined in terms of continuous paths and optimization problems, represents a very powerful tool to measure the significance of extrema in a gray-level image. Thanks to dynamics, we can construct efficient markers of objects belonging to an image which do not depend on the size or on the shape of the object we want to segment (to compute watershed transforms [15,20] and proceed to image segmentation). This contrasts with convolution filters very often used in digital signal processing or morphological filters [16–18] where geometrical properties do matter.

Selecting components of high dynamics in an image is a way to filter objects depending on their contrast, whatever the scale of the objects. In *Persistent Homology* [7,9] well-known in *Computational Topology* [8], we can find the same paradigm: topological features whose *persistence* is high are "true" when the ones whose persistence is low are considered as sampling artifacts, whatever their scale. An example of application of persistence is the filtering of *Morse-Smale complexes* used in *Morse Theory* [10,13] where pairs of extrema of low persistence are canceled for simplification purpose. This way, we obtain simplified topological representations of *Morse functions*.

© Springer Nature Switzerland AG 2019
B. Burgeth et al. (Eds.): ISMM 2019, LNCS 11564, pp. 57–68, 2019.
https://doi.org/10.1007/978-3-030-20867-7_5

In this paper, we prove that the relation between Mathematical Morphology and Persistent Homology is strong in the sense that pairing by dynamics and pairing by persistence are equivalent (and then dynamics and persistence are equal), at least in 1D, when we work with $\mathfrak{D}$-Morse functions (we define $\mathfrak{D}$-Morse functions as Morse functions which tend to $\pm\infty$ when the 2-norm of their argument tends to $+\infty$).

The plan of the paper is the following: Sect. 2 recalls the mathematical background needed in this paper, Sect. 3 proves the equivalence between pairing by dynamics and pairing by persistence and Sect. 4 concludes the paper.

2   Mathematical Background

A 1D Morse function is a function $f : \mathbb{R} \to \mathbb{R}$ which belongs to $\mathcal{C}^2(\mathbb{R})$ and whose second derivative $f''(x^*)$ at each critical point $x^* \in \mathbb{R}$ is different from 0. A consequence of this property is that the critical points of a Morse function are isolated. A 1D $\mathfrak{D}$-Morse function $f : \mathbb{R} \to \mathbb{R}$ is a Morse function verifying that $f(x)$ tends to $\pm\infty$ when x tends to $\pm\infty$.

In this paper, we work with one-dimensional $\mathfrak{D}$-Morse functions $f : \mathbb{R} \to \mathbb{R}$ with the additional property that for any two local extrema x_1 and x_2 of f, $x_1 \neq x_2$ implies that $f(x_1) \neq f(x_2)$. In other words, critical values of f are "unique".

Even if it does not seem realistic to assume that the critical values are unique, we can easily obtain this property by perturbing slightly the given function while preserving its topology.

Let us define the *lower threshold sets*: the set $[f \leq \lambda]$ for any $\lambda \in \mathbb{R}$ is defined as the set $\{x \in \mathbb{R} \; ; \; f(x) \leq \lambda\}$. Then, we define the *connected component* of a set $X \subseteq \mathbb{R}$ containing $x \in X$ as the greatest interval (in the inclusion sense) contained in X and containing x and we denote it $\mathcal{CC}(X, x)$.

We denote as usual $\overline{\mathbb{R}} := \mathbb{R} \cup \{-\infty, +\infty\}$. For a, b two elements of $\overline{\mathbb{R}}$, $\mathrm{iv}(a, b)$ is defined as the *interval value*

$$[\min(a, b), \max(a, b)].$$

Also, for a given function $f : \mathbb{R} \to \mathbb{R}$ and for $(a, b) \in \overline{\mathbb{R}}^2$ verifying $a < b$, we denote:

$$\mathrm{Rep}([a, b], f) := \arg \min_{x \in \mathrm{locmin}([a,b],f)} f(x),$$

where $\mathrm{locmin}(A, f)$ is defined as the set of local minima of f belonging to the set $A \subseteq \overline{\mathbb{R}}$. When f is a $\mathfrak{D}$-Morse function with unique critical values and when $\mathrm{locmin}([a,b], f)$ is non-empty, $\mathrm{Rep}([a, b], f)$ is a degenerate set. Then, $\mathrm{Rep}([a, b], f)$ is abusively said to be the *representative* [7] of the interval $[a, b]$ relatively to f.

Finally, we denote by $\varepsilon \to 0^+$ the fact that ε tends to 0 with the constraint $\varepsilon > 0$.

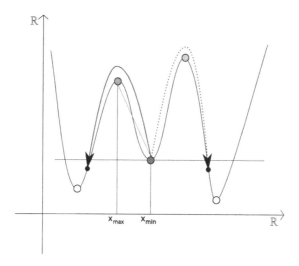

Fig. 1. Example of pairing by dynamics: the abscissa $x_{\min}$ of the red point is paired by dynamics relatively to f with the abscissa $x_{\max}$ of the green point on its left because the "effort" needed to reach a point of lower height than $f(x_{\min})$ (like the two black points) following the graph of f is minimal on the left. (Color figure online)

2.1 Pairing by Dynamics

Let $f : \mathbb{R} \rightarrow \mathbb{R}$ be a $\mathfrak{D}$-Morse function with unique critical values. For $x_{\min} \in \mathbb{R}$ a local minimum of f, if there exists at least one abscissa $x'_{\min} \in \mathbb{R}$ of f such that $f(x'_{\min}) < f(x_{\min})$, then we define the *dynamics* [12] of $x_{\min}$ by:

$$\mathrm{dyn}(x_{\min}) := \min_{\gamma \in C} \max_{s \in [0,1]} f(\gamma(s)) - f(x_{\min}),$$

where C is the set of paths $\gamma : [0,1] \rightarrow \mathbb{R}$ verifying $\gamma(0) := x_{\min}$ and verifying that there exists some $s \in]0,1]$ such that $f(\gamma(s)) < f(x_{\min})$.

Let us now define γ^* as a path of C verifying:

$$\max_{s \in [0,1]} f(\gamma^*(s)) - f(x_{\min}) = \min_{\gamma \in C} \max_{s \in [0,1]} f(\gamma(s)) - f(x_{\min}),$$

then we say that this path is *optimal*. The real value $x_{\max}$ *paired by dynamics* to $x_{\min}$ (relatively to f) is the local maximum of f characterized by:

$$x_{\max} := \gamma^*(s^*),$$

with $f(\gamma^*(s^*)) = \max_{s \in [0,1]} f(\gamma^*(s))$. We obtain then:

$$f(x_{\max}) - f(x_{\min}) = \mathrm{dyn}(x_{\min}).$$

Note that the local maximum $x_{\max}$ of f does not depend on the path γ^* (see Fig. 1), and its value is unique (by hypothesis on f), which shows that in some way $x_{\max}$ and $x_{\min}$ are "naturally" paired by dynamics.

2.2 Pairing by Persistence

From now on, we will denote by $\mathrm{cl}_{\overline{\mathbb{R}}}(A)$ the closure in $\overline{\mathbb{R}}$ of the set $A \subseteq \mathbb{R}$.

Algorithm 1. Pairing by persistence of $x_{\max}$.

$x_{\min} := \emptyset;$
$[x_{\max}^-, x_{\max}^+] := \mathrm{cl}_{\overline{\mathbb{R}}}(\mathcal{CC}([f \leq f(x_{\max})], x_{\max}));$
if $x_{\max}^- > -\infty \parallel x_{\max}^+ < +\infty$ **then**
$\quad x_{\min}^- := \mathrm{Rep}([x_{\max}^-, x_{\max}], f);$
$\quad x_{\min}^+ := \mathrm{Rep}([x_{\max}, x_{\max}^+], f);$
$\quad$ **if** $x_{\max}^- > -\infty \,\&\&\, x_{\max}^+ < +\infty$ **then**
$\quad\quad\lfloor\ x_{\min} := \arg\max_{x \in \{x_{\min}^-, x_{\min}^+\}} f(x);$
$\quad$ **if** $x_{\max}^- > -\infty \,\&\&\, x_{\max}^+ = +\infty$ **then**
$\quad\quad\lfloor\ x_{\min} := x_{\min}^-;$
$\quad$ **if** $x_{\max}^- = -\infty \,\&\&\, x_{\max}^+ < +\infty$ **then**
$\quad\quad\lfloor\ x_{\min} := x_{\min}^+;$

return $x_{\min}$;

Let $f : \mathbb{R} \to \mathbb{R}$ be a $\mathfrak{D}$-Morse function with unique critical values, and let $x_{\max}$ be a local maximum of f. Let us recall the 1D procedure [7] which pairs (relatively to f) local maxima to local minima (see Algorithm 1). Roughly speaking, the representatives $x_{\min}^-$ and $x_{\min}^+$ are the abscissas where connected components of respectively $[f \leq (f(x_{\min}^-)]$ and $[f \leq (f(x_{\min}^+)]$ "emerge" (see Fig. 2), when $x_{\max}$ is the abscissa where two connected components of $[f < f(x_{\max})]$ "merge" into a single component of $[f \leq f(x_{\max})]$. Pairing by persistence associates then $x_{\max}$ to the value $x_{\min}$ belonging to $\{x_{\min}^-, x_{\min}^+\}$ which maximizes $f(x_{\min})$. The *persistence* of $x_{\max}$ relatively to f is then equal to $\mathrm{per}(x_{\max}) := f(x_{\max}) - f(x_{\min})$.

3 Pairings by Dynamics and by Persistence are Equivalent in 1D

In this section, we prove that under some constraints, pairings by dynamics and by persistence are equivalent in the 1D case.

Proposition 1. *Let* $f : \mathbb{R} \to \mathbb{R}$ *be a* $\mathfrak{D}$-*Morse function with unique critical values. Now let us assume that a local minimum* $x_{\min} \in \mathbb{R}$ *of* f *is paired with a local maximum* $x_{\max}$ *of* f *by dynamics. We assume without constraints that* $x_{\min} < x_{\max}$ *(the reasoning is the same for the opposite assumption). Also, we denote by* $(x_{\max}^-, x_{\max}^+) \in \overline{\mathbb{R}}^2$ *the two values verifying:*

$$[x_{\max}^-, x_{\max}^+] = \mathrm{cl}_{\overline{\mathbb{R}}}(\mathcal{CC}([f \leq f(x_{\max})], x_{\max})).$$

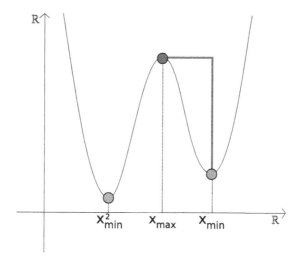

Fig. 2. Example of pairing by persistence: the abscissa $x_{\max}$ of the local maximum in red is paired by persistence relatively to f with the abscissa $x_{\min}$ of the local minimum in green since its image by f is greater than the image by f of the abscissa $x^2_{\min}$ of the local minimum drawn in pink. (Color figure online)

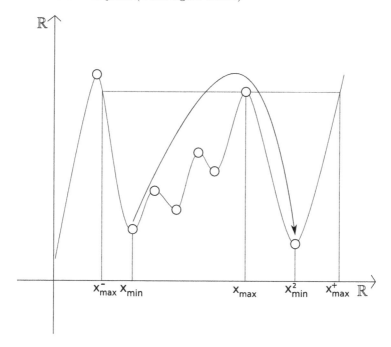

Fig. 3. A $\mathfrak{D}$-Morse function where the local extrema $x_{\min}$ and $x_{\max}$ are paired by dynamics.

Then the following properties are true:

(P1) $x_{\min} = \mathrm{Rep}([x_{\max}^-, x_{\max}], f)$,

(P2) When $x_{\max}^+$ is finite, $x_{\min}^2 := \mathrm{Rep}([x_{\max}, x_{\max}^+], f)$ verifies $f(x_{\min}^2) < f(x_{\min})$,

(P3) $x_{\max}$ and $x_{\min}$ are paired by persistence.

Proof: Figure 3 depicts an example of $\mathfrak{D}$-Morse function where $x_{\min}$ and $x_{\max}$ are paired by dynamics.

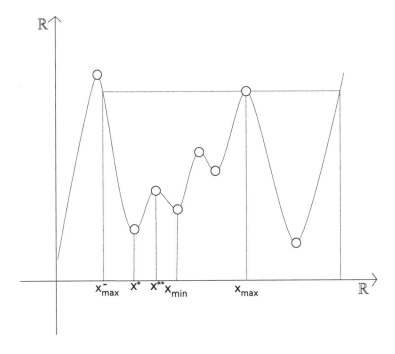

Fig. 4. Proof of (P1).

Let us prove (P1); we proceed by *reductio ad absurdum*. When $x_{\min}$ is not the lowest local minimum of f on the interval $[x_{\max}^-, x_{\max}]$, then there exists another local minimum $x^* \in [x_{\max}^-, x_{\max}]$ of f (see Fig. 4) which verifies $f(x^*) < f(x_{\min})$ (x^* and $x_{\min}$ being distinct local extrema of f, their images by f are not equal). Then, because the path joining $x_{\min}$ and x^* belongs to C (defined in Subsect. 2.1), we have:

$$\mathrm{dyn}(x_{\min}) \leq \max\{f(x) - f(x_{\min}) \; ; \; x \in \mathrm{iv}(x^*, x_{\min})\}.$$

Let us call $x^{**} := \arg\max_{x \in [\mathrm{iv}(x_{\min}, x^*)]} f(x)$, we can deduce that $f(x^{**}) < f(x_{\max})$ since $x^{**} \in \mathrm{iv}(x^*, x_{\min}) \subseteq]x_{\max}^-, x_{\max}[$. This way,

$$\mathrm{dyn}(x_{\min}) \leq f(x^{**}) - f(x_{\min}),$$

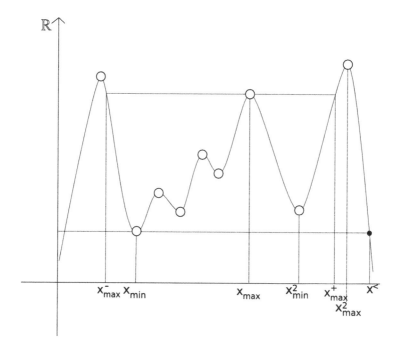

Fig. 5. Proof of $(P2)$ in the case where x_{max}^+ is finite.

which is lower than $f(x_{\mathrm{max}}) - f(x_{\mathrm{min}})$; this is a contradiction since x_{min} and x_{max} are paired by dynamics. $(P1)$ is then proved.

Now let us prove $(P2)$. Let us assume that x_{max}^+ is finite and let x_{min}^2 be the representative of $[x_{\mathrm{max}}, x_{\mathrm{max}}^+]$ relatively to f. Let us assume that $f(x_{\mathrm{min}}^2) > f(x_{\mathrm{min}})$. Note that we cannot have equality of $f(x_{\mathrm{min}}^2)$ and $f(x_{\mathrm{min}})$ since x_{min} and x_{min}^2 are both local extrema of f. Then we obtain Fig. 5. Since with $x \in [x_{\mathrm{max}}, x_{\mathrm{max}}^+]$, we have $f(x) \geq f(x_{\mathrm{min}}^2) > f(x_{\mathrm{min}})$, and because x_{min} is paired with x_{max} by dynamics with $x_{\mathrm{min}} < x_{\mathrm{max}}$, then there exists a value x on the right of x_{max} where $f(x)$ is lower than $f(x_{\mathrm{min}})$. In other words, there exists:

$$x^< := \inf\{x \in [x_{\mathrm{max}}, +\infty[\; ; \; f(x) < f(x_{\mathrm{min}})\}$$

such that for any $\varepsilon \to 0^+$, $f(x^< + \varepsilon) < f(x_{\mathrm{min}})$. Since $x^< > x_{\mathrm{max}}^+$, every path γ joining x_{min} to $x^<$ goes through a local maximum x_{max}^2 defined by

$$x_{\mathrm{max}}^2 := \arg \max_{x \in [x_{\mathrm{max}}^+, x^<]} f(x)$$

which verifies $f(x_{\mathrm{max}}^2) > f(x_{\mathrm{max}}^+)$. Then the dynamics of x_{min} is greater than or equal to $f(x_{\mathrm{max}}^2) - f(x_{\mathrm{min}})$ which is greater than $f(x_{\mathrm{max}}) - f(x_{\mathrm{min}})$. We obtain a contradiction. Then we have $f(x_{\mathrm{min}}^2) < f(x_{\mathrm{min}})$. The proof of $(P2)$ is done.

Thanks to $(P1)$ and $(P2)$, we obtain directly $(P3)$ by applying Algorithm 1. □

Proposition 2. *Let $f : \mathbb{R} \to \mathbb{R}$ be a $\mathfrak{D}$-Morse function with unique critical values. Now let us assume that a local minimum $x_{\min} \in \mathbb{R}$ of f is paired with a local maximum $x_{\max}$ of f by persistence. We assume without constraints that $x_{\min} < x_{\max}$ (the reasoning is the same for the opposite assumption). Then, $x_{\max}$ and $x_{\min}$ are paired by dynamics.*

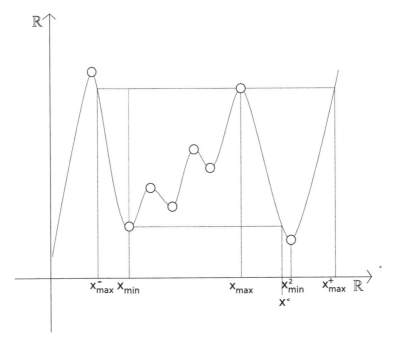

Fig. 6. A $\mathfrak{D}$-Morse function $f : \mathbb{R} \to \mathbb{R}$ where the local extrema $x_{\min}$ and $x_{\max}$ are paired by persistence relatively to f.

Proof: We denote by $(x_{\max}^-, x_{\max}^+) \in \overline{\mathbb{R}}^2$ the two values verifying:

$$[x_{\max}^-, x_{\max}^+] = \mathrm{cl}_{\overline{\mathbb{R}}}(\mathcal{CC}([f \leq f(x_{\max})], x_{\max})).$$

Since $x_{\min}$ is paired by persistence to $x_{\max}$ with $x_{\min} < x_{\max}$ (see Fig. 6), then:

$$x_{\min} = \mathrm{Rep}([x_{\max}^-, x_{\max}], f) \in \mathbb{R},$$

and, by Algorithm 1, we know that $x_{\max}^- > -\infty$ (then $x_{\max}^-$ is finite).

When $x_{\max}^+ < +\infty$ (Case 1), the representative $x_{\min}^2$ of $[x_{\max}, x_{\max}^+]$ relatively to f is exists in $]x_{\max}, x_{\max}^+[$ and is unique, and its image by f is lower than $f(x_{\min})$. When $x_{\max}^+ = +\infty$ (Case 2), $\lim_{x \to +\infty} f(x) = -\infty$, and then there exists one more time an abscissa $x_{\min}^2 \in \mathbb{R}$ whose image by f is lower than $f(x_{\min})$. So, in both cases, there exists a (finite) value $x_{\min}^2 \in]x_{\max}, x_{\max}^+[$

verifying $f(x_{\min}^2) < f(x_{\min})$. This way, we know that $x_{\min}$ is paired with some abcissa in $\mathbb{R}$ by dynamics.

In Case 1, we know that the path defined as:

$$\gamma : \lambda \in [0,1] \rightarrow \gamma(\lambda) := (1-\lambda)x_{\min} + \lambda x_{\min}^2$$

belongs to the set of paths C defining the dynamics of $x_{\min}$ (see Subsect. 2.1). Then,

$$\mathrm{dyn}(x_{\min}) \leq \max\{f(x) - f(x_{\min}) \; ; \; x \in \gamma([0,1])\},$$

which is lower than or equal to $f(x_{\max}) - f(x_{\min})$ since f is maximal at $x_{\max}$ on $[x_{\max}^-, x_{\max}^+]$. Then we have the following property:

$$\mathrm{dyn}(x_{\min}) \leq f(x_{\max}) - f(x_{\min}).$$

In Case 2, since $f(x)$ is lower than $f(x_{\max})$ for $x \in]x_{\max}, +\infty[$, then one more time we get $\mathrm{dyn}(x_{\min}) \leq f(x_{\max}) - f(x_{\min})$. Let us call this property (P).

Even if we know that there exists some local maximum of f which is paired with $x_{\min}$ by dynamics, we do not know whether the abscissa of this local maximum is lower than or greater than $x_{\min}$. Then, let us assume that there exists a local maximum $x^* < x_{\min}$ (lower case) which is associated to $x_{\min}$ by dynamics. We denote this property (H) and we depict it in Fig. 7. Since $f(x)$ is greater than or equal to $f(x_{\min})$ for $x \in [x_{\max}^-, x_{\min}]$, (H) implies that $x^* < x_{\max}^-$. Then, we can observe that the local maximum x^1 of f of maximal abscissa in $[x^*, x_{\max}^-]$ verifies $f(x^1) > f(x_{\max})$, which implies that $\mathrm{dyn}(x_{\min}) \geq f(x^1) - f(x_{\min}) > f(x_{\max}) - f(x_{\min})$ (since we go through x^1 to reach x^*), which contradicts (P). (H) is then false. In other words, we are in the upper case: the local maximum paired by dynamics to $x_{\min}$ belongs to $]x_{\min}, +\infty[$, let us call this property (P').

Now let us define:

$$x^< := \inf\{x > x_{\min} \; ; \; f(x) < f(x_{\min})\},$$

(see again Fig. 6) and let us remark that $x^< > x_{\max}$ (because $x_{\min}$ is the representative of f on $[x_{\max}^-, x_{\max}]$). Since we know by (P') that a local maximum $x > x_{\min}$ of f is paired by dynamics with $x_{\min}$, then the image of every optimal path belonging to C contains $\{x^<\}$, and then contains $[x_{\min}, x^<]$. Indeed, an optimal path in C whose image would not contain $\{x^<\}$ would then contain an abscissa $x < x_{\max}^-$ and then we would obtain $\mathrm{dyn}(x_{\min}) > f(x_{\max}) - f(x_{\min})$, which would contradict (P).

Now, the maximal value of f on $[x_{\min}, x^<]$ is equal to $f(x_{\max})$, then $\mathrm{dyn}(x_{\min}) = f(x_{\max}) - f(x_{\min})$. The only local maximum of f whose value is $f(x_{\max})$ is $x_{\max}$, then $x_{\max}$ is paired with $x_{\min}$ by dynamics relatively to f. $\square$

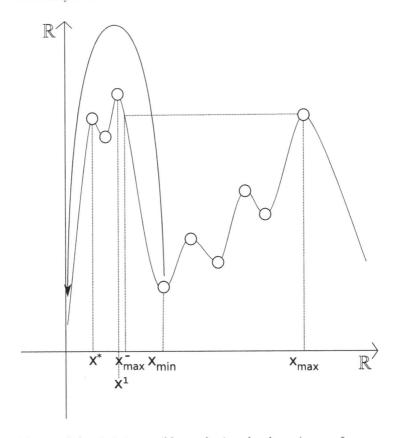

Fig. 7. The proof that it is impossible to obtain a local maximum $x^* < x_{\min}$ paired with $x_{\min}$ by dynamics when $x_{\min}$ is paired with $x_{\max} > x_{\min}$ by persistence.

Theorem 1. *Let $f : \mathbb{R} \to \mathbb{R}$ be a $\mathfrak{D}$-Morse function with a finite number of local extrema and unique critical values. A local minimum $x_{\min} \in \mathbb{R}$ of f is paired by dynamics to a local maximum $x_{\max} \in \mathbb{R}$ of f iff $x_{\max}$ is paired by persistence to $x_{\min}$. In other words, pairings by dynamics and by persistence lead to the same result. Furthermore, we obtain $\mathrm{per}(x_{\max}) = \mathrm{dyn}(x_{\min})$.*

Proof: This theorem results from Propositions 1 and 2.   □

Note that pairing by persistence has been proved to be *symmetric* in [1] for Morse functions defined on manifolds: the pairing is the same for a Morse function and its negative.

4   Conclusion

In this paper, we prove the equivalence between pairing by dynamics and pairing by persistence for 1D $\mathfrak{D}$-Morse functions. As future work, we plan to study their relation in the n-D case, $n \geq 2$. Another interesting issue is to explore how ideas steaming from Morse Theory can infuse Mathematical Morphology. Conversely, since the watershed is clearly linked to the topology of the surfaces [4,5,14], it is definitely worthwhile to search how such ideas can contribute to Discrete Morse Theory. This can be done along the same lines as what is proposed in [2,3,6].

References

1. Cohen-Steiner, D., Edelsbrunner, H., Harer, J.: Extending persistence using Poincaré and Lefschetz duality. Found. Comput. Math. **9**(1), 79–103 (2009)
2. Čomić, L., De Floriani, L., Iuricich, F., Magillo, P.: Computing a discrete Morse gradient from a watershed decomposition. Comput. Graph. **58**, 43–52 (2016)
3. Čomić, L., De Floriani, L., Magillo, P., Iuricich, F.: Morphological Modeling of Terrains and Volume Data. SCS. Springer, New York (2014). https://doi.org/10.1007/978-1-4939-2149-2
4. Cousty, J., Bertrand, G., Couprie, M., Najman, L.: Collapses and watersheds in pseudomanifolds of arbitrary dimension. J. Math. Imaging Vis. **50**(3), 261–285 (2014)
5. Cousty, J., Bertrand, G., Najman, L., Couprie, M.: Watershed cuts: minimum spanning forests and the drop of water principle. IEEE Trans. Pattern Anal. Mach. Intell. **31**(8), 1362–1374 (2009)
6. De Floriani, L., Iuricich, F., Magillo, P., Simari, P.: Discrete morse versus watershed decompositions of tessellated manifolds. In: Petrosino, A. (ed.) ICIAP 2013. LNCS, vol. 8157, pp. 339–348. Springer, Heidelberg (2013). https://doi.org/10.1007/978-3-642-41184-7_35
7. Edelsbrunner, H., Harer, J.: Persistent homology - a survey. Contemp. Math. **453**, 257–282 (2008)
8. Edelsbrunner, H., Harer, J.: Computational Topology: An Introduction. American Mathematical Society, New York (2010)
9. Edelsbrunner, H., Letscher, D., Zomorodian, A.: Topological persistence and simplification. In: Foundations of Computer Science, pp. 454–463. IEEE (2000)
10. Forman, R.: A user's guide to Discrete Morse Theory. Séminaire Lotharingien de Combinatoire **48**, 1–35 (2002)
11. Grimaud, M.: La géodésie numérique en Morphologie Mathématique. Application à la détection automatique des microcalcifications en mammographie numérique. Ph.D. thesis, École des Mines de Paris (1991)
12. Grimaud, M.: New measure of contrast: the dynamics. In: Image Algebra and Morphological Image Processing III, vol. 1769, pp. 292–306. International Society for Optics and Photonics (1992)
13. Milnor, J.W., Spivak, M., Wells, R.: Morse Theory. Princeton University Press, Princeton (1963)
14. Najman, L., Schmitt, M.: Watershed of a continuous function. Signal Process. **38**(1), 99–112 (1994)
15. Najman, L., Schmitt, M.: Geodesic saliency of watershed contours and hierarchical segmentation. IEEE Trans. Pattern Anal. Mach. Intell. **18**(12), 1163–1173 (1996)
16. Najman, L., Talbot, H.: Mathematical Morphology: From Theory to Applications. Wiley, Hoboken (2013)
17. Serra, J.: Introduction to mathematical morphology. Comput. Vis. Graph. Image Process. **35**(3), 283–305 (1986)

18. Serra, J., Soille, P.: Mathematical Morphology and its Applications to Image Processing, vol. 2. Springer Science & Business Media, Heidelberg (2012)
19. Vachier, C.: Extraction de caractéristiques, segmentation d'image et Morphologie Mathématique. Ph.D. thesis, École Nationale Supérieure des Mines de Paris (1995)
20. Vincent, L., Soille, P.: Watersheds in digital spaces: an efficient algorithm based on immersion simulations. IEEE Trans. Pattern Anal. Mach. Intell. **13**(6), 583–598 (1991)

Dual-Primal Skeleton: A Thinning Scheme for Vertex Sets Lying on a Surface Mesh

Ricardo Uribe Lobello[✉] and Jean-Luc Mari

Aix Marseille Université, Université de Toulon, CNRS, LIS, Marseille, France
ricardo.uribe-lobello@univ-amu.fr

Abstract. We present a new algorithm for the skeletonization of shapes lying on surface meshes, which is based on a thinning scheme with a granularity that is twice as fine as that of other thinning methods, since the proposed approach uses dual-primal iterations in the region of interest to perform the skeleton extraction. This dual operator is built on specific construction rules, and it is applied until idempotency, which provides a better geometric positioning of the skeleton compared to other thinning methods. Moreover, the skeleton has the property of ensuring the same topological guarantees as other homotopic thinning approaches: the skeleton is thin, connected and can include Y-branches and cycles if the input region contains holes.

Keywords: Skeletonization · Surface mesh · Homotopic thinning · Shape description · Topology preservation

1 Introduction

1.1 Context, Problem and Related Work

The skeleton is a well-known shape descriptor. It is an entity that is globally centered in a 2D or 3D object, and it characterizes its topology and its geometry. This structure is widely used in various applications (video tracking [4], shape recognition [13], surface sketching [9], etc.). Several techniques exist in order to extract the skeleton from binary 2D images [14], 3D closed volume meshes [2], or 3D cubic grids [8].

Nevertheless, very few approaches have been dedicated to the extraction of skeletons from binary information located on an arbitrary triangulated mesh (see [3,11,12] for the state of the art). Therefore, the task of computing the skeleton of the subset of a discrete surface embedded in $\mathbb{R}^3$ remains. Rössl *et al.* [10] have presented the first method that uses the elementary mathematical morphology *opening* operator, ported to triangulated meshes. However, the operator's definition is incomplete, and the underlying algorithm presents some issues. Therefore, several drawbacks have been pointed out, which mainly lead to unexpectedly

© Springer Nature Switzerland AG 2019
B. Burgeth et al. (Eds.): ISMM 2019, LNCS 11564, pp. 69–83, 2019.
https://doi.org/10.1007/978-3-030-20867-7_6

disconnected skeletons [5]. Kudelski *et al.* have later proposed a modified algorithm that produces topologically robust skeletons by generalizing the notion of *morphological erosion* to arbitrary meshes [6,7]. This approach takes, as an input, a subset lying on a triangulated surface mesh in 3D, and as outputs, thin lines corresponding to the skeleton obtained by homotopic thinning. The main idea is to transpose the notion of neighborhood from the classical thinning algorithms where the adjacency is constant (e.g., 26-adjacency in digital volumes, 8-adjacency in 2D grids) to the mesh domain where the neighborhood is variable due to the adjacency of each vertex.

1.2   Contribution

Our work continues the idea of homotopic thinning using a generalized adjacency described in [6]. Instead of iteratively removing nonrelevant vertices of the subset (topologically speaking, i.e., the *simple vertices*), the erosion step is replaced by a *dual-primal* operation. The area of interest is converted to the dual space, and all semi-infinite edges are removed from the structure. This process is repeated (dual to primal to dual) until idempotency. It produces a lineal skeleton with a resolution increased by a factor 2, as the resulting structure is not only composed of initial vertices and edges but also composed of dual vertices and edges. This skeletonization is more general because it can address nondevelopable surfaces (the operations are local, and there is no need to have a $[i, j]$ indexing such as in 2D grids). Moreover, the resulting skeleton preserves the topology of the original shape lying on the surface.

This paper is divided as follows: Sect. 2 briefly develops some basic notions and definitions. Then, Sect. 3 explains in detail our approach of dual-primal skeletonization using a specific thinning scheme on surface meshes. Section 4 is dedicated to the validation of the method, including tests on irregular meshes and an application to the extraction of feature lines.

2   Basic Notions

Let $\mathcal{M}$ be an unstructured mesh patch representing an arbitrary manifold surface $\mathcal{S}$, such as $\mathcal{M} = (\mathcal{V}, \mathcal{E}, \mathcal{T})$. The sets $\mathcal{V}$, $\mathcal{E}$ and $\mathcal{T}$ correspond, respectively, to the vertices, the edges, and the triangles composing $\mathcal{M}$, a piecewise linear approximation of $\mathcal{S}$. We denote p_i the vertices, with $i \in [0; n[$ and $n = |\mathcal{V}|$ being the number of vertices in $\mathcal{M}$. The neighborhood $\mathcal{N}$ of a vertex p_i is defined as follows:

$$\mathcal{N}(p_i) = \{q_j \mid \exists \, (p_i, q_j) \in \mathcal{E}\}. \tag{1}$$

In such a case, $m_i = |\mathcal{N}(p_i)|$ represents the total number of neighbors of p_i. As we consider obtaining a skeleton of a subset of $\mathcal{M}$, let us now define a binary attribute F on each vertex of $\mathcal{V}$. The set $R \subseteq \mathcal{V}$ is then written as follows:

$$\forall p_i \in R \iff F(p_i) = 1. \tag{2}$$

The attribute F may be defined from a previous process such as manual selection, thresholding based on geometric properties (triangle area, principal curvatures, *etc.*) or any binarization process. Then, an edge $e = (p, q)$ belongs to R if and only if $p, q \in R$. Similarly, a triangle $t = (p, q, r)$ belongs to R if and only if $p, q, r \in R$.

3   Dual-Primal Skeletonization

Our approach is based on a robust dual-primal erosion algorithm that preserves the topology of the original mesh. In the next sections, we present an overview of our algorithm and a detailed explanation of each part of our algorithm.

3.1   Overview of the Approach

The input to our algorithm can be any triangular mesh $\mathcal{M}$ that is a set of triangular faces connected by edges and vertices as defined in Sect. 2. As in Sect. 2, this mesh must have the vertices marked with a function in order to define regions of interest. $\mathcal{M}$ can contain several connected components, and it can have one or several borders. Hereafter, we will call the input mesh the *initial primal mesh (IPM)*. Our approach is straightforward; it starts by extracting the dual representation of the *IPM*. Then, it detects the faces that can be eliminated by exploring the dual mesh. Finally, it computes the intermediate primal representation of the mesh by eliminating the primal faces that are not going to affect the topology of the final skeleton. Then, we repeat the dual-primal iteration over this new primal mesh. Finally, it stops when the primal mesh in iteration N is equal to the primal mesh in iteration $N+1$. The main steps of our approach are illustrated in Fig. 1.

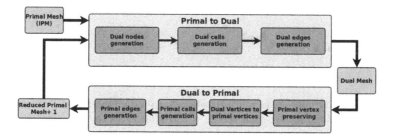

Fig. 1. General structure of the workflow of our approach. The initial mesh is provided as input. A dual representation is computed. Then, a primal mesh is obtained by eliminating those dual cells that are not totally contained in the areas of interest. The new primal mesh can contain edges representing the thin features of the input mesh. Finally, the reduced mesh is provided as input to the next iteration until idempotency.

The primal-to-dual and dual-to-primal transformations are performed by following a well-defined set of rules that guarantee that the final mesh will preserve the topology of the *IPM*.

3.2   Primal-Dual-Primal Computation

In classic dual extraction algorithms, the dual mesh is extracted based on an edge adjacency relationship between cells. For each edge in the primal mesh, a dual edge is generated. However, this narrow definition is not sufficient to obtain a 1-dimensional skeleton because it does not consider vertex adjacencies between edges and vertex adjacency between faces. Vertex adjacency is necessary to detect all connected components in the area of interest in the original mesh. Furthermore, as our algorithm applies an erosion to the input mesh, it will eventually lead to the creation of mixed meshes containing 2-dimensional (faces) and 1-dimensional (edges) cells that are uniquely connected by vertices. Our approach addresses these cases in order to keep the consistency and fidelity of the final mesh with respect to the *IPM*.

3.3   General Rules for Dual Mesh Generation

These rules follow classic definitions of dual meshes extracted from polygonal meshes.

1. Each primal cell is replaced by a dual vertex. It includes primal faces and primal edges. Initially, we always place the dual vertex in the barycenter of the primal cell as illustrated in Figs. 2a and b.
2. Each primal vertex is replaced by a dual cell. In the case of vertices surrounded by 2-dimensional cells, the dual cell is a polygon. In the case of a vertex surrounded by edges, it is replaced by a dual edge. These cases are illustrated in Figs. 2c and d.

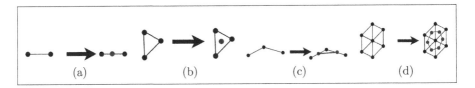

Fig. 2. Basic transformations from a primal mesh to a dual mesh. (a) Primal edge to dual vertex. (b) Primal face to dual vertex. (c) Primal vertex to dual edge. (d) Primal vertex to dual cell (a polygon).

As a consequence of the previous rules, starting from a primal mesh $\mathcal{M}_k = \{\mathcal{V}, \mathcal{E}, \mathcal{C}\}$, its dual mesh $\mathcal{D}_k = \{\mathcal{V}_d, \mathcal{E}_d, \mathcal{C}_d\}$ (in the general case) is defined by equations:

$$\mathcal{V}_d = \{v_{d_i} \mid \exists c_i \in \mathcal{C} \text{ so that } v_{d_i} \text{ is located at the barycenter of } f_i\} \qquad (3)$$

$$\mathcal{E}_d = \{e_{d_i} \mid \exists e_i \in \mathcal{E}\} \qquad (4)$$

$$\mathcal{C}_d = \{c_{d_i} \mid \exists v_i \in \mathcal{V}\} \qquad (5)$$

These definitions apply, as mentioned, in the general case of a closed mesh without borders. The *IPM* can have borders, and thus, the intermediate meshes can contain borders. Therefore, the previous definitions have to be expanded to consider these special cases. For example, the region $\mathcal{R}$ can contain thin structures; hence, it is not possible to generate the classic dual cell, but it is possible to generate linear dual cells. We will explain these cases in more detail in the next section.

3.4   Generation of Dual Cells in Particular Cases

To address all possible cases in a region with borders and to prove its topological correctness, it is necessary to expand our definition of a dual mesh.

Dual Vertex Generation. Our algorithm simply traverses each cell in the primal mesh $\mathcal{M}$ and computes a dual vertex placed in the barycenter of the current cell. This algorithm is easily applied over 1-dimensional and 2-dimensional cells. There is no need to extend the current definitions.

Dual Cell Generation. In the general case, it is possible to build the dual cell d_i around a primal vertex c_i by extracting its face neighborhood as defined in Eq. 6

$$\mathcal{N}_f(c_i) = \{f_j \mid c_i \in \mathcal{V}(f_j) \text{ where } \mathcal{V}(f_j) \text{ is the set of vertices of cell } f_j\}. \quad (6)$$

From $\mathcal{N}_f(c_i)$, it is possible to generate the dual vertices that will be connected to generate the dual cell. However, it is necessary to ensure that the faces are ordered counterclockwise in order to obtain a well oriented dual mesh. To do this, our algorithm extracts the ordered neighborhood of the primal vertices around c_i by traversing all faces incidents to c_i. Then, we extract all edges in those faces not containing c_i. Finally, we order these edges to create a closed path. This algorithm only works in well oriented surfaces. Thus, this vertex neighborhood can be redefined as follows.

$$\mathcal{N}_v(c_i) = \{c_j \in \mathcal{V}(f_k) \mid \exists f_k \in \mathcal{N}_f(c_i) \text{ and } c_j \neq c_i\}. \quad (7)$$

The vertices contained in $\mathcal{N}_v(c_i)$ can be ordered to check if the current vertex c_i is actually surrounded by a closed path of vertices. If it is the case, it is possible to generate the dual cell. If it is not the case, the current vertex belongs to a thin structure or a border. Our algorithm to process these cases will be explained in the next section.

It is necessary to also handle the cases where 1-dimensional and 2-dimensional cells are adjacent. In these cases, it is necessary to make a decision about the way how these cells have to be connected, which occurs in both the primal space and in the dual space, and the method used to solve this is presented in the next section.

3.5   Handling Mixed Regions

Mixed regions usually appear when large components encounter thin structures. It can happen in the transition from primal to dual and in the dual-to-primal transformation. Therefore, we develop these two transitions separately.

Primal to Dual Transformation. Once the dual vertices have been generated, it is necessary to detect thin structures to avoid the separation of components connected by a strip of triangles or a set of vertices. To detect if a primal face f is contained in a thin structure, we traverse each of its primal vertices checking if a ring can be built around each one. If this is not possible for any of them, this primal face is marked as comprising part of a thin structure. Then, each dual vertex in every face is connected to the dual vertex in the edge-adjacent cell in order to generate a dual edge. This process is repeated until all dual vertices in the thin structure are connected. It is important to clarify that edges generated in thin structures are 1-dimensional dual cells and are stored explicitly, in contrast to edges comprising part of 2-dimensional dual cells, which are stored implicitly. This process is illustrated in Fig. 3.

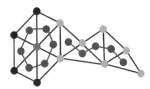

Fig. 3. Detection of thin structures. For each primal vertex, we check the 1-ring built from its incident faces. In red, a primal vertex in a larger structure. In yellow, these primal vertices do not have a 1-ring around them. Consequently, all primal faces containing only these kinds of vertices are part of a thin structure. In blue are dual vertices, dual faces and dual edges. (Color figure online)

As seen in Fig. 3, dual vertices in thin structures are still not connected to dual vertices in the larger structure. Our algorithm traverses all dual vertices belonging to thin structures to detect end points, which is done by checking if the dual vertex is only adjacent to one 1-dimensional cell. Then, for each end point $v_{endpoint}$, we obtain its primal cell c_p, and we check if c_p is an edge-adjacent to another primal cell c_a so that it contains a dual vertex belonging to a 2-dimensional dual cell. We do this only once in order to avoid connecting the thin structure with the larger structure multiple times. This procedure is listed in the Algorithm 1.

As mentioned previously, our algorithm connects vertex-adjacent regions in order to preserve the topology of a region of interest defined with marked vertices, which is implemented in the primal-to-dual phase. Dual vertices in vertex-adjacent primal vertices are always connected by an edge. It is important to

Algorithm 1. Connecting the dual vertices in a thin structure with the dual vertices in larger structures.

Data: Dual mesh with thin and large structures disconnected.
Result: Dual mesh with thin structures connected to large structures.

```
dualVertices ← GetDualVerticesInThinStructure(...)
forall the dualVertices as v do
    if IsEndPoint(v) then
        cp ← GetPrimalCell(v)
        adjacentCells ← GetAdjacentCells(cp)
        forall the adjacentCells as c do
            vd ← GetDualVertex(c)
            if BelongTo2DDualCell(vd ) then
                GenerateDualEdge(v, vd)
                Break
            end
        end
    end
end
```

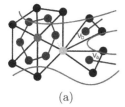   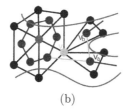   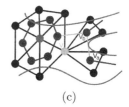

(a)   (b)   (c)

Fig. 4. In red, we illustrate potential borders for the structure approximated by the area of interest. (a) Vertex-adjacent components in the Dual cells and dual vertices are in blue. The large structure is composed of a dual cell in blue. This structure is connected to two thin structures by the primal vertex in orange. (b) The three components are connected by the orange edges from v_a. It is clear, however, that we can obtain different meshes depending on the initial dual vertex (c). As explained later, we do not consider this as a limitation for our approach because the topology of the area of interest is preserved. (Color figure online)

remember that a primal cell can be edge-adjacent to several structures. Consequently, it is necessary to detect all components that have to be connected with the current dual vertex, which is illustrated in Fig. 4a.

To connect these structures, it is necessary to detect one of the end points v_a or v_b. Then, the algorithm extracts the set of primal cells C that are adjacent to the shared primal vertex v. Next, we extract the connected components of C. Finally, we choose one of the primal cells belonging to each connected component that contains a dual vertex, and we connect v_a (or v_b) with each one of these dual vertices. This procedure is listed in Algorithm 2 and illustrated in Fig. 4b.

By using this method, it is possible to obtain different meshes from the same input depending on the dual vertex that is first detected (v_a or v_b). However, as this choice does not change the topology (but the connectivity) of the final mesh. As our objective is to preserve the topology in the sense of connected components, holes and voids and not in terms of connectivity of the mesh, this result is correct.

Algorithm 2. Connecting the dual vertices of vertex-adjacent primal cells. This algorithm can connect multiple dual vertices from multiple vertex-adjacent primal cells (see Fig. 4).

Data: Dual mesh with vertex-adjacent structures disconnected.
Result: Dual mesh with vertex-adjacent structures connected.

```
dualVertices ← GetDualVerticesInEdges(...)
forall the dualVertices as v do
    if IsEndPoint(v) then
        c_p ← GetPrimalCell(v)
        v_shared ← GetSharedVertex(c_p)
        vertexAdjacentCells ← GetAdjacentCellsByVertex(v_shared )
        connectedComponents ← ExtractConnectedComponents(vertexAdjacentCells)
        forall the connectedComponent as cc do
            v_d ← GetOneDualVertexInComponent(cc)
            GenerateDualEdge(v, v_d)
        end
    end
end
```

Dual-to-Primal Transformation. Once a dual mesh has been generated, our approach detects which primal cells can be eliminated without affecting the topology of the final linear structure. To do this, we apply the following rule:

Rule 1. *A primal cell must be preserved if and only if its dual vertex is shared by three 2D dual cells.*

The rule 1 is intended to preserve primal cells that are part of large structures. This rule also implies that all primal cells corresponding to a dual vertex shared uniquely by one or two 2D dual cells must be eliminated. These primal cells are, evidently, always located in the frontiers of the structure. Therefore, the application of rule 1 results in an erosion of the larger structures from the border to the inside, similar to the prairie fire algorithm used in topological thinning methods. It is implemented by traversing all dual vertices and obtaining its adjacent dual cells.

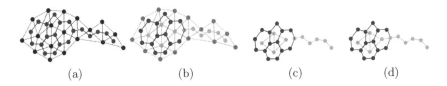

(a)   (b)   (c)   (d)

Fig. 5. General illustration of the transition from dual to primal. This example shows a large structure connected to a thin structure.

The rule 1 is only intended to preserve interior faces. The connectivity between large and thin structures is represented as a dual edge sharing a dual vertex with a 2-dimensional dual cell, as explained in the previous section. To preserve this topology in the next primal mesh, we proceed as follows. In the case of primal faces belonging to thin structures and, as consequence, represented in the dual mesh by dual vertices, they are not preserved; however, each one is replaced by its dual vertex, which is added to the next primal mesh. Dual edges inside these thin structures are all preserved as primal edges. Finally, dual edges connecting to 2-dimensional dual cells are also preserved by adding the shared dual vertex to the primal vertices and connecting it to the primal vertex at the origin of the primal 2-dimensional cell.

This process is illustrated in Fig. 5 with the original mesh in black and the initial dual mesh in blue. In Fig. 5b, green faces and vertices are preserved, all primal faces and vertices that are going to be eliminated (by using the complement of rule 1) are in red. In Fig. 5c, thin and large structure are still not connected. In Fig. 5d, These structures are connected by using the primal vertex at the origin of the 2-dimensional dual cell and the dual vertex in the same dual cell connected to the thin structure.

The method previously described is sufficiently general to address several thin structures or branches converging toward a single 2-dimensional dual cell and can generate multiple junctions at the same point; it preserves the topology of the original area of interest.

As evident from the primal mesh obtained in Fig. 5d, our approach generates intermediary meshes that are a mixture of faces and edges as illustrated in Fig. 6a. Therefore, this kind of configuration has to be addressed in the primal-to-dual transition. In these cases, the algorithm connects the dual vertex in the primal edge with the dual vertex belonging to the adjacent primal face, forming a junction as shown in Fig. 6b.

(a)   (b)

Fig. 6. General illustration of the extraction of a dual mesh in areas of the mesh where primal faces and edges are connected. (a) This primal mesh has been generated from a previous iteration. It generates mixed meshes with edges and faces. (b) The dual mesh in blue is generated mostly by using the standard rules explained earlier. However, in mixed areas, dual vertices in edges have to be connected to dual vertices in primal faces generating a junction.

The previous primal-to-dual and dual-to-primal processes are applied iteratively until idempotency. In this final stage, we obtain a mesh containing edges only. These edges are positioned approximately at the center of the topological structures of the original area of interest. This fact will help to better understand the next section.

3.6   Geometric Positioning

The algorithm generates linear meshes well positioned in the center of the main structures of the original object, which is possible because of the symmetrical erosion procedure that is used to progressively eliminate the most external primal faces in the mesh. As explained above, our algorithm advances from the borders of the mesh to the interior. Thus, the final structure converges slowly towards the center of the structure. Additionally, we use the primal vertices if possible; however, if necessary, we generate and add dual vertices to the next primal mesh. These dual vertices are placed in the barycenter of primal faces, thus increasing the centering of the final structure, especially in thin structures.

Concerning the geometric position with respect to the original surface, we use, if possible, the original primal vertices. If dual vertices are used in the final mesh, they are strictly located over the primal faces. Therefore, the vertices of the final mesh are always located over the original mesh. By contrast, final edges are not always located on the surface. It is only the case when the edge is relying two initial primal meshes. In any other case, all the points of the edge are not necessarily on the surface.

3.7   Topological Guarantees

The topological guarantees offered by our approach can be proven by analyzing the different cases that can appear in the primal-to-dual and dual-to-primal transitions.

Lemma 1. *In the transition from the primal mesh to the dual mesh, all topological structures are represented through their dual counterparts.*

Proof. Let $A_p = \{V_p, F_p, E_p\}$ with V_p be a set of primal vertices, F_p be a set of primal faces and E_p be a set of edges. We consider S_p a *conformal* mesh so that N-dimensional cells only intersect on (N-1)-dimensional cells. Triangles only intersect in an edge or a vertex and edges only intersect in a vertex. Therefore, we consider that A_p is also conformal.

First, let the dual mesh be defined equivalently as $A_d = \{V_d, F_d, E_d\}$. As our algorithm generates one dual vertex for each primal cell or edge, we find that the set of dual vertices V_d is composed of dual vertices centered at their corresponding primal cell, as illustrated in Fig. 2a and 2b. Hence, each face $f_p^i \in F_p$ or edge $e_p^i \in E_p$ in the region A_p is represented by a dual vertex $v_d^i \in V_d$.
Uniform regions: In the case of 2 edges connected by a primal vertex, dual vertices are connected by a dual edge (see Fig. 2c). In the case of primal vertices surrounded by faces, dual vertices are connected as illustrated in Fig. 2d. This construction guarantees that linear structures will still be represented by linear (but dual) structures. In the case of triangulations, they will be represented by at least one dual cell.
Mixed regions: These are regions where a triangulation representing a large structure L_p is connected to a thin structure T_p by at least a primal vertex.

These thin structures can be a strip of triangles or a set of edges. Two cases can be considered:

- **If $L_p \cup T_p$ is an edge:** The dual vertices belonging to the edge-adjacent triangles are connected by a dual edge, thus connecting the two regions.
- **If $L_p \cup T_p$ is a primal vertex:** L_p and T_p are either two primal faces or a primal face and a primal edge. Then, the dual vertex in T_p is connected to the first primal cell of each vertex-adjacent connected component as illustrated in Fig. 4.

Mixed regions in the primal mesh only transform into dual edges in the dual mesh. As we have assumed that the area of interest is a conformal surface, the cases explained above connect all cells that are at least connected by a primal vertex. Consequently, this algorithm preserves the topology of the original surface.

Lemma 2. *In the transition from the dual mesh to the primal mesh, all topological structures are preserved.*

Proof. To prove Lemma 2, we need to proceed by case:

- **Dual edges:** Dual edges are only converted to primal edges. Therefore, thin structures are preserved.
- **Large structures:** As mentioned previously, all internal dual vertices are transformed to primal cells. All noninternal dual vertices and their corresponding primal faces are eliminated. If no internal dual vertices exist, it means it is a thin structure; in any other case, primal faces will be conserved, and this structure will be present in the next iteration.
- **Mixed structures:** A 2-dimensional dual cell connected to a dual edge. In this case, the connection between the components is always conserved because the primal vertex at the center of the dual edge will be connected to the primal vertex at the origin of the 2-dimensional dual cell.

In conclusion, no components that have been connected during the primal-to-dual process are separated in the dual-to-primal transition. All connections are preserved in the edge and vertex adjacency. Consequently, the topology of the original region of interest is preserved.

3.8   Post-processing

Pruning. Our algorithm uses a simple criterion to eliminate parasite branches. It consists of the detection of ending vertices of the structure. For each of these ending vertices, it checks if it belongs to the original mesh. If this is the case, this branch is preserved because it can represent an important feature of the region. If the ending vertex does not belong to the initial mesh, it means that it has been created during the primal-dual-primal iteration process. In this case, it is not sure that the vertex reflects a thin structure in the initial mesh. In this article, we have decided to keep these branches if they have a minimum length (a parameter) but more research is needed to establish if they really have to be conserved.

Smoothing. As our approach is based entirely on the geometrical information of the mesh, it can produce highly oscillating structures. To improve these results, we apply a simple smoothing algorithm that moves every vertex to the barycenter of the vertices composing its topological neighborhood similar to the "Beatifying" algorithm proposed in [1].

Evidently, as this process can affect the approximation quality of the final skeleton, we bound the maximum approximation error from the original mesh.

4   Results and Validation

These results have been obtained in the Ubuntu 16.04 LTS 64-bit environment with the processor, Intel i7-7820HQ CPU @ 2.90 GHz × 8 and 32 GBytes of internal memory. It has been implemented in C++, and the mesh data structure used was the vtkPolyData. We used VTK built-in methods to traverse the mesh and access to its properties. Our approach is intended to work in 2-dimensional simplex meshes that can be mapped in 3D in order to represent volumes. For this, we have started applying it to 2D planar meshes and the results are discussed in the next section.

4.1   Application on Planar Meshes

In Fig. 7, our implementation is applied to four different meshes. These meshes are regular meshes with relatively well-formed faces. In the figure, we show the original mesh and the extracted skeleton after cleaning and some iterations of the smoothing process. Table 1 shows relevant information about the examples presented in Fig. 7.

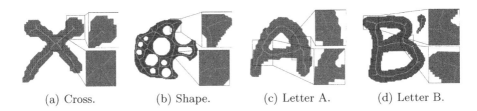

(a) Cross.   (b) Shape.   (c) Letter A.   (d) Letter B.

Fig. 7. Application of our algorithm to 2D meshes. The cleaned skeleton is shown after a smoothing process.

From Fig. 7, it is clear that the algorithm is able to extract skeletons from arbitrary triangular meshes while preserving the topology of the original mesh. It keeps junctions and holes, and places the final skeleton at approximately the center of the topological structure. However, the method is entirely based on the geometry and the connectivity of the mesh, and consequently, a different triangulation of the same object will produce a different skeleton.

Table 1. Information on the execution of the algorithm in the four previous planar meshes. This table shows the number of primal-to-dual and dual-to-primal iterations, the execution time and the total number of edges in the final skeleton.

Data set	# Triangles	# Iterations	Exec. time (in ms)	Lines final skeleton
Cross	819	5	246	108
Shape	8556	13	4529	1009
A letter	812	6	261	103
B letter	4307	9	2049	361

Concerning the execution time, our implementation is not optimal because of the choice of the vtkPolyData structure. The vtkPolyData depends on a logarithmic data structure to traverse neighboring cells and adjacent vertices. Having into account the vtkPolyData data structure, the complexity of the dual vertex generation process is $\mathcal{O}(N\frac{1}{K}(F + V))$ and including the complexity of the dual cell generation, the final complexity of our method is $\mathcal{O}(N[\frac{1}{K}[(F+V)+V\log F])$. As seen, the execution time of our algorithm is highly dependent of the size and connectivity of the input mesh. As optimization and with a half-edge data structure, the $\frac{1}{k}V\log F$ part of our approach can be optimized to a constant factor.

4.2   Application on Surfaces

We have applied our approach to surfaces mapped in 3D that represent volumes. These surfaces have to be marked in order to define areas of interest. In Fig. 8, we show two surface meshes where areas of maximum curvature are marked. Our approach has extracted skeletons close to the center of interest areas.

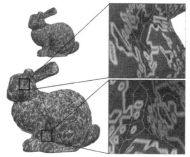

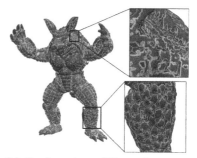

(a) Surface Bunny with 69451 faces and 35947 vertices. The final skeleton has 9053 edges.

(b) Surface Armadillo with 345944 faces and 172974 vertices. The final skeleton has 46725 edges.

Fig. 8. Application to the high curvature areas of a surface (in red). Two closeups show how the skeleton (in white) is located in the middle of the region of interest. (Color figure online)

As seen in Fig. 8, our algorithm is robust enough to be applied to any 2D mesh with complex topology. It is able to connect connected components with edge and vertex connectivity, allowing us to preserve the topology of thin structures connected only by 1-dimensional cells or edges. However, our current implementation does not guarantee that the skeleton is located in the center of the connected component but tends to place it close to it. To offer guarantees on the location of the skeleton, it will likely be necessary to follow closely the position of dual vertices, thus minimizing its distance to the center of mass of the current component. This kind of implementation can be used to extract the main features of surface meshes in the future.

5   Conclusion and Future Work

We have tested our approach and we have confirmed that the algorithm is robust and capable of extracting a 1-dimensional structure from any region lying on a mesh with arbitrary topology. However, this approach is highly dependent on the size and connectivity of the mesh. Thus, we think that a more efficient mesh data structure can strongly improve the complexity and the execution time of our method.

As future work, we consider that our approach is not limited to 2D triangular meshes, and it can be easily extended to simplicial meshes in higher dimensions.

References

1. Arcelli, C., di Baja, G.S.: Euclidean skeleton via centre-of-maximal-disc extraction. Image Vis. Comput. **11**(3), 163–173 (1993)
2. Au, O.K.-C., Tai, C.-L., Chu, H.-K., Cohen-Or, D., Lee, T.-Y.: Skeleton extraction by mesh contraction. ACM Trans. Graph. **27**(3), 1–10 (2008)
3. Delame, T., Kustra, J., Telea, A.: Structuring 3D medial skeletons: a comparative study. In: Symposium on Vision, Modeling and Visualization (2016)
4. Gall, J., Stoll, C., De Aguiar, E., Theobalt, C., Rosenhahn, B., Seidel, H.P.: Motion capture using joint skeleton tracking and surface estimation. In: IEEE Computer Society Conference on Computer Vision and Pattern Recognition (CVPR 2009), pp. 1746–1753. IEEE Computer Society (2009)
5. Kudelski, D., Mari, J.-L., Viseur, S.: Extraction of feature lines with connectivity preservation. In: Computer Graphics International (CGI 2011). Electronic Proceedings (2011)
6. Kudelski, D., Viseur, S., Mari, J.-L.: Skeleton extraction of vertex sets lying on arbitrary triangulated 3D meshes. In: Gonzalez-Diaz, R., Jimenez, M.-J., Medrano, B. (eds.) DGCI 2013. LNCS, vol. 7749, pp. 203–214. Springer, Heidelberg (2013). https://doi.org/10.1007/978-3-642-37067-0_18
7. Kudelski, D., Viseur, S., Scrofani, G., Mari, J.-L.: Feature line extraction on meshes through vertex marking and 2D topological operators. Int. J. Image Graph. **11**(4), 531–548 (2011)
8. Lee, T.C., Kashyap, R.L., Chu, C.N.: Building skeleton models via 3-D medial surface/axis thinning algorithms. Graph. Models Image Process. **56**(6), 462–478 (1994)

9. Mari, J.-L.: Surface sketching with a voxel-based skeleton. In: Brlek, S., Reutenauer, C., Provençal, X. (eds.) DGCI 2009. LNCS, vol. 5810, pp. 325–336. Springer, Heidelberg (2009). https://doi.org/10.1007/978-3-642-04397-0_28

10. Rössl, C., Kobbelt, L., Seidel, H.-P.: Extraction of feature lines on triangulated surfaces using morphological operators. In: AAAI Spring Symposium on Smart Graphics, vol. 00–04, pp. 71–75 (2000)

11. Saha, P.K., Borgefors, G., di Baja, G.S.: A survey on skeletonization algorithms and their applications. Pattern Recogn. Lett. **76**, 3–12 (2016)

12. Tagliasacchi, A., Delame, T., Spagnuolo, M., Amenta, N., Telea, A.: 3D skeletons: a state-of-the-art report. Comput. Graph. Forum **35**, 573–597 (2016). Wiley Online Library

13. Yu, K., Wu, J., Zhuang, Y.: Skeleton-Based Recognition of Chinese Calligraphic Character Image. In: Huang, Y.-M.R., et al. (eds.) PCM 2008. LNCS, vol. 5353, pp. 228–237. Springer, Heidelberg (2008). https://doi.org/10.1007/978-3-540-89796-5_24

14. Zhang, T.Y., Suen, C.Y.: A fast parallel algorithm for thinning digital patterns. Commun. ACM **27**(3), 236–239 (1984)

Topological Mapper for 3D Volumetric Images

Daniel H. Chitwood, Mitchell Eithun$^{(\boxtimes)}$, Elizabeth Munch, and Tim Ophelders

Michigan State University, East Lansing, MI, USA
{chitwoo9,eithunmi,muncheli,ophelder}@msu.edu

Abstract. Mapper is a topological construction similar to a Reeb graph, and is used to summarize the shape of a dataset as a (generalized) graph. Formally, mapper can be constructed for any connected space and algorithms have been developed to compute mapper for point clouds and 2D images. In this paper, we extend mapper to 3D volumetric images. We use our algorithm to compute mapper for scans of barley generated using computed tomography. We demonstrate the flexibility of the construction by highlighting different aspects of the morphology through different choices of starting parameters. Applying mapper to this type of data provides an integrated means of visualization, segmentation and clustering, and can thus be used to study the topology of any 3D object.

Keywords: Topological mapper · Image processing ·
Computed tomography · Topological data analysis

1 Introduction

Mapper is a construction from topological data analysis [3,10,15] that generates an abstract simplicial complex from data. Traditionally, mapper bins a point cloud based on a filter function, clusters the points inside each bin and builds a simplicial complex based on how the clusters of different bins overlap [27]. Statistically, mapper is an optimal estimator of a Reeb graph [4,16], which summarizes connected components of functions on manifolds [22,26]. For surfaces with genus 0, a Reeb graph reduces to a contour tree [7].

While mapper was originally defined for point-clouds, the idea can be applied to any dataset in a metric space. Some applications of mapper include visualization, clustering and classification. For example, mapper has identified a new subtype of breast cancer [18]. Mapper has also been used to mine data related to spinal chord and traumatic brain injuries [19]. The more general concept of a Reeb graph has been used for phenotyping root architectures [14] and comparing cortical surfaces in brain scans [25]. Mapper graphs can be constructed from Reeb graphs, which can be constructed in $O(n \log n)$ time [7].

This project was supported by the USDA National Institute of Food and Agriculture, and by Michigan State University AgBioResearch. The work of EM was supported in part by NSF grants DMS-1800446 and CMMI-1800466.

B. Burgeth et al. (Eds.): ISMM 2019, LNCS 11564, pp. 84–95, 2019.
https://doi.org/10.1007/978-3-030-20867-7_7

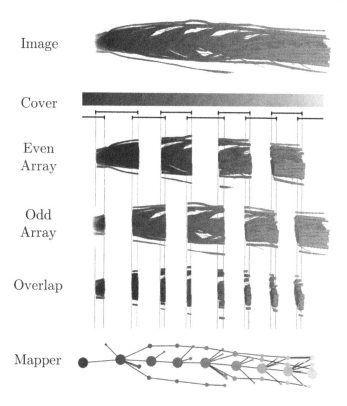

Image

Cover

Even
Array

Odd
Array

Overlap

Mapper

Fig. 1. Mapper applied to a 3D volumetric image, filtered by height. The preimages of cover elements along the range are clustered to form vertices in the mapper graph. Edges are defined by overlap between clusters.

Recently, a version of mapper was developed for 2D images [24]. We extend this algorithm to voxel-based images in 3D, called volumetric images. In particular we clarify how the domain of an image is connected after imposing a lattice graph, which naturally defines clusters as connected components. We also explain different parameters used to generate an open interval cover. To our knowledge, this is the first time mapper has been applied to volumetric images.

As an example we apply mapper to a volumetric image obtained through X-ray computed tomography (CT) on a barley spike. Different choices of filter functions and cover elements reveal different aspects of the morphology and structure of barley.

2   Background

Mapper separates data using a filter function, performs clustering, and outputs a *simplicial complex* describing how clusters overlap. A simplicial complex can be thought of as a generalization of a graph; that is, a combinatorial object with vertices and edges, but also triangles, tetrahedra, and higher dimensional ana-logues. However, the reader not familiar with this concept can internally replace "simplicial complex" with "graph" for this paper as we will immediately restrict our constructions of interest so that all simplicial complexes are 1-dimensional.

The general mapper construction uses the concept of the nerve of a cover [13, §3.3]. Given an open cover of a topological space (a collection of open sets whose union contains the space), the nerve is a skeleton that summarizes the intersections of the cover elements. For ease of notation throughout, define $[N] := \{1, 2, \ldots, N\}$.

Definition 1. *Let the collection* $\mathcal{U} = \{U_\alpha\}_{\alpha=1}^N$ *be an open cover of a topological space* $\mathbb{X}$. *The* **nerve** *of* $\mathcal{U}$, *written* $Nrv(\mathcal{U})$, *is the simplicial complex with vertex set* $[N]$, *for which* $\{\alpha_1, \alpha_2, \ldots, \alpha_k\} \subseteq [N]$ *spans a* k-*simplex in* $Nrv(\mathcal{U})$ *if and only if* $\bigcap_{i=1}^k U_{\alpha_i} \neq \emptyset$.

To define mapper, consider a real-valued function f on some topological space (this is commonly called a filter function) and choose an open cover of the range of f. Mapper is the nerve of the components of the preimages (under f) of the cover elements. Note that if there are no triple intersections between these preimages, then all of the simplices in the nerve will be of degree 0 or 1, i.e. the nerve is a graph.

Definition 2. *Let* $\mathbb{X}$ *be a connected topological space and let* $f \colon \mathbb{X} \to \mathbb{R}$ *be a continuous function. Choose an open interval cover* $\mathcal{U} = \{U_\alpha\}_{\alpha=1}^N$ *of* $f(\mathbb{X})$ *such that* $U_\alpha \cap U_{\alpha'} \cap U_{\alpha''} = \emptyset$ *for all distinct* $U_\alpha, U_{\alpha'}, U_{\alpha''} \in \mathcal{U}$. *Let* $\mathcal{V}$ *be the cover of the domain of* f *whose elements are the connected components of* $f^{-1}(U)$ *for all* $U \in \mathcal{U}$. *The* **mapper graph** *is*

$$\mathcal{M}(\mathbb{X}, f, \mathcal{U}) := Nrv(\mathcal{V}).$$

One important difference to note between our definition and the original definition from [27] is that we assume our input data is a continuous function on a topological space, while the original definition was given for point cloud data (or more generally, finite metric space data). This actually simplifies the construction as it means that no choice of clustering method is required; this will be further discussed in Sect. 3.1.

3   Topological Mapper for 3D Images

Imaging methods such as magnetic resonance imaging (MRI) and CT generate 3D volumetric images. Volumetric images are 3D arrays of intensity values, which roughly correspond to the density of the object in a particular spot. Each cubical cell in the image is called a *voxel*, which is analogous to the concept of a *pixel* in a 2D image. See Fig. 1 for an example of the full construction.

Applying mapper to volumetric images follows the basic construction in Sect. 2, except that the structure of the domain warrants a careful approach

6-connectivity   18-connectivity   26-connectivity

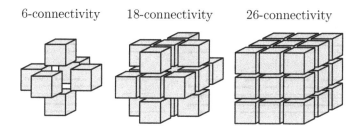

Fig. 2. Standard structuring elements for 3D lattices.

to connectivity. Denote an $m \times n \times p$ cubical lattice by $L_{m,n,p} := [m] \times [n] \times [p]$. An $m \times n \times p$ volumetric image $\mathcal{P}$ is a function $\mathcal{P} \colon L_{m,n,p} \to \mathbb{R}$ and a set of input coordinates (i,j,k) is called a *voxel coordinate*. Elements in the range of $\mathcal{P}$ are *intensity values* given by the method used to create the scan (e.g. CT or MRI).

3.1   Structuring Element

There are various ways in which one can define connectivity between the voxel coordinates. We capture such a model of connectivity using a graph on $L_{m,n,p}$ that specifies the pairs of points that are adjacent. In the models of connectivity we consider, the edges of the graphs are based on the relative coordinates between points. That is, there is an edge between two points if the difference between those points lies in a given set D. Following the usual convention in 2D [17], we refer to D as the *structuring element*, and define the *voxel connectivity graph* $G = (V, E)$ by

$$V := L_{m,n,p},$$
$$E := \{(u, v) \in V \times V \mid u - v \in D\}.$$

The L_p-norm of a point $x = (x_1, \ldots, x_k) \in \mathbb{R}^k$ is given by $\|x\|_p = (\sum_{i=1}^{k} |x_i|^p)^{1/p}$. Standard structuring elements for a k-dimensional lattice include the sets $D = \{x \mid 0 < \|x\|_2 \le \sqrt{r}\}$ for $r \in [k]$. Further natural structuring elements are the sets $D = \{x \mid \|x\|_1 \le r \text{ and } \|x\|_\infty = 1\}$ of points simultaneously at distance 1 in the L_∞-norm and at distance at most r from the origin in the L_1-norm for $r \in [k]$. For $k \le 3$, these two models of connectivity are identical for each r. For $r = 1, 2$, and 3, these structuring elements are illustrated in Fig. 2, and are called 6-, 18-, and 26-connectivity, respectively.

In practice, the vertex set V is chosen to be a subset of $L_{m,n,p}$ that only includes the foreground of the image. This is commonly done by thresholding the image (setting intensity values in some range equal to 0). In this paper we use 26-connectivity ("full connectivity") on interior voxel coordinates (Fig. 3).

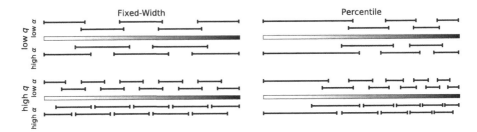

Fig. 3. Examples of fixed-width and percentile intervals over the range of the filter $f: [0,1] \rightarrow [0,1]$ given by $f(x) = \sqrt{x}$. We consider three parameters for cover elements: the type of scheme (evenly-spaced or percentile), the number of cover elements $(q+1)$ and the amount of overlap, or gain (α).

3.2   Cover Elements

We suggest two heuristics for choosing intervals $\{(a_0, b_0), (a_1, b_1), \ldots, (a_q, b_q)\}$ to cover the range of a filter function $f: \mathbb{X} \rightarrow \mathbb{R}$. See Fig. 4 for examples of these constructions.

First we define evenly-spaced intervals over the range. Suppose that $f(\mathbb{X}) \subseteq (a, b)$ and let $a_0 := a$ and $b_q := b$. Let $r = (b-a)/q$ be the interval length and choose $\alpha \in (0, 1]$ to be the proportion of overlap between cover elements. Now for $1 \leq i \leq q-1$, let

$$a_i = a + (i + \alpha)r,$$
$$b_i = a + (i + 1 - \alpha)r.$$

This rule generates $q+1$ overlapping intervals. The parameter r is called the resolution and controls the diameter of the elements; the gain α captures the proportion of overlap [4].

The second scheme is based on the underlying distribution of intensity values. Specifically, define $\mathrm{Perc}(x): [0, 1] \rightarrow \mathbb{R}$ so that $\mathrm{Perc}(x)$ is the $(100x)^{\mathrm{th}}$ percentile of the range of f. Let $a_0 = \mathrm{Perc}(0)$ and $b_q = \mathrm{Perc}(1)$ and for $1 \leq i \leq q-1$, let

$$a_i = \mathrm{Perc}(i/q) - \alpha(\mathrm{Perc}(i/q) - \mathrm{Perc}((i-1)/q)),$$
$$b_i = \mathrm{Perc}((i+1)/q) + \alpha(\mathrm{Perc}((i+1)/q) - \mathrm{Perc}(i/q)).$$

The cover intervals defined here correspond to subsets of the same size in the underlying dataset.

3.3   Filtering

A filter function used to construct the mapper graph is a choice of real valued function on the vertices. The most natural filter function to use for a volumetric image $\mathcal{P}$ is the input data itself, i.e $f(i, j, k) = \mathcal{P}(i, j, k)$. Filters may also be constructed by viewing the volumetric image as a collection of foreground voxel coordinates. For example, the z-filter $f_z: V(G) \rightarrow \mathbb{R}$ is defined by

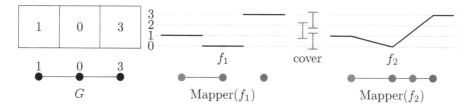

Fig. 4. An example of a 2D image with three voxels. The resulting mapper is disconnected if the function is piecewise-constant, but connected if the function value is interpolated along edges.

$f_z((i, j, k)) = k$. Similar filters can be constructed in the x- and y-directions and, more generally, from sweeping a plane with an arbitrary direction through the finite lattice $L_{m,n,p}$.

There are various ways we can extend a filter function to the voxel connectivity graph. A natural way to construct a filter on a 3D volumetric image is to extend a function $f \colon L_{m,n,p} \to \mathbb{R}$ defined on the vertices $V(G)$ to the points interior to edges of G. This can be done through linear interpolation between the function values at the end points of the edge, but also by assigning the function value of the closest end point of that edge, resulting in a piecewise-constant function. Figure 4 illustrates how these two choices may result in a different mapper graph. However, if adjacent voxels always lie in a common element of the cover of the range of f, the resulting mapper is the same.

For our algorithm, we choose the piecewise-constant method of extending the function to edges. In the context of CT images this choice makes sense because the intensity values arise from the X-ray densities of the scanned materials. The overlap between cover elements is often chosen to capture adjacency between certain materials. For instance, we might be interested when a material of intensity 0 is adjacent to a material of intensity 1. However, if we use linear interpolation, adjacent materials of intensities 0 and 3 would (falsely) appear to be separated by a material of intensity 1.

In practice the filtering step takes advantage of the fact that no three cover elements intersect. Consider a filter function f with the cover intervals

$$\{(a_0, b_0), (a_1, b_1), \ldots, (a_q, b_q)\}.$$

Assume that $a_0 \le a_1 \le \cdots \le a_q$, as in our construction of fixed-width or percentile intervals. By avoiding triple intersections between cover intervals we can assume that $f^{-1}((a_\ell, b_\ell))$ and $f^{-1}((a_{\ell+2}, b_{\ell+2}))$ are disjoint. Furthermore, the preimages of intervals with even index are disjoint; similarly the preimage of intervals with odd index are disjoint. Following [24], we exploit this fact to construct two 3D label arrays to filter the data: L_0 and L_1, called the even and odd arrays respectively. For $b \in \{0, 1\}$, define

$$L_b(i, j, k) = \begin{cases} 2\ell + b & \text{if } (i, j, k) \in f^{-1}((a_{2\ell+b}, b_{2\ell+b})) \text{ for some } \ell \in \mathbb{Z} \\ 0 & \text{otherwise} \end{cases}.$$

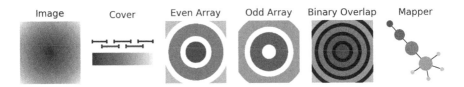

Fig. 5. Mapper on a 2D circular gradient image. The even (odd) array contains the voxels corresponding to cover elements with even (odd) indices. Each region (black) in the overlap between the two arrays suggests an edge in the mapper graph.

In other words, a non-zero value in L_0 or L_1 at voxel coordinate (i, j, k) is the index of the cover interval that contains $\mathcal{P}(i, j, k)$. Examples of even and odd arrays are shown in Fig. 5.

3.4 Clustering

In the point-cloud version of mapper, a clustering algorithm (e.g. k-means) is required to separate points within the pullback of each cover element. In this version for 3D volumetric images, clusters are defined by connected components in the voxel connectivity graph. These components are dependent on the chosen structuring element, as described in Sect. 3.1. Further, they are sensitive to the choice of filter function interpolation as described in Sect. 3.3.

In practice, label the disjoint regions in L_0 and L_1 to generate arrays L_0' and L_1'. These arrays simply choose a labeling for each connected component in L_i and give each voxel in the connected component the same entry. The connected components in these images are what will be used to determine the vertices of the mapper graph. Note that the preimage of a cover element may have several connected components.

3.5 Mapper Graph Construction

To construct the mapper graph, define an overlap array $\mathcal{O}$ as the logical AND of L_0' and L_1'. The disjoint regions in this image represent the overlap between clusters. To identify the edges in the mapper graph let (i, j, k) be a voxel coordinate in a connected component of $\mathcal{O}$. The mapper graph has the edge $(L_0'(i, j, k), L_1'(i, j, k))$.

3.6 Implementation

While there is available open and closed-source code for point cloud input data, to the best of our knowledge, no previous code is available for applying mapper to 3D images. Open-source software to apply mapper to point clouds includes an implementation in R [21], Kepler Mapper in Scikit-TDA for Python 3 [2], and Python Mapper, which has a GUI for Python 2 [1]. The private company

Ayasdi also offers closed-source software to generate mapper graphs for point clouds, but not volumetric images.

Our implementation for 3D images is done in Python 3, using `numpy` [20] and `skimage` [28]. It is open-source and available on Github [8]. This algorithm is based on extending the ideas of [24] which is restricted to 2D images.

4   Example: The Shape of Barley

We apply our 3D mapper algorithm to a CT reconstruction of barley (*Hordeum vulgare* L.). The specimen, a reconstructed spike, is shown in Fig. 1. A spike is an inflorescence bearing sessile flowers. Spikelets consist of single flowers and are attached to the rachis (the backbone of the inflorescence) in groups of three at each node, alternately arranged. Subtending each spikelet are two glumes (bracts), and enclosing the seed of each flower are upper (palea) and lower (lemma) bracts. From each lemma extends a prominent awn (a bristle-like appendage) [23]. The seed of each spikelet has a conspicuously high X-ray density. The surrounding organs of each seed, the awns, and rachis are less X-ray dense. We use this example to show how different choices of parameters in the mapper construction highlight different aspects of the data in question. An interactive visualization of mapper applied to this barley scan is available at egr.msu.edu/~eithunmi/js_mapper.

4.1   Parameters

Our construction of mapper for images requires three main parameters: a choice of structuring element, a filter function and an open cover of its range. In this example we adopt 26-connectivity and give examples using both the intensity filter and the z-filter. For each filter function, we consider different numbers of evenly-spaced intervals and percentile intervals with different levels of overlap.

4.2 z-filter

The output of mapper with the z-filter is shown in Fig. 6. Following [27], the graphs are drawn to reflect the underlying data. Vertices are colored according to the average value of the clusters they represent and their size is proportional to the square root of the size of each cluster. Additionally, vertices are plotted at the centroids of their underlying clusters in a 2D projection of the volumetric image, which suggests the overall structure of the spike. The bristle-like awns extend out from the nodes of the spike, and the individual spikelets from which the awns arise are segmented along the central axis representing the rachis. The correspondence of the mapper graph to the architecture of the barley spike is because of the choice of direction of the z-axis along the rachis, which is sensitive to the separation of the spikelets and awns from the main body of the inflorescence.

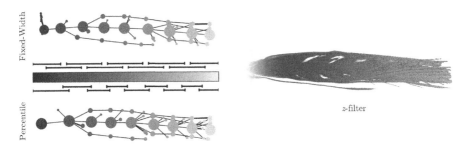

Fig. 6. Mapper graphs generated using the z-filter. The color bar shows the cover elements on the range of the filter. The nodes are drawn in line with the average function value of vertices represented. (Color figure online)

4.3 Intensity Filter

The output of mapper with the intensity filter is shown in Fig. 7. Like the z-filter, the intensity filter summarizes the anatomical features of the barley spike, but as defined by X-ray density, yielding a different graph. Clusters corresponding to X-ray dense seeds all connects to a central node, representing the rachis. Similarly, there are separate hubs connected to numerous edges corresponding to awns and anatomical features surrounding the seed.

5   Discussion

New data analysis tools from the field of applied topology have provided previously unavailable insights on the shape of data. Applying mapper to 3D volumetric images has implications for studying the shape and structure of 3D images. In the future we plan to study other intricate plant structures including maize and citrus using X-ray CT. Extending this algorithm to 4D images or using time-varying mapper graphs could also be used to track morphological changes over time or across species. Further structural information (at the expense of computational cost) could be also constructed by removing the restriction on triple intersections and producing a higher-dimensional simplicial complex. Recently proposed parallelization schemes might make this more feasible [11,12]. Multiscale mapper approaches may also be useful for capturing structural details at different scales [6].

Another important area for future work is parameter selection, which may reduce noise in resultant mapper graphs. Recently a bootstrap method was proposed to automatically tune mapper parameters by estimating the distance between mapper graph and the underlying Reeb graph [4]. Along with the rise and the gain parameters used to generate a fixed-width interval cover, a δ parameter is introduced which corresponds to our choice of structuring element. To our knowledge no process has been proposed to choose parameters for a percentile-based interval scheme. Parameters could also be chosen based on some desirable outcome, such as the connectivity or diameter of the output graph.

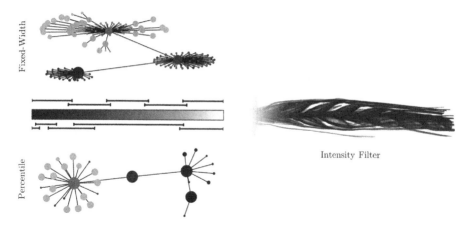

Fig. 7. Mapper graphs for the barley shown in Fig. 6 generated using the intensity filter with two different choices of cover. The locations of the vertices are drawn with the Fruchterman-Reingold spring layout [9].

Concerning the input data, while the scheme proposed here solves a key problem related to connectivity by imposing a graph structure on the data, the clustering scheme is highly susceptible to noise. For example, if a connected graph component in a preimage of a cover element has connectivity 1, the removal of one edge would disconnect the graph. A way to combat this might be to use graph connectivity measures to test each cluster. If a cluster is not connected by sufficiently many edges it can be cut to form multiple clusters.

Acknowledgments. The authors thank Jacob Landis and Daniel Koenig for providing the barley spike and X-ray Computed Tomography data. The data set is available on the figshare repository [5].

References

1. Python mapper. http://danifold.net/mapper/
2. Sckit-tda. https://github.com/scikit-tda/scikit-tda
3. Carlsson, G.: Topological pattern recognition for point cloud data. Acta Numer. **23**, 289–368 (2014). https://doi.org/10.1017/S0962492914000051, http://journals. cambridge.org/article_S0962492914000051, survey
4. Carrière, M., Michel, B., Oudot, S.: Statistical analysis and parameter selection for mapper (2017). arXiv:1706.00204
5. Chitwood, D.H., Eithun, M., Koenig, D., Landis, J., Munch, E., Ophelders, T.: CT scan of barley (Hordeum vulgare L.) (2019). https://doi.org/10.6084/m9.figshare. 7590833
6. Dey, T., Mémoli, F., Wang, Y.: Multiscale mapper: topological summarization via codomain covers. In: Proceedings of the Twenty-Seventh Annual ACM-SIAM Symposium on Discrete Algorithms, pp. 997–1013. Society for Industrial and Applied Mathematics, December 2015. https://doi.org/10.1137/1.9781611974331.ch71

7. Doraiswamy, H., Natarajan, V.: Computing Reeb graphs as a union of contour trees. IEEE Trans. Vis. Comput. Graph. **2**, 249–262 (2013)
8. Eithun, M.: eithun/3D-mapper: topological mapper for 3D volumetric images (2019). https://doi.org/10.5281/zenodo.2602059
9. Fruchterman, T.M., Reingold, E.M.: Graph drawing by force-directed placement. Softw. Pract. Exp. **21**(11), 1129–1164 (1991)
10. Ghrist, R.: Elementary Applied Topology. CreateSpace Independent Publishing Platform, London (2014)
11. Hajij, M., Assiri, B., Rosen, P.: Distributed mapper, December 2017. arXiv:1712.03660 [cs, stat]
12. Hajij, M., Rosen, P.: An efficient data retrieval parallel Reeb graph algorithm, October 2018. arXiv:1810.08310 [cs]
13. Hatcher, A.: Algebraic Topology. Cambridge University Press, Cambridge (2001)
14. Janusch, I.: Reeb graph based image representation for phenotyping of plants. Master's thesis, Vienna University of Technology (2014)
15. Munch, E.: A user's guide to topological data analysis. J. Learn. Anal. **4**(2) (2017). https://doi.org/10.18608/jla.2017.42.6
16. Munch, E., Wang, B.: Convergence between categorical representations of Reeb space and Mapper. In: Fekete, S., Lubiw, A. (eds.) 32nd International Symposium on Computational Geometry (SoCG 2016). Leibniz International Proceedings in Informatics (LIPIcs), vol. 51, pp. 53:1–53:16. Schloss Dagstuhl-Leibniz-Zentrum fuer Informatik, Dagstuhl (2016). https://doi.org/10.4230/LIPIcs.SoCG.2016.53
17. Najman, L., Cousty, J.: A graph-based mathematical morphology reader. Pattern Recogn. Lett. **47**, 3–17 (2014). https://doi.org/10.1016/j.patrec.2014.05.007
18. Nicolau, M., Levine, A.J., Carlsson, G.: Topology based data analysis identifies a subgroup of breast cancers with a unique mutational profile and excellent survival. Proc. Natl. Acad. Sci., 201102826 (2011). https://doi.org/10.1073/pnas.1102826108, https://www.pnas.org/content/early/2011/04/07/1102826108
19. Nielson, J.L., et al.: Topological data analysis for discovery in preclinical spinal cord injury and traumatic brain injury. Nat. Commun. **6** (2015). https://doi.org/10.1038/ncomms9581, https://www.ncbi.nlm.nih.gov/pmc/articles/PMC4634208/
20. Oliphant, T.E.: Guide to numpy (2006)
21. Piekenbrock, M., Doran, D., Kram, R.: Efficient multi-scale simplicial complex generation for mapper (2018). https://peekxc.github.io/resources/indexed_mapper.pdf
22. Reeb, G.: Sur les points singuliers d'une forme de pfaff complèment intégrable ou d'une fonction numérique. C. R. Acad. Séances **222**, 847–849 (1946)
23. Reid, D.A.: Morphology and anatomy of the barley plant. Barley (Barley) **26**, 73–101 (1985)
24. Robles, A., Hajij, M., Rosen, P.: The shape of an image - a study of mapper on images. In: Proceedings of the 13th International Joint Conference on Computer Vision, Imaging and Computer Graphics Theory and Applications. VISAPP, vol. 4, pp. 339–347. INSTICC, SciTePress (2018). https://doi.org/10.5220/0006574803390347
25. Shi, Y., Li, J., Toga, A.W.: Persistent Reeb graph matching for fast brain search. In: Wu, G., Zhang, D., Zhou, L. (eds.) MLMI 2014. LNCS, vol. 8679, pp. 306–313. Springer, Cham (2014). https://doi.org/10.1007/978-3-319-10581-9_38
26. de Silva, V., Munch, E., Patel, A.: Categorified Reeb graphs. Discrete Comput. Geom. **55**(4), 854–906 (2016). https://doi.org/10.1007/s00454-016-9763-9

27. Singh, G., Memoli, F., Carlsson, G.: Topological methods for the analysis of high dimensional data sets and 3D object recognition. In: Botsch, M., Pajarola, R., Chen, B., Zwicker, M. (eds.) Eurographics Symposium on Point-Based Graphics. The Eurographics Association (2007). https://doi.org/10.2312/SPBG/SPBG07/091-100

28. van der Walt, S., et al.: Scikit-image: image processing in Python. PeerJ **2**, e453 (2014). https://doi.org/10.7717/peerj.453

Local Uniqueness Under Two Directions in Discrete Tomography: A Graph-Theoretical Approach

Silvia M. C. Pagani[(✉)] [iD]

Università Cattolica del Sacro Cuore, via Musei 41, 25121 Brescia, Italy
silvia.pagani@unicatt.it

Abstract. The goal of discrete tomography is to reconstruct an image, seen as a finite set of pixels, by knowing its projections along given directions. Uniqueness of reconstruction cannot be guaranteed in general, because of the existence of the switching components. Therefore, instead of considering the uniqueness problem for the whole image, in this paper we focus on local uniqueness, i.e., we seek what pixels have uniquely determined value. Two different kinds of local uniqueness are presented: one related to the structure of the directions and of the grid supporting the image, having as a sub-case the region of uniqueness (ROU), and the other one depending on the available projections. In the case when projections are taken along two lattice directions, both kinds of uniqueness have been characterized in a graph-theoretical reformulation. This paper is intended to be a starting point in the construction of connections between pixels with uniquely determined value and graphs.

Keywords: Discrete tomography · Graph · Lattice direction · Region of uniqueness · Uniqueness of reconstruction

1 Introduction

The reconstruction (namely, the recovery) of the interior of an object, modeled as a density function, from the knowledge of its projections along given directions is the aim of tomography. When the object is regarded as a finite set of pixels and the directions' slopes are rational numbers, as in our investigation, one is dealing with *discrete tomography*. In particular, our framework is the two-dimensional lattice $\mathbb{Z}^2$, so that the object to be reconstructed is also called an image.

One of the main tasks of tomographic reconstruction is to ensure that the obtained solution equals the original object. It cannot be achieved in general, as the tomographic problem is ill-posed. Ambiguities in the reconstruction are due to the presence of the so-called *switching functions* (and in particular their nonzero pixels, constituting the *switching components*) related to a set of directions. Switching functions are images that can be added to a solution of the tomographic problem in order to obtain another admissible solution. By [11], uniqueness of reconstruction is equivalent to the absence of switching functions.

© Springer Nature Switzerland AG 2019
B. Burgeth et al. (Eds.): ISMM 2019, LNCS 11564, pp. 96–107, 2019.
https://doi.org/10.1007/978-3-030-20867-7_8

Different further constraints may be imposed on the image in order to prevent switching functions. One is to confine the object in a finite subset $\mathcal{A}$ of the lattice $\mathbb{Z}^2$, so that no switching component can be contained in $\mathcal{A}$, meaning that every image defined on $\mathcal{A}$ is uniquely determined by projections along the considered directions. Another extra condition which is commonly exploited is an upper bound on the number of the values the unknown function can take. In particular, *binary* (namely, 0-1-) images have been extensively studied. Combining the binary property with confining the object in a finite grid, in [5] a uniqueness result has been obtained for projections taken along four suitable directions. On the other side, the binary constraint, together with the *convexity* property for the image, enables uniqueness of reconstruction along every set of seven nonparallel directions, or even along four suitable directions, in the whole lattice $\mathbb{Z}^2$ ([13]). There is, however, a fundamental issue which prevents a direct implementation of such results: the reconstruction problem for more than two directions is NP-hard ([14]).

In the present paper we want to exploit a connection between tomography and graphs which has been investigated since [12]. Basing on [4], such a link works when projections from two lattice directions are considered. We therefore restrict to the case involving two directions.

We move from considering uniqueness of reconstruction for the whole image and focus on a *local* perspective. In other words, we are interested in detecting what pixels have uniquely determined value under the given directions. Such an issue is investigated from two viewpoints: a *structural* one, i.e., for all tomographic problems defined on a given subset $\mathcal{A}$ of the lattice and for given directions, independently of the projection data; and a *data-dependent* perspective, namely, for a given set of projections. In particular, in the structural case a remarkable subset is highlighted: that of *region of uniqueness* (ROU), first introduced in [7]. Structural and data-dependent uniqueness, for two directions, are defined and characterized in graph-theoretical terms: pixels of the grid are seen as edges in a (di)graph and their value is uniquely determined if and only if they are not on cycles. This paper is intended to be a starting point in relating pixels with uniquely determined value and graphs.

The paper is organized as follows. Section 2 reports the basic notions of discrete tomography, together with the definitions of the various kinds of uniqueness and a couple of examples. Moreover, in Subsect. 2.1 the link between tomographic problem under two directions and graphs is presented. In Sect. 3 the main results are stated and proven, and strategies to retrieve uniquely determined pixel values are suggested. Section 4 provides possible further work and concludes the paper.

2   Definitions and Known Results

A common assumption in the tomographic investigation is to know the size of the object to be reconstructed, so that positive integers m, n are set to be the number of the columns and the number of the rows, respectively, of the minimal grid containing the object. Up to translation, we can set the origin of the integer lattice as the bottom-left corner of the grid. We therefore fix the grid $\mathcal{A} = \{(z_1, z_2) \in \mathbb{Z}^2 : 0 \leq z_1 < m, 0 \leq z_2 < n\}$ throughout the paper. Elements of $\mathcal{A}$ are called *pixels*. In what follows, we will identify a pixel $P = (z_1, z_2)$ with the unit square $[z_1, z_1 + 1) \times [z_2, z_2 + 1)$, having P as bottom-left corner. An *image* is a map $f : \mathcal{A} \to \mathbb{R}$, where pixels are sent to real numbers representing colors, or grey levels. If $f \in \{0, 1\}^{\mathcal{A}}$, we say that f is a *binary image*, whose 0-valued pixels are called, for instance, *white pixels* and 1-valued ones are the *black pixels*.

A *(lattice) direction* is a pair $[(a, b)]$ of coprime integers. A *(lattice) line* with direction $[(a, b)]$ is the set $\ell_t = \{(x, y) \in \mathbb{Z}^2 : ay = bx + t\}$, for $t \in \mathbb{Z}$. The *projection* (or *line sum*) p_ℓ of f along a line ℓ is the sum of values of the pixels lying on ℓ:

$$p_\ell = \sum_{z \in \mathcal{A} \cap \ell} f(z).$$

Projections, ordered in some way, are collected in the *projection vector* $\mathbf{p}$; so, for a finite set $U = \{[(a_i, b_i)] : i = 1, \ldots, d\}$ of distinct directions, we define a *projection operator* $\pi_U : \mathbb{R}^{\mathcal{A}} \longrightarrow \mathbb{R}^{\kappa(U, \mathcal{A})}$, where

$$\kappa(U, \mathcal{A}) = \sum_{i=1}^{d} (|a_i| \, n + |b_i| \, m - |a_i| \, |b_i|)$$

is the number of projections (i.e., the size of $\mathbf{p}$), mapping an image to the corresponding projection vector.

The *tomographic problem* we deal with can be described as follows: given a grid $\mathcal{A}$, a vector $\mathbf{p}$ and a set U of directions, find an image f whose line sums satisfy $\mathbf{p}$ (namely, such that $\pi_U(f) = \mathbf{p}$). Sometimes, by an abuse of notation, we will say that we reconstruct a pixel instead of its value, i.e., instead of an image defined on that pixel.

In general, for a given set U of directions, the presence of the so-called *switching functions*, namely, nonzero images with null projections along the directions in U, prevents uniqueness of reconstruction. If a switching function is added to an image which is a solution of the tomographic problem, another image, different from the previous one and still consistent with the projections, is obtained. The support of a switching function is called *switching domain*. Every minimal subset of the switching domain with the property to be a switching domain itself is said to form a *switching component*. By [11], an image is uniquely determined by its projections along U if and only if the switching domain is empty. This happens when no switching function with respect to U can be defined inside the grid $\mathcal{A}$, i.e., when the switching domain exceeds the sides of the grid. In this case uniqueness of reconstruction is guaranteed, and the focus moves on other issues, such as speeding up the reconstruction algorithms (see for instance [17]). We are however interested in detecting whether a pixel value is uniquely recoverable even when switching functions can be defined. Therefore, we focus on the case when

$$\sum_{i=1}^{d} |a_i| < m \quad \text{and} \quad \sum_{i=1}^{d} |b_i| < n \tag{1}$$

hold for $U = \{[(a_i, b_i)] : i = 1, \ldots, d\}$. Sets satisfying (1) are said to be *valid* for the grid $\mathcal{A}$. Note that (1) is the negation of the Katz condition (see [16]).

Given a set of lattice directions, valid for a grid $\mathcal{A}$, one can construct switching components inside $\mathcal{A}$. Therefore, the investigation moves from looking for uniqueness of reconstruction in the whole grid to a notion of *local* uniqueness, namely, the focus is on what parts of the grid can be reconstructed with no ambiguities. Note that this kind of problem is related only to the size of the grid and on the set of directions, and it is independent of the projection vector (i.e., of a specific tomographic problem). The issue is therefore related to the structure of the switching domains (see for instance [15]). Since uniqueness of reconstruction is equivalent to the absence of switching components, uniquely determined parts of the grid are sought among pixels which are not part of a switching component.

Definition 1. *Let $\mathcal{A}$ be a finite grid, U a valid sets of directions for $\mathcal{A}$. A pixel z is said to be* structurally unique *if it is not in a switching domain. Equivalently, z is structurally unique if its value is zero in every switching function associated to U.*

Lemma 1. *Let $z \in \mathcal{A}$, U a set of directions. The following statements are equivalent:*

(i) z is a structurally unique pixel;
(ii) $\forall f \in \ker(\pi_U) : f(z) = 0$;
(iii) for all projections vectors $\mathbf{p}$ related to U there exists $b \in \mathbb{R}$ such that

$$\forall f \in \mathbb{R}^{\mathcal{A}} : \pi_U(f) = \mathbf{p} \Longrightarrow f(z) = b.$$

Proof. Statements (i) and (ii) are equivalent by Definition 1. Equivalence of statements (ii) and (iii) comes from the linearity of π_U. □

If m, n are large enough, structurally unique pixels are located in the corners of the grid (see [18]). Among structurally unique pixels, those belonging to the so-called *region of uniqueness*, first introduced in [7], deserve special attention. It is defined recursively as follows.

Definition 2. *The region of uniqueness (ROU) related to a set U of directions is the final set of pixels $z \in \mathcal{A}$ such that*

(i) z is the only pixel on a line parallel to a direction of U, or
(ii) z is on a line with direction in U, whose other pixels are already in the ROU.

Note that Definitions 1 and 2 do not involve peculiar features of the set of values ($\mathbb{R}$) which can actually be replaced by any set.

As proven in [8], the ROU is the set of structurally unique pixels that can be reconstructed in linear time with respect to the sizes of the grid. The shape of the ROU has been described for two directions in [8], for special sets of three directions in [9,10] and for other special sets in [18]. When the ROU does not coincide with the set of structurally unique pixels, we say that there is *erosion*. The following examples show cases when the two sets are not equal.

Example 1. Consider the grid $\mathcal{A} = \{(z_1, z_2) \in \mathbb{Z}^2 : 0 \leq z_1 < 3, 0 \leq z_2 < 3\}$ and the set $U = \{[(1,0)], [(0,1)], [(1,-1)]\}$. Pixels $(0,0)$ and $(2,2)$ belong to the ROU, since both of them are the only pixels on a line parallel to direction $[(1,-1)]$ (see Fig. 1). No other pixel is in the ROU, while pixel $P = (1,1)$ is a structurally unique one. Its value, say p, can be computed as

$$p = \frac{h + v - d_1 - d_2}{2},$$

being h (resp., v) the value of the horizontal (resp., vertical) projection along P, and d_1 (resp., d_2) the projection along the direction $[(1,-1)]$ through $(1,0)$ (resp., $(2,1)$). Therefore, the value p is uniquely determined for all tomographic problems defined for $\mathcal{A}$ and U. The pixel P is not in the ROU because all lines through P parallel to directions of U have further intersections with $\mathcal{A}$ which are not in the ROU.

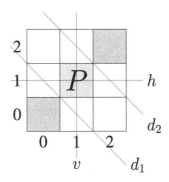

Fig. 1. Structurally unique pixels (grey and yellow ones) for a 3×3 grid $\mathcal{A}$ and $U = \{[(1,0)], [(0,1)], [(1,-1)]\}$. Pixel P (in yellow) is not in the ROU. (Color figure online)

In the following example, taken from [8], the ROU is concentrated in the corners of the grid.

Example 2 (Example 6 of [8]). Let $\mathcal{A} = \{(z_1, z_2) \in \mathbb{Z}^2 : 0 \leq z_1 < 8, 0 \leq z_2 < 7\}$, $U = \{u_1 = [(4,-3)], u_2 = [(3,-2)]\}$. The ROU, computed as explained in [8], consists of the grey pixels in Fig. 2. Pixels $(1,5), (2,4), (5,2), (6,1)$ are structurally unique, but do not satisfy Definition 2. Consider for instance pixel $P = (1,5)$ and denote by $x_{(z_1,z_2)}$ the value of pixel (z_1, z_2). Consider moreover the equations

$$x_{(0,6)} + x_{(4,3)} = \lambda_1, \tag{2}$$

$$x_{(3,4)} + x_{(7,1)} = \lambda_2, \tag{3}$$

$$x_{(1,5)} + x_{(4,3)} + x_{(7,1)} = \lambda_3, \tag{4}$$

$$x_{(0,6)} + x_{(3,4)} + x_{(6,2)} = \lambda_4, \tag{5}$$

where (2) and (3) refer to lines parallel to u_1, (4) and (5) are related to lines with direction u_2, λ_i-s, $i = 1, 2, 3, 4$, are the corresponding projections and $x_{(6,2)}$ is known, since pixel $(6, 2)$ is in the ROU. The value of P can be computed as $(4) + (5) - (2) - (3)$, namely,

$$x_{(1,5)} = \lambda_3 + \lambda_4 - \lambda_1 - \lambda_2 - x_{(6,2)}.$$

Therefore, P is not in the ROU (other pixels on lines through P and parallel to either u_1 or u_2 are not uniquely determined), even if it is structurally unique.

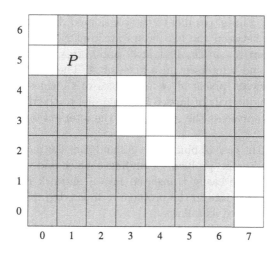

Fig. 2. Pixels in the ROU (in grey) for a 8×7 grid and $U = \{[(4, -3)], [(3, -2)]\}$, and structurally unique pixels not in the ROU (in yellow). White pixels belong to two distinct switching components. (Color figure online)

We now introduce a wider notion of uniqueness, which exploits a further constraint on the class of admissible solutions, namely, the specification of the set of allowed grey levels for the image.

Definition 3. *Let U be a set of directions, $\mathbf{p}_U$ a projection vector, $B \subseteq \mathbb{R}$. A pixel z is* data-dependent unique *if there exists $b \in B$ such that, for all $f \in B^{\mathcal{A}}$ with $\pi_U(f) = \mathbf{p}_U$, it results $f(z) = b$.*

As an example, if $B = \mathbb{N}$ and the projection along a line is zero, then we know that all pixels on that line must get zero value. Note that a structurally unique pixel is also a data-dependent unique one. Roughly speaking, we say that, given a grid $\mathcal{A}$ and a set U of directions, a pixel is structurally unique if its value is constant among all solutions of all possible tomographic problems (namely, for all possible projection vectors $\mathbf{p}$), while a pixel is data-dependent unique if, for a given projection vector $\mathbf{p}$, its value does not change in all B-valued solutions of that particular tomographic problem. In what follows, we will employ $B = \{0, 1\}$. Note that in this case a pixel in a switching domain only takes either the value 1 or -1.

2.1   The Graph Theoretical Formulation

When U consists of two lattice directions, one can build a correspondence between tomographic problems and graphs (first described in [12]; see also [21]). Since the tomographic problem needs the knowledge of a projection vector, we first introduce the link between undirected bipartite graphs and the *structure* of the grid, where pixels are seen as intersections of lines with direction in U.

Given a grid $\mathcal{A}$ and a set U of two lattice directions, an undirected bipartite graph G can be constructed, consisting of two layers of vertices, each one corresponding to a direction. Each layer contains a number of vertices equal to the number of lattice lines, parallel to the considered direction, intersecting the grid $\mathcal{A}$. The edges between the two layers are constructed as follows: there is an edge between vertex i in the first layer and vertex j in the second layer if and only if lines i and j intersect in a pixel of $\mathcal{A}$. There is then an one-to-one correspondence between pixels in $\mathcal{A}$ and edges of G, since every pixel is the intersection of a unique pair of lattice lines parallel to the directions of U. The graph G therefore represents the structure of the intersections among lines with directions in U.

If we restrict to the case of binary images and consider moreover a projection vector $\mathbf{p}$, the tomographic problem of finding a black-and-white image f such that $\pi_U(f) = \mathbf{p}$ can be easily translated into a network flow model (see for instance [2,4]). We associate to the problem a directed graph $\tilde{G}$, whose vertices and edges are constructed as follows.

The set of vertices of $\tilde{G}$ consists of the two layers as for G, together with two more vertices, called the *source* and the *sink* respectively. There are three kinds of edges in $\tilde{G}$: edges from the vertices in the first layer to vertices in the second layer, defined as before; edges from the source to the vertices in the first layer, whose capacity is the value of the corresponding projection; similarly, edges from the second layer to the sink. Then there is a solution to the reconstruction problem if and only if a maximum flows fills all the edges from the source and to the sink to capacity. If there is a solution, there is an one-to-one correspondence between reconstructions and maximum flows. For a given maximum flow, the corresponding reconstruction can be obtained by giving a pixel value 1 if the related edge is used by the flow, and value 0 otherwise (see for instance Fig. 3 for a 3×3 grid and U consisting of the coordinate directions).

3   Results

By means of the graph formulation, both the structurally and data-dependent unique pixels are characterized. In the first case, only the undirected graph G described in the previous section is necessary.

Theorem 1. *Let $\mathcal{A}$ be a finite grid, U a set of two directions, G the undirected graph defined as in Subsect. 2.1. A pixel $z \in \mathcal{A}$ is structurally unique if and only if its corresponding edge $e(z)$ does not lie on a cycle of G.*

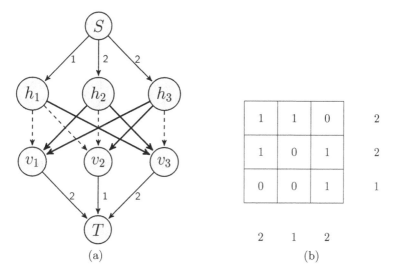

Fig. 3. (a) The network flow representation of a 3×3 grid, where projections are taken along the horizontal and the vertical direction. Vertices S and T are the source and the sink, respectively. Edges between the first and the second layer represent a possible solution to the problem. Dashed edges correspond to zero-valued pixels. (b) A grid representation of the solution found in (a).

Proof. If $e(z)$ belongs to a cycle of G, it means that z is in a switching component related to U, since G shows the relations among pixels regardless of the projections. It results that z is not structurally unique.

On the other hand, assume that z is not structurally unique. This is equivalent to saying that z is in a switching domain, then on a cycle in the undirected graph G. □

Pixels in the ROU can be detected even in the graph formulation.

Theorem 2. *If a vertex of G has degree one, then the pixel corresponding to its incident edge belongs to the ROU. Moreover, if one removes edges incident with vertices of degree one until no such vertices remain, the ROU is characterized as the set of pixels whose corresponding edges have been removed.*

Proof. If a vertex v has degree one, this means that the corresponding line has just one intersection with the grid. The pixel corresponding to the edge incident with v is therefore in the ROU by Definition 2, case (i).

Assume now to remove edges from G corresponding to pixels falling in the above case. If a vertex in one of the two layers has now degree one, the only edge incident with it corresponds to a pixel in the case (ii) of Definition 2. By proceeding to remove edges incident with vertices of degree one, exactly all edges corresponding to pixels in the ROU are taken. □

Example 3 (Continuation of Example 2). Figure 4 reports (part of) the graph reformulation of an 8×7 grid $\mathcal{A}$ and a set $U = \{u_1 = [(4, -3)], u_2 = [(3, -2)]\}$. Vertices in the upper (respectively, lower) layer correspond to u_1 (respectively, u_2). The number attached to each vertex in the upper (respectively, lower) layer represents the value q assumes in the equation $y = -\frac{3}{4}x + q$ (respectively, $y = -\frac{2}{3}x + q$) of the line corresponding to that vertex. Black and blue edges represent structurally unique pixels, while red edges correspond to pixels of the switching domain (in particular, the pixels split into two distinct switching components).

Blue edges, related to pixels $(1, 5), (2, 4), (5, 2), (6, 1)$, are not on cycles, so they are structurally unique. However, when removing edges corresponding to pixels in the ROU as in Theorem 2, such edges are not taken. This is due to the fact that edges corresponding to pixels $(1, 5), (2, 4), (5, 2), (6, 1)$ link vertices whose other incident edges are either such edges or edges related to switching components.

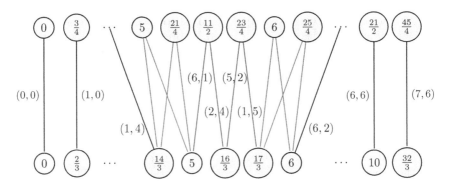

Fig. 4. (Part of) the graph translation of the grid in Example 2. Black edges correspond to pixels in the ROU, blue edges refer to structurally unique pixels not in the ROU, red edges are related to pixels on switching components. (Color figure online)

In order to exploit the network flow construction, we treat the recovery of data-dependent unique pixels only in the binary case. The value of data-dependent unique pixels can be computed only when we have a solution of the problem; so, assume that an image $\tilde{f}$ satisfying the projections corresponding to two of the directions has been obtained. We translate it into a directed graph $\tilde{G}$, defined differently from Sect. 2.1: the edges between the two layers are upwards (namely, from the layer related to the second direction to the layer of the first direction) if the corresponding pixel has value 0 and downwards if the corresponding pixel has value 1 (see Fig. 5 for an example). Such a new way of defining the edges enables us to construct the cycles we will need later.

The following result can be found, without a proof, in [1], which in turn refers to [6].

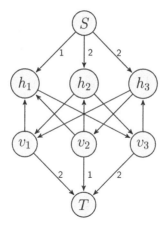

Fig. 5. The new graph associated to the image represented in Fig. 3(b). Ascending edges refer to 0-valued pixels, descending edges to 1-valued ones.

Theorem 3. *Let $\mathcal{A}$ be a finite grid, U a set of two directions, $\mathbf{p}$ a projection vector, $f \in \{0,1\}^{\mathcal{A}}$ a solution, $\tilde{G}$ the graph associated to f, defined as before. A pixel $z \in \mathcal{A}$ is data-dependent unique if and only if its corresponding edge $e(z)$ does not lie on a cycle of $\tilde{G}$.*

Proof. If $e(z)$ is on a cycle, then its value can be switched along such a cycle without changing the values of the projections, so the value of the pixel z is not uniquely determined.

Conversely, if the value of z is not uniquely determined, there exist two solutions, say f_1, f_2, of the related tomographic problem such that $f_1(z) \neq f_2(z)$. We associate graphs $\tilde{G}_1, \tilde{G}_2$ to f_1, f_2, respectively, built as before. Consider the switching function $h = f_1 - f_2$; pixels of h have values in the set $\{-1, 0, 1\}$. We build the corresponding digraph $\tilde{H}$ by skipping drawing edges with value zero and represent minus ones as ascending edges and ones as descending edges. Being $h(z) \neq 0$, $e(z)$ is drawn. Since all projections are zero, the incoming degree of a vertex must equal its outgoing degree. We start a walk from one of the two vertices incident with $e(z)$: each time an edge is crossed, it is removed. The only way to preserve zero line sums is that the walk ends in the same vertex as it started; this means that there is a cycle including $e(z)$ in the graph $\tilde{H}$. Since a -1 in $\tilde{H}$ corresponds to a 0 in $\tilde{G}_1$ and an 1 in $\tilde{G}_2$, and an 1 in $\tilde{H}$ corresponds to an 1 in $\tilde{G}_1$ and a 0 in $\tilde{G}_2$, also in the solutions there exist cycles including $e(z)$.   □

The above characterizations allow to employ tools from graph theory to retrieve both structurally and data-dependent unique pixels.

The ROU is constructed iteratively, by adding pixels corresponding to edges incident with a vertex with degree one, as long as such edges are removed from the graph. When no more pixels are added, the remaining edges which are not on cycles correspond to structurally unique pixels which do not satisfy Definition 2.

On the other side, data-dependent unique pixels correspond to edges not on cycles in the digraph $\tilde{G}$. In order to construct $\tilde{G}$, we first need a solution of the tomographic problem. Being $|U| = 2$, the reconstruction problem can be solved in polynomial time. For instance, if U consists of the coordinate directions, Ryser's algorithm ([19]) can be employed. Suppose therefore to have a binary image f satisfying the projections, and construct the digraph $\tilde{G}$ as before. The recovery of cycles in a digraph is one of the application of the search of strongly connected components, which can be solved in linear time by Tarjan's algorithm (see [20]).

4   Conclusions

In this paper we address the problem of recovering the pixels of a grid, whose value is uniquely determined by projections taken along a set of lattice directions. The instance is investigated both in general, namely, for all tomographic problems, and for assigned projections. The two cases lead to the definition of structurally unique pixels, with the special sub-case of the region of uniqueness (ROU), and data-dependent unique pixels, respectively. All kinds of uniquely determined pixels have been characterized in graph-theoretical terms as edges not on cycles, and strategies for their detection have been suggested.

The correspondence between tomographic problems and graphs considers only the case when $|U| = 2$. It would be worth to extend it to consider more than two lattice directions. It is not equivalent to take into account a pair of directions per time, since there could be some extra information which is not captured in this case (for instance, when considering all projections together). Moreover, concerning data-dependent unique pixels, a comparison between the presented approach and a 2SAT one as in [3], where clauses involving a pair of pixels per time are generated, would be worthwhile.

References

1. Aharoni, R., Herman, G., Kuba, A.: Binary vectors partially determined by linear equation systems. Discrete Math. **171**(1–3), 1–16 (1997). https://doi.org/10.1016/S0012-365X(96)00068-4

2. Batenburg, K.J.: Network flow algorithms for discrete tomography. In: Herman, G.T., Kuba, A. (eds.) Advances in discrete Tomography and Its Applications, pp. 175–205. Springer, Boston (2007). https://doi.org/10.1007/978-0-8176-4543-4_9

3. Batenburg, K.J., Kosters, W.: Solving nonograms by combining relaxations. Pattern Recogn. **42**(8), 1672–1683 (2009). https://doi.org/10.1016/j.patcog.2008.12.003

4. Batenburg, K.J., Sijbers, J.: Generic iterative subset algorithms for discrete tomography. Discrete Appl. Math. **157**(3), 438–451 (2009). https://doi.org/10.1016/j.dam.2008.05.033

5. Brunetti, S., Dulio, P., Peri, C.: Discrete tomography determination of bounded lattice sets from four X-rays. Discrete Appl. Math. **161**(15), 2281–2292 (2013). https://doi.org/10.1016/j.dam.2012.09.010

6. Chen, W.: Integral matrices with given row and column sums. J. Combin. Theory Ser. A **61**(2), 153–172 (1992). https://doi.org/10.1016/0097-3165(92)90015-M

7. Dulio, P., Frosini, A., Pagani, S.M.C.: Uniqueness regions under sets of generic projections in discrete tomography. In: Barcucci, E., Frosini, A., Rinaldi, S. (eds.) DGCI 2014. LNCS, vol. 8668, pp. 285–296. Springer, Cham (2014). https://doi.org/10.1007/978-3-319-09955-2_24

8. Dulio, P., Frosini, A., Pagani, S.M.C.: A geometrical characterization of regions of uniqueness and applications to discrete tomography. Inverse Prob. **31**(12), 125011 (2015). https://doi.org/10.1088/0266-5611/31/12/125011

9. Dulio, P., Frosini, A., Pagani, S.M.C.: Geometrical characterization of the uniqueness regions under special sets of three directions in discrete tomography. In: Normand, N., Guédon, J., Autrusseau, F. (eds.) DGCI 2016. LNCS, vol. 9647, pp. 105–116. Springer, Cham (2016). https://doi.org/10.1007/978-3-319-32360-2_8

10. Dulio, P., Frosini, A., Pagani, S.M.C.: Regions of uniqueness quickly reconstructed by three directions in discrete tomography. Fund. Inform. **155**(4), 407–423 (2017). https://doi.org/10.3233/FI-2017-1592

11. Fishburn, P., Lagarias, J., Reeds, J., Shepp, L.: Sets uniquely determined by projections on axes. II. Discrete case. Discrete Math. **91**(2), 149–159 (1991). https://doi.org/10.1016/0012-365X(91)90106-C

12. Gale, D.: A theorem on flows in networks. Pacific J. Math. **7**(2), 1073–1082 (1957). https://doi.org/10.2140/pjm.1957.7.1073

13. Gardner, R.J., Gritzmann, P.: Discrete tomography: determination of finite sets by X-rays. Trans. Amer. Math. Soc. **349**(6), 2271–2295 (1997). https://doi.org/10.1090/S0002-9947-97-01741-8

14. Gardner, R.J., Gritzmann, P., Prangenberg, D.: On the computational complexity of reconstructing lattice sets from their X-rays. Discrete Math. **202**(1–3), 45–71 (1999). https://doi.org/10.1016/S0012-365X(98)00347-1

15. Hajdu, L., Tijdeman, R.: Algebraic aspects of discrete tomography. J. Reine Angew. Math. **534**, 119–128 (2001). https://doi.org/10.1515/crll.2001.037

16. Katz, M.: Questions of Uniqueness and Resolution in Reconstruction from Projections. Lecture Notes in Biomathematics. Springer, Heidelberg (1978). https://doi.org/10.1007/978-3-642-45507-0

17. Normand, N., Kingston, A., Évenou, P.: A geometry driven reconstruction algorithm for the mojette transform. In: Kuba, A., Nyúl, L.G., Palágyi, K. (eds.) DGCI 2006. LNCS, vol. 4245, pp. 122–133. Springer, Heidelberg (2006). https://doi.org/10.1007/11907350_11

18. Pagani, S.M.C., Tijdeman, R.: Algorithms for fast reconstruction by discrete tomography. Researchgate preprint. https://doi.org/10.13140/RG.2.2.20108.56969

19. Ryser, H.: Combinatorial properties of matrices of zeros and ones. Canad. J. Math. **9**, 371–377 (1957). https://doi.org/10.4153/CJM-1957-044-3

20. Tarjan, R.: Depth-first search and linear graph algorithms. SIAM J. Comput. **1**(2), 146–160 (1972). https://doi.org/10.1137/0201010

21. de Werra, D., Costa, M., Picouleau, C., Ries, B.: On the use of graphs in discrete tomography. 4OR **6**(2), 101–123 (2008). https://doi.org/10.1007/s10288-008-0077-5

Trees and Hierarchies

Constructing a Braid of Partitions
from Hierarchies of Partitions

Guillaume Tochon[1(✉)], Mauro Dalla Mura[2,3], and Jocelyn Chanussot[2]

[1] EPITA Research and Development Laboratory (LRDE),
Le Kremlin-Bicêtre, France
guillaume.tochon@lrde.epita.fr
[2] GIPSA-lab, Grenoble Institute of Technology, Saint Martin d'Hères, France
{mauro.dalla-mura,jocelyn.chanussot}@gipsa-lab.grenoble-inp.fr
[3] Tokyo Tech World Research Hub Initiative (WRHI), School of Computing,
Tokyo Institute of Technology, Tokyo, Japan

Abstract. Braids of partitions have been introduced in a theoretical framework as a generalization of hierarchies of partitions, but practical guidelines to derive such structures remained an open question. In a previous work, we proposed a methodology to build a braid of partitions by experimentally composing cuts extracted from two hierarchies of partitions, notably paving the way for the hierarchical representation of multimodal images. However, we did not provide the formal proof that our proposed methodology was yielding a braid structure. We remedy to this point in the present paper and give a brief insight on the structural properties of the resulting braid of partitions.

Keywords: Braid of partitions · Hierarchy of partitions ·
h-Equivalence · Multimodal images

1 Introduction

Hierarchical representations are a well suited tool to handle the multi-scale nature of images, since they allow to encompass all potential scales of interest in a single structure. The hierarchical representation can be constructed once and regardless of the application, and its scale of analysis can then be tuned afterward to comply with the pursued task. The component tree (also called min-tree and max-tree) [14], the inclusion tree (also called tree of shapes (ToS)) [7], the α-tree (also called the hierarchy of quasi-flat zones) [16] and the binary partition tree (BPT) [13] constitute a non-exhaustive list of the most known hierarchical representations in the mathematical morphology literature. Reviews can be found in [3,8]. Hierarchical representations have proven to be useful for many image processing and computer vision tasks, such as image filtering [20] and simplification [21], image segmentation [11] as well as object recognition [19].

The most common framework is to build and process a single hierarchical representation for a given input image. In some cases however, it could be interesting to associate one image with multiple hierarchical representations (each

© Springer Nature Switzerland AG 2019
B. Burgeth et al. (Eds.): ISMM 2019, LNCS 11564, pp. 111–123, 2019.
https://doi.org/10.1007/978-3-030-20867-7_9

one focusing on a particular feature of the image for example), or, on the contrary, to build a common hierarchical representation for multiple input images (each being a single modality of a multimodal image for instance). While these largely remain open questions, some recent works have been devoted to such fusion issues. The fusion of multiple hierarchical representation is for instance addressed in [5], where hierarchies of watersheds (see [2]) driven by area and dynamics attributes are combined through the composition by infimum, supremum or averaging of their saliency maps. The representation of multimodal images (*i.e.* several images acquired over the same scene with different setups, such as different sensor types or acquisition dates) with a single hierarchical structure is another challenge studied in the literature [17]. The ToS structure has for instance been extended to multivariate images in [4], where univariate ToS are first computed for each individual modality and then further merged into a graph from which is derived the final multivariate ToS representation (note that a similar idea is presented in [10] to extend component trees to multivariate images, but the final result is a graph and no longer a tree-based representation). In [9], a single BPT is built over a whole video sequence by integrating motion cues during the construction stage, allowing to perform some object tracking by simply identifying nodes of interest in the resulting trajectory BPT structure. Another approach for the construction of a multi-feature BPT has been introduced in [12], where all modalities of the input multimodal image cooperate in a consensus framework to allow for the construction of a single tree structure. Finally, braids of partitions were proposed in [6] as a generalization of hierarchies of partitions for a theoretical standpoint, and we actually sketched in a previous work [18] the potentiality of such braid structures to act as suited hierarchical representations of multimodal images. In [18], we proposed to build the braid structure (and its associated monitor hierarchy) by experimentally combining cuts extracted from two hierarchies of partitions, and showed the interest of the resulting structure within the framework of multimodal image segmentation. However, we did not provide the formal proof that the proposed methodology yielded a braid structure.

We remedy to this point in this present article, whose organization is as follows: Sect. 2 introduces all used notations and formally defines the notions of hierarchies and braids of partitions. Section 3 recalls the construction procedure introduced in [18] and formally proves that the resulting structure is a braid. Section 4 gives a few insights on the structural properties of braids obtained with the presented construction procedure while Sect. 5 concludes and presents the perspectives of the actual work.

2   From Hierarchies to Braids of Partitions

2.1   Hierarchies of Partitions

Let E be the spatial support of a generic image, *i.e.*, its pixel grid (in which case $E \subseteq \mathbb{Z}^2$ although there is no requirement for E to be discrete in the following). A partition π of E is a collection of regions $\{\mathcal{R}_i \subseteq E\}$ of E such that $\mathcal{R}_i \cap \mathcal{R}_{j \neq i} = \emptyset$

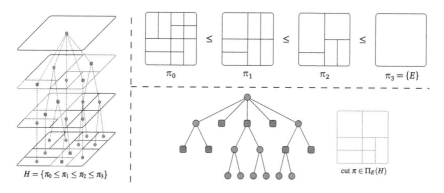

Fig. 1. Example (left) of a hierarchy of partitions H of E, represented as (top) a sequence of partitions ordered by refinement and (bottom) its corresponding dendogram. A particular cut $\pi \in \Pi_E(H)$ is represented with red squared nodes. (Color figure online)

and $\bigcup_i \mathcal{R}_i = E$. The set of all possible partitions of E is denoted Π_E. For any two partitions $\pi_i, \pi_j \in \Pi_E$, $\pi_i \leq \pi_j$ when for each region $\mathcal{R}_i \in \pi_i$, there exists a region $\mathcal{R}_j \in \pi_j$ such that $\mathcal{R}_i \subseteq \mathcal{R}_j$. π_i is said to refine π_j in such case. Π_E is a complete lattice for the refinement (partial) ordering $\leq$. In particular, it is possible to define the refinement supremum $\pi_i \vee \pi_j$ of two partitions π_i and π_j as the smallest partition of Π_E that is refined by both π_i and π_j.

A hierarchy of partitions H of E is a collection of partitions $\{\pi_i \in \Pi_E\}_{i=0}^n$ ordered by refinement, that is $H = \{\pi_0 \leq \pi_1 \leq \cdots \leq \pi_n\}$. π_0 is called the *leaf* partition (its regions are the *leaves* of H) and $\pi_n = \{E\}$ is the *root* of the hierarchy. A hierarchy of partitions is often represented as a tree graph (also called dendogram), where the nodes of the graph correspond to the various regions contained in the partitions of the sequence, and the vertices denote the inclusion between these regions. Alternatively, H can be equivalently defined as a collection of regions $H = \{\mathcal{R} \subseteq E\}$ that satisfy the following 3 properties:

1. $\emptyset \notin H$, $E \in H$.
2. $\forall \mathcal{R}_i, \mathcal{R}_j \in H$, $\mathcal{R}_i \cap \mathcal{R}_j \in \{\emptyset, \mathcal{R}_i, \mathcal{R}_j\}$. Any two regions belonging to H are either disjoint or nested.
3. $\forall \mathcal{R} \in H, \mathcal{R} \notin \pi_0 \Rightarrow \mathcal{R} = \bigcup_{r \in \pi_0}\{r \,|\, r \subset \mathcal{R}\}$. Any non leaf region $\mathcal{R}$ is exactly recovered by the union of all leaves of H that are included in $\mathcal{R}$.

Note that considering only items 1 and 2 allows to define tree-based representations such as the ToS, but item 3 is mandatory to define hierarchies or partitions. A *cut* of H is a partition π of E whose regions belong to H. The set of all cuts of a hierarchy H is denoted $\Pi_E(H)$, and is a sub-lattice of Π_E. All those notions related to hierarchies of partitions are summarized by Fig. 1.

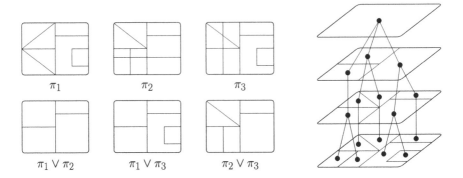

Fig. 2. Example of braid of partitions $B = \{\pi_1, \pi_2, \pi_3\}$. On the right is a monitor hierarchy of B since the pairwise refinement suprema $\pi_i \vee \pi_j, i, j \in \{1, 2, 3\}, i \neq j$ define cuts of this hierarchy different from the whole space E.

2.2 Braids of Partitions

Braids of partitions have been introduced in [6] as a more general structure than hierarchies of partitions, and are defined as follows:

Definition 1 (Braid of partitions). *A family of partitions $B = \{\pi_i \in \Pi_E\}$ is called a braid of partitions whenever there exists some hierarchy of partitions H_m such that:*

$$\forall \pi_i, \pi_j \in B, \pi_i \vee \pi_{j \neq i} \in \Pi_E(H_m) \backslash \{E\} \tag{1}$$

Braids of partitions generalize hierarchies of partitions in the sense that the refinement ordering between the partitions composing the braid no longer needs to exist. However, there must exist some hierarchy of partition H_m, called the *monitor hierarchy*, such that the refinement supremum of any two partitions in the braid defines a cut of this hierarchy H_m. It is also worth noting that this refinement supremum must differ from the whole space $\{E\}$. Otherwise, any family of arbitrary partitions would form a braid with $\{E\}$ as a supremum, thus loosing any interesting structure. An example of braid of partitions B composed of three partitions $B = \{\pi_1, \pi_2, \pi_3\}$ as well as one possible monitor hierarchy H_m of B are displayed by Fig. 2.

As we pointed out in our previous work [18], the structure of a braid B and its monitor hierarchy H_m are particulary suited for the hierarchical representation of multimodal images. As it can be observed in Fig. 2, the monitor hierarchy H_m encodes all regions that are common to at least two different partitions contained in B. Assuming that the partitions composing B originate from different modalities, H_m encodes in a hierarchical manner the information that is shared by those modalities. On the other hand, all regions contained in B but not in H_m belong to a single modality, and are thus responsible for some exclusive information. Jointly considering the braid B and its monitor hierarchy H_m therefore leads to a hierarchical representation of the multimodal image that relies both on the complementary and redundant information contained in the data.

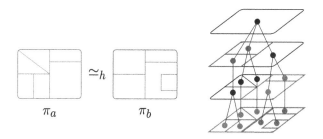

Fig. 3. Illustration of the h-equivalence relation: π_a and π_b are h-equivalent (left), they define two different cuts of the same hierarchy (right).

2.3 The h-Equivalence as Starting Point to Compose a Braid

As pointed out in [6], two issues arise when working with braids of partitions:

- validating that a given family of partitions has a braid structure (that is, Eq. (1) is satisfied).
- defining general procedures that generate braids of partitions.

It is straightforward to compose a braid using a single hierarchy since the supremum of two cuts of a hierarchy also defines a cut of this hierarchy. For this reason, any set of cuts coming from a single hierarchy is a braid of partitions. However, this guarantee is lost when one wants to compose a braid from cuts coming from multiple hierarchies, or, even further, with arbitrary partitions (note in that respect that, although tempting to think so, the family of partitions generated by the stochastic watershed [1] has not a braid structure in general). As a matter of fact, all those cuts must be sufficiently related so their pairwise refinement suprema define cuts of the same monitor hierarchy H_m. To analyze the relationships which must be holding between the cuts of various hierarchies to form a braid, we introduced in [18] the property of *h-equivalence* (h standing for *hierarchical*):

Definition 2 (h-equivalence). *Two partitions π_a and π_b are said to be h-equivalent, and one notes $\pi_a \simeq_h \pi_b$ if and only if*

$$\forall \mathcal{R}_a \in \pi_a, \, \forall \mathcal{R}_b \in \pi_b, \mathcal{R}_a \cap \mathcal{R}_b \in \{\emptyset, \mathcal{R}_a, \mathcal{R}_b\}. \tag{2}$$

In other words, a region in π_a either refines or is a refinement of a region in π_b. Partitions π_a and π_b may not be globally comparable for the refinement ordering, but they locally are, as displayed by Fig. 3. Obviously, if $\pi_a \leq \pi_b$ or $\pi_a \geq \pi_b$, then $\pi_a \simeq_h \pi_b$. All cuts of a hierarchy H are h-equivalent: $\forall \pi_1, \pi_2 \in \Pi_E(H), \pi_1 \simeq_h \pi_2$. Conversely, if two partitions are h-equivalent, they define two cuts of the same hierarchy. Despite a somewhat misleading name, $\simeq_h$ is not an equivalence relation but only a tolerance relation: it is reflexive and symmetric but not transitive in general.

Following, we aim to build a braid B using cuts extracted from several hierarchical representations. To do so, we must investigate what kind of relationship must be holding between those cuts in order to guarantee the braid structure (that is, Eq. (1) is satisfied). Let the family of partitions $B = \{\pi_i \in \Pi_E\}$ be a braid, and let H_m be a monitor hierarchy of B.

Proposition 1. *If there exist $\pi_i, \pi_j \in B$ such that $\pi_i \leq \pi_j$, then $\pi_j \in \Pi_E(H_m)$.*

Proof. As $\pi_i \leq \pi_j$, it follows that $\pi_i \vee \pi_j = \pi_j$. And from the definition (1) of a braid, $\pi_i \vee \pi_j \in \Pi_E(H_m)$, so $\pi_j \in \Pi_E(H_m)$. $\qquad\square$

Thus, if the braid B has two partitions ordered by refinement (two cuts extracted from the same hierarchy for instance), the coarest of them also belongs to the set of cuts $\Pi_E(H_m)$ of the monitor hierarchy H_m.

Proposition 2. *If there exist $\pi_i, \pi_j, \pi_k, \pi_l \in B$ such that $\pi_i \leq \pi_j$ and $\pi_k \leq \pi_l$, then $\pi_j \simeq_h \pi_l$.*

Proof. Using Proposition (1) for both $\pi_i \leq \pi_j$ and $\pi_k \leq \pi_l$, it follows that $\pi_j, \pi_l \in \Pi_E(H_m)$. Thus $\pi_j \simeq_h \pi_l$ using the property of h-equivalence. $\qquad\square$

Therefore, if the braid B has two pairs partitions ordered by refinement, the coarest of both pairs are necessarily h-equivalent to each other since they both belong to the set of cuts $\Pi_E(H_m)$ of the monitor hierarchy H_m.

3   The Braid Construction Procedure

Given some hierarchy H and a partition $\pi_* \in \Pi_E$, we denote by $H^{\simeq_h}(\pi_*)$ the set of cuts of H that are h-equivalent to π_*: $H^{\simeq_h}(\pi_*) \subseteq \Pi_E(H)$ with equality if and only if $\pi_* \in \Pi_E(H)$. Similarly, we denote by $H^{\leq}(\pi_*)$ the set of cuts of H that are a refinement of π_*.

Now, let H_1 and H_2 be two hierarchies of partitions built over the same space E. We aim to extract two cuts $\pi_i^1, \pi_i^2 \in \Pi_E(H_i)$ from each of those two hierarchies $H_i, i \in \{1, 2\}$ in order for the family $B = \{\pi_1^1, \pi_1^2, \pi_2^1, \pi_2^2\}$ to be a braid. For this purpose, we propose the following iterative procedure:

1. First select arbitrarily some cut $\pi_1^1 \in \Pi_E(H_1)$.
2. Then choose a cut π_2^1 in the constrained set $H_2^{\simeq_h}(\pi_1^1) \backslash \{E\}$, that is, a cut from H_2 which is h-equivalent to π_1^1 and different from the whole space $\{E\}$.
3. Finally, complete by taking a cut in each hierarchy that is a refinement of the cut previously extracted from the other hierarchy, that is $\pi_i^2 \in \Pi_E(H_i), i \in \{1, 2\}$ such that $\pi_1^2 \leq \pi_2^1$ and $\pi_2^2 \leq \pi_1^1$.

This procedure is summarized by Fig. 4.

Proposition 3. *Under this configuration, $B = \{\pi_1^1, \pi_1^2, \pi_2^1, \pi_2^2\}$ has a braid structure.*

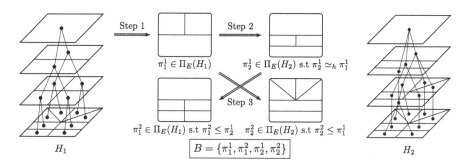

Fig. 4. Composing a braid B with cuts from two hierarchies H_1 and H_2.

Proof. Let $B = \{\pi_1^1, \pi_1^2, \pi_2^1, \pi_2^2\}$ be a family of partitions composed following the previously described procedure, and let $\pi_{i,j}^{k,l} = \pi_i^k \vee \pi_j^l$ denote the pairwise refinement suprema of partitions in B. In particular, the 4 partitions composing B generates $\binom{4}{2} = 6$ different pairwise refinement suprema $\pi_{1,1}^{1,2}, \pi_{1,2}^{1,1}, \pi_{1,2}^{1,2}, \pi_{1,2}^{2,1}, \pi_{1,2}^{2,2}, \pi_{2,2}^{1,2}$. Checking that B is a braid amounts to verify whether the $\pi_{i,j}^{k,l}$ all defines cuts of the same monitor hierarchy H_m, which is equivalent to showing that they are (at least) all h-equivalent to each other. In order to show the braid structure of B, we first demonstrate the following result:

Lemma 1. *Let* $\pi_1, \pi_2, \pi_3 \in \Pi_E$ *be some partitions of E such that $\pi_1 \simeq_h \pi_3$ and $\pi_2 \leq \pi_3$. Then $\pi_1 \vee \pi_2 \simeq_h \pi_3$.*

Proof. If $\pi_1 \leq \pi_3$, then $\pi_1 \vee \pi_2 \leq \pi_3$ by definition of the refinement supremum, and so $\pi_1 \vee \pi_2 \simeq_h \pi_3$ since any two ordered partitions are also h-equivalent.

On the other hand, if $\pi_1 \geq \pi_3$, then $\pi_1 \geq \pi_2$, hence $\pi_1 \vee \pi_2 = \pi_1$ and so $\pi_1 \vee \pi_2 \simeq_h \pi_3$ for the same reason as above.

In the most general case where π_1 and π_3 are h-equivalent but can nonetheless not be ordered, it means that π_1 is a refinement of π_3 in some parts of E, and is refined by π_3 in the other parts. In the former case, let $\mathcal{R}_3$ be a region of π_3 and $\pi_1(\mathcal{R}_3), \pi_2(\mathcal{R}_3)$ be the refinements (partial partitions) of $\mathcal{R}_3$ in π_1 and π_2. Then, $\pi_1(\mathcal{R}_3) \vee \pi_2(\mathcal{R}_3)$ is also a refinement of $\mathcal{R}_3$, implying that $\pi_1 \vee \pi_2$ refines π_3 in the part of E covered by $\mathcal{R}_3$. In the case where π_3 is locally a refinement of π_1, then given $\mathcal{R}_1 \in \pi_1$, there exists a refinement $\pi_3(\mathcal{R}_1)$ of $\mathcal{R}_1$ in π_3, and therefore a refinement $\pi_2(\mathcal{R}_1)$ of $\mathcal{R}_1$ in π_2 since $\pi_2 \leq \pi_3$. Therefore, $\{\mathcal{R}_1\} \vee \pi_2(\mathcal{R}_1) = \{\mathcal{R}_1\}$ and thus π_3 refines $\pi_1 \vee \pi_2$ in the part of E covered by $\mathcal{R}_1$. Finally, $\pi_1 \vee \pi_2$ either refines or is refined by π_3 in all parts of E, hence $\pi_1 \vee \pi_2 \simeq_h \pi_3$. □

To ease the reading of the proof of Proposition (3), we first recall the relations holding between the various partitions composing the braid B:

- $\pi_1^1 \simeq_h \pi_2^1$ by construction.
- $\pi_1^2 \leq \pi_2^1$ and $\pi_2^2 \leq \pi_1^1$ by construction.
- $\pi_1^1 \simeq_h \pi_1^2$ because they are both cuts of the same hierarchy H_1. Similarly, $\pi_2^1 \simeq_h \pi_2^2$.

Following, we prove that all the pairwise refinement suprema of B are at least all h-equivalent to each other. Their relationships are summarized in Table 1.

Table 1. Summary of the relationships holding between all pairwise refinement suprema of B with their corresponding item in the proof.

	$\pi_{1,1}^{1,2}$	$\pi_{1,2}^{1,1}$	$\pi_{1,2}^{1,2}$	$\pi_{1,2}^{2,1}$	$\pi_{1,2}^{2,2}$	$\pi_{2,2}^{1,2}$
$\pi_{1,1}^{1,2}$	X	1. $\leq$	2. $\leq$	3. $\simeq_h$	4. $\leq$	5. $\simeq_h$
$\pi_{1,2}^{1,1}$		X	6. $\leq$	7. $\leq$	8. $\leq$	9. $\leq$
$\pi_{1,2}^{1,2}$			X	10. $\simeq_h$	11. $\simeq_h$	12. $\simeq_h$
$\pi_{1,2}^{2,1}$				X	13. $\simeq_h$	14. $\leq$
$\pi_{1,2}^{2,2}$					X	15. $\leq$
$\pi_{2,2}^{1,2}$						X

1. $\pi_{1,1}^{1,2} = \pi_1^1 \vee \pi_1^2$. As $\pi_1^2 \leq \pi_2^1$ by construction of B, it follows that $\pi_1^1 \vee \pi_1^2 \leq \pi_1^1 \vee \pi_2^1$, hence $\pi_{1,1}^{1,2} \leq \pi_{1,2}^{1,1}$.

2. $\pi_{1,2}^{1,2} = \pi_1^1 \vee \pi_2^2 = \pi_1^1$ as $\pi_2^2 \leq \pi_1^1$ by construction of B. By property of the refinement supremum, one has $\pi_1^1 \leq \pi_1^1 \vee \pi_1^2 = \pi_{1,1}^{1,2}$, hence $\pi_{1,2}^{1,2} \leq \pi_{1,1}^{1,2}$.

3. By construction of B, one has $\pi_1^1 \simeq_h \pi_2^1$ and $\pi_1^2 \leq \pi_2^1 = \pi_{1,2}^{2,1}$. Using Lemma 1, it follows that $\pi_1^1 \vee \pi_1^2 = \pi_{1,1}^{1,2} \simeq_h \pi_{1,2}^{2,1}$.

4. $\pi_2^2 \leq \pi_1^1$ by construction of B, meaning that $\pi_1^1 \vee \pi_2^2 \leq \pi_1^1 \vee \pi_1^1$, hence $\pi_{1,2}^{2,2} \leq \pi_{1,1}^{1,2}$.

5. Using item 3, we first have $\pi_{1,1}^{1,2} \simeq_h \pi_2^1 = \pi_{1,2}^{2,1}$. In addition, $\pi_2^2 \leq \pi_1^1$ by construction of B, implying that $\pi_2^2 \leq \pi_1^1 \vee \pi_1^2 = \pi_{1,1}^{1,2}$. Using Lemma 1 finally leads to $\pi_{1,1}^{1,2} \simeq_h \pi_{2,2}^{1,2}$.

6. $\pi_{1,2}^{1,1} = \pi_1^1$ as $\pi_2^2 \leq \pi_1^1$ by construction of B. The basic property of the refinement supremum allows to conclude that $\pi_1^1 \leq \pi_1^1 \vee \pi_2^1$, hence $\pi_{1,2}^{1,1} \leq \pi_{1,2}^{1,2}$.

7. The exact same reasoning as item 6 applied to $\pi_{1,2}^{2,1} = \pi_2^1$ leads to $\pi_{1,2}^{2,1} \leq \pi_{1,2}^{1,1}$.

8. $\pi_1^2 \leq \pi_2^1$ and $\pi_2^2 \leq \pi_1^1$, both by construction of B. It immediately follows that $\pi_1^2 \vee \pi_2^2 \leq \pi_1^1 \vee \pi_2^1$, hence $\pi_{1,2}^{2,2} \leq \pi_{1,2}^{1,1}$.

9. The same reasoning as item 1 applies to $\pi_{2,2}^{1,2} = \pi_2^1 \vee \pi_2^2$, leading to $\pi_{2,2}^{1,2} \leq \pi_{1,2}^{1,1}$.

10. By construction of B, one has $\pi_1^1 = \pi_{1,2}^{1,2} \simeq_h \pi_{1,2}^{2,1} = \pi_2^1$, hence the result.

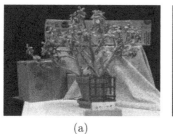 

(a)   (b)

Fig. 5. (a) RGB modality and (b) Depth map of a RGB/Depth multimodal image from the Middlebury Stereo Dataset.

11. $\pi_{1,2}^{1,2} = \pi_1^1$ as $\pi_2^2 \le \pi_1^1$ by construction of B. In addition, $\pi_1^1 \simeq_h \pi_1^2$ as they are both cuts of the same hierarchy H_1. Using Lemma 1, it follows that $\pi_1^1 = \pi_{1,2}^{1,2} \simeq_h \pi_{1,2}^{2,2} = \pi_1^2 \vee \pi_2^2$.

12. The same reasoning as item 3 applies to $\pi_{2,2}^{1,2}$ and $\pi_{1,2}^{1,2} = \pi_1^1$ and, relying upon Lemma 1, leads to $\pi_{1,2}^{1,2} \simeq_h \pi_{2,2}^{1,2}$.

13. The same reasoning as item 11 applies to $\pi_{1,2}^{2,1} = \pi_2^1$ and $\pi_{1,2}^{2,2}$, leading to $\pi_{1,2}^{2,1} \simeq_h \pi_{1,2}^{2,2}$.

14. The same reasoning as item 2 applies to $\pi_{2,2}^{1,2}$ and $\pi_{1,2}^{2,1} = \pi_2^1$, leading to $\pi_{1,2}^{2,1} \le \pi_{2,2}^{1,2}$.

15. The same reasoning as item 4 applies to $\pi_{1,2}^{2,2}$ and $\pi_{2,2}^{1,2}$, leading to $\pi_{1,2}^{2,2} \le \pi_{2,2}^{1,2}$.

Finally, all the pairwise refinement supremum $\pi_{i,j}^{k,l} = \pi_i^k \vee \pi_j^l$ that can be formed using the partitions belonging to B are (at least) all h-equivalent to each other. Therefore, there exists some hierarchy H_m such that all $\pi_{i,j}^{k,l} \in \Pi_E(H_m)$, which proves that B has a braid structure when constructed following the proposed procedure.   □

4   A Quick Look into the Braid Structure

The braid construction procedure presented in Sect. 3 and summarized by Fig. 4 is only operable when two cuts are extracted from two hierarchies of partitions. While this may appear quite restrictive from an applicative point of view, we are up to now only able to provide the proposed procedure in this context. As a matter of fact, the braid structure is guaranteed whenever the refinement suprema of all partitions composing the braid are all h-equivalent to each other, but some additional efforts are still needed to understand how this h-equivalence constraint can be formulated as a constraint on the partitions of the braid.

In order to give a brief insight on the structural properties of the braid of partitions and its monitor hierarchy when the proposed procedure is applied on a real multimodal scenario, we consider the RGB/Depth multimodal image presented by Fig. 5, originating from the 2014 database of the Middlebury Stereo Dataset [15]. Both modalities are co-registered and comprise 400×300 pixels.

The two hierarchies of partitions H_1 and H_2 are obtained by means of two BPTs [13] with standard parameters (mean value for the region model and Euclidean distance for the merging criterion). We use the same initial partition for both BPTs, namely the intersection of two mean shift clustering procedures ran independently on each modality (yielding a total of 2148 leaf regions). The first cut π_1^1 is extracted from H_1 as the partition composed of the $NbReg$ last regions before completion of the region merging process of the BPT. It defines the extraction of the three remaining cuts $\pi_1^2, \pi_2^1, \pi_2^2$ through the application of the procedure presented in Sect. 3 as follows: $\pi_2^1 = \bigvee\{H_2^{\simeq_h}(\pi_1^1)\backslash\{E\}\}$, $\pi_1^2 = \bigvee\{H_1^{\leq}(\pi_2^1)\}$ and $\pi_2^2 = \bigvee\{H_2^{\leq}(\pi_1^2)\}$. The three constrained set of cuts $H_2^{\simeq_h}(\pi_1^1)\backslash\{E\}$, $H_1^{\leq}(\pi_2^1)$ and $H_2^{\leq}(\pi_1^2)$ are non empty since both hierarchies H_1 and H_2 have the same leaf partition (all three sets thus contain at least the leaf partition). We set $NbReg$ (that is, the number of regions in π_1^1) to $10, 20, 50, 100, 200, 500$ and 1000. Tables 2 and 3 give the number of regions composing partitions π_1^2, π_2^1 and π_2^2 when H_1 is the BPT built on the RGB and Depth modality, respectively. As it can be seen, there is in both cases a big gap between the number of regions in π_1^1 and π_2^1. Even though π_2^1 is defined to be h-equivalent to π_1^1, it turns out in practice to be a refinement of π_1^1 since $\pi_2^1 \leq \pi_1^1$ has the same number of regions as π_2^1 (except for $NbReg = 500, 1000$). The reason is that when π_1^1 has a relatively small number of regions, there are no regions in H_2 that are refined by those of π_1^1, hence the largest cut in $H_2^{\simeq_h}(\pi_1^1)\backslash\{E\}$ is equivalent to the one in $H_2^{\leq}(\pi_1^1)$. There is an equivalently big gap between the number of regions in π_2^1 and π_1^2. While this can be explained by the fact that π_1^2 is defined as a refinement of π_2^1, it also means that it is very difficult to find "intermediary" regions in H_1 (that is, regions that are not close from the root or the leaves of H_1, and that are more likely to be associated with semantic objects in the image) that correspond to those in H_2. While this is out of the scope of this paper, this interpretation might be a potential issue in a practical scenario of braid-based hierarchical representation of multimodal images and will be further investigated.

Table 2. Number of regions $NbReg$ in all partitions composing the braid B when the first cut is extracted from H_1 built on RGB modality.

π_1^1	10	20	50	100	200	500	1000
$\pi_2^1 \simeq_h \pi_1^1$	544	822	1095	1267	1415	1609	1684
$\pi_1^2 \leq \pi_2^1$	1358	1527	1680	1750	1800	1860	1876
$\pi_2^2 \leq \pi_1^1$	544	822	1095	1267	1415	1620	1801

Regarding the influence of the hierarchy from which is extracted the first cut π_1^1, it can be appreciated by comparing Tables 2 and 3 that the proposed procedure yields cuts π_1^2, π_2^1 and π_2^2 whose number of regions remains relatively stable whether H_1 was defined to be the BPT constructed on the RGB modality or the Depth modality. This observation will have to be validated with more

Table 3. Number of regions *NbReg* in all partitions composing the braid B when the first cut is extracted from H_1 built on Depth modality.

π_1^1	10	20	50	100	200	500	1000
$\pi_2^1 \simeq_h \pi_1^1$	633	829	994	1138	1315	1569	1699
$\pi_1^2 \leq \pi_2^1$	1530	1612	1657	1707	1785	1864	1865
$\pi_2^2 \leq \pi_1^1$	633	829	994	1138	1315	1578	1824

in-depth experiments, and the influence of the choice of the modality associated with H_1 on practical image processing tasks (such as segmentation or object recognition) will also need to be evaluated (since the semantic content of those regions will depend on the identity of the modality associated with H_1 and H_2). It could also be interesting to study whether the number of regions *NbReg* of the first partition π_1^1 can vary within a given range of values without impacting the structure of the monitor hierarchy. Similarly, the influence of the choice of π_1^2, π_2^1 and π_2^2 in their respective constrained sets of cuts will also have to be investigated. While these considerations are probably data dependent, they could nevertheless give a deeper insight on the stability of the braid structure generated by the proposed procedure.

5   Conclusion

Braids of partitions were defined as a generalization of hierarchies of partitions, in the sense that all partitions composing the braid do not need to be ordered by refinement. This more permissive property opens the door to potentially several applications of interest for the braid structure, but it also brings difficulties to build such structure in practice. In our previous work [18], we experimentally provided a procedure to build a braid of partitions as a combination of cuts coming from hierarchies, and intuited the potential of braids to perform multimodal image segmentation.

In this paper, we formally demonstrated that the procedure proposed in [18] was indeed yielding a braid of partitions. While we did not focus here on a more thorough evaluation of the usefulness of the braid structure for multimodal image analysis, this is obviously an important future research avenue. In addition, we are up to now bound to build a braid by using only two hierarchies of partitions, and two cuts per hierarchy. In that respect, future work will investigate theoretical aspects related to the construction of the braid of partitions, namely how to extract more cuts coming from various hierarchies and still maintain the braid structure.

References

1. Angulo, J., Jeulin, D.: Stochastic watershed segmentation. In: Proceedings of the 8th International Symposium on Mathematical Morphology, pp. 265–276 (2007)
2. Beucher, S.: Watershed, hierarchical segmentation and waterfall algorithm. In: Serra, J., Soille, P. (eds.) Mathematical Morphology and its Applications to Image Processing, pp. 69–76. Springer, Dordrecht (1994). https://doi.org/10.1007/978-94-011-1040-2_10
3. Bosilj, P., Kijak, E., Lefèvre, S.: Partition and inclusion hierarchies of images: a comprehensive survey. J. Imaging **4**(2), 33 (2018)
4. Carlinet, E., Géraud, T.: MToS: a tree of shapes for multivariate images. IEEE Trans. Image Process. **24**(12), 5330–5342 (2015)
5. Cousty, J., Najman, L., Kenmochi, Y., Guimarães, S.: Hierarchical segmentations with graphs: quasi-flat zones, minimum spanning trees, and saliency maps. J. Math. Imaging Vis. **60**(4), 479–502 (2018)
6. Kiran, B.R., Serra, J.: Braids of partitions. In: Benediktsson, J.A., Chanussot, J., Najman, L., Talbot, H. (eds.) ISMM 2015. LNCS, vol. 9082, pp. 217–228. Springer, Cham (2015). https://doi.org/10.1007/978-3-319-18720-4_19
7. Monasse, P., Guichard, F.: Fast computation of a contrast-invariant image representation. IEEE Trans. Image Process. **9**(5), 860–872 (2000)
8. Najman, L., Cousty, J.: A graph-based mathematical morphology reader. Pattern Recogn. Lett. **47**, 3–17 (2014)
9. Palou, G., Salembier, P.: Hierarchical video representation with trajectory binary partition tree. In: Proceedings of the IEEE Conference on Computer Vision and Pattern Recognition, pp. 2099–2106 (2013)
10. Passat, N., Naegel, B.: Component-trees and multivalued images: structural properties. J. Math. Imaging Vis. **49**(1), 37–50 (2014)
11. Perret, B., Cousty, J., Guimaraes, S.J.F., Maia, D.S.: Evaluation of hierarchical watersheds. IEEE Trans. Image Process. **27**(4), 1676–1688 (2018)
12. Randrianasoa, J.F., Kurtz, C., Desjardin, E., Passat, N.: Binary partition tree construction from multiple features for image segmentation. Pattern Recogn. **84**, 237–250 (2018)
13. Salembier, P., Garrido, L.: Binary partition tree as an efficient representation for image processing, segmentation, and information retrieval. IEEE Trans. Image Process. **9**(4), 561–576 (2000)
14. Salembier, P., Oliveras, A., Garrido, L.: Antiextensive connected operators for image and sequence processing. IEEE Trans. Image Process. **7**(4), 555–570 (1998)
15. Scharstein, D., et al.: High-resolution stereo datasets with subpixel-accurate ground truth. In: Jiang, X., Hornegger, J., Koch, R. (eds.) GCPR 2014. LNCS, vol. 8753, pp. 31–42. Springer, Cham (2014). https://doi.org/10.1007/978-3-319-11752-2_3
16. Soille, P.: Constrained connectivity for hierarchical image partitioning and simplification. IEEE Trans. Pattern Anal. Mach. Intell. **30**(7), 1132–1145 (2008)
17. Tochon, G.: Hierarchical analysis of multimodal images. Ph.D. thesis, Université Grenoble Alpes (2015)
18. Tochon, G., Dalla Mura, M., Chanussot, J.: Segmentation of multimodal images based on hierarchies of partitions. In: Benediktsson, J.A., Chanussot, J., Najman, L., Talbot, H. (eds.) ISMM 2015. LNCS, vol. 9082, pp. 241–252. Springer, Cham (2015). https://doi.org/10.1007/978-3-319-18720-4_21
19. Vilaplana, V., Marques, F., Salembier, P.: Binary partition trees for object detection. IEEE Trans. Image Process. **17**(11), 2201–2216 (2008)

20. Xu, Y., Géraud, T., Najman, L.: Connected filtering on tree-based shape-spaces. IEEE Trans. Pattern Anal. Mach. Intell. **38**(6), 1126–1140 (2016)
21. Xu, Y., Géraud, T., Najman, L.: Hierarchical image simplification and segmentation based on Mumford-Shah-salient level line selection. Pattern Recogn. Lett. **83**, 278–286 (2016)

Watersheding Hierarchies

Deise Santana Maia$^{(\boxtimes)}$, Jean Cousty, Laurent Najman, and Benjamin Perret

Université Paris-Est, LIGM (UMR 8049), CNRS, ENPC,
ESIEE Paris, UPEM, 93162 Noisy-le-Grand, France
deisesantanamaia@gmail.com

Abstract. The computation of hierarchies of partitions from the watershed transform is a well-established segmentation technique in mathematical morphology. In this article, we introduce the *watersheding operator*, which maps any hierarchy into a hierarchical watershed. The hierarchical watersheds are the only hierarchies that remain unchanged under the action of this operator. After defining the watersheding operator, we present its main properties, namely its relation with extinction values and sequences of minima of weighted graphs. Finally, we discuss practical applications of the watersheding operator.

1 Introduction

In the context of image segmentation, hierarchies (of partitions) are sequences of nested partitions of image pixels. At the highest levels of a hierarchy, we have the most representative regions according to a given criterion, such as size and contrast. Hierarchies can be equivalently represented by *saliency maps* [1,6,18], in which the contours between regions are weighted according to their level of disappearance in the hierarchy. Thank to the bijection between hierarchies and saliency maps [6], we work indifferently with any of those notions in this study.

In mathematical morphology, hierarchies are often obtained from the watershed transform [2,4]. With the definitions proposed in [4], hierarchical watersheds are optimal in the sense of minimum spanning forests. Furthermore, efficient algorithms to compute those hierarchies have been designed [5,17]. The competitive performance of hierarchical watersheds is attested by the quantitative evaluation presented in [19]. Moreover, watersheds can be linked to a broader family of combinatorial optimization problems, such as random walkers and graph cuts, as demonstrated in [3].

In this article, we propose the *watersheding operator*, which, given an edge-weighted graph $((V, E), w)$, transforms (the saliency map of) any hierarchy on V into (the saliency map of) a hierarchical watershed of $((V, E), w)$. Figure 1 illustrates an application of the watersheding operator. The goal is to obtain, from the image gradient G, a hierarchical watershed of G that highlights the circular regions of the image I. A straightforward method to achieve this goal is to

This research is partly funded by the Bézout Labex, funded by ANR, reference ANR-10-LABX-58.

B. Burgeth et al. (Eds.): ISMM 2019, LNCS 11564, pp. 124–136, 2019.
https://doi.org/10.1007/978-3-030-20867-7_10

Fig. 1. An image I, the gradient G of I computed using the structured edge detector introduced in [8], the hierarchical watershed $\mathcal{H}_c$ of G based on a regularized circularity attribute, a circularity based hierarchy $\mathcal{H}_{cc}$ and the watersheding $\mathcal{H}_w$ of $\mathcal{H}_{cc}$ (for G).

compute the hierarchical watershed of G based on a (regularized) circularity attribute, which results in the hierarchy $\mathcal{H}_c$. We can observe that the hierarchy $\mathcal{H}_c$ does not succeed at highlighting all the main circular regions of the image I. Alternatively, we can compute another hierarchy $\mathcal{H}_{cc}$ by simply weighing each contour as the maximum circularity value among the regions that share this contour. The hierarchy $\mathcal{H}_{cc}$ brings to the fore the main circular regions of I, however it is not a hierarchical watershed of G. By applying the watersheding operator on $\mathcal{H}_{cc}$, we obtain the hierarchical watershed $\mathcal{H}_w$. We can see that the circular regions of the image I are better highlighted in the result of the watersheding operator $\mathcal{H}_w$ when compared to the straightforward approach $\mathcal{H}_c$.

This article is organized as follows. We review basic notions on graphs, hierarchies and saliency maps in Sect. 2. Then, we present the watersheding operator and some properties of this operator in Sect. 3. Finally, we discuss applications of the watersheding of hierarchies in Sect. 4.

2   Background Notions

In this section, we first introduce hierarchies of partitions. Then, we review the definition of graphs, connected hierarchies and saliency maps. Subsequently, we define hierarchical watersheds.

2.1   Hierarchies of Partitions

Let V be a set. A *partition* of V is a set $\mathbf{P}$ of non empty disjoint subsets of V whose union is V. If $\mathbf{P}$ is a partition of V, any element of $\mathbf{P}$ is called a *region of* $\mathbf{P}$. Let V be a set and let $\mathbf{P}_1$ and $\mathbf{P}_2$ be two partitions of V. We say that $\mathbf{P}_1$ is a *refinement* of $\mathbf{P}_2$ if every element of $\mathbf{P}_1$ is included in an element of $\mathbf{P}_2$. A *hierarchy (of partitions on V)* is a sequence $\mathcal{H} = (\mathbf{P}_0, \ldots, \mathbf{P}_n)$ of partitions of V such that $\mathbf{P}_0$ is an arbitrary partition of V, the partition $\mathbf{P}_n$ is the set $\{V\}$ and $\mathbf{P}_{i-1}$ is a refinement of $\mathbf{P}_i$, for any $i \in \{1, \ldots, n\}$. For any i in $\{0, \ldots, n\}$, any region of the partition $\mathbf{P}_i$ is called a *region of* $\mathcal{H}$.

Hierarchies of partitions can be represented as trees whose nodes correspond to regions, as shown in Fig. 2. Given a hierarchy $\mathcal{H}$ and two regions X and Y of $\mathcal{H}$, we say that *X is a parent of Y (or that Y is a child of X)* if $Y \subset X$ and X

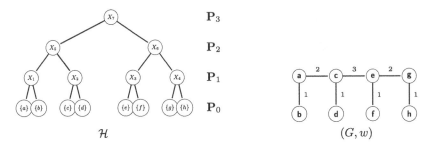

Fig. 2. A representation of a hierarchy of partitions $\mathcal{H} = (\mathbf{P}_0, \mathbf{P}_1, \mathbf{P}_2, \mathbf{P}_3)$ on the set $\{a, b, c, d, e, f, g, h\}$ and a weighted graph (G, w).

is minimal for this property. In other words, if X is a parent of Y and if there is a region Z such that $Y \subseteq Z \subset X$, then $Y = Z$. It can be seen that any region X of $\mathcal{H}$ such that $X \neq V$ has exactly one parent. Thus, for any region X such that $X \neq V$, we write $parent(X) = Y$ where Y is the unique parent of X. For any region R of $\mathcal{H}$, if R is not the parent of any region of $\mathcal{H}$, we say that R is a *leaf region of* $\mathcal{H}$. Otherwise, we say that R is a *non-leaf region of* $\mathcal{H}$.

In Fig. 2, the regions of a hierarchy $\mathcal{H}$ are linked to their parents and children by straight lines. The partition $\mathbf{P}_0$ of $\mathcal{H}$ contains all the leaf regions of $\mathcal{H}$.

2.2   Graphs, Connected Hierarchies and Saliency Maps

A *graph* is a pair $G = (V, E)$, where V is a finite set and E is a set of pairs of distinct elements of V, *i.e.*, $E \subseteq \{\{x, y\} \subseteq V | x \neq y\}$. Each element of V is called a *vertex (of G)*, and each element of E is called an *edge (of G)*. To simplify the notations, the set of vertices and edges of a graph G will be also denoted by $V(G)$ and $E(G)$, respectively.

Let $G = (V, E)$ be a graph and let X be a subset of V. A sequence $\pi = (x_0, \ldots, x_n)$ of elements of X is a *path (in X) from* x_0 *to* x_n if $\{x_{i-1}, x_i\}$ is an edge of G for any i in $\{1, \ldots, n\}$. The subset X of V is said to be *connected for G* if, for any x and y in X, there exists a path from x to y. The subset X is a *connected component* of G if X is connected and if, for any connected subset Y of V such that $X \subseteq Y$, we have $X = Y$. In the following, we denote by $CC(G)$ the set of all connected components of G. It is well known that this set $CC(G)$ of all connected components of G is a partition of the set V.

Let $G = (V, E)$ be a graph. *A partition of V is connected for G* if each of its regions is connected for G and *a hierarchy on V is connected (for G)* if each of its partitions is connected. For example, in Fig. 2, the hierarchy $\mathcal{H}$ is connected for the graph (G, w).

Let G be a graph. If w is a map from the edge set of G to the set $\mathbb{R}^+$ of positive real numbers, then the pair (G, w) is called an *(edge) weighted graph*. If (G, w) is a weighted graph, for any edge u of G, the value $w(u)$ is called the *weight of u (for w)*.

As established in [6], a connected hierarchy can be equivalently treated by means of a weighted graph through the notion of a saliency map. Given a graph G and a hierarchy $\mathcal{H} = (\mathbf{P}_0, \ldots, \mathbf{P}_n)$ connected for G, the *saliency map of* $\mathcal{H}$ (also known as ultrametric watershed [16] and ultrametric contour map [1]) is the map from $E(G)$ to $\{0, \ldots, n\}$, denoted by $\Phi(\mathcal{H})$, such that, for any edge $u = \{x, y\}$ in $E(G)$, the value $\Phi(\mathcal{H})(u)$ is the smallest value i in $\{0, \ldots, n\}$ such that x and y belong to a same region of $\mathbf{P}_i$. Therefore, the weight $\Phi(\mathcal{H})(u)$ of any edge $u = \{x, y\}$ is the ultrametric distance between x and y on the hierarchy $\mathcal{H}$. It follows that any connected hierarchy has a unique saliency map. Moreover, we can recover any hierarchy $\mathcal{H}$ connected for G from the saliency map $\Phi(\mathcal{H})$ of $\mathcal{H}$ through the notion of quasi-flat zones [13,15], as established in [6]. For instance, in Fig. 2, the weight map w is the saliency map of the hierarchy $\mathcal{H}$.

2.3 Hierarchies of Minimum Spanning Forests and Watersheds

The watershed segmentation [2,4] derives from the topographic notion of watersheds lines and catchment basins. A catchment basin is a region whose collected precipitation drains to the same body of water, as a sea, and the watershed lines are the separating lines between neighbouring catchment basins. In [4], the authors formalize watersheds in the framework of weighted graphs and show the optimality of watersheds in the sense of minimum spanning forests. In this section, we present hierarchical watersheds following the definition of hierarchies of minimum spanning forests presented in [7].

Let G be a graph. We say that G *is a forest* if, for any edge u in $E(G)$, the number of connected components of the graph $(V(G), E(G) \setminus \{u\})$ is larger than the number of connected components of G. Given another graph G', we say that G' *is a subgraph of* G, denoted by $G' \sqsubseteq G$, if $V(G') \subseteq V(G)$ and $E(G') \subseteq E(G)$. Let (G, w) be a weighted graph and let G' be a subgraph of G. A graph G'' is a *Minimum Spanning Forest (MSF) of* G *rooted in* G' if:

1. the graphs G and G'' have the same set of vertices, *i.e.*, $V(G'') = V(G)$; and
2. the graph G' is a subgraph of G''; and
3. each connected component of G'' includes exactly one connected component of G'; and
4. the sum of the weight of the edges of G'' is minimal among all graphs for which the above conditions 1, 2 and 3 hold true.

Important Notations and Notions: in the sequel of this article, the symbol G denotes a tree, *i.e.*, a forest with a unique connected component. This implies that any map from $E(G)$ into $\mathbb{Z}^+$ is the saliency map of a hierarchy which is connected for G, as established by Properties 9 and 10 of [6]. To shorten the notation, the vertex set of G is denoted by V and its edge set is denoted by E. The symbol w denotes a map from E into $\mathbb{R}^+$ such that, for any pair of distinct edges u and v in E, we have $w(u) \neq w(v)$. Thus, the pair (G, w) is a weighted graph. Every hierarchy considered in this article is connected for G and therefore, for the sake of simplicity, we use the term *hierarchy* instead of *hierarchy which is connected for* G.

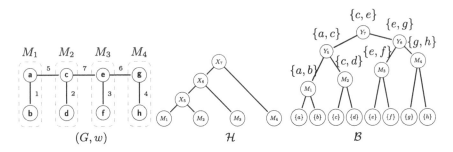

Fig. 3. A weighted graph (G, w) with four minima delimited by the dashed rectangles, a hierarchical watershed $\mathcal{H}$ of (G, w), and the binary partition hierarchy by altitude ordering $\mathcal{B}$ of (G, w).

Intuitively, a drop of water on a topographic surface drains in the direction of a local minimum and there is a correspondence between the catchment basins of a surface and its local minima. As established in [4], in the context of cuts, a notion of watershed in the framework of edge-weighted graphs is characterized as a (graph) cut induced by a minimum spanning forest rooted in the minima of this graph. Let k be a value in $\mathbb{R}^+$. A connected subgraph G' of G is a *(regional) minimum* (for w) at level k if:

1. for any edge u in $E(G')$, the weight of u is equal to k; and
2. for any edge $\{x, y\}$ in $E(G) \setminus E(G')$ such that $|\{x, y\} \cap V(G')| \geq 1$, the weight of $\{x, y\}$ is strictly greater than k.

Let $\{G_1, \ldots, G_\ell\}$ be a set of graphs. We denote by $\sqcup\{G_1, \ldots, G_\ell\}$ the graph $(\cup\{V(G_j) \mid j \in \{1, \ldots, \ell\}\}, \cup\{E(G_j) \mid j \in \{1, \ldots, \ell\}\})$. In the following, we define hierarchical watersheds which are optimal in the sense of minimum spanning forests [5].

Definition 1 (hierarchical watershed). *Let $\mathcal{S} = (M_1, \ldots, M_n)$ be a sequence of pairwise distinct minima of w such that $\{M_1, \ldots, M_n\}$ is the set of all minima of w. Let $(G_0, \ldots, G_{n-1})$ be a sequence of subgraphs of G such that:*

1. *for any $i \in \{0, \ldots, n-1\}$, the graph G_i is a MSF of G rooted in $\sqcup\{M_j \mid j \in \{i+1, \ldots, n\}\}$; and*
2. *for any $i \in \{1, \ldots, n-1\}$, $G_{i-1} \sqsubseteq G_i$.*

The sequence $\mathcal{T} = (CC(G_0), \ldots, CC(G_{n-1}))$ is called a hierarchical watershed of (G,w) for $\mathcal{S}$. Given a hierarchy $\mathcal{H}$, we say that $\mathcal{H}$ is a hierarchical watershed of (G,w) if there exists a sequence $\mathcal{S} = (M_1, \ldots, M_n)$ of pairwise distinct minima of w such that $\{M_1, \ldots, M_n\}$ is the set of all minima of w and such that $\mathcal{H}$ is a hierarchical watershed for $\mathcal{S}$.

A weighted graph (G, w) and a hierarchical watershed $\mathcal{H}$ of (G, w) are illustrated in Fig. 3. We can see that $\mathcal{H}$ is the hierarchical watershed of (G, w) for the sequence $\mathcal{S} = (M_1, M_2, M_3, M_4)$ of minima of w.

Important Notations and Notions: since the weight of the edges of G are pairwise distinct, it follows that any minimum of G is a graph with a single edge. The number of minima of w is denoted by n. Moreover, for any sequence $\mathcal{S}$ of pairwise distinct minima of w, the hierarchical watershed of (G, w) for $\mathcal{S}$ is unique. Every sequence of minima of w considered in this article is a sequence of n pairwise distinct minima of w and, therefore, we use the term *sequence of minima of w* instead of *sequence of n pairwise distinct minima of w*.

3   Watersheding

In this section, we introduce the watersheding (operator), which maps any saliency map into the saliency map of a hierarchical watershed of (G, w). To present the watersheding, we first introduce binary partition hierarchies by altitude ordering, whose link with hierarchical watersheds is established in [7].

3.1   Binary Partition Hierarchies (by Altitude Ordering)

Binary partition trees [20] are widely used for hierarchical image representation. In this section, we describe the particular case where the merging process is guided by the edge weights [7].

Given any set X, we denote by $|X|$ the number of elements of X. Let k be any element in $\{1, \ldots, |E|\}$. We denote by u_k the edge in E such that there are $k-1$ edges in E of weight strictly smaller than $w(u_k)$. We set $\mathbf{B}_0 = \{\{x\}|x \in V\}$. The *k-partition of V* is defined by $\mathbf{B}_k = (\mathbf{B}_{k-1} \backslash \{\mathbf{B}_{k-1}^x, \mathbf{B}_{k-1}^y\}) \cup \{\mathbf{B}_{k-1}^y \cup \mathbf{B}_{k-1}^x\}$, where $u_k = \{x, y\}$ and $\mathbf{B}_{k-1}^x$ and $\mathbf{B}_{k-1}^y$ are the regions of $\mathbf{B}_{k-1}$ that contain x and y, respectively. The *binary partition hierarchy (by altitude ordering) of (G, w)*, denoted by $\mathcal{B}$, is the hierarchy $(\mathbf{B}_0, \ldots, \mathbf{B}_{|E|})$.

Let $\mathcal{B} = (\mathbf{B}_0, \ldots, \mathbf{B}_{|E|})$ be the binary partition hierarchy of (G, w) and let k be a value in $\{1, \ldots, |E|\}$. Since G is a tree, given $u_k = \{x, y\}$, it can be seen that $\mathbf{B}_{k-1}^x$ and $\mathbf{B}_{k-1}^y$ are disjoint. Thus $\mathbf{B}_k$ is different from $\mathbf{B}_{k-1}$. We can affirm that $\mathbf{B}_{k-1}^x \cup \mathbf{B}_{k-1}^y$ is not a region of $\mathbf{B}_{k'}$ for any $k' < k$. Since $\mathcal{B}$ is a hierarchy, we say that u_k *is the building edge of the region* $\{\mathbf{B}_{k-1}^y \cup \mathbf{B}_{k-1}^x\}$ of $\mathcal{B}$. Given any edge u in E, we denote by R_u the region of $\mathcal{B}$ whose building edge is u.

In Fig. 3, we present the binary partition hierarchy $\mathcal{B}$ of the graph (G, w). The building edge of each non-leaf region R of $\mathcal{B}$ is shown above the node that represents R.

Important Notation: in the sequel of this article, the binary partition hierarchy by altitude ordering of (G, w) is denoted by $\mathcal{B}$.

Let X and Y be two distinct regions of $\mathcal{B}$. If the parent of X is equal to the parent of Y, we say that X is a sibling of Y, that Y is a sibling of X and that X and Y are siblings. It can be seen that X has exactly one sibling and we denote this unique sibling of X by *sibling(X)*.

3.2   Watersheding Operator

The idea underlying the watersheding operator is to invert the method to compute hierarchical watersheds proposed in [7,17]. In order to present this method, we first present the definition of extinction values and extinction maps.

Definition 2 (extinction value and extinction map). *Let $\mathcal{S}$ be a sequence of minima of w and let R be any region of $\mathcal{B}$. The* extinction value *of R for $\mathcal{S}$ is zero if there is no minimum of w included in R and, otherwise, it is the maximum value i in $\{1, \ldots, n\}$ such that the minimum M_i is included in R. Let P be a map from the regions of $\mathcal{B}$ into $\mathbb{R}^+$. We say that P is the* extinction map *for $\mathcal{S}$ if, for any region R of $\mathcal{B}$, the value $P(R)$ is the extinction value of R for $\mathcal{S}$. We say that P is an* extinction map *if there exists a sequence of minima $\mathcal{S}$ such that P is the extinction map for $\mathcal{S}$.*

The following property characterizes extinction maps.[1]

Property 3 (extinction map). *Let P be a map from the regions of $\mathcal{B}$ into $\mathbb{R}^+$. The map P is an* extinction map *for w if and only if the following statements hold true:*

- *$\cup\{P(R) \mid R$ is a region of $\mathcal{B}\} = \{0, \ldots, n\}$;*
- *for any two distinct minima M_1 and M_2, $P(M_1) \neq P(M_2)$; and*
- *for any region R of $\mathcal{B}$, we have $P(R) = \vee\{P(M)$ such that M is a minimum of w included in $R\}$.*

Let $\mathcal{S}$ be a sequence of minima of w. As established in [7], the saliency map f of the hierarchical watershed of (G, w) for $\mathcal{S}$ can be obtained through the following steps:

1. computation of the extinction map P for $\mathcal{S}$; and
2. computation of f: for any edge u in E, the weight of u for f is $min\{P(R) \mid R$ is a child of $R_u\}$.

Inspired by this simple and efficient method to compute hierarchical watersheds, we propose the watersheding of hierarchies. From any saliency map f, we find an *approximated* extinction map P such that, for any edge u of G, we have $f(u) = min\{P(R) \mid R$ is a child of $R_u\}$. Then, we define a sequence $\mathcal{S}$ of minima of w ordered in increasing order for P. The watersheding of f is then defined as the saliency map of the hierarchical watershed of (G, w) for $\mathcal{S}$.

In order to introduce approximated extinction maps, we first present the auxiliary notions of supremum descendant values, non-leaf orderings and dominant regions.

Definition 4 (supremum descendant value). *Let f be a map from E into $\mathbb{R}^+$ and let R be a region of $\mathcal{B}$. The* supremum descendant value *of R for f, denoted by $\vee^f(R)$, is defined as $\vee^f(R) = \vee\{f(v) \mid v \in E, R_v \subseteq R\}$.*

---

[1] The proofs of the lemmas, properties and theorem presented in this article can be found in [10].

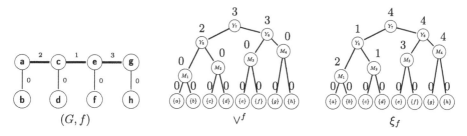

Fig. 4. A weighted graph (G, f), the suppremum descendant values $\vee^f$ for f, and the approximated extinction map ξ_f of f for the binary partition hierarchy by altitude ordering $\mathcal{B}$ of Fig. 3.

Definition 5 (non-leaf ordering). *Let f be a map from E into $\mathbb{R}^+$ and let R be a region of $\mathcal{B}$. A non-leaf ordering by f (for $\mathcal{B}$) is a total ordering on the non-leaf regions of $\mathcal{B}$, denoted by $\prec^{(f,w)}$, such that, for any two non-leaf regions X and Y whose building edges are u and v, respectively, if $\vee^f(X) < \vee^f(Y)$ or if $\vee^f(X) = \vee^f(Y)$ and $w(u) < w(v)$, then $X \prec^{(f,w)} Y$.*

Since the edge weights for w are pairwise distinct, we can conclude that, for any map f, there is a unique non-leaf ordering $\prec^{(f,w)}$ by f.

Definition 6 (dominant region). *Let f be a map from E into $\mathbb{R}^+$ and let $\prec^{(f,w)}$ be the non-leaf ordering by f. Given any non-leaf region R different from V of $\mathcal{B}$, we say that R is a dominant region for f if:*

- *$sibling(R) \prec^{(f,w)} R$; or*
- *$sibling(R)$ is a leaf region of $\mathcal{B}$.*

For instance, consider the weighted graph (G, w) and the binary partition hierarchy $\mathcal{B}$ of (G, w) shown in Fig. 3, and the weighted graph (G, f) shown in Fig. 4. The non-leaf ordering $\prec^{(f,w)}$ by f for $\mathcal{B}$ is such that $M_1 \prec^{(f,w)} M_2 \prec^{(f,w)} M_3 \prec^{(f,w)} M_4 \prec^{(f,w)} Y_5 \prec^{(f,w)} Y_6 \prec^{(f,w)} Y_7$. The dominant regions of $\mathcal{B}$ for f are the regions M_2, M_4, Y_6.

Definition 7 (approximated extinction map). *Let f be a map from E into $\mathbb{R}^+$. The approximated extinction map of f (for $\mathcal{B}$) is the map ξ_f from the set of regions of $\mathcal{B}$ into $\mathbb{R}^+$ such that:*

1. *$\xi_f(R) = \vee^f(R) + 1$ if R is the vertex set V of G; and*
2. *$\xi_f(R) = \xi_f(parent(R))$ if R is a dominant region of $\mathcal{B}$; and*
3. *$\xi_f(R) = f(u)$, where u is the building edge of $parent(R)$, otherwise.*

In Fig. 4, we show the approximated extinction map ξ_f of a map f for the hierarchy $\mathcal{B}$ shown in Fig. 3. The next property establishes that ξ_f is an extinction map if and only if f is the saliency map of a hierarchical watershed.

Property 8. *Let f be a map from E into $\mathbb{R}^+$ and let ξ_f be the approximated extinction map of f. The map f is the saliency map of a hierarchical watershed of (G, w) if and only if the map ξ_f is an extinction map.*

By Property 3, we can verify that the map ξ_f of Fig. 4 is not an extinction map because $\vee\{P(M_1), P(M_2)\} = 2$ is different from $P(Y_5) = 1$. By Property 8, we may conclude that f is not the saliency map of a hierarchical watershed.

In the next definition, we introduce estimated sequences of minima obtained through approximated extinction maps.

Definition 9 (estimated sequence of minima). *Let f be a map from E into $\mathbb{R}^+$ and let ξ_f be the approximated extinction map of f. The estimated sequence of minima (of w) for f is the sequence $S_f = (M_1, \ldots, M_n)$ such that for i and j in $\{1, \ldots, n\}$, if $i < j$, then:*

- $\xi_f(M_i) < \xi_f(M_j)$; or
- $\xi_f(M_i) = \xi_f(M_j)$ and $M_i \prec^{(f,w)} M_j$.

For instance, in Fig. 4, we can see that $\xi_f(M_2) < \xi_f(M_1) < \xi_f(M_3) < \xi_f(M_4)$. Therefore, the estimated sequence of minima for f is $S_f = (M_2, M_1, M_3, M_4)$. The next property establishes that, if f is a hierarchical watershed of (G, w), then f is the hierarchical watershed of (G, w) for S_f.

Property 10. *Let f be a map from E into $\mathbb{R}^+$ and let S_f be the estimated sequence of minima of f. If f is a hierarchical watershed of (G, w), then f is the hierarchical watershed of (G, w) for S_f.*

Having defined approximated extinction maps and estimated sequences of minima, we formalize the watersheding operator in the following definition.

Definition 11 (watersheding). *Given a map f from E into $\mathbb{R}^+$, let S_f be the estimated sequence of minima of f and let P be the extinction map for S_f. The watersheding of f is the map $\omega(f)$ from E into $\mathbb{R}^+$ such that, for any edge u:*

$$\omega(f)(u) = min\{P(R) \mid R \text{ is a child of } R_u\}.$$

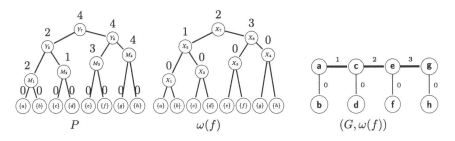

Fig. 5. The extinction map P for the estimated sequence of minima $S_f = (M_2, M_1, M_3, M_4)$ for the map f (Fig. 4) and watersheding $\omega(f)$ of f.

In Fig. 5, we show the watersheding $\omega(f)$ of the map f of Fig. 4. After obtaining the estimated sequence of minima $\mathcal{S}_f = (M_2, M_1, M_3, M_4)$ for f (shown in Fig. 4), we compute the extinction map P for $\mathcal{S}_f$. Then, the watersheding of f is obtained according to Definition 11. We can verify that $\omega(f)$ is the saliency map of a hierarchical watershed of (G, w) of Fig. 3. Indeed, the map $\omega(f)$ is the saliency map of the hierarchical watershed $\mathcal{H}$ of Fig. 3.

In the following theorem, we establish that the watersheding of any map is the saliency map of a hierarchical watershed of (G, w).

Theorem 12. *Let f be a map from E into $\mathbb{R}^+$. The watersheding $\omega(f)$ of f is the saliency map of the hierarchical watershed of (G, w) for the estimated sequence of minima for f.*

The following properties establish that the watersheding operator is idempotent and that the saliency maps of the hierarchical watersheds of (G, w) are the fixed points of the watersheding operator.

Property 13. *Let f be a map from E into $\mathbb{R}^+$. The watersheding $\omega(\omega(f))$ of $\omega(f)$ is equal to $\omega(f)$.*

Property 14. *Let f be the saliency map of a hierarchical watershed of (G, w). The watersheding $\omega(f)$ of f is equal to f.*

From Theorem 12 and from Properties 13 and 14, we may conclude the following property, which links the watersheding operator to the problem of recognition of hierarchical watersheds discussed in [11].

Property 15. *Let $\mathcal{H}$ be a hierarchy and let f be the saliency map of $\mathcal{H}$. The hierarchy $\mathcal{H}$ is a hierarchical watershed of (G, w) if and only if the watersheding $\omega(f)$ of f is equal to f.*

4   Discussions and Perspectives

In this article, we propose the watersheding operator, which converts any map into the saliency map of a hierarchical watershed. The genericity of the watersheding operator opens a range of potential applications:

– *Regularization of hierarchies based on non-increasing attributes.* In a hierarchical watershed, the order in which catchment basins are merged are often defined by extinction values associated with increasing regional attributes, such as area and volume [14]. Given an attribute A, we say that A is increasing if, given any hierarchy $\mathcal{H}$, for any pair of regions R_1 and R_2 of $\mathcal{H}$, if $R_1 \subset R_2$ then $A(R_1) < A(R_2)$. To compute a hierarchical watershed of (G, w) based on the attribute A, we first obtain extinction values based on A [22]. Then, we compute the hierarchical watershed $\mathcal{H}$ of (G, w) for the sequence $\mathcal{S}$ of minima w ordered by their extinction values. The hierarchy $\mathcal{H}$ corresponds to a sequence of filterings of the watershed of (G, w) in which the least important

I G $\mathcal{H}_c$ $\mathcal{H}_{cc}$ $\mathcal{H}_w$

I' G' $\mathcal{H}'_{cob}$ $\mathcal{H}'_w$

Fig. 6. First row: an image I, a gradient G of I [8], a hierarchical watershed $\mathcal{H}_c$ of G based on regularized circularity attribute values, a circularity based hierarchy $\mathcal{H}_{cc}$ and the watersheding $\mathcal{H}_w$ of $\mathcal{H}_{cc}$ for G. Second row: an image I', a gradient G' of I' [8], the hierarchy $\mathcal{H}'_{cob}$ obtained through the method described in [12] and the watersheding $\mathcal{H}'_w$ of $\mathcal{H}'_{cob}$ for G'.

regions according to A are the first regions to be suppressed. However, this property is not guaranteed when A is not increasing. When dealing with non-increasing attributes, *e.g.* circularity and perimeter, we can obtain extinction values by applying a regularization rule [21] on the attribute values. Alternatively, instead of computing hierarchical watersheds from regularized attribute values, we can compute the watersheding of any hierarchy based on a non-increasing attribute. This is illustrated in Fig. 6. We can see that, in the watersheding $\mathcal{H}_w$ of the circularity based hierarchy $\mathcal{H}_{cc}$, the circular regions are more highlighted when compared to the hierarchical watershed $\mathcal{H}_c$ computed from regularized circularity values.

- *Refinement of coarse hierarchies.* In [12], the authors propose a high-quality method (COB) to compute hierarchies. However, in some cases, fine regions are not included in the resulting hierarchies. This is the case of the hierarchy $\mathcal{H}'_{cob}$ of Fig. 6. The watersheding $\mathcal{H}'_w$ of $\mathcal{H}'_{cob}$ allows fine regions to appear in the hierarchy while taking into consideration coarse levels of $\mathcal{H}'_{cob}$.
- *Learning of optimal attributes to compute hierarchical watersheds.* In [9], the authors showed that combinations of hierarchical watersheds improve the performance of individual hierarchical watersheds. However, there is no guarantee that the resulting combinations are hierarchical watersheds themselves. By using the watersheding operator, we can compute hierarchical watersheds from the combinations of hierarchies. If we obtain similar results, those could be used to learn an attribute A such that the watersheding of the combinations corresponds to the hierarchical watersheds based on A.

In future work, we will extend the watersheding operator to generic graphs. We may also perform the experiments suggested here, namely testing the watersheding operator on hierarchies based on non-increasing attributes, on combinations of hierarchical watersheds and on other coarse hierarchies.

References

1. Arbelaez, P., Maire, M., Fowlkes, C., Malik, J.: Contour detection and hierarchical image segmentation. PAMI **33**(5), 898–916 (2011)
2. Beucher, S., Meyer, F.: The morphological approach to segmentation: the watershed transformation. Opt. Eng. New York-Marcel Dekker Inc. **34**, 433–481 (1992)
3. Couprie, C., Grady, L., Najman, L., Talbot, H.: Power watersheds: a new image segmentation framework extending graph cuts, random walker and optimal spanning forest. In: ICCV, pp. 731–738. IEEE (2009)
4. Cousty, J., Bertrand, G., Najman, L., Couprie, M.: Watershed cuts: minimum spanning forests and the drop of water principle. PAMI **31**(8), 1362–1374 (2009)
5. Cousty, J., Najman, L.: Incremental algorithm for hierarchical minimum spanning forests and saliency of watershed cuts. In: Soille, P., Pesaresi, M., Ouzounis, G.K. (eds.) ISMM 2011. LNCS, vol. 6671, pp. 272–283. Springer, Heidelberg (2011). https://doi.org/10.1007/978-3-642-21569-8_24
6. Cousty, J., Najman, L., Kenmochi, Y., Guimarães, S.: Hierarchical segmentations with graphs: quasi-flat zones, minimum spanning trees, and saliency maps. JMIV **60**(4), 479–502 (2018)
7. Cousty, J., Najman, L., Perret, B.: Constructive links between some morphological hierarchies on edge-weighted graphs. In: Hendriks, C.L.L., Borgefors, G., Strand, R. (eds.) ISMM 2013. LNCS, vol. 7883, pp. 86–97. Springer, Heidelberg (2013). https://doi.org/10.1007/978-3-642-38294-9_8
8. Dollár, P., Zitnick, C.L.: Structured forests for fast edge detection. In: ICCV, pp. 1841–1848. IEEE (2013)
9. Santana Maia, D., de Albuquerque Araujo, A., Cousty, J., Najman, L., Perret, B., Talbot, H.: Evaluation of combinations of watershed hierarchies. In: Angulo, J., Velasco-Forero, S., Meyer, F. (eds.) ISMM 2017. LNCS, vol. 10225, pp. 133–145. Springer, Cham (2017). https://doi.org/10.1007/978-3-319-57240-6_11
10. Maia, D.S., Cousty, J., Najman, L., Perret, B.: Proofs of the properties presented in the article "Watersheding hierarchies". Research report, April 2019
11. Maia, D.S., Cousty, J., Najman, L., Perret, B.: Recognizing hierarchical watersheds. In: Couprie, M., Cousty, J., Kenmochi, Y., Mustafa, N. (eds.) DGCI 2019. LNCS, vol. 11414, pp. 300–313. Springer, Cham (2019). https://doi.org/10.1007/978-3-030-14085-4_24
12. Maninis, K.-K., Pont-Tuset, J., Arbeláez, P., Van Gool, L.: Convolutional oriented boundaries: from image segmentation to high-level tasks. PAMI **40**(4), 819–833 (2018)
13. Meyer, F., Maragos, P.: Morphological scale-space representation with levelings. In: Nielsen, M., Johansen, P., Olsen, O.F., Weickert, J. (eds.) Scale-Space 1999. LNCS, vol. 1682, pp. 187–198. Springer, Heidelberg (1999). https://doi.org/10.1007/3-540-48236-9_17
14. Meyer, F., Vachier, C., Oliveras, A., Salembier, P.: Morphological tools for segmentation: connected filters and watersheds. Annales des télécommunications **52**, 367–379 (1997). Springer

15. Nagao, M., Matsuyama, T., Ikeda, Y.: Region extraction and shape analysis in aerial photographs. CGIP **10**(3), 195–223 (1979)
16. Najman, L.: On the equivalence between hierarchical segmentations and ultrametric watersheds. JMIV **40**(3), 231–247 (2011)
17. Najman, L., Cousty, J., Perret, B.: Playing with kruskal: algorithms for morphological trees in edge-weighted graphs. In: Hendriks, C.L.L., Borgefors, G., Strand, R. (eds.) ISMM 2013. LNCS, vol. 7883, pp. 135–146. Springer, Heidelberg (2013). https://doi.org/10.1007/978-3-642-38294-9_12
18. Najman, L., Schmitt, M.: Geodesic saliency of watershed contours and hierarchical segmentation. PAMI **18**(12), 1163–1173 (1996)
19. Perret, B., Cousty, J., Guimaraes, S.J.F., Maia, D.S.: Evaluation of hierarchical watersheds. TIP **27**(4), 1676–1688 (2018)
20. Salembier, P., Garrido, L.: Binary partition tree as an efficient representation for image processing, segmentation, and information retrieval. TIP **9**(4), 561–576 (2000)
21. Salembier, P., Oliveras, A., Garrido, L.: Antiextensive connected operators for image and sequence processing. TIP **7**(4), 555–570 (1998)
22. Vachier, C., Meyer, F.: Extinction value: a new measurement of persistence. In: Workshop on nonlinear signal and image processing, pp. 254–257 (1995)

On the Probabilities of Hierarchical Watersheds

Deise Santana Maia[(⊠)], Jean Cousty, Laurent Najman, and Benjamin Perret

Université Paris-Est, LIGM (UMR 8049), CNRS, ENPC, ESIEE Paris, UPEM,
93162 Noisy-le-Grand, France
deisesantanamaia@gmail.com

Abstract. Hierarchical watersheds are obtained by iteratively merging the regions of a watershed segmentation. In the watershed segmentation of an image, each region contains exactly one (local) minimum of the original image. Therefore, the construction of a hierarchical watershed of any image I can be guided by a total order $\prec$ on the set of minima of I. The regions that contain the least minima according to the order $\prec$ are the first regions to be merged in the hierarchy. In fact, given any image I, for any hierarchical watershed $\mathcal{H}$ of I, there exists more than one total order on the set of minima of I which could be used to obtain $\mathcal{H}$. In this article, we define the probability of a hierarchical watershed $\mathcal{H}$ as the probability of $\mathcal{H}$ to be the hierarchical watershed of I for an arbitrary total order on the set of minima of I. We introduce an efficient method to obtain the probability of hierarchical watersheds and we provide a characterization of the most probable hierarchical watersheds.

1 Introduction

Watershed [3,4] is a well established segmentation technique in the field of mathematical morphology. The idea underlying this technique comes from the topographic definition of watersheds: dividing lines between neighboring catchment basins, *i.e.*, regions whose collected water drains to a common point. We say that the point (or region) of lowest altitude of a catchment basin is a (local) minimum of a topographic surface. In the context of digital image processing, gray-level images can be treated as topographic surfaces whose altitudes are determined by the pixel gray-levels. The local minima of an image are the regions of uniform grey-level surrounded by pixels of strictly higher gray-levels. We show the representation of a gray-scale image with four local minima and a watershed segmentation in Fig. 1(a) and (b), respectively.

Hierarchical watersheds are sequences of nested segmentations equivalent to filterings of an initial watershed segmentation. Let I be an image. The construction of a hierarchical watershed of I is often based on a criterion used to order the minima of I, as the area and the dynamics [8,13]. More specifically, given

This research is partly funded by the Bézout Labex, funded by ANR, reference ANR-10-LABX-58.

B. Burgeth et al. (Eds.): ISMM 2019, LNCS 11564, pp. 137–149, 2019.
https://doi.org/10.1007/978-3-030-20867-7_11

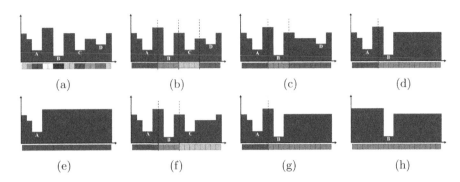

Fig. 1. (a): A gray-scale image I with four minima. (b) A watershed segmentation of I: the vertical dashed lines represent the watershed lines. (c), (d) and (e): The watershed segmentations resulting from iteratively flooding the minima C, D and B, respectively. (f), (g) and (h): The watershed segmentations resulting from iteratively flooding the minima D, C and A, respectively.

any total order $\prec$ on the set of minima of I, the hierarchical watershed of I for $\prec$ is constructed by iteratively "flooding" the minima of I according to $\prec$. For instance, let us consider the total order $\prec$ on the set of minima $\{A, B, C, D\}$ of the image I of Fig. 1(a) such that $C \prec D \prec B \prec A$. In Fig. 1(b), (c), (d) and (e), we show the sequence of floodings of the minima of I for $\prec$. The watershed segmentation of those floodings are the segmentations of the hierarchical watershed of I for $\prec$.

In fact, we may obtain the same hierarchical watershed for several total orders on the set of minima of an image. For example, we show in Fig. 1(b), (f), (g) and (h) the floodings of the minima of the image I for another total order $\prec'$ such that $D \prec' C \prec' A \prec' B$. We can observe that the floodings for the total orders $\prec'$ and $\prec$ induce the same sequence of watershed segmentations. Indeed, given any image I and any hierarchical watershed $\mathcal{H}$ of I, there may exist several total orders on the set of minima of I whose hierarchical watersheds correspond to $\mathcal{H}$. In other words, it is possible to order the minima of I according to distinct criteria and still obtain the same hierarchical watershed.

In this article, (the gradients of) images are represented as weighted graphs. We define the probability of a hierarchical watershed $\mathcal{H}$ as the probability of $\mathcal{H}$ to be the hierarchical watershed of a given weighted graph (G, w) for an arbitrary sequence of minima of w. Our contributions are twofold: (1) an efficient method to obtain the probability of hierarchical watersheds; and (2) a characterization of the most probable hierarchical watersheds of any weighted graph.

Other studies related to probability and (watershed) segmentations are found in [1,2,9,12]. In [1], a stochastic watershed segmentation based on random markers is introduced. In [2], the definitions of watersheds with multiple solutions for a single image are unified in the definition of tie-zone watersheds, which returns a unique solution. In [12], the authors propose a method to list the k-minimum spanning trees that induce distinct segmentations for a given set of markers.

In [9], the authors estimate the probability that any two regions of a watershed segmentation have the same texture, which is further used to build hierarchies of segmentations.

This paper is organized as follows. In Sect. 2, we review graphs, hierarchies of partitions and saliency maps. In Sect. 3, we propose an efficient method to compute the probability of hierarchical watersheds and we characterize the most probable hierarchical watersheds of any weighted graph.

2   Background Notions

In this section, we first introduce hierarchies of partitions. Then, we review the definition of graphs, connected hierarchies and saliency maps. Subsequently, we define hierarchical watersheds.

2.1   Hierarchies of Partitions

Let V be a set. A *partition* of V is a set $\mathbf{P}$ of non empty disjoint subsets of V whose union is V. If $\mathbf{P}$ is a partition of V, any element of $\mathbf{P}$ is called a *region of* $\mathbf{P}$. Let V be a set and let $\mathbf{P}_1$ and $\mathbf{P}_2$ be two partitions of V. We say that $\mathbf{P}_1$ is a *refinement* of $\mathbf{P}_2$ if every element of $\mathbf{P}_1$ is included in an element of $\mathbf{P}_2$. A *hierarchy (of partitions on V)* is a sequence $\mathcal{H} = (\mathbf{P}_0, \dots, \mathbf{P}_n)$ of partitions of V such that $\mathbf{P}_{i-1}$ is a refinement of $\mathbf{P}_i$, for any $i \in \{1, \dots, n\}$ and such that $\mathbf{P}_n = \{V\}$. For any i in $\{0, \dots, n\}$, any region of the partition $\mathbf{P}_i$ is called a *region of* $\mathcal{H}$.

Hierarchies of partitions can be represented as trees whose nodes correspond to regions, as shown in Fig. 2(a). Given a hierarchy $\mathcal{H}$ and two regions X and Y of $\mathcal{H}$, we say that X *is a parent of* Y *(or that* Y *is a child of X)* if $Y \subset X$ and if X is minimal for this property. In other words, if X is a parent of Y and if there is a region Z such that $Y \subseteq Z \subset X$, then $Y = Z$. It can be seen that any region X of $\mathcal{H}$ such that $X \neq V$ has exactly one parent. Thus, for any region X such that $X \neq V$, we write $parent(X) = Y$ where Y is the unique parent of X. Given any region R of $\mathcal{H}$, if R is not the parent of any region of $\mathcal{H}$, we say that R *is a leaf region of* $\mathcal{H}$. Otherwise, we say that R *is a non-leaf region of* $\mathcal{H}$.

In Fig. 2(a), the regions of a hierarchy $\mathcal{H}$ are linked to their parents and children by straight lines. The partition $\mathbf{P}_0$ of $\mathcal{H}$ contains all the leaf regions of $\mathcal{H}$.

2.2   Graphs, Connected Hierarchies and Saliency Maps

A *graph* is a pair $G = (V, E)$, where V is a finite set and E is a set of pairs of distinct elements of V, *i.e.*, $E \subseteq \{\{x, y\} \subseteq V \mid x \neq y\}$. Each element of V is called a *vertex (of G)*, and each element of E is called an *edge (of G)*. To simplify the notations, the set of vertices and edges of a graph G will be also denoted by $V(G)$ and $E(G)$, respectively.

Let $G = (V, E)$ be a graph and let X be a subset of V. A sequence $\pi = (x_0, \dots, x_n)$ of elements of X is a *path (in X) from* x_0 *to* x_n if $\{x_{i-1}, x_i\}$ is an

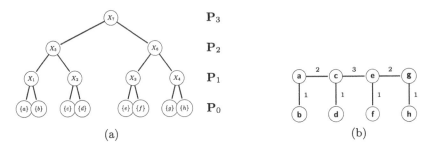

Fig. 2. (a): A representation of a hierarchy of partitions $\mathcal{H} = (\mathbf{P}_0, \mathbf{P}_1, \mathbf{P}_2, \mathbf{P}_3)$ on the set $\{a, b, c, d, e, f, g, h\}$. (b): A weighted graph (G, w).

edge of G for any i in $\{1, \ldots, n\}$. The subset X of V is said to be *connected for G* if, for any x and y in X, there exists a path from x to y. The subset X of V is a *connected component* of G if X is connected and if, for any connected subset Y of V, if $X \subseteq Y$, then we have $X = Y$. In the following, we denote by $CC(G)$ the set of all connected components of G. It is well known that this set $CC(G)$ of all connected components of G is a partition of the set V.

Let $G = (V, E)$ be a graph. *A partition of V is connected for G* if each of its regions is connected for G and *a hierarchy on V is connected (for G)* if any of its partitions is connected. For example, the hierarchy of Fig. 2(a) is connected for the graph of Fig. 2(b).

Let G be a graph. If w is a map from the edge set of G to the set $\mathbb{R}^+$ of positive real numbers, then the pair (G, w) is called an *(edge) weighted graph*. If (G, w) is a weighted graph, for any edge u of G, the value $w(u)$ is called the *weight of u (for w)*.

As established in [6], a connected hierarchy can be equivalently defined with a weighted graph through the notion of a saliency map. Given a weighted graph (G, w) and a hierarchy $\mathcal{H} = (\mathbf{P}_0, \ldots, \mathbf{P}_n)$ connected for G, the *saliency map of $\mathcal{H}$* is the map from $E(G)$ to $\{0, \ldots, n\}$, denoted by $\Phi(\mathcal{H})$, such that, for any edge $u = \{x, y\}$ in $E(G)$, the value $\Phi(\mathcal{H})(u)$ is the smallest value i in $\{0, \ldots, n\}$ such that x and y belong to a same region of $\mathbf{P}_i$. Therefore, the weight $\Phi(\mathcal{H})(u)$ of any edge $u = \{x, y\}$ is the ultrametric distance between x and y on the hierarchy $\mathcal{H}$. It follows that any connected hierarchy has a unique saliency map. Moreover, we can recover any hierarchy $\mathcal{H}$ connected for G from the saliency map $\Phi(\mathcal{H})$ of $\mathcal{H}$. For instance, the weight map depicted in Fig. 2(b) is the saliency map of the hierarchy of Fig. 2(a).

2.3   Hierarchies of Minimum Spanning Forests and Watersheds

In [4], the authors formalize watersheds in the framework of weighted graphs and show the optimality of watersheds in the sense of minimum spanning forests. In this section, we present hierarchical watersheds following the definition of hierarchies of minimum spanning forests presented in [7].

Let G be a graph. We say that G *is a forest* if, for any edge u in $E(G)$, the number of connected components of the graph $(V(G), E(G) \setminus \{u\})$ is greater than the number of connected components of G. Given another graph G', we say that G' *is a subgraph of* G, denoted by $G' \sqsubseteq G$, if $V(G') \subseteq V(G)$ and $E(G') \subseteq E(G)$. Let (G, w) be a weighted graph and let G' be a subgraph of G. A graph G'' is a *Minimum Spanning Forest (MSF) of G rooted in G'* if:

1. G' is a subgraph of G''; and
2. the graphs G and G'' have the same set of vertices, *i.e.*, $V(G'') = V(G)$; and
3. each connected component of G'' includes exactly one connected component of G'; and
4. the sum of the weight of the edges of G'' is minimal among all graphs for which the above conditions 1, 2 and 3 hold true.

Important Notations and Notions: In the sequel of this article, the symbol G denotes a tree, *i.e.*, a forest with a unique connected component. This implies that any map from $E(G)$ into $\mathbb{Z}^+$ is the saliency map of a hierarchy which is connected for G. To shorten the notation, the vertex set of G is denoted by V and its edge set is denoted by E. The symbol w denotes a map from E into $\mathbb{R}^+$ such that, for any pair of distinct edges u and v in E, we have $w(u) \neq w(v)$. Thus, the pair (G, w) is a weighted graph. Every hierarchy considered in this article is connected for G and therefore, for the sake of simplicity, we use the term *hierarchy* instead of *hierarchy which is connected for G*.

Intuitively, a drop of water on a topographic surface drains in the direction of a local minimum and there is a correspondence between the catchment basins of a surface and its local minima. As established in [4], in the context of cuts, a notion of watershed in the framework of edge-weighted graphs is characterized as a (graph) cut induced by a minimum spanning forest rooted in the minima of this graph. Let k be a value in $\mathbb{R}^+$. A connected subgraph G' of G is a *(regional) minimum* (for w) at level k if:

1. G' has at least one edge: $E(G') \neq \emptyset$; and
2. for any edge u in $E(G')$, the weight of u is equal to k; and
3. for any edge $\{x, y\}$ in $E(G) \setminus E(G')$ such that $|\{x, y\} \cap V(G')| \geq 1$, $\{x, y\}$ has a weight greater than k.

One can note that, since the weights of the edges of G are pairwise distinct, it follows that any minimum of G is a graph with a single edge.

Then, we follow the definition of hierarchical watersheds proposed in [5] which are optimal in the sense of minimum spanning forests. The sequence $(M_1, \ldots, M_n)$ of pairwise distinct subgraphs of G is *a sequence of minima of w* if M_i is a minimum of w for any $i \in \{1, \ldots, n\}$ and if n is equal to the number of minima of w. In other words, a sequence of minima of w is equivalent to a total order on the set of minima of w. We denote by $\mathcal{M}_w$ the set of all sequences of minima of w.

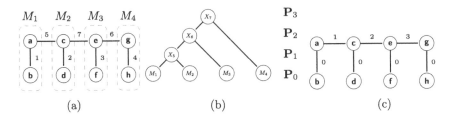

Fig. 3. (a): A weighted graph (G, w) with four minima delimited by the dashed rectangles. (b): A hierarchical watershed $\mathcal{H} = (\mathbf{P}_0, \mathbf{P}_1, \mathbf{P}_2, \mathbf{P}_3)$ of (G, w). (c): The saliency map $\Phi(\mathcal{H})$ of $\mathcal{H}$.

Definition 1 (hierarchical watershed). *Let* $\mathcal{S} = (M_1, \ldots, M_n)$ *be a sequence of minima of* w*. Let* $(G_0, \ldots, G_{n-1})$ *be a sequence of subgraphs of* G *such that:*

1. *for any* $i \in \{0, \ldots, n-1\}$*, the graph* G_i *is a MSF of* G *rooted in* $(\cup \{V(M_j) \mid j \in \{i+1, \ldots, n\}\}, \cup \{E(M_j) \mid j \in \{i+1, \ldots, n\}\})$*; and*
2. *for any* $i \in \{1, \ldots, n-1\}$*,* $G_{i-1} \sqsubseteq G_i$*.*

The sequence $\mathcal{T} = (CC(G_0), \ldots, CC(G_{n-1}))$ *is called a* hierarchical watershed *of* (G, w) *for* $\mathcal{S}$*. Given a hierarchy* $\mathcal{H}$*, we say that* $\mathcal{H}$ *is a hierarchical watershed of* (G, w) *if there exists a sequence* $\mathcal{S}$ *of minima of* w *such that* $\mathcal{H}$ *is a hierarchical watershed for* $\mathcal{S}$*.*

A weighted graph (G, w) and a hierarchical watershed $\mathcal{H}$ of (G, w) are illustrated in Fig. 3(a) and (b), respectively. We can see that $\mathcal{H}$ is the hierarchical watershed of (G, w) for the sequence $\mathcal{S} = (M_1, M_2, M_3, M_4)$ of minima of w.

Since we assumed that the edge weights for w are pairwise distinct, for any sequence $\mathcal{S}$ of minima of w, the hierarchical watershed of (G, w) for $\mathcal{S}$ is unique.

3   Studying Probabilities of Hierarchical Watersheds

Let $\mathcal{H}$ be a hierarchical watershed of (G, w). In this study, we tackle the following problems: (P_1) Find the probability of $\mathcal{H}$ to be the hierarchical watershed of (G, w) for an arbitrary sequence of minima of w; and (P_2) Characterize the most probable hierarchical watersheds of (G, w).

In this section, we first introduce binary partition hierarchies by altitude ordering, which are closely linked to hierarchical watersheds, as established in [7]. Then, we present our solution to Problem (P_1) and a quasi-linear time algorithm to compute probabilities of hierarchical watersheds. Finally, we propose a solution to Problem (P_2).

3.1   Binary Partition Hierarchies (by Altitude Ordering)

Binary partition trees are widely used for hierarchical image representation [11]. In this section, we describe the particular case where the merging process is guided by the edge weights [7].

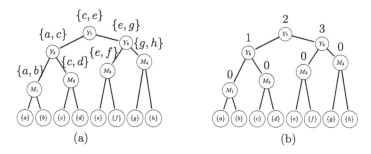

Fig. 4. (a) The binary partition hierarchy $\mathcal{B}$ of (G, w) of Fig. 3(a). (b) A representation of the saliency map $\Phi(\mathcal{H})$ of Fig. 3(c).

Given any set X, we denote by $|X|$ the number of elements of X. Let k be any element in $\{1, \ldots, |E|\}$. We denote by u_k the edge in E such that there are $k-1$ edges in E of weight strictly smaller than $w(u_k)$. We set $\mathbf{B}_0 = \{\{x\}|x \in V\}$. The k-*partition of V* is defined by $\mathbf{B}_k = (\mathbf{B}_{k-1} \setminus \{\mathbf{B}_{k-1}^x, \mathbf{B}_{k-1}^y\}) \cup \{\mathbf{B}_{k-1}^y \cup \mathbf{B}_{k-1}^x\}$, where $u_k = \{x, y\}$ and $\mathbf{B}_{k-1}^x$ and $\mathbf{B}_{k-1}^y$ are the regions of $\mathbf{B}_{k-1}$ that contain x and y, respectively. The *binary partition hierarchy (by altitude ordering)* of (G, w), denoted by $\mathcal{B}$, is the hierarchy $(\mathbf{B}_0, \ldots, \mathbf{B}_{|E|})$.

Let $\mathcal{B} = (\mathbf{B}_0, \ldots, \mathbf{B}_{|E|})$ be the binary partition hierarchy of (G, w) and let k be a value in $\{1, \ldots, |E|\}$. Since G is a tree, given $u_k = \{x, y\}$, it can be seen that $\mathbf{B}_{k-1}^x$ and $\mathbf{B}_{k-1}^y$ are disjoint. Thus $\mathbf{B}_k$ is different from $\mathbf{B}_{k-1}$. We can affirm that $\mathbf{B}_{k-1}^x \cup \mathbf{B}_{k-1}^y$ is not a region of $\mathbf{B}_{k'}$ for any $k' < k$. Since $\mathcal{B}$ is a hierarchy, we say that u_k is *the building edge of the region* $\{\mathbf{B}_{k-1}^y \cup \mathbf{B}_{k-1}^x\}$ of $\mathcal{B}$. Given any edge u in E, we denote by R_u the region of $\mathcal{B}$ whose building edge is u.

In Fig. 4(a), we present the binary partition hierarchy $\mathcal{B}$ of the graph (G, w) of Fig. 3(a). The building edge of each non-leaf region R of $\mathcal{B}$ is shown above the node that represents R.

3.2   Finding Probabilities of Hierarchical Watersheds

Let $\mathcal{H}$ be a hierarchical watershed of (G, w). As defined previously, there is a sequence $\mathcal{S}$ of minima of w such that $\mathcal{H}$ is the hierarchical watershed of (G, w) for $\mathcal{S}$. Indeed, as illustrated in Fig. 1, the hierarchy $\mathcal{H}$ may be the hierarchical watershed of (G, w) for several sequences of minima of w. Thus, for any hierarchical watershed $\mathcal{H}$ of (G, w), we denote by $S_w(\mathcal{H})$ the set which contains every sequence S of minima such that the hierarchical watershed of (G, w) for S is $\mathcal{H}$.

Definition 2 (probability of a hierarchical watershed). *Let $\mathcal{H}$ be a hierarchical watershed of (G, w). Let S be uniformly distributed on the set $\mathcal{M}_w$ of all sequences of minima of w. We define the probability of $\mathcal{H}$ knowing w, denoted by $p(\mathcal{H}|w)$, as the probability that the hierarchical watershed of (G, w) for S is equal to $\mathcal{H}$.*

Property 3. *Let* $\mathcal{H}$ *be a hierarchical watershed of* (G, w)*. The probability* $p(\mathcal{H}|w)$ *of* $\mathcal{H}$ *knowing* w *is the ratio* k/n *where* k *and* n *are the numbers of elements of* $S_w(\mathcal{H})$ *and of* $\mathcal{M}_w$, *respectively.*

In order to solve the problem of finding the probability of hierarchical watersheds, we first introduce watershed-cut edges and maximal regions.

Definition 4 (watershed-cut edge). *Let* $\mathcal{H}$ *be a hierarchical watershed of* (G, w) *and let* $\mathbf{P}$ *be the set of leaf regions of* $\mathcal{H}$*. Let* $u = \{x, y\}$ *be an edge of* G*. If* x *and* y *belong to distinct regions of* $\mathbf{P}$, *we say that* u *is a watershed-cut edge of* w.

We can observe that any hierarchical watershed of (G, w) has the same set of leaf regions. For example, the set of watershed-cut edges of the graph (G, w) of Fig. 3(a) is $\{\{a, c\}, \{c, e\}, \{e, g\}\}$, which is the set of edges whose vertices are in distinct leaf regions of the hierarchical watershed $\mathcal{H}$ of Fig. 3(b).

Definition 5 (maximal region). *Let* u *be a watershed-cut edge of* w *and let* f *be a map from* E *into* $\mathbb{R}^+$*. Let* R_u *be the region of* $\mathcal{B}$ *whose building edge is* u*. We say that the region* R_u *is a maximal region of* $\mathcal{B}$ *for* f *if the value of* f *on the building edge* u *of* R_u *is greater than the value of* f *on the building edge of any region of* $\mathcal{B}$ *included in* R_u*: i.e., if* $f(u) > max\{f(v), v \in E \mid R_v \subset R_u\}$.

In Fig. 4(b), we represent the saliency map $\Phi(\mathcal{H})$ of Fig. 3(c) on the binary partition hierarchy by altitude ordering $\mathcal{B}$ of Fig. 4(a). The weight above each region R of $\mathcal{B}$ is the weight of the building edge of R for $\Phi(\mathcal{H})$. We can see that the only maximal regions of $\mathcal{B}$ for $\Phi(\mathcal{H})$ are the regions Y_5 and Y_6.

The following property establishes that, given any hierarchical watershed $\mathcal{H}$ of (G, w), the probability of $\mathcal{H}$ knowing w can be defined through the number of maximal regions of $\mathcal{B}$ for the saliency map $\Phi(\mathcal{H})$ of $\mathcal{H}$.

Property 6. *Let* $\mathcal{H}$ *be a hierarchical watershed of* (G, w) *and let* m *be the number of maximal regions of* $\mathcal{B}$ *for the saliency map* $\Phi(\mathcal{H})$ *of* $\mathcal{H}$*. The probability of* $\mathcal{H}$ *knowing* w *is*

$$p(\mathcal{H} \mid w) = \frac{2^m}{|\mathcal{M}_w|}. \tag{1}$$

For instance, let us consider the hierarchical watershed $\mathcal{H}$ of Fig. 3(b). As stated previously, the hierarchy $\mathcal{B}$ has two maximal regions (Y_5 and Y_6) for $\Phi(\mathcal{H})$. Since the graph (G, w) of Fig. 3 has four minima, there are 4! sequences of minima of w. By Property 6, we may conclude that the probability of $\mathcal{H}$ knowing w is $\frac{2^2}{4!}$. Indeed, $\mathcal{H}$ is the hierarchical watershed of (G, w) for four sequences of minima: (M_1, M_2, M_3, M_4), (M_1, M_2, M_4, M_3), (M_2, M_1, M_3, M_4) and (M_2, M_1, M_4, M_3).

From Property 6, we derive a quasi-linear time algorithm (Algorithm 1) to compute the probability of hierarchical watersheds. The inputs are a weighted graph $((V, E), w)$ and the saliency map f of a hierarchical watershed of $((V, E), w)$. First, the binary partition hierarchy $\mathcal{B}$ of $((V, E), w)$ is computed

---

Algorithm 1. Probability of hierarchical watersheds

    Data: $((V, E), w)$: a weighted graph whose edges are already sorted in increasing order according to w

          f: the saliency map of a hierarchical watershed $\mathcal{H}$ of $((V, E), w)$

    Result: the probability of $\mathcal{H}$ knowing w

    // We consider that all variables are implicitly initialized to 0

1: Compute the binary partition hierarchy $\mathcal{B}$ of $((V, E), w)$

    // Identification of watershed edges in WS and counting of minima in $numMin$

2: $numMin := 1$

3: Declare WS as an array of $|E|$ integers

4: **for** each edge u in E **do**

5:     **if** none of the children of R_u is a leaf region of $\mathcal{B}$ **then**

6:         $WS[u] := 1$

7:         $numMin = numMin + 1$

    // Identification of maximal region by computing, for any edge u in E, $MaxF[u] = \vee\{f(v) \mid R_v \subset R_u\}$

8: Declare $MaxF$ as an array of $|E|$ real numbers

9: **for** each edge u in increasing order of weights for w **do**

10:     **for** each child X of R_u **do**

11:         **if** X is not a leaf region of $\mathcal{B}$ **then**

12:             $v :=$ the building edge of X

13:             $MaxF[u] := max(MaxF[u], MaxF[v])$

    // Find the number m of maximal regions of $\mathcal{B}$ for f

14: $m := 0$

15: **for** each edge u in E **do**

16:     **if** $WS[u] = 1$ **then**

17:         $v_1 :=$ the building edge of a child of R_u

18:         $v_2 :=$ the building edge of another child of R_u

19:         **if** $f[u] > Max[v_1]$ and $f[u] > Max[v_2]$ **then**

20:             $m := m + 1$

    return $\dfrac{2^m}{numMin!}$

---

in quasi-linear time with respect to $|E|$ [10]. At lines 2–7, we compute the number of minima of w and the set of watershed-cut edges of w in linear time with respect to $|E|$ using the method proposed in [10]. At lines 8–20, we find the number m of maximal regions of $\mathcal{B}$ for f. Both **for** loops at lines 8 and 15 can be computed in linear time with respect to $|E|$. Therefore, the overall time complexity of Algorithm 1 is quasi-linear with respect to $|E|$. The output of Algorithm 1 is the probability of $\mathcal{H}$ knowing w, which is $\frac{2^m}{numMin!}$ by Property 6.

3.3 Most Probable Hierarchical Watersheds

Let ℓ be the number of watershed-cut edges of w. We can affirm that there are at most ℓ maximal regions of $\mathcal{B}$ for the saliency map of any hierarchical watershed of (G, w). Thus, we can derive the following Corollary 7, which establishes the tight upper bound on the probability of any hierarchical watershed of (G, w).

Corollary 7. *Let ℓ be the number of watershed-cut edges of w and let $\mathcal{H}$ be a hierarchical watershed of (G, w). The tight upper bound on the probability of $\mathcal{H}$ knowing w is $\frac{2^\ell}{|\mathcal{M}_w|}$.*

Let u be a watershed-cut edge of w. We say that R_u *is a primary region of* $\mathcal{B}$ if there is no watershed-cut edge v of w such that $R_v \subset R_u$. Let f be the saliency map of a hierarchical watershed of (G, w). One can note that the value of f is null on non watershed-cut edges of w and non null on the watershed-cut edges of w. Therefore, given a watershed-cut edge u of w, if R_u is a primary region of $\mathcal{B}$, then $f(u) > max\{f(v), v \in E \mid R_v \subset R_u\} = 0$ and, consequently, R_u is a maximal region of $\mathcal{B}$ for f. We conclude that each primary region of $\mathcal{B}$ is a maximal region of $\mathcal{B}$ for every saliency map of a hierarchical watershed of (G, w). We can now define the tight lower bound on the probability of hierarchical watersheds.

Corollary 8. *Let k be the number of primary regions of $\mathcal{B}$ and let $\mathcal{H}$ be a hierarchical watershed of (G, w). The tight lower bound on the probability of $\mathcal{H}$ knowing w is $\frac{2^k}{|\mathcal{M}_w|}$.*

If the map w has more than two minima, then there is at least one watershed-cut edge u of w such that R_u is not a primary region of $\mathcal{B}$. Therefore, the tight lower bound and the tight upper bound on the probabilities of hierarchical watersheds of (G, w) are not equal. This justifies the following definition of most probable hierarchical watersheds.

Definition 9 (most probable hierarchical watersheds). *Let $\mathcal{H}$ be a hierarchical watershed of (G,w). We say that $\mathcal{H}$ is a* most probable hierarchical watershed of (G, w), *if, for any hierarchical watershed $\mathcal{H}'$ for (G, w), we have $p(\mathcal{H} \mid w) \geq p(\mathcal{H}' \mid w)$.*

Let f be the saliency map of a hierarchical watershed of (G, w). Let R be a non-leaf region of $\mathcal{B}$ and let u be the building edge of R. We define the *persistence value of R for f* as the weight $f(u)$ of u. Let $\mathcal{H}$ be a hierarchical watershed of (G, w). By Corollary 7, the probability of $\mathcal{H}$ knowing w is maximal when, for every watershed-cut edge u of w, R_u is a maximal region of $\mathcal{B}$ for $\Phi(\mathcal{H})$. By the definition of maximal regions, we can establish the following characterization of the most probable hierarchical watersheds of (G, w).

Property 10. *Let $\mathcal{H}$ be a hierarchical watershed of (G, w). $\mathcal{H}$ is a most probable hierarchical watershed of (G, w) if and only if the persistence values for $\Phi(\mathcal{H})$ are increasing on the hierarchy $\mathcal{B}$.*

Let $\mathcal{H}$ be a most probable hierarchical watershed of (G, w). By Property 10, we may conclude that the order in which the regions of $\mathcal{H}$ are merged along the partitions of $\mathcal{H}$ are constrained by the hierarchy $\mathcal{B}$. Thus, we can deduce the following corollary from Property 10.

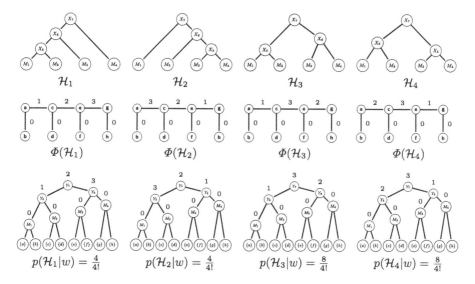

Fig. 5. The hierarchical watersheds for the weighted graph (G, w) of Fig. 3(a), their saliency maps represented on the graph (G, w) and on the binary partition hierarchy $\mathcal{B}$. The probability of each hierarchical watershed knowing w is presented under $\mathcal{B}$.

Corollary 11. *Let $\mathcal{H}$ be a hierarchical watershed of (G, w). $\mathcal{H}$ is a most probable hierarchical watershed of (G, w) if and only if each non-leaf region of $\mathcal{H}$ is a region of $\mathcal{B}$.*

In Fig. 5, we present four hierarchical watersheds for the weighted graph of Fig. 3(a). Indeed, those are the only hierarchical watersheds for (G, w). For each hierarchy, we present its saliency map represented on G and on the binary partition hierarchy by altitude ordering $\mathcal{B}$. Since w has four minima, the number of sequences of minima of w is 4!. The probability of the hierarchies $\mathcal{H}_1$, $\mathcal{H}_2$, $\mathcal{H}_3$ and $\mathcal{H}_4$ knowing w are $\frac{4}{4!}, \frac{4}{4!}, \frac{8}{4!}$ and $\frac{8}{4!}$, respectively. Therefore, the set of most probable hierarchical watersheds of (G, w) is $\{\mathcal{H}_3, \mathcal{H}_4\}$. We can verify that each non-leaf region of $\mathcal{H}_3$ and $\mathcal{H}_4$ is a region of $\mathcal{B}$, as established by Corollary 11, which is not the case for $\mathcal{H}_1$ and $\mathcal{H}_2$.

From Property 10, we can deduce a recursive method to find the saliency map f of a most probable hierarchical watershed of (G, w). Let ℓ be the number of watershed-cut edges of w and let L be the set $\{1, \ldots, \ell\}$. Let R be the root of the tree representing $\mathcal{B}$ and let u be the building edge of R. First, we assign $f(u)$ to $max\{1, \ldots, \ell\}$. Then, we divide the set $\{1, \ldots, \ell - 1\}$ into two subsets L' and L'' with $n' - 1$ and $n'' - 1$ elements, respectively, where n' and n'' are the number of minima included in the children R' and R'' of R. Subsequently, the sets L' and L'' are propagated to R' and R'', respectively. The subtrees rooted in R' and R'' are treated separately. This process is performed until the weight of all watershed-cut edges of w have been assigned. An illustration of this method is presented in Fig. 6, where we show two of the most probable hierarchical watersheds of an image.

I $Grad$ f_1 f_2

Fig. 6. An image I, a gradient $Grad$ of I and the saliency maps f_1 and f_2 of two of the most probable hierarchical watersheds of $Grad$.

4   Conclusion

We proposed an efficient method to obtain the probability of hierarchical watersheds in the framework of weighted graphs. We also provided a characterization of the most probable hierarchical watersheds of any weighted graph. In future work, we will extend the notions presented here to arbitrary graphs, *i.e.* graphs that are not trees and graphs whose edge weights are not pairwise distinct.

References

1. Angulo, J., Jeulin, D.: Stochastic watershed segmentation. In: ISMM. Springer (2007)
2. Audigier, R., Lotufo, R.: Uniquely-determined thinning of the tie-zone watershed based on label frequency. JMIV **27**(2), 157–173 (2007)
3. Beucher, S., Meyer, F.: The morphological approach to segmentation: the watershed transformation. Opt. Eng. N.Y. Marcel Dekker Inc. **34**, 433–481 (1992)
4. Cousty, J., Bertrand, G., Najman, L., Couprie, M.: Watershed cuts: minimum spanning forests and the drop of water principle. IEEE PAMI **31**(8), 1362–1374 (2009)
5. Cousty, J., Najman, L.: Incremental algorithm for hierarchical minimum spanning forests and saliency of watershed cuts. In: Soille, P., Pesaresi, M., Ouzounis, G.K. (eds.) ISMM 2011. LNCS, vol. 6671, pp. 272–283. Springer, Heidelberg (2011). https://doi.org/10.1007/978-3-642-21569-8_24
6. Cousty, J., Najman, L., Kenmochi, Y., Guimarães, S.: Hierarchical segmentations with graphs: quasi-flat zones, minimum spanning trees, and saliency maps. JMIV **60**(4), 479–502 (2018)
7. Cousty, J., Najman, L., Perret, B.: Constructive links between some morphological hierarchies on edge-weighted graphs. In: Hendriks, C.L.L., Borgefors, G., Strand, R. (eds.) ISMM 2013. LNCS, vol. 7883, pp. 86–97. Springer, Heidelberg (2013). https://doi.org/10.1007/978-3-642-38294-9_8
8. Grimaud, M.: New measure of contrast: the dynamics. In: Image Algebra and Morphological Image Processing III, vol. 1769, pp. 292–306. International Society for Optics and Photonics (1992)
9. Jeulin, D.: Morphological probabilistic hierarchies for texture segmentation. Math. Morphol.-Theory Appl. **1**(1), 216–324 (2016)
10. Najman, L., Cousty, J., Perret, B.: Playing with kruskal: algorithms for morphological trees in edge-weighted graphs. In: Hendriks, C.L.L., Borgefors, G., Strand, R. (eds.) ISMM 2013. LNCS, vol. 7883, pp. 135–146. Springer, Heidelberg (2013). https://doi.org/10.1007/978-3-642-38294-9_12

11. Salembier, P., Garrido, L.: Binary partition tree as an efficient representation for image processing, segmentation, and information retrieval. IEEE Trans. Image Process. **9**(4), 561–576 (2000)

12. Straehle, C., Peter, S., Köthe, U., Hamprecht, F.A.: K-smallest spanning tree segmentations. In: Weickert, J., Hein, M., Schiele, B. (eds.) GCPR 2013. LNCS, vol. 8142, pp. 375–384. Springer, Heidelberg (2013). https://doi.org/10.1007/978-3-642-40602-7_40

13. Vachier, C., Meyer, F.: Extinction value: a new measurement of persistence. In IEEE Workshop on Nonlinear Signal and Image Processing, pp. 254–257 (1995)

Incremental Attribute Computation
in Component-Hypertrees

Alexandre Morimitsu[1(✉)], Wonder Alexandre Luz Alves[2],
Dennis José da Silva[1], Charles Ferreira Gobber[2],
and Ronaldo Fumio Hashimoto[1(✉)]

[1] Instituto de Matemática e Estatística, Universidade de São Paulo, São Paulo, Brazil
`alexandre.morimitsu@usp.br`,`{dennis,ronaldo}@ime.usp.br`
[2] Universidade Nove de Julho, São Paulo, Brazil
`wonder@uni9.pro.br`

Abstract. Component-hypertrees are structures that store nodes of multiple component trees built with increasing neighborhoods, meaning they retain the same desirable properties of component trees but also store nodes from multiple scales, at the cost of increasing time and memory consumption for building, storing and processing the structure. In recent years, algorithmic advances resulted in optimization for both building and storing hypertrees. In this paper, we intend to further extend advances in this field, by presenting algorithms for efficient attribute computation and statistical measures that analyze how attribute values vary when nodes are merged in bigger scales. To validate the efficiency of our method, we present complexity and time consumption analyses, as well as a simple application to show the usefulness of the statistical measurements.

Keywords: Component tree · Component-hypertree ·
Connected operators · Attribute computation

1 Introduction

In Image Processing applications, characterization and feature extraction of shapes in images are important steps towards understanding their contents. For this reason, studies related to shapes on digital images date from many decades ago, such as the one of Gray [3], which investigated local properties of discrete binary images. Using these properties, efficient ways of computing attributes such as area and perimeter of binary shapes were developed. Gray's paper also emphasizes the importance of connectivity, showing how it may affect the way objects connect and how it may impact the computation of these attributes [3].

Over time, advances in theory and analysis of shapes led to the development of connected filtering and, in the late 1990s, to the emergence of component trees [4,15], where connected components (CCs) of consecutive level sets of

© Springer Nature Switzerland AG 2019
B. Burgeth et al. (Eds.): ISMM 2019, LNCS 11564, pp. 150–161, 2019.
https://doi.org/10.1007/978-3-030-20867-7_12

grayscale images are hierarchically stored in a tree based on the inclusion relation. This structure allows fast attribute computation since, given an incremental attribute, it is possible to obtain the value of a given node by using previously computed attribute values of all its children nodes. Many authors have developed efficient methods for attribute computation in component trees including, but not limited to, Passat et al. [13], which proposed a way to calculate a pseudo-distance that can be used to compare the similarity of each node to a given binary shape in order to perform segmentation; Neumann and Matas [10], which showed a way to compute attributes such as perimeter (using 4-connectivity) and horizontal and vertical crossings; Silva et al. [17], that extended Gray's study to work directly on component trees and Kiwakuna et al. [6], which quickly computes surface area from 3D-images.

In parallel with the development of connected filtering, theoretical advances expanded the concept of connectivity for digital topology, leading to the development of second-generation connectivities [1,15,16]. Merging these two fields led to the development of even more generalized structures, such as component trees built with richer connectivities [11] and component-hypertrees [12], which can be considered a generalization of component trees for multiple increasing connectivities.

Although having the advantage of storing nodes of multiple trees, hypertrees have not gained widespread adoption. Simplistic approaches for building them are slow and are not efficient in terms of storage. However, recent studies have reduced both time complexity [7] and memory consumption [8] for building hypertrees, making them a viable option.

In order to continue the studies on efficient algorithms, we show in this paper how to efficiently perform attribute computation in component-hypertrees, by both extending known algorithms used in component trees and also presenting new attributes and statistical measurements that take advantage of the inclusion relation of CCs among different connectivities.

For that, this paper is organized as follows: in Sect. 2 we review the background related to images, component trees and previous studies in hypertrees. In Sect. 3, we review attribute computation in component trees and show how to extend these ideas to hypertrees. We also show some other types of measurements that can be extracted by taking advantage of the fact that nodes of hypertrees may consist of clusters of nodes of previous trees. To show the efficiency of our proposed method, experimental results are presented in Sect. 4. Finally, Sect. 5 shows our final thoughts and possible future works related to this study.

2 Background

A gray-level image is represented as a function $f : \mathcal{D}_f \rightarrow \mathbb{K}$, where $\mathcal{D}_f \subset \mathbb{Z}^2$ and $\mathbb{K} = \{0, 1, \ldots, K - 1\}$. Images with $K = 2$ are also called *binary images*. Elements of the domain $\mathcal{D}_f$ of an image are called *pixels*. Storing only pixels $p \in \mathcal{D}_f$ such that $f(p) = 1$ suffices to represent binary images. Any gray-level image f can be transformed into a binary image by *thresholding*. In this sense, we define the set $X_\lambda(f) = \{p : f(p) \geq \lambda\}$. This set of pixels defines a binary image usually referred as the *upper level set* of f with level λ, where $\lambda \in \mathbb{K}$.

Pixels of an image are related using the notion of *neighborhood*. In this paper, neighborhoods are defined using structuring elements (SEs), which are sets $\mathcal{S} \subset \mathbb{Z}^2$. Given a set of pixels $\mathcal{P}$ and a SE $\mathcal{S}$, the dilation of $\mathcal{P}$ by $\mathcal{S}$ is defined as $\mathcal{P} \oplus \mathcal{S} = \{p + s : p \in \mathcal{P}, s \in \mathcal{S}\}$. Using this definition, the $\mathcal{S}$−neighbors of a pixel p are in the set $\{p\} \oplus \mathcal{S}$. Given an image f and a SE $\mathcal{S}$, we define $\mathcal{A}(\mathcal{S}) = \{\{p, q\} : q \in (\{p\} \oplus \mathcal{S}) \text{ and } p, q \in \mathcal{D}_f\}$ as the set of $\mathcal{S}$-neighbors in f. If $\mathcal{S}$ is clear from the context, $\mathcal{S}$-neighbors may be simply referred as neighbors.

Using this definition, we can define connectedness in binary images. Let X be a binary image, $\mathcal{S}$ a SE and $\mathcal{A} := \mathcal{A}(\mathcal{S})$ its set of neighbors. Then, two pixels $p, q \in X$ are said to be $\mathcal{A}$-connected if and only if there exists a sequence of m pixels $(p_1, p_2, \ldots, p_m)$, such that the following conditions hold: (a) $p_1 = p$; (b) $p_m = q$; (c) $\{p_j, p_{j+1}\} \in \mathcal{A}$ for all $1 \leq j < m$. Additionally, an $\mathcal{A}$-*connected component* ($\mathcal{A}$-CC) is defined as a maximal set of $\mathcal{A}$-connected pixels in X. The set of $\mathcal{A}$-CCs of X is denoted by $CC(X, \mathcal{A})$. An $\mathcal{A}$-CC may be also be simply written CC for brevity.

Given a gray-level image f, multiple upper level sets can be extracted by using different thresholds. Upper level sets are decreasing, i.e., if two thresholds α and β satisfying $0 \leq \alpha < \beta < K$ are given, then their respective upper level sets satisfy $X_\alpha(f) \supseteq X_\beta(f)$. The same property holds to the CCs of these level sets. This means that if we build a graph with CCs as nodes and inclusion relation of consecutive level sets as arcs, this entire hierarchy of the CCs of level sets can be represented using a tree structure. This structure is commonly called *component tree*. An example is given in Fig. 1.

Component trees can be efficiently computed and stored using disjoint sets (union-find). This structure is also commonly referred as the *max-tree*. In this representation, each CCs is represented by a *representative pixel*, and these CCs can be reconstructed by looking at all pixels that point towards the representative pixel. A graphical illustration of a max-tree is given in Fig. 1 (rightmost column), with each representative pixel marked in bold.

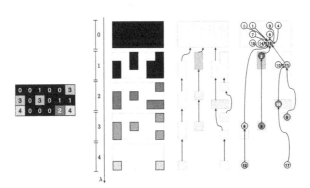

Fig. 1. From left to right: an example of a gray-level image; its upper level sets; its component tree with the shape of each node indicating its respective CC; and the max-tree, with representative pixels marked in bold and the shape of each node showing the reconstruction of its CC.

An extension of component trees can be obtained by considering multiple neighborhoods. Let $\mathbb{A}$ be a sequence of n set of neighboring pixels $(\mathcal{A}_1, \ldots, \mathcal{A}_n)$. If these neighborhoods are increasing (i.e., if $1 \leq i < j \leq n$, then $\mathcal{A}_i \subseteq \mathcal{A}_j$), then CCs of component trees built with consecutive neighborhoods also satisfy the inclusion relation. If CCs of all these trees are used as nodes and the inclusion relations of both level set and neighborhoods are used as arcs, a directed acyclic graph (DAG) is obtained. This DAG is referred as the *component-hypertree*. An example is given in Fig. 2 (left).

More formally, supposing an image f and a sequence $\mathbb{A} = (\mathcal{A}_1, \ldots, \mathcal{A}_n)$ are given, the *complete component-hypertree* of f using $\mathbb{A}$, $HT(f, \mathbb{A})$, is defined as the graph $HT(f, \mathbb{A}) = (N(f, \mathbb{A}), E(f, \mathbb{A}))$, where

$$N(f, \mathbb{A}) = \{(\mathcal{C}, \lambda, i) : \mathcal{C} \in CC(X_\lambda(f), \mathcal{A}_i), \lambda \in \mathbb{K}, 1 \leq i \leq n\} \qquad (1)$$

$$E(f, \mathbb{A}) = \{((\mathcal{C}, \alpha, i), (\mathcal{C}', \beta, j)) \in N(f, \mathbb{A}) \times N(f, \mathbb{A}) : \mathcal{C} \subseteq \mathcal{C}' \text{ and either} \\ (\alpha = \beta + 1 \text{ and } i = j) \text{ or } (\alpha = \beta \text{ and } i = j - 1)\}. \qquad (2)$$

An example of a component-hypertree is given in Fig. 2 (left).

In terms of nomenclature, given an arc $e = (N, N') \in E(f, \mathbb{A})$ with N and N' having the same gray-level but different neighborhood indices, we say N is a *partial node* of N' and N' is a *composite node* of N. In the case they differ only by gray-level, then the nomenclature of component trees is used: N is a *child* of N' and N' is a *parent* of N. If there is a path from a node N to N', we say that N is a *subnode* of N' and the set of all subnodes of N' form the *sub-DAG* of N'.

In terms of storage, similarly to max-trees, one can optimize how a hypertree is represented by storing pixels only in the smallest CC that contain them. In this sense, a good strategy is to store these pixels only in the nodes $(\mathcal{C}, \lambda, i)$ such that $i = 1$. All nodes with $i > 1$ can be reconstructed by using all pixels in its sub-DAG.

Since some nodes do not store pixels, a single array can no longer be used. In this sense, a good strategy consists of explicitly allocating each node. A node $N = (\mathcal{C}, \lambda, i)$ is represented as a data structure with at least the following information: (a) N.rep, the representative pixel of $\mathcal{C}$; (b) N.level, the gray-level λ; (c) N.adjIndex, the neighborhood index i; (d) N.pixels, the set of stored pixels in N; (e) N.parents, a list of parent nodes of N; (f) N.children, a list of children nodes of N; (g) N.composites, a list of composite nodes of N; (h) N.partials, a list of partial nodes of N; and (i) N.attributes, a list of values of some attributes, which will be explained in more details in the next section.

Another optimization in memory that can be performed is to store nodes representing repeated CCs only once (we refer to the remaining nodes as *compact nodes*). It is also possible to remove most arcs that give redundant information in terms of inclusion relation. A structure that performs such optimization consists of the *minimal component hypertree* [8]. An example of this structure is shown in Fig. 2 (right). Theory and algorithms used to build the minimal hypertree are presented in [8].

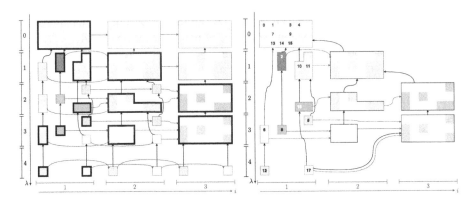

Fig. 2. Left: a complete hypertree, with compact nodes drawn with a thick border. Right: its respective minimal representation, with numbers inside nodes indicating pixels stored in each node.

3   Attributes

3.1   Simple Attributes

Attributes are functions $\kappa : CC(X, \mathbb{A}) \to \mathbb{R}$ that, given a CC $\mathcal{C}$, return a numeric value that gives quantitative information about the shape of $\mathcal{C}$. For a given attribute *att*, we denote its value in $\mathcal{C}$ as $\kappa_{att}(\mathcal{C})$. For example, area of a CC $\mathcal{C}$ is denoted by $\kappa_{area}(\mathcal{C})$. Some examples of attributes and their definitions are given in the list below:

- $\kappa_{area}(\mathcal{C}) = |\mathcal{C}|$, i.e., the number of pixels $p \in \mathcal{C}$;
- $\kappa_{minX}(\mathcal{C})$: the lowest x coordinate of all pixels $p = (x, y) \in \mathcal{C}$;
- $\kappa_{maxX}(\mathcal{C})$: the highest x coordinate of all pixels $p = (x, y) \in \mathcal{C}$;
- $\kappa_{width}(\mathcal{C}) = \kappa_{maxX}(\mathcal{C}) - \kappa_{minX}(\mathcal{C}) + 1$.

There are attributes that can take advantage of the inherently hierarchical structure of component trees and hypertrees and be efficiently computed. These attributes are known as *incremental attributes*. In this case, given a node N of a max-tree, an incremental attribute can be computed using only pixels stored in N and the attribute values stored in all children of N. For instance, all attributes presented above are incremental and can be computed in max-trees using the formulas below:

- $\kappa_{area}(N) = |N.\mathsf{pixels}| + \displaystyle\sum_{N_C \in N.\mathsf{children}} \kappa_{area}(N_C)$;
- $\kappa_{minX}(N) = \min\{x : p = (x, y) \in N.\mathsf{pixels}, \displaystyle\min_{N_C \in N.\mathsf{children}} \kappa_{minX}(N_C)\}$;

For component-hypertrees, analogous strategies can be used for some attributes. Given a node N with $N.\mathsf{adjIndex} > 1$ (if $N.\mathsf{adjIndex} = 1$, the same strategies used in max-tree can be used, since $N.\mathsf{partials} = \emptyset$), for attributes such as $minX$, $maxX$ and $width$, these attributes can be efficiently computed as follows:

$$- \; \kappa_{minX}(N) = \min_{N' \in (N.\text{children} \cup N.\text{partials})} \{\kappa_{minX}(N')\}.$$

For other attributes, a better strategy consists of taking into consideration all the partial nodes. Since every node is a cluster of nodes obtained from a smaller neighborhood, attributes of nodes of hypertrees can be efficiently computed by only looking at their partial nodes. For example, assuming that all partial nodes of N are stored in N.partials, an alternative way of incrementally computing the values of some attributes for nodes N of component-hypertrees with neighborhood index higher than 1 is shown below:

$$- \; \kappa_{area}(N) = \sum_{N_P \in N.\text{partials}} \kappa_{area}(N_P);$$
$$- \; \kappa_{minX}(N) = \min_{N_P \in N.\text{partials}} \kappa_{minX}(N_P);$$

Due to the storage optimization of minimal hypertrees, some of these partial nodes N_P may not be in N.partials, in case the arc that links N_P to N is redundant, i.e., there is another longer path from N_P to N that gives the same information. However, these redundant arcs can be inferred from the structure of the minimal hypertree: given a node N and its composite N_C in N.composites, missing composite arcs from N to ancestors of N_C can be found using Algorithm 1.

Algorithm 1. Algorithm that, given two nodes N, N_C with N_C a composite of N, finds nodes N' such that $e = (N, N')$ is a missing redundant composite arc in a minimal hypertree.

1: **procedure** GETCOMPOSITES(N, N_C)
2:    $composites \leftarrow \emptyset$;
3:    $k \leftarrow N_P$.level, where $N_P \in N$.parents is the parent of N with highest N_P.adjIndex satisfying N_P.adjIndex $< N_C$.adjIndex;
4:    $N' \leftarrow$ parent of N_C with highest neighborhood index $\leq N_C$.adjIndex;
5:    **while** N'.level $> k$ **do**
6:        $composites \leftarrow composites \cup \{N'\}$;
7:        $N' \leftarrow$ parent of N' with highest neighborhood index $\leq N_C$.adjIndex;
8:    **return** $composites$;

An example illustrating Algorithm 1 is given in Fig. 3. The idea of the algorithm is straightforward: for all gray-levels between N.level and $k + 1$ (where k is obtained at Line 3), there are nodes representing the same CC of N. In the complete representation, these repeated nodes would point to nodes on the $composites$ set, and these are the missing arcs.

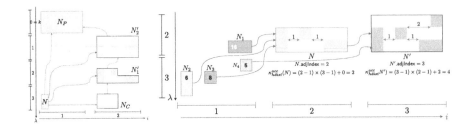

Fig. 3. Left: illustration of Algorithm 1. Nodes N'_1 and N'_2 will be added to the *composites* set, indicating there are missing redundant composite arcs from N to N'_1 and N'_2 (blue dashed arcs). Right: computation of horizontal distance. The computed value is equivalent to the sum of all horizontal spacing (in red) between the isolated components. (Color figure online)

After using Algorithm 1, for all pairs of nodes (N, N_C), all the missing redundant arcs are found. In this way, we can consider running Algorithm 1 as preprocessing step for attribute computation. From now on, we assume that N.partials contains all the partials nodes of N.

It is also worth noting that, since incremental computation depends on the attribute value of subnodes, the order of execution matters. Thus, given a node N, it is important to update the attributes of subnodes of N before computing the value of the attribute in N.

3.2   Statistical Measures of Clusters of Nodes

Since nodes with index $i > 1$ are clusters of nodes of previous trees, we can compute additional statistics that measure how the values of an attribute vary in partial nodes. Some of these statistical measurements, such as average and variance, can also be incrementally computed by storing some additional information: let N be a node of a component-hypertree with N.adjIndex > 1 and att be any attribute. Then, we will define the accumulated and squared values of the attribute att in N as, respectively:

$$\kappa_{att}^{acc}(N) = \sum_{N_P \in N.\text{partials}} \kappa_{att}(N_P); \tag{3}$$

$$\kappa_{att}^{sq}(N) = \sum_{N_P \in N.\text{partials}} \left(\kappa_{att}(N_P)^2 \right). \tag{4}$$

Then, we can compute the mean and variance of the values of κ_{att} in the partial nodes of N using the following statistical properties:

$$\overline{\kappa}_{att}(N) = \frac{\kappa_{att}^{acc}(N)}{|N.\text{partials}|}; \tag{5}$$

$$\kappa_{att}^{var}(N) = \frac{\kappa_{att}^{sq}(N)}{|N.\text{partials}|} - \overline{\kappa}_{att}(N)^2. \tag{6}$$

It is important to note that we can change the denominator of these function to compute mean and variance according to some other rules. For example, given a node N, instead of computing the mean of an attribute in N according to their direct partial nodes, we can compute the mean according to the number of CCs that are parts of N in any given neighborhood index. For example, in Fig. 3 (right), the node N' has two direct partial nodes, but it is composed of 4 nodes with neighborhood index 1 (N_1 to N_4).

In particular, the number of CCs with neighborhood index 1 will be referred as $partials_1$. It can be computed as an incremental attribute: given a node N of a component-hypertree, we can efficiently compute $\kappa_{partials_1}(N)$ according to the following rules:

$$\kappa_{partials1}(N) = \begin{cases} 1, & \text{if } N.\text{adjIndex} = 1; \\ \sum_{N_C \in N.\text{partials}} \kappa_{partials_1}(N_C), & \text{otherwise.} \end{cases} \quad (7)$$

3.3   Attributes Between Nodes

For nodes of component-hypertrees that have more than one partial node, we can also compute attributes that give an insight of how nodes are merged or some values quantifying information between merged nodes.

For example, assuming $\mathbb{A} = (\mathcal{A}_1, \ldots, \mathcal{A}_n)$ is defined such that $\mathcal{A}_i = \mathcal{A}(\mathcal{S}_i)$ and $\mathcal{S}_i$ is a $(2i+1) \times 3$ SE for $1 \leq i \leq n$, the neighborhood index will give the size of the horizontal gap between partial nodes. We define this attribute as *horizontal distance*. Analogously, we can obtain the *vertical distance* by changing the SEs to a sequence of $3 \times (2i+1)$ rectangles; and the *Chebyschev distance* by changing SEs to a sequence of squares with side $2i+1$, for $1 \leq i \leq n$.

All of these distances can be computed as incremental attributes. For example, let N be a node and assume $\mathbb{A} = (\mathcal{A}_1, \ldots, \mathcal{A}_n)$ with $\mathcal{A}_i = (\mathcal{A}(\mathcal{S}_i))$ and $\mathcal{S}_i$ as the $(2i+1) \times 3$ SE, for $1 \leq i \leq n$, we can compute $\kappa_{hdist}^{acc}(N)$, the value of the accumulated horizontal distance in the partial nodes, as follows:

$$\kappa_{hdist}^{acc}(N) = \begin{cases} 0, & \text{if } adj = 1; \\ (adj-1)(|Parts|-1) + \sum_{N_C \in Parts} \kappa_{hdist}^{acc}(N_C), & \text{otherwise;} \end{cases} \quad (8)$$

where $adj = N.\text{adjIndex}$ and $Parts = N.\text{partials}$. An example is provided in Fig. 3 (right).

4   Experiments

4.1   Text Extraction

In this section, we present a simple application to show the usefulness of the computed attributes. Note that it is not our objective to show that this is the best method for text extraction.

Since texts usually consist of groups of letters with consistent height, color and spacing, component-hypertrees are suitable structures for text extraction. Lines of text can be extracted using a sequence $\mathbb{A} = (\mathcal{A}(\mathcal{S}_1), \ldots, \mathcal{A}(\mathcal{S}_n))$ considering: (a) Assuming letters are connected, we choose $\mathcal{S}_1$ as the 3×3 SE. (b) Assuming texts are written horizontally, we choose $\mathcal{S}_i$ as $(2i+1) \times 3$, $1 \leq i \leq n$. (c) We assume texts have letters with similar heights, i.e., they are nodes N such that $normVar_{height}(N) \leq 0.3$, where $normVar_{att}(N) = \frac{\kappa_{att}^{var}(N)}{(\overline{\kappa}_{att}(N))^2}$. This is used to add scale invariance. Variance is computed using $partials_1$, since we want the variance of the letters that are extracted using the first SE; (d) We assume texts have relatively consistent spaces between letters and words, i.e., they are nodes N such that $normVar_{spacing}(N) \leq 0.5$. ($e$) A line of text must contain at least 2 letters, with each letter being at least 8 pixels tall and with a height of at least half of height average of the entire line.

Figure 4 shows some results using the 2013 ICDAR born-digital database [5] which contains images with text from webpages and e-mail attachments.

Fig. 4. Text extraction using our method. At the top, we show original images and, at the bottom, the results. Letters with same color belong to the same node. (Color figure online)

A main difference between this method compared to other studies using connected components [2,14] is that they use an approach that consists of recognizing each letter and grouping them to form the text. Differently, our method has a *top-down* approach, i.e., texts are extracted without explicitly recognizing the letters. Naturally, since letters are also stored in the hypertree, a bottom-up approach is still possible. Also, there is no need to use additional image processing techniques to group letters into words, since these words are already stored as nodes of the hypertree.

Finally, since component-hypertrees are inherently multiscale structures, changing thresholds allows extraction of different scales of clusters. An example is shown in Fig. 5.

4.2 Computational Complexity and Time Consumption

In this section, we compare the computational complexity and time consumption of attribute computation of all nodes of a component-hypertree with and without using incremental strategies. For these experiments, 50 images were

SensoMotoric Instruments SensoMotoric M
     0.358   0.125   0

Fig. 5. Extraction of different scales of objects, with their respective $normVar_{spacing}$ values shown below each image. Once a horizontal text is found (left), words can be extracted by restricting the values of $normVar$ (middle). Letters can be obtained by selecting nodes extracted using $\mathcal{A}(\mathcal{S}_1)$, assuming they are 8-connected.

randomly chosen from the ICDAR2017 Robust Reading Challenge Dataset [9]. These images contains scene images with text in multiple languages and have sizes ranging from 0.05 to 12 megapixels (MP). Hypertrees are built using a sequence $\mathbb{A} = (\mathcal{A}_1, \dots, \mathcal{A}_{10})$, with $\mathcal{A}_i := \mathcal{A}(\mathcal{S}_i)$ and $\mathcal{S}_i$ defined as $(2i+1) \times (2i+1)$ SE, for $1 \le i \le 10$.

Asymptotic computational complexity of incremental approach depends on the number of partial nodes of each node, while the non-incremental approach depends on the number of subnodes. When considering individual images, the average number of partial nodes usually varies between 1 and 2 (Fig. 6, blue bars in the left graph), but this amount is not related to the total number of nodes of hypertree, i.e., the average number of partial nodes tend to stay around 1.4 (red line) regardless of the number of nodes of the input image. The number of subnodes also varies considerably according to the input image (blue bars on the right in Fig. 6), but, on average, the number of subnodes tends to slightly increase as the number of nodes of hypertree increases (red line). In particular, it reaches about 920 for images with almost 2 million nodes. This means that the incremental attribute computation is linear in the number of nodes, while the non-incremental approach grows faster and is multiplied by a bigger constant.

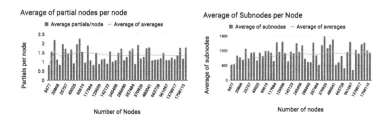

Fig. 6. Left: each bar shows the average number of partial nodes per node for hypertrees with varying number of nodes and the red line shows the average of the values in the blue bars. Right: each bar represents the average number of subnodes per node and the red line shows the average of these numbers. (Color figure online)

To validate these observation, we computed, for the selected images, the area attribute for all nodes of the hypertree using both strategies. For the incremental approach, we compute the area of the nodes of $\mathcal{A}_1$ using the same strategies for max-trees and then use the ideas presented in Sect. 3 for the nodes of other neighborhood, while in the non-incremental approach, each node is reconstructed entirely by visiting all nodes in its sub-DAG, and the area is computed by accumulating the number of pixels stored in the visited nodes.

These experiments are written in Java and executed in a notebook with i7 6700HQ processor and 12 GB of RAM. Time spent for hypertree construction and allocation of nodes was discarded but, for reference, images with about 0.1 MP usually takes less than half a second to be built, while big images (about 12 MP) takes from 30 to about 55 s to be constructed using the algorithm in [7]. The results for attribute computation are depicted in Fig. 7. As we can see, incremental approach has a huge gain compared to the non-incremental one: about 100 times faster for simple images and becoming about 300 times for more complex images with a higher amount of nodes.

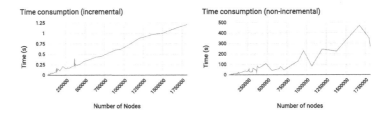

Fig. 7. Time consumption for the incremental approach (left) compared to the non-incremental approach (right).

5   Conclusion

In this paper, an efficient way of computing incremental attributes in minimal component-hypertrees was presented. Asymptotic complexity analysis showed that the proposed approach is linear in the number of nodes of the hypertree and that the asymptotic complexity grows slower than a naive approach that reconstructs the nodes. Moreover, a simple word extraction application was developed to show that the computed attributes may be useful in problems that deal with clustering of small objects.

Acklowledgements. This study was financed in part by the CAPES - Coordenação de Aperfeiçoamento de Pessoal de Nível Superior (Finance Code 001); FAPESP - Fundação de Amparo a Pesquisa do Estado de São Paulo (Proc. 2015/01587-0 and 2018/15652-7); CNPq - Conselho Nacional de Desenvolvimento Científico e Tecnológico (Proc. 428720/2018-8).

References

1. Braga-Neto, U., Goutsias, J.: Connectivity on complete lattices: new results. Comput. Vis. Image Underst. **85**(1), 22–53 (2002)
2. Epshtein, B., Ofek, E., Wexler, Y.: Detecting text in natural scenes with stroke width transform. In: 2010 IEEE Computer Society Conference on Computer Vision and Pattern Recognition, pp. 2963–2970 (2010)

3. Gray, S.B.: Local properties of binary images in two dimensions. IEEE Trans. Comput. **C–20**(5), 551–561 (1971)
4. Jones, R.: Connected filtering and segmentation using component trees. Comput. Vis. Image Underst. **75**(3), 215–228 (1999)
5. Karatzas, D., Mestre, S.R., Mas, J., Nourbakhsh, F., Roy, P.P.: ICDAR 2011 robust reading competition - challenge 1: reading text in born-digital images (web and email). In: 2011 International Conference on Document Analysis and Recognition, pp. 1485–1490 (2011)
6. Kiwanuka, F.N., Ouzounis, G.K., Wilkinson, M.H.F.: Surface-area-based attribute filtering in 3D. In: Wilkinson, M.H.F., Roerdink, J.B.T.M. (eds.) ISMM 2009. LNCS, vol. 5720, pp. 70–81. Springer, Heidelberg (2009). https://doi.org/10.1007/978-3-642-03613-2_7
7. Morimitsu, A., Alves, W.A.L., Hashimoto, R.F.: Incremental and efficient computation of families of component trees. In: Benediktsson, J.A., Chanussot, J., Najman, L., Talbot, H. (eds.) ISMM 2015. LNCS, vol. 9082, pp. 681–692. Springer, Cham (2015). https://doi.org/10.1007/978-3-319-18720-4_57
8. Morimitsu, A., Alves, W.A.L., Silva, D.J., Gobber, C.F., Hashimoto, R.F.: Minimal component-hypertrees. In: Couprie, M., Cousty, J., Kenmochi, Y., Mustafa, N. (eds.) DGCI 2019. LNCS, vol. 11414, pp. 276–287. Springer, Cham (2019). https://doi.org/10.1007/978-3-030-14085-4_22
9. Nayef, N., et al.: ICDAR 2017 robust reading challenge on multi-lingual scene text detection and script identification - RRC-MLT. In: 2017 14th IAPR International Conference on Document Analysis and Recognition (ICDAR), vol. 01, pp. 1454–1459 (2017)
10. Neumann, L., Matas, J.: Real-time scene text localization and recognition. In: 2012 IEEE Conference on Computer Vision and Pattern Recognition, pp. 3538–3545 (2012)
11. Ouzounis, G.K., Wilkinson, M.H.F.: Mask-based second-generation connectivity and attribute filters. IEEE Trans. Pattern Anal. Mach. Intell. **29**(6), 990–1004 (2007)
12. Passat, N., Naegel, B.: Component-hypertrees for image segmentation. In: Soille, P., Pesaresi, M., Ouzounis, G.K. (eds.) ISMM 2011. LNCS, vol. 6671, pp. 284–295. Springer, Heidelberg (2011). https://doi.org/10.1007/978-3-642-21569-8_25
13. Passat, N., Naegel, B., Rousseau, F., Koob, M., Dietemann, J.L.: Interactive segmentation based on component-trees. Pattern Recogn. **44**(10), 2539–2554 (2011). semi-Supervised Learning for Visual Content Analysis and Understanding
14. Retornaz, T., Marcotegui, B.: Scene text localization based on the ultimate opening. In: Proceedings of the 8 th International Symposium on Mathematical Morphology, October 10–13, 2007, MCT/INPE, vol. 1, pp. 177–188. Rio de Janeiro (2007)
15. Salembier, P., Oliveras, A., Garrido, L.: Antiextensive connected operators for image and sequence processing. IEEE Trans. Image Process. **7**(4), 555–570 (1998)
16. Serra, J.: Connectivity on complete lattices. J. Math. Imaging Vis. **9**(3), 231–251 (1998)
17. Silva, D.J., Alves, W.A.L., Morimitsu, A., Hashimoto, R.F.: Efficient incremental computation of attributes based on locally countable patterns in component trees. In: 2016 IEEE International Conference on Image Processing (ICIP), pp. 3738–3742 (2016)

Incremental Bit-Quads Count in Tree of Shapes

Dennis José da Silva$^{1(\boxtimes)}$, Wonder Alexandre Luz Alves^{2},
Alexandre Morimitsu^{1}, Charles Ferreira Gobber^{2},
and Ronaldo Fumio Hashimoto$^{1(\boxtimes)}$

^{1} Institute of Mathematics and Statistics,
Instituto de Matemática e Estatística, São Paulo, Brazil
`alexandre.morimitsu@usp.br`, `{dennis,ronaldo}@ime.usp.br`
^{2} Universidade Nove de Julho, São Paulo, Brazil

Abstract. Bit-quads are 2×2 binary patterns which are counted within a binary image and can be used to compute attributes. Based on previous works which proposed an efficient algorithm to count bit-quads in component trees, in this paper, we discuss how we can count these patterns in tree of shapes by presenting two approaches. In the first one, we show how counting quads in component trees can be used to count them in tree of shapes by using the depth of the node as the value of pixels in a larger and interpolated image representation (used in an algorithm for constructing tree of shapes). In the second approach, we propose a novel algorithm which uses this larger image representation, but, the resulting quad counts are for the input image. In this way, our approach gives exactly the counts for the original image. We also provide experimental results showing that our algorithm is much faster than the non-incremental naive approach.

Keywords: Tree of shapes · Attributes · Bit-quads

1 Introduction

In Image Processing and Mathematical Morphology (MM) applications, a crucial step consists of *feature extraction* of the underlying structures of an image. These features, or attributes, can be used to develop algorithms that can obtain information about the contents of the input image.

One early work related to this problem consists of the study done by Gray [1]. Gray has shown that some attributes, such as area, perimeter and Euler number, can be extracted using local information. This local information is extracted by counting occurrences of 2×2 binary patterns (called bit-quads) in the input image. These bit-quads can be combined to compute attributes that describe the whole image. However, Gray's approach is restricted only to binary images.

Afterwards, many richer ways of representing gray-level images have been introduced. In particular, over the last two decades, hierarchical representations

© Springer Nature Switzerland AG 2019
B. Burgeth et al. (Eds.): ISMM 2019, LNCS 11564, pp. 162–173, 2019.
https://doi.org/10.1007/978-3-030-20867-7_13

have become popular. Such representations include the so-called morphological trees, such as component trees [2] and tree of shapes [3]. In these trees, each node is a connected component or a shape of the input image. Thus, the extraction of attributes for each node can be used to perform connected filtering [2,4–6].

Based on these ideas, Climent and Oliveira [7] adapted Gray's bit-quads to 2×2 patterns in order to incrementally compute the number of holes of each node in component trees. However, since there is no bijection between some of Gray's and Climent and Oliveira's patterns, computation of area and perimeter using this approach is not possible. To fill this gap, Silva et al. [8] extended Climent and Oliveira's pattern sets so that all Gray's bit-quads can be counted, allowing us to incrementally compute, in every node of component trees, all possible attributes that can be obtained from Gray's bit-quads.

Despite the known incremental algorithm to count bit-quads in component trees, there is no such algorithm to be used in tree of shapes. In this paper, we discuss how we can use the algorithm proposed in [8] and the enlarged image representation given in [9–11] to incrementally count bit-quads in all nodes of tree of shapes. Note that the image representation given in [9] enlarges considerably the input image, consequently its tree of shapes is different (as well as the bit-quad counts). Therefore, we approach this issue by proposing a novel method which uses the image representation and tree of shapes built with the algorithm in [11] to incrementally count bit-quads in tree of shapes considering only pixels of the input image. In this way, our approach gives exactly the counts for the original image. In addition, we also present experimental results showing that our algorithm is much faster than the non-incremental naive approach.

This paper is organized as follows: in Sect. 2, we recall some concepts on morphological trees and tree of shapes building methods. Then, in Sect. 3, we propose methods to incrementally count bit-quads in tree of shapes with both enlarged shapes (with additional pixels from the tree building algorithm [11]) and original shapes (only pixels from the input image). Lastly, we present an example of connected filtering using our proposed algorithm and its execution times in Sect. 4 and, finally, conclusions in Sect. 5.

2   Background

In this section, we shortly recall the definition and properties of the tree of shapes and its relation with max-tree and min-tree. We also discuss the interpolation and immersion steps of the tree of shapes building algorithm proposed in [11].

2.1   Definition and Properties of the Tree of Shapes

In the following, we consider images as mappings from a regular rectangular grid $\mathcal{D} \subset \mathbb{Z}^2$ to a discrete set of $k \geq 2$ integers $\mathbb{K} = \{0, 1, \ldots, k-1\}$ (precisely, $f : \mathcal{D} \to \mathbb{K}$). These mappings can be decomposed into *lower* (*strict*) and *upper* (*large*) level sets, i.e., for any $\lambda \in \mathbb{K}$, $\mathcal{X}_{\downarrow}^{\lambda}(f) = \{p \in \mathcal{D} : f(p) < \lambda\}$ and $\mathcal{X}_{\lambda}^{\uparrow}(f) = \{p \in \mathcal{D} : f(p) \geq \lambda\}$. One important property of these level sets is that they are nested, i.e., $\mathcal{X}_{\downarrow}^{0}(f) \subseteq \mathcal{X}_{\downarrow}^{1}(f) \subseteq \ldots \subseteq \mathcal{X}_{\downarrow}^{k-1}(f)$ and $\mathcal{X}_{k-1}^{\uparrow}(f) \subseteq \mathcal{X}_{k-2}^{\uparrow}(f) \subseteq \ldots \subseteq \mathcal{X}_{0}^{\uparrow}(f)$. From all these level sets, two other sets $\mathcal{L}(f)$ and $\mathcal{U}(f)$ composed of the connected components (CCs) of, respectively, for all $\lambda \in \mathbb{K}$, $\mathcal{X}^\lambda_\downarrow(f)$ and $\mathcal{X}^\uparrow_\lambda(f)$ are defined, i.e., $\mathcal{L}(f) = \{C \in \mathcal{CC}(\mathcal{X}^\lambda_\downarrow(f)) : \lambda \in \mathbb{K}\}$ and $\mathcal{U}(f) = \{C \in \mathcal{CC}(\mathcal{X}^\uparrow_\lambda(f)) : \lambda \in \mathbb{K}\}$, where $\mathcal{CC}(\mathcal{X})$ denotes the sets of either 4 or 8 CCs of $\mathcal{X}$, respectively. Using the usual set inclusion relation, the ordered pairs $(\mathcal{L}(f), \subseteq)$ and $(\mathcal{U}(f), \subseteq)$ define two dual component trees [12,13]. It is possible to combine them into a single tree in order to obtain the so-called *tree of shapes*. In fact, if $C \in \mathcal{L}(f)$, then CCs of $\mathcal{D} \backslash C$ are in $\mathcal{U}(f)$ (or vice versa, $C \in \mathcal{U}(f) \Rightarrow \mathcal{CC}(\mathcal{D} \backslash C) \subseteq \mathcal{L}(f)$). Then, let $\mathcal{P}(\mathcal{D})$ denote the *powerset* of $\mathcal{D}$ and let $sat : \mathcal{P}(\mathcal{D}) \rightarrow \mathcal{P}(\mathcal{D})$ be the operator of saturation [14] (or filling holes) and $\mathcal{SAT}(f) = \{sat(C) : C \in \mathcal{L}(f) \cup \mathcal{U}(f)\}$ be the family of CCs of upper and lower level sets with filled holes. The elements of $\mathcal{SAT}(f)$, called *shapes*, are nested by the inclusion relation and thus the pair $(\mathcal{SAT}(f), \subseteq)$, also defines a tree called *tree of shapes* [13,14].

Tree of shapes and component trees are complete representations of images and they can be stored using compact and non-redundant data structures. In the case of component trees $(\mathcal{L}(f), \subseteq)$ and $(\mathcal{U}(f), \subseteq)$, these structures are known, respectively, as min-tree and max-tree [2,15]. In these structures, each pixel $p \in \mathcal{D}$ is associated only (or belongs) to the smallest CC of the tree containing it; and through parenthood relationship, it is also associated (or also belongs) to all its ancestor nodes. Then, we denote by $\mathcal{SC}(\mathcal{T}, p)$ the smallest CC containing p in tree $\mathcal{T}$. Similarly, we say $p \in \mathcal{D}$ is a *compact node pixel* (CNP) of a given CC $C \in \mathcal{T}$ if and only if C is the smallest CC containing p, i.e., $C = \mathcal{SC}(\mathcal{T}, p)$. These compact and non-redundant representations for component trees and tree of shapes are called generically as *morphological tree*. Figure 1 shows examples of min-tree, max-tree, and the tree of shapes, where CNPs are highlighted in red.

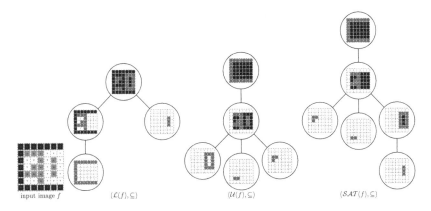

Fig. 1. Min-tree, max-tree and tree of shapes as compact representations of $(\mathcal{L}(f), \subseteq)$ and $(\mathcal{U}(f), \subseteq)$, and $(\mathcal{SAT}(f), \subseteq)$, respectively. CNPs are highlighted in red. (Color figure online)

2.2   Immersion, Interpolation and Tree of Shapes Construction

Carlinet et al. [11] have proposed an efficient algorithm for building tree of shapes based on the theory proposed in [9] and in a previous algorithm presented in [10]. Their algorithm can be summarized in the following four steps:

1. **Interpolation:** For each consecutive pair of pixel columns, a new pixel column is inserted between them; and for each consecutive pair of pixel rows, a new pixel row is inserted between them. The value of each inserted pixel is computed by applying an operator (max, min, or median) on the values of its neighboring pixels from the original image. This step defines the background/foreground connectivity and grants the tree of shapes uniqueness.
2. **Immersion:** Then, the resulting image is immersed in the Khalimsky grid and represented by an interval-valued map [11], where new intermediate pixels are inserted and represent level lines.
3. **Sort and topological order:** The generated interval-valued map is used to compute: (i) an image, where each pixel value corresponds to the depth of the smallest CC that contains it in the tree of shapes (we call it *depth image*), and (ii) a lookup table that gives the pixel order used by the algorithm to build the tree of shapes.
4. **Tree building:** The algorithm builds the max-tree of the depth image using any known efficient method [16]. The computed pixel order can be used to avoid sorting in the max-tree building algorithm.

It is worth noting that the interpolation step doubles the size of the input image domain, and thereby, nodes of tree of shapes built with this algorithm represent shapes larger than the ones of the input image. However, shapes with only pixels of the input image can be easily computed by taking only pixels with even x- and y-coordinates on the larger shapes.

2.3   Quads

Gray proposed a method to compute attributes such as perimeter, number of Euler, and area of binary images by locally counting 2×2 binary patterns named *bit-quads* [1]. These bit-quads are organized in sets based on the number of foreground pixels and their shapes. Silva et al. [8] redefined theses bit-quads by inserting an origin o in each bit-quad and called it *quads* as follows:

$$Q_1 = \left\{ \begin{smallmatrix} o\ 0 \\ 0\ 0 \end{smallmatrix}, \begin{smallmatrix} 0\ o \\ 0\ 0 \end{smallmatrix}, \begin{smallmatrix} 0\ 0 \\ o\ 0 \end{smallmatrix}, \begin{smallmatrix} 0\ 0 \\ 0\ o \end{smallmatrix} \right\}, Q_2 = \left\{ \begin{smallmatrix} o\ 1 \\ 0\ 0 \end{smallmatrix}, \begin{smallmatrix} 0\ o \\ 0\ 1 \end{smallmatrix}, \begin{smallmatrix} 0\ 0 \\ 1\ o \end{smallmatrix}, \begin{smallmatrix} 1\ 0 \\ o\ 0 \end{smallmatrix} \right\},$$

$$Q_D = \left\{ \begin{smallmatrix} 0\ 0 \\ 0\ 1 \end{smallmatrix}, \begin{smallmatrix} 0\ 1 \\ o\ 0 \end{smallmatrix} \right\}, Q_3 = \left\{ \begin{smallmatrix} 1\ 1 \\ o\ 0 \end{smallmatrix}, \begin{smallmatrix} 0\ o \\ 1\ 1 \end{smallmatrix}, \begin{smallmatrix} o\ 0 \\ 1\ 1 \end{smallmatrix}, \begin{smallmatrix} 1\ 1 \\ o\ 0 \end{smallmatrix} \right\}, Q_4 = \left\{ \begin{smallmatrix} o\ 1 \\ 1\ 1 \end{smallmatrix} \right\}.$$

Attributes are computed by counting quads for each set within a binary image. This count is performed by translating each quad for each foreground pixel of the input image and checking whether the foreground and background pixels match the foreground and background of the quad. Thus, if X is a binary image and $n_{Q_i}(X)$ is the number of quads in Q_i (with $i \in \{1, 2, 3, 4, D\}$), then one may compute area A, perimeter P, and Euler number E_4 for 4-connectivity and E_8 for 8-connectivity:

$$A(X) = (n_{Q_1}(X) + 2n_{Q_2}(X) + 2n_{Q_D}(X) + 3n_{Q_3}(X) + 4n_{Q_4}(X)) / 4,$$
$$P(X) = n_{Q_1}(X) + n_{Q_2}(X) + 2n_{Q_D}(X) + n_{Q_3}(X),$$
$$E_8(X) = (n_{Q_1}(X) - n_{Q_3}(X) - 2n_{Q_D}(X)) / 4,$$
$$E_4(X) = (n_{Q_1}(X) - n_{Q_3}(X) + 2n_{Q_D}(X)) / 4.$$

Gray also presented continuous approximation of area $\hat{A}$ and perimeter $\hat{P}$:

$$\hat{A}(X) = ((n_{Q_1}(X)/2) + n_{Q_2}(X) + n_{Q_D}(X) + ((7/2)n_{Q_3}(X)) + 4n + Q_4(X)) / 4,$$
$$\hat{P}(X) = n_{Q_2}(X) + (n_{Q_1}(X) + n_{Q_3}(X)) / \sqrt{2}.$$

In one hand, quads can be used to efficiently compute attributes in binary images, in the other hand, they may take very long to count in morphological trees due to repeated counts on the same region of the nested CCs or shapes. Thus, given a morphological tree node N, one may try to reuse the count of its descendant nodes to count the quads in N, but, some quads which were counted in the descendant nodes may not be present in N. Figure 2 shows a case which the counted quad can be reused and another case which it cannot.

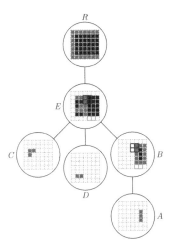

Fig. 2. Tree of shapes with examples of quads within nodes which can and cannot be reused in its ancestor. Quad in Q_2 highlighted in blue at node B was transformed into a quad in Q_4 at node E, therefore, it cannot be reused. Quad in Q_2 highlighted in green at node B is not changed at node E, therefore, it can be reused. (Color figure online)

When a quad in Q_i is counted at a node but it is changed into another quad in Q_j, for $j \neq i$, at its ancestor node due to background pixel changing into foreground pixel (blue quad in Fig. 2), we say quad Q_i is transformed into quad Q_j and denote it as Q_{iT}. Using these ideas and the ones in Climent and Oliveira's work [7], Silva et al. [8] proposed a method which incrementally counts quads in max-trees and min-trees. It reuses quad counts of descendant nodes and updates quad transformations.

3   Proposed Method

In this section, we present two approaches to efficiently count quads in tree of shapes. In the first one, we show how algorithm in [8] can be used to count quads on tree of shapes nodes of interpolated images. In the second one, we present a novel algorithm to count quads in tree of shapes but considering only pixels of the input image.

3.1   Counting Bit-Quads in Interpolated Images

First, let $f : \mathcal{D}_f \rightarrow \mathbb{K}_f$ be the input image, $d : \mathcal{D}_d \rightarrow \mathbb{K}_d$ be its depth image, $\mathfrak{S}(f)$ be the tree of shapes of f, and $\mathcal{M}(d)$ be the max-tree of d. As discussed in [9–11], there exists a function τ such that $\tau(\mathcal{M}(d)) = \mathfrak{S}(f)$. This function τ is obtained by removing the interpolated pixels from all nodes of tree $\mathcal{M}(d)$.

Since the state of art algorithm to construct tree of shapes first builds $\mathcal{M}(d)$, there are some applications which, by simply considering the relation between $\mathfrak{S}(f)$ and $\mathcal{M}(d)$, use $\mathcal{M}(d)$ as $\mathfrak{S}(f)$. Using $\mathcal{M}(d)$ as the representation of $\mathfrak{S}(f)$ has the advantage of having fast algorithms to compute attributes, since there are efficient algorithms to compute many different attributes in max-trees found in the literature [2,8,17]. Thus, approximations of bit-quad counts in tree of shapes can be obtained by applying the algorithm published in [8] in $\mathcal{M}(d)$. It is worth noting that, since $|\mathcal{D}_f| < |\mathcal{D}_d|$, the number of quad counts in $\mathcal{M}(d)$ is larger than the ones in $\mathfrak{S}(f)$ and, consequently, extracted attributes from them describe the CCs in $\mathcal{M}(d)$ instead of $\mathfrak{S}(f)$.

3.2   Counting Bit-Quads in Tree of Shapes Using Only Input Image Pixels

Even though using $\mathcal{M}(d)$ instead of $\mathfrak{S}(f)$ has advantages, it enlarges the CCs. In this section, we propose an incremental algorithm that uses the image d and the tree $\mathcal{M}(d)$ to count quads in $\mathfrak{S}(f)$.

Before presenting the algorithm, we define useful concepts for our proposed algorithm. Given a pixel $p \in \mathbb{Z}^2$, we call the set $W_p = \{p + r : r \in \{0,1,2\}^2\}$ as the *window of size* 3×3 at p (or *window* at p - note that p is at upper-left corner of the window). We also use the counting map $M(N, Q_i) \rightarrow \mathbb{N}$, where $N \in \mathcal{M}(d)$ and $i \in \{1, 2, 3, 4, D, 1T, 2T, 3T, DT\}$, which gives the number of quads Q_i of node N. Figure 3 graphically shows a depth image, a window of the depth image, and a data structure for the counting map.

Algorithm 1 presents our method to count all quads in $\mathfrak{S}(f)$. It starts initializing the counting map M with zeros and then it calls two functions: (*i*) *Pre-Count-Quads* (see Algorithms 2 and 3); (*ii*) *Update-Quads-Counting*

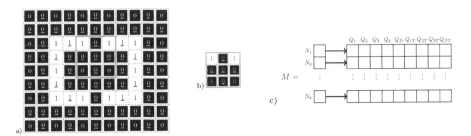

Fig. 3. (a) depth image, (b) window at pixel $p = (4, 2)$ of the depth image, (c) possible data structure for counting map, where $N_i \in \mathcal{M}(d)$, for $i = 1, \ldots, k$. The interpolated pixels are underlined.

Algorithm 1. Algorithm for counting all quads in $\mathfrak{S}(f)$

---

1 **Function** *Count-Quads*(f, d, $\mathcal{M}(d)$, $\underline{M}$)
2   **foreach** *Node* $N \in \mathcal{M}(d)$ **do**
3       $\lfloor$ **foreach** $i \in \{1, 2, 3, 4, D, 1T, 2T, 3T, DT\}$ **do** $M(N, Q_i) \leftarrow 0$;
4   *Pre-Count-Quads*($f, d, \mathcal{M}(d), \underline{M}$);
5   *Update-Quads-Counting*($\underline{M}, \mathcal{M}(d)$);

(see Algorithm 4). At the end of Algorithm 1, the counting map M contains all the counts for all quads and for all nodes of the tree of shapes of the input image. The underlined parameter indicates that it is modified during the function call.

Algorithms 2 and 3 work together. Algorithm 2 transforms each pixel $p \in D_f$ (represented by $T : D_f \rightarrow D_d$) into its corresponding pixel $p' \in D_d$ and calls Algorithm 3. Algorithm 3 first computes the window at p' (Line 2), then it builds the max-tree of this window (Line 3) and, for each node $\hat{N}$ of this max-tree, computes its size considering only pixels of the input image (interpolated pixels are ignored). Afterwards, given pixels $q \in \hat{N}$, $r \in parent(\hat{N})$ and nodes $N = \mathcal{SC}(\mathcal{M}(d), q)$, $P = \mathcal{SC}(\mathcal{M}(d), r)$, it updates the quad counts of Q_i and Q_{iT} for N and P (if node $\hat{N}$ is not root) considering the index i as the size of node $\hat{N}$ (computed in previous step) and its shape (to differentiate Q_2 and Q_D) (Lines 4–8). Note that within the window at pixel $p' \in D_d$, with $p' = T(p)$ and $p \in D_f$, there exists exactly one quad for each possible node $\hat{N} \in \mathcal{M}(W_{p'})$ and if this quad is not $\begin{smallmatrix} 1 & 1 \\ 1 & 1 \end{smallmatrix}$ (Q_4), then $\hat{N}$ is not the root node and this quad will be transformed into another quad in its parent node. In addition, the quads in the windows which intersect D_f but are not contained in D_f must also be counted. We omit these counts in the algorithm for space saving. They can be obtained by generating windows of size 3×1 and 1×3 for every border pixel (pixel which has a 4-connected neighbor outside D_f). Figure 4 shows the operations performed by Algorithm 3.

Algorithm 4 adjusts, for each node $N \in \mathcal{M}(d)$, their quad counts by summing them with its parent counts (Line 4) and removing the counts of transformed quads (Line 5). See Fig. 5 for an example.

Algorithm 2. Initialization of the number of quads.

1 **Function** *Pre-Count-Quads(f, d, $\mathcal{M}(d)$, $\underline{M}$)*
2   **foreach** $p \in \mathcal{D}_f$ **do**
3        $p' \leftarrow T(p)$;
4        $Count(M, p', \mathcal{M}(d))$;

Algorithm 3. Precomputation of the number of quads in a window.

1 **Function** *Count($\underline{M}$, p', $\mathcal{M}(d)$)*
2   Let $W_{p'}$ be the window at p';
3   Build the max-tree $\mathcal{M}(W_{p'})$ of $W_{p'}$;
4   **foreach** $\hat{N} \in \mathcal{M}(W_{p'})$ *from leaves to root* **do**
5        $i \leftarrow S_Z(\hat{N}) \leftarrow$ Size (#pixels from the input image) of $\hat{N}$;
6   Let q be a pixel of $\hat{N}$; $N \leftarrow \mathcal{SC}(\mathcal{M}(d), q)$;
7        $M(N, Q_i)$++ ;   // $S_Z(\hat{N}) = 2 \Rightarrow$ **set** Q_2 or Q_D
8   **if** $\hat{N}$ *is not root* **then**
9   Let r be a pixel of $parent(\hat{N})$; $P \leftarrow \mathcal{SC}(\mathcal{M}(d), r)$;
10           $M(P, Q_{iT})$++;

We state Proposition 1 which guarantees the correctness of our algorithm:

Proposition 1. *After the counting map M has been computed by the Pre-Count-Quads (see Algorithms 2 and 3), the following equality holds for any node $N \in \mathfrak{S}(f)$:*

$$n_{Q_i}(N) = \sum_{C \in \mathfrak{S}(f): C \subseteq N} (M(C, Q_i) - M(C, Q_{iT})).$$

The method asymptotic execution time is determined by the max-tree building at line 3 of Algorithm 3. Since the line builds a max-tree for a 3×3 window, it runs in $\mathcal{O}(1)$ time. It runs for $|D_f|$ windows, therefore, the algorithm runs in $\mathcal{O}(|D_f|)$ which is experimentally (experiments described Sect. 4) showed in Fig. 7.

Algorithm 4. Algorithm to merge quad counts of nodes with their parent.

1 **Function** *Update-Quads-Counting($\underline{M}$, $\mathcal{M}(d)$)*
2   **foreach** $N \in \mathcal{M}(d)$ *from leaves to root* **do**
3   **foreach** $i \in \{1, 2, 3, D\}$ **do**
4           $M(N, Q_i) \leftarrow M(N, Q_i) - M(N, Q_{iT})$;
5           $M(parent(N), Q_i) \leftarrow M(parent(N), Q_i) + M(N, Q_i)$;
6        $M(parent(N), Q_4) \leftarrow M(parent(N), Q_4) + M(N, Q_4)$;

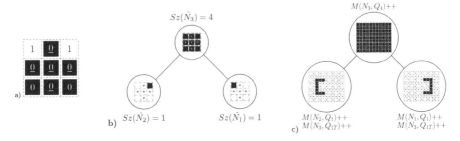

Fig. 4. Example for Algorithm 3. **(a)** An example of window used at Line 2; **(b)** Max-tree and its node sizes (S_z) computed at Line 5; **(c)** update quad counts using node size computed in previous step as the index of quad and quad transformation (Lines 5–8). Nodes using hat represent nodes in the max-tree of the window that is related (by a pixel of the window) to the node (without hat) in $\mathcal{M}(d)$ (Line 6). For example, $\hat{N}_1$ is the node of the max-tree of the window related to $N_1 \in \mathcal{M}(d)$ by a common pixel $q \in \hat{N}_1$, i.e., $N_1 \leftarrow \mathcal{SC}(\mathcal{M}(d), q)$.

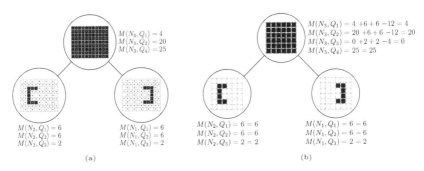

Fig. 5. **(a)** M after execution of Algorithm 2 on $\mathcal{M}(d)$; **(b)** updating of M performed in Algorithm 4 on $\mathfrak{S}(f)$, where green denotes current values, blue denotes values which came from children nodes, and red denotes counts from transformations. (Color figure online)

4 Experimental Analysis

In this section, we experimentally analyze the proposed method by (i) implementing a connected filter using attributes computed by our approach and (ii) comparing its execution time with the one that uses a naive approach to count quads.

4.1 Connected Filter

Based on Léon et al. [17], we developed a connected filter for text segmentation that uses the proposed algorithm to compute the attributes of area A and perimeter P, which in turn are used to compute, for any node $N \in \mathfrak{S}(f)$, its complexity $CX(N)$ and compactness $CP(N)$:

$$CX(N) = \frac{P(N)}{A(N)} \quad \text{and} \quad CP(N) = \frac{A(N)}{P(N)^2}.$$

Then, we use a filter that keeps nodes which hold the following criterion:

$$0.0025 < CP(N) < 0.060 \text{ and } 1.25 < CX(N) < 3.85 \text{ and } 10 < A(N) < 95.$$

This criterion tries to find nodes which represent letters. The area term is used to remove large or small noisy. Figure 6 shows an example of input image and the resulting image computed by this connected filter.

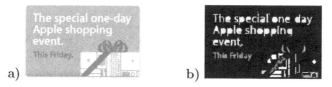

a)   b)

Fig. 6. (a) input image; (b) resulting image computed by the connected filter.

It is worth noting that the resulting image can still have undesired artifacts (see Fig. 6), but, text has been successfully separated from them.

4.2 Execution Time

In order to measure the execution time of the proposed method, we compare it with the naive approach which counts the quads in each binary image reconstructed from nodes of the tree of shapes. The Pylene tree of shapes implementation (https://gitlab.lrde.epita.fr/olena/pylene) was used.

The steps to take the measurements are: (i) read the input image, (ii) compute its tree of shapes, (iii) take the current time t_1 using the C++ standard library (chronos), (iv) execute the method which is being measured (either the proposed or the naive method), (v) take the current time t_2, and (vi) compute the spent time $t = t_2 - t_1$. Some observations about the experiment settings: (1) since Pylene does not store the computed depth image for the tree of shapes construction, we compute it right after the tree of shapes construction and consider this computation in the time analysis; (2) the naive method considers also the time to reconstruct the binary image from the nodes of the tree of shapes; (3) the naive approach visits all pixels of the binary images, but it does not count quads in background pixels.

In our experiment, we use a subset of the dataset available at http://www.lrde.epita.fr/Olena/MaxtreeReview (images larger than 1381×1297 pixels were not considered due to the very high execution time of the non-incremental approach). The images were grouped by their dimensions and the mean execution time for the group was registered in the table. The experiments were executed in a computer with Ubuntu 18.10, Processor Intel i5-430M, and 4 GB of memory.

Table 1 presents the mean execution time for the naive and the proposed methods; and also the the corresponding speed-up. It shows that our proposed method is faster than the non-incremental method in all tested images giving the speed-up ranging from 362 and 11244 (7995.20 in average) for image sizes ranging from 256×256 to 1381×1297. In addition, Fig. 7 gives a plot showing that the execution time of the proposed method is linear.

Table 1. The execution time comparison between the naive and proposed methods.

Dimension	Execution time in milliseconds		Speed-up
	Naive method	Proposed method	
256×256	2.3×10^4	6.3×10^1	362
512×512	3.9×10^5	2.6×10^2	1479
480×640	6.8×10^5	3.2×10^2	2133
498×621	5.4×10^5	3.0×10^2	1821
1381×1297	2.0×10^7	1.8×10^3	11244
Mean	4.3×10^6	5.5×10^2	7995

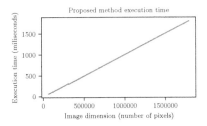

Fig. 7. Experimental execution time of the proposed method.

5   Conclusion

In this paper, we proposed a novel algorithm that uses the depth image [9–11] to incrementally count quads (bit-quads) for all nodes of tree of shapes of the input image. Although these quads can be efficiently counted in the depth image [11] using the algorithm presented in [8], the resulting counts are for the depth image which is larger than the input image. In this way, our approach gives exactly the counts for the original image. In addition, the proposed algorithm is linear on the number of pixels of the input image and it is much faster than a non-incremental naive approach. We also presented an application which implements a self-dual connected operator (used to segment texts [17]) based on attributes computed by using the quad counts provided by our approach. As future work, we intend to compare our approach to other incremental area and perimeter computations as the ones proposed in [18].

Acknowledgements. This study was financed in part by the CNPq - Conselho Nacional de Desenvolvimento Científico e Tecnológico (Proc. 141422/2018-1 and 428720/2018-8); CAPES - Coordenação de Aperfeiçoamento de Pessoal de Nível Superior (Finance Code 001); FAPESP - Fundação de Amparo a Pesquisa do Estado de São Paulo (Proc. 2015/01587-0 and 2018/15652-7).

References

1. Gray, S.B.: Local properties of binary images in two dimensions. IEEE Trans. Comput. **C–20**(5), 551–561 (1971)
2. Salembier, P., Oliveras, A., Garrido, L.: Antiextensive connected operators for image and sequence processing. IEEE Trans. Image Process. **7**(4), 555–570 (1998)

3. Monasse, P., Guichard, F.: Fast computation of a contrast-invariant image representation. IEEE Trans. Image Process. **9**(5), 860–872 (2000)
4. Salembier, P., Wilkinson, M.H.F.: Connected operators. IEEE Signal Process. Mag. **26**(6), 136–157 (2009)
5. Xu, Y., Géraud, T., Najman, L.: Morphological filtering in shape spaces: applications using tree-based image representations. In: International Conference on Pattern Recognition (ICPR), pp. 485–488 (2012)
6. Alves, W.A.L., Hashimoto, R.F.: Ultimate grain filter. In: IEEE International Conference on Image Processing (ICIP), pp. 2953–2957 (2014)
7. Climent, J., Oliveira, L.S.: A new algorithm for number of holes attribute filtering of grey-level images. Pattern Recogn. Lett. **53**, 24–30 (2015)
8. Silva, D.J., Alves, W.A.L., Morimitsu, A., Hashimoto, R.F.: Efficient incremental computation of attributes based on locally countable patterns in component trees. In: IEEE International Conference on Image Processing (ICIP), pp. 3738–3742 (2016)
9. Najman, L., Géraud, T.: Discrete set-valued continuity and interpolation. In: Hendriks, C.L.L., Borgefors, G., Strand, R. (eds.) ISMM 2013. LNCS, vol. 7883, pp. 37–48. Springer, Heidelberg (2013). https://doi.org/10.1007/978-3-642-38294-9_4
10. Géraud, T., Carlinet, E., Crozet, S., Najman, L.: A quasi-linear algorithm to compute the tree of shapes of nD images. In: Hendriks, C.L.L., Borgefors, G., Strand, R. (eds.) ISMM 2013. LNCS, vol. 7883, pp. 98–110. Springer, Heidelberg (2013). https://doi.org/10.1007/978-3-642-38294-9_9
11. Carlinet, E., Crozet, S., Géraud, T.: The tree of shapes turned into a max-tree: a simple and efficient linear algorithm. In: IEEE International Conference on Image Processing (ICIP), pp. 1488–1492 (2018)
12. Alves, W.A.L., Morimitsu, A., Hashimoto, R.F.: Scale-space representation based on levelings through hierarchies of level sets. In: Benediktsson, J.A., Chanussot, J., Najman, L., Talbot, H. (eds.) ISMM 2015. LNCS, vol. 9082, pp. 265–276. Springer, Cham (2015). https://doi.org/10.1007/978-3-319-18720-4_23
13. Caselles, V., Meinhardt, E., Monasse, P.: Constructing the tree of shapes of an image by fusion of the trees of connected components of upper and lower level sets. Positivity **12**(1), 55–73 (2008)
14. Caselles, V., Monasse, P.: Geometric Description of Images as Topographic Maps, 1st edn. Springer, Heidelberg (2009). https://doi.org/10.1007/978-3-642-04611-7
15. Berger, C., Geraud, T., Levillain, R., Widynski, N., Baillard, A., Bertin, E.: Effective component tree computation with application to pattern recognition in astronomical imaging. In: IEEE International Conference on Image Processing, vol. 4, pp. IV-41–IV-44 (2007)
16. Carlinet, E., Géraud, T.: A comparative review of component tree computation algorithms. IEEE Trans. Image Process. **23**(9), 3885–3895 (2014)
17. León, M., Mallo, S., Gasull, A.: A tree structured-based caption text detection approach. In: Fifth IASTED VIIP, pp. 220–225 (2005)
18. Xu, Y., Carlinet, E., Géraud, T., Najman, L.: Efficient computation of attributes and saliency maps on tree-based image representations. In: Benediktsson, J.A., Chanussot, J., Najman, L., Talbot, H. (eds.) ISMM 2015. LNCS, vol. 9082, pp. 693–704. Springer, Cham (2015). https://doi.org/10.1007/978-3-319-18720-4_58

Multivariate Morphology

Matrix Morphology with Extremum Principle

Martin Welk[1(✉)], Michael Breuß[2], and Vivek Sridhar[2]

[1] Institute of Biomedical Image Analysis,
Private University of Health Sciences, Medical Informatics and Technology,
Eduard-Wallnöfer-Zentrum 1, 6060 Hall/Tyrol, Austria
martin.welk@umit.at
[2] Brandenburg University of Technology Cottbus–Senftenberg,
Platz der Deutschen Einheit 1, 03046 Cottbus, Germany
{breuss,sridharvivek95}@b-tu.de

Abstract. The fundamental operations of mathematical morphology are dilation and erosion. In previous works, these operations have been generalised in a discrete setting to work with fields of symmetric matrices, and also corresponding methods based on partial differential equations have been constructed. However, the existing methods for dilation and erosion in the matrix-valued setting are not overall satisfying. By construction they may violate a discrete extremum principle, which means that results may leave the convex hull of the matrices that participate in the computation. This may not be desirable from the theoretical point of view, as the corresponding property is fundamental for discrete and continuous-scale formulations of dilation and erosion in the scalar setting. Moreover, if such a principle could be established in the matrix-valued framework, this would help to make computed solutions more interpretable.

In our paper we address this issue. We show how to construct a method for matrix-valued morphological dilation and erosion that satisfies a discrete extremum principle. We validate the construction by showing experimental results on synthetic data as well as colour images, as the latter can be cast as fields of symmetric matrices.

Keywords: Dilation · Erosion · Matrix valued images · Extremum principle

1 Introduction

Images that take symmetric matrices as values arise in several ways in image acquisition and processing. For example, in diffusion tensor MRI [17] they result from the measurement of second-order diffusion tensors. Fields of structure tensors [13] arise as derived quantity within many image processing methods such as anisotropic diffusion [20] or variational optic flow computation [5]. By a transform introduced in [6,7], see also [8], colour images can be transformed into fields of symmetric matrices.

© Springer Nature Switzerland AG 2019
B. Burgeth et al. (Eds.): ISMM 2019, LNCS 11564, pp. 177–188, 2019.
https://doi.org/10.1007/978-3-030-20867-7_14

Over the last two decades, many image processing methods have been devised for (symmetric) matrix-valued images, see e.g. the edited volumes [15,21,24]. In particular, a framework of matrix-valued morphology has been built up, starting from the seminal work of Burgeth et al. [10,11] in which matrix-valued dilation and erosion operations were introduced. Continuous-scale matrix-valued morphology was established in [9]. Applications to colour images are found in [3,6,7,14].

In their classical formulation morphological dilation and erosion are local filters in which image values from a neighbourhood of each image location are selected by a mask (or structuring element) and aggregated by taking the maximum and minimum, respectively. Since the application of the structuring element does not depend on the range of the image values being selected, the essential step in extending dilation and erosion to matrix-valued images is an appropriate generalisation of the maximum and minimum operations.

The concept of matrix-valued dilation and erosion from [10] relies on the combination of the Loewner ordering [16] as a partial order on symmetric matrices with the concept of total ordering by the trace of occurring matrices. In order to define the supremum of a set of symmetric matrices, one considers the set of all matrices that are greater than or equal to all given matrices w.r.t. the Loewner order, and chooses from this upper bound set the minimal element w.r.t. the trace total order. The criterion for choosing the minimal element from the upper bound set may be varied, leading to variants of the dilation and erosion operations, see [22, Sect. 2.4].

Clearly, the so-defined supremum and infimum generalise the scalar maximum and minimum operations underlying classical grey-value morphology in the sense that also the maximum of grey-values can be understood as the smallest value that is greater or equal to all grey-values from an input set. On the other hand, a sacrifice is made for this: In classical morphology, dilation and erosion and their compositions never extend the range of grey-values of an input set. This is of particular importance when the continuous limit of morphological operations is considered, giving rise to the partial differential equations (PDEs) of continuous-scale morphology [1,4,19]: These PDEs then fulfil an extremum principle. Unfortunately, this is no longer true for the matrix-valued dilation and erosion based on [10]; in general the supremum of a set of matrices will have a strictly greater trace than each of them; likewise for the infimum. We are therefore led to ask whether variants of matrix-valued dilation and erosion can be devised that allow for an extremum principle.

Taking a more principled approach to the scalar operations of classical morphology, we can identify two essential features of the maximum (and analogously for the minimum):

(i) The maximum of given input values is greater or equal to each of them.
(ii) The maximum of given input values is contained in their convex hull.

We remark that the scalar-valued maximum even happens to coincide with one of the input values. On one hand, in the context of multivariate data such a requirement would lead to discontinuous dependence on the input data and often result in values that do not represent the input data adequately. One might compare also the situation for median filtering of multivariate images, see [23], where filters that select only among the input values have turned out too restrictive. On the other hand, individual values of an image normally result from a sampling process, and are just representatives of a larger set. By admitting convex combination (averaging) which occurs in sampling in a natural way anyway, (ii) provides a conservative estimate of the underlying set of values. We are thus convinced that (ii) is the essential property of the maximum in this context.

The matrix dilation from [10] guarantees property (i) at the expense of giving up (ii). The trace criterion or its alternatives serve as a way to minimise the degree of violation of (ii). In turn, the exact fulfilment of (ii) is what underlies the extremum principle.

As for a given set of symmetric matrices, its upper bound set will in general be disjoint from its convex hull, so that (ii) can be enforced only at the expense of tolerating violations of (i). We aim therefore at finding matrix-valued replacements for the scalar maximum/minimum operations that satisfy (ii), while minimising the degree of violation of (i) in a suitable sense.

Our Contribution and Outline of the Paper. The structure of the paper is adapted to the goals formulated above. In Sect. 2 we show how to construct an optimisation procedure based on an interior point method that satisfies property (ii) while minimising the violation of property (i). In doing this, our method is to our best knowledge the first one proposed intentionally with the purpose to fulfil the mentioned essential aim. With the help of experiments on synthetic data sets as well as dilation and erosion of colour images, we validate our proceeding as discussed in Sect. 3. As indicated we conjecture that the results obtained with our method are more intuitive than those obtained with previous methods that violate the extremum principle. We end our paper by a conclusion which indicates the potential of the proposed method for future developments.

2   Matrix Pseudomaximum and Pseudominimum with Extremum Principle

In this section we proceed by giving technical details of the underlying model for decomposing symmetric matrices, see e.g. [23] for related developments in addition to the works mentioned above. After that we give as indicated details on the algorithmic realisation of property (ii).

2.1   Theoretical Background of the Model

For a given matrix $Y \in \text{Sym}(n)$ we define $N(Y)$ as the square sum of its negative eigenvalues,

$$N(Y) := \frac{1}{2} \sum_{j=1}^{n} [\lambda_j(Y)]_{-}^{2} \tag{1}$$

where $\lambda_j(\boldsymbol{Y})$ is the j-th-largest eigenvalue of $\boldsymbol{Y}$ and $[z]_- := \frac{1}{2}(|z| - z)$ for $z \in \mathbb{R}$. This can be seen as a penaliser for the degree of violation of the relation $\boldsymbol{Y} \succeq \boldsymbol{0}$ where $\succeq$ is the Loewner ordering and $\boldsymbol{0} \in \mathrm{Sym}(n)$ the zero matrix.

For a given (multi-) set

$$\mathcal{X} := (\boldsymbol{X}_1, \ldots, \boldsymbol{X}_m), \quad \boldsymbol{X}_i \in \mathrm{Sym}(n) \tag{2}$$

we define the pseudomaximum $\bigvee(\mathcal{X})$ as the matrix from the convex hull $\mathrm{conv}(\mathcal{X})$ of $\mathcal{X}$ for which the total measure of violations of $\boldsymbol{Y} \succeq \boldsymbol{X}_i$ is minimal:

$$\bigvee(\mathcal{X}) := \underset{\boldsymbol{Y} \in \mathrm{conv}(\mathcal{X})}{\mathrm{argmin}} \; E_{\mathcal{X}}(\boldsymbol{Y}), \quad E_{\mathcal{X}}(\boldsymbol{Y}) := \sum_{i=1}^{m} N(\boldsymbol{Y} - \boldsymbol{X}_i). \tag{3}$$

Abbreviating $-\mathcal{X} := (-\boldsymbol{X}_1, \ldots, -\boldsymbol{X}_m)$, the pseudominimum $\bigwedge(\mathcal{X})$ is defined as

$$\bigwedge(\mathcal{X}) := -\bigvee(-\mathcal{X}). \tag{4}$$

2.2   Analysis

If $\boldsymbol{Y} \in \mathrm{Sym}(n)$ has the spectral decomposition

$$\boldsymbol{Y} = \sum_{j=1}^{n} \lambda_j \boldsymbol{w}_j \boldsymbol{w}_j^{\mathrm{T}} \tag{5}$$

with eigenvalues λ_j and unit eigenvectors $\boldsymbol{w}_j$, then the directional derivative of the j-th-largest eigenvalue in direction of a perturbation matrix $\boldsymbol{Z} \in \mathrm{Sym}(n)$ is

$$\left. \frac{\mathrm{d}\lambda_j(\boldsymbol{Y} + \varepsilon\boldsymbol{Z})}{\mathrm{d}\varepsilon} \right|_{\varepsilon=0} = \boldsymbol{w}_j^{\mathrm{T}} \boldsymbol{Z} \boldsymbol{w}_j. \tag{6}$$

From this it follows that the one-sided derivative of $N(\boldsymbol{Y})$ w.r.t. $\boldsymbol{Z}$ is

$$\left. \frac{\mathrm{d}N(\boldsymbol{Y} + \varepsilon\boldsymbol{Z})}{\mathrm{d}\varepsilon} \right|_{\varepsilon=0^+} = \sum_{j \in \mathcal{T}^-} \lambda_j \boldsymbol{w}_j^{\mathrm{T}} \boldsymbol{Z} \boldsymbol{w}_j \tag{7}$$

where $\mathcal{T}^- := \{j \in \{1, \ldots, n\} \mid \lambda_j < 0\}$.

The objective function $E_{\mathcal{X}}$ in (3) is convex. Whereas we have to defer a detailed proof to a future paper, we mention an important observation which is used in the proof: Let $N^* > 0$ be given, and dilate the cone of positive semidefinite matrices from $\mathrm{Sym}(n)$ with the Euclidean ball of radius $\sqrt{N^*}$ as structuring element. This results in a convex set in $\mathrm{Sym}(n)$, the boundary of which (a hypersurface) is the set of all $\boldsymbol{Y}$ that fulfil $N(\boldsymbol{Y}) = N^*$. For example, in $\mathrm{Sym}(2)$ this set is a cone (open toward the direction of increasing trace) with its tip truncated and replaced with a sphere segment. With additional calculations it follows that $N(\boldsymbol{Y})$ is convex. The same is true for $E_{\mathcal{X}}$ which is the sum of translated copies of $N(\boldsymbol{Y})$. By refining the argument, it can be shown that the convexity is even strict within the convex hull of input data.

Therefore (7) can be used to construct a gradient descent algorithm similar to an interior point method [12] for the pseudomaximum: One starts at some location in $\mathrm{conv}(\mathcal{X})$ and continues by update steps within the convex hull as long as $E_{\mathcal{X}}$ can be reduced. As initialisation, one might simply choose the $\boldsymbol{X}_i$ with maximal trace; update steps for $\boldsymbol{Y}$ within the convex hull can be devised as moving towards any $\boldsymbol{X}_i \neq \boldsymbol{Y}$. This can be realised with $\boldsymbol{Z} := \boldsymbol{X}_i - \boldsymbol{Y}$ and a suitable step size which should be chosen small enough so that the sign of no relevant eigenvalue of any $\boldsymbol{Y} - \boldsymbol{X}_i$ changes within the update step.

The pseudomaximum (3) is a weighted average of some Loewner-maximal input matrices, i.e. those which are not Loewner-less or equal to any other input matrix. If there is only one Loewner-maximal input matrix, which is then Loewner-greater or equal to all other input matrices, it is the pseudomaximum (and in this case also the matrix supremum as defined in [10]).

2.3 Exposition on the Algorithm

Given $\mathcal{X}$ as above, our algorithm for finding $\bigvee(\mathcal{X})$ proceeds as follows.

Initialisation. Let $\boldsymbol{Y}_0 := \mathrm{argmax}_{\boldsymbol{Y} \in \mathcal{X}}\, \mathrm{trace}(\boldsymbol{Y})$.
Iteration. For $k = 0, 1, \ldots$:

1. For each $i = 1, \ldots, m$:
   Compute the spectral decomposition of $\boldsymbol{D}_{k,i} := \boldsymbol{Y}_k - \boldsymbol{X}_i$:

$$\boldsymbol{D}_{k,i} = \sum_{r=1}^{n} \lambda_{k,i,r} \boldsymbol{w}_{k,i,r} \boldsymbol{w}_{k,i,r}^{\mathrm{T}} \ . \tag{8}$$

2. Determine the index set

$$\mathcal{T}_k^- := \{(i,r) \mid \lambda_{k,i,r} < 0\} \ . \tag{9}$$

3. For each $j = 1, \ldots, m$:
   Let $\boldsymbol{D}_{k,j}$ be defined as in Step 1. For $(i,r) \in \mathcal{T}_k^-$ let

$$d_{k,j,i,r} := \lambda_{k,i,r} \boldsymbol{w}_{k,i,r}^{\mathrm{T}} \boldsymbol{D}_{k,j} \boldsymbol{w}_{k,i,r} \ . \tag{10}$$

   Let

$$g_{k,j} := - \sum_{(i,r) \in \mathcal{T}_k^-} d_{k,j,i,r} \ . \tag{11}$$

4. Let

$$(j^*(k), g_k) := (\mathrm{argmin}, \min)_{j=1,\ldots,m}\, g_{k,j} \ . \tag{12}$$

5. If $g_k \geq 0$, stop; $\boldsymbol{Y}_k$ is the sought minimiser.
   Otherwise choose a step size τ_k which fulfils

$$2\tau_k d_{k,j^*(k),i,r} \leq |\lambda_{k,i,r}| \quad \text{for all } (i,r) \in \mathcal{T}_k^- , \qquad \tau_k \leq 1 \tag{13}$$

   and let

$$\boldsymbol{Y}_{k+1} := \boldsymbol{Y}_k - \tau_k \boldsymbol{D}_{k,j^*(k)} \ . \tag{14}$$

   Check whether $E(\boldsymbol{Y}_{k+1}) < E(\boldsymbol{Y}_k)$; if this is not the case, choose a smaller value for τ_k and repeat (14).

6. *Numerical stopping criterion:* If $|E(\boldsymbol{Y}_{k+1}) - E(\boldsymbol{Y}_k)|$ is below a predefined threshold, stop; $\boldsymbol{Y}_{k+1}$ is an approximation to the sought minimiser.

3   Experiments

In this section we show the effect of morphological dilation and erosion using the pseudomaximum and pseudominimum of symmetric matrices introduced in Sect. 2. We will shortly denote these operations as X-dilation and X-erosion (X standing for "obeying extremum principle").

For comparison, we use morphological dilation using the matrix supremum from [10] which we will denote as L-dilation (L indicating the strict Loewner order between the supremum and the input data guaranteed by this approach), and two versions of the morphological erosion using matrix infima as defined in [10] and [7]. In [10], the infimum of positive definite matrices is defined as the matrix inverse of the supremum of the matrix inverses of the input matrices; this definition is suitable for positive definite matrix data as it is designed to preserve positive definiteness. In contrast, [7] uses an infimum that is minus the supremum of the sign-inverted input matrices. The latter definition cannot guarantee positive definite results even for positive definite input matrices; it is suitable for matrix data the eigenvalues of which can take either sign. We will denote the first variant as LP-erosion (P standing for "positive definite"), and the second one as LI-erosion (I for "indefinite"). Note that no such distinction is needed for X-erosion because by virtue of the extremum principle (4) can be used also for positive definite matrices.

Experiment 1: Synthetic Data. We start with an experiment on synthetic data. Figure 1 a shows an array of 100 symmetric positive definite 2×2 matrices $\boldsymbol{A}$ depicted by ellipses $\boldsymbol{x}^{\mathrm{T}} \boldsymbol{A}^{-1} \boldsymbol{x} = 1$. For the subsequent morphological operations we use a structuring element containing all pixels with distance ≤ 2 from the centre (as shown for one exemplary pixel in the figure), and reflecting boundary conditions.

Frame b shows the result of L-dilation. Whereas in regions with well-aligned eigensystems (as near the lower boundary of the array) larger values are nicely propagated, the dilation creates matrices exceeding all contributing input matrices when these are not well aligned (as near the top boundary). Frame c shows the result of X-dilation, i.e. obtained using the proposed framework. In regions with well-aligned eigensystems the result is similar to that of L-dilation. In non-aligned regions, still larger eigenvalues are propagated but by the restriction to the convex hull of contributing input matrices no amplification of values is observed. Frames d and e in the bottom row of Fig. 1 show a similar effect for LP-erosion and X-erosion.

Let us note that the amplification of values as observable within the results of L-dilation could be interpreted as a potential trace of instability in the context of a PDE-based formulation.

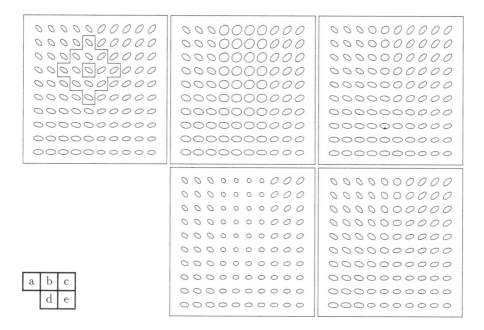

Fig. 1. Synthetic matrix morphology example. **a** Original data set consisting of symmetric positive definite 2×2 matrices $\boldsymbol{A}$ (depicted by ellipses $\boldsymbol{x}^{\mathrm{T}}\boldsymbol{A}^{-1}\boldsymbol{x} = 1$). Thin lines delineate one exemplary pixel and its corresponding structuring element. – **b** L-dilation following [10]. – **c** X-dilation. – **d** LP-erosion following [10]. – **e** X-erosion.

Experiment 2: Colour Imagery. In our second experiment we apply the matrix-valued morphological operations to filter colour images. A correspondence between colour images and fields of Sym(2) matrices was established in [6]. This correspondence is mediated by the HCL (hue–chroma–luminance) colour space. From a given RGB triple (r, g, b), chroma c is obtained by $c := M - m$ where $M := \max\{r, g, b\}$, $m := \min\{r, g, b\}$, luminance l by $l := \frac{1}{2}(M - m)$ and hue h by $D + \frac{1}{6}d/M \bmod 1$ where $d := g - b$, $D := 0$ for $r \geq g, b$, $d := b - r$, $D := 1/3$ for $g \geq r, b$, $d := r - g$, $D := 2/3$ for $b \geq r, b$. A symmetric 2×2 matrix $\boldsymbol{A} = \boldsymbol{A}(r, g, b)$ is then obtained by

$$\boldsymbol{A} := \frac{2l - 1}{\sqrt{2}}\begin{pmatrix} 1 & 0 \\ 0 & 1 \end{pmatrix} + \frac{c}{\sqrt{2}}\begin{pmatrix} -\sin(2\pi h) & \cos(2\pi h) \\ \cos(2\pi h) & \sin(2\pi h) \end{pmatrix} . \tag{15}$$

For further details see [6, 8, 14]. By this transformation a bijection between the RGB colour space and a compact convex set of symmetric matrices (namely, a bi-cone) is established, see Fig. 2.

Following the procedure from [6], we can now wrap matrix-valued dilations and erosions in the RGB–Sym(2) transform (15) and its inverse to obtain dilations and erosions for colour images. In this case, the infimum-based erosion is chosen as LI-erosion because the bi-cone of matrices is symmetric about zero. Our comparison therefore includes L-dilation, LI-erosion, X-dilation and X-erosion.

However, as pointed out in [6], a difficulty arises for L-dilation and LI-erosion as the supremum and infimum of matrices may generate values outside the bi-cone to which RGB colours are mapped. In [6,7], an additional transform is therefore proposed to map back the supremum and infimum into the bi-cone.

For better comparison with X-dilation and X-erosion which do not require such a transform due to their built-in extremum principle, we omit the additional transform also for L-dilation and LI-erosion. Effectively, overshooting matrix values are just projected to the admissible colour range (sacrificing invertibility for these values).

Our colour test image is shown in Fig. 3a; two zoom-ins are shown in Fig. 4a, f. As in the previous experiment, we use the 13-pixel structuring element shown in Fig. 1, and reflecting boundary conditions. Frames b and c of Fig. 3 show the results of L-dilation and X-dilation, respectively, see also the clippings in Fig. 4b, c and g, h.

As expected, both operations extend bright structures. However, at locations where regions of comparable brightness but different colours meet, as in the region of Fig. 4a, the supremum-based dilation generates artificial colours brighter than their surrounds. This is not the case for X-dilation; instead,

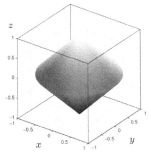

Fig. 2. Color bi-cone, figure adapted from [7]

colours of similar brightness mutually retard their propagation. The difference image in Fig. 4k confirms this effect. In contrast, the clipping in Fig. 4f has fairly similar dilation results, see also the difference image in frame l. Here, adjacent colours are sufficiently similar or differ substantially in brightness such that the brighter colour can be propagated without generating artificial colours or exaggerating brightness.

Analogous observations can be made for erosion, see Fig. 3d, e as well as Fig. 4d, e, i, j (clippings) and l, m (difference images). Again, LI-erosion leads to brightness undershoots and artificial colours (albeit visually less pronounced due to their dark appearance) which are safely avoided by X-erosion.

Experiment 3: Discontinuous Synthetic Colour Images. Making use of the same framework for colour images as in the previous paragraph, we now consider a particularly simple setting for colour images in which the interaction of colours during filtering is easily observed. Modifying the type of our experiments, we will now compare the proposed method with the PDE-based scheme built upon the discretisation of Rouy and Tourin [18], which was used as a building block in [2], and with the lattice-based dilation/erosion procedure from [6,7], this time without any modification. We denote by RT-dilation and RT-erosion, respectively, the PDE-based results.

Let us note that the method of Rouy and Tourin is given by a first-order scheme. On one hand, this means that in the scalar case it is known to introduce blurry artefacts, compare again [2]. On the other hand, this is why it satisfies by construction in the scalar case the extremum principle as discussed in this work.

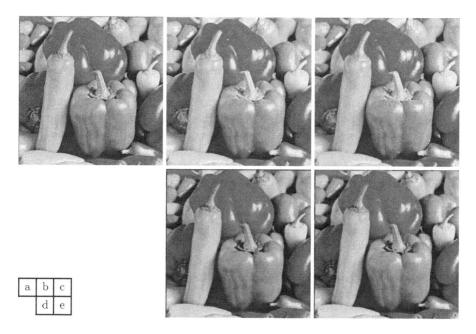

Fig. 3. Colour morphology example. **a** Original colour image *peppers*, 512×512 pixels. – **b** L-dilation similar to [7]. – **c** X-dilation. – **d** LI-erosion similar to [7]. – **e** X-erosion.

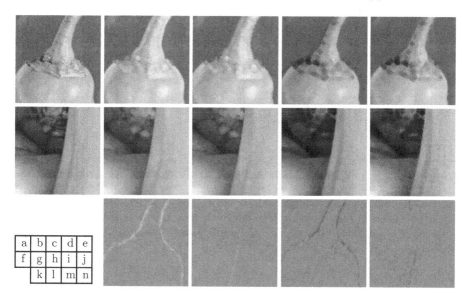

Fig. 4. Colour morphology example, continued. **a, f** Two zoomed details from *peppers* image, 100×100 pixels each. – **b, g** L-dilation. – **c, h** X-dilation. – **d, i** LI-erosion. – **e, j** X-erosion. – **k** Difference of b and c. Middle grey represents zero, brighter colours represent positive differences, darker colours negative differences. – **l** Difference of g and h. – **m** Difference of d and e. – **n** Difference of i and j.

Thus, any potential over-/undershoots that could be observed in results are due to the maximum respectively minimum construction as used in previous work.

In Fig. 5 we show the results of the experiment. Let us note that images are of very small size so that we see in practice a zoom on results. We have done just one step of dilation/erosion with the discrete methods using a small 3×3 structuring element, which corresponds to two steps with the RT method. The results by the original lattice-based method displayed in the second column clearly show the effect of leaving the convex hull of values. The remarkable colour mixture is due to the colours chosen in the experiment and their arrangement in the color bicone. The RT-scheme, the results of which are shown in the second column, displays some blurry artefacts, yet one also observes clearly similar effects of leaving the convex hull of colours as with the previous method. The results obtained by the proposed method as given in the third column are obviously much more intuitive for interpretation. Still we observe some mixing of the colours, which is due to the choice of our objective function within the optimisation: if there exist multiple input values not dominated by others in the underlying partial order, none of which can therefore well represent the input set, a compromise between them is found.

Fig. 5. Colour morphology example with simple images. This test is designed to observe in detail the effect of the extremum principle. **a** Original colour image, 8×8 pixels. – **b** L-dilation of [7]. – **c** RT-dilation. – **d** X-dilation. – **e** LI-erosion of [7]. – **f** RT-erosion. – **g** X-erosion.

4   Summary and Conclusion

We have shown that the use of the convex hull of matrices within the structuring element of matrix-valued dilation/erosion appears to be a suitable generalisation of the corresponding property in the scalar setting. The computational results confirm favourable stability properties in comparison to other methods in the field and encourage further investigation. In this context, also generalisations of the objective function used in our optimisation will be considered. It will also be of interest to which extent algebraic properties of classical scalar-valued dilation and erosion, such as adjunction, can be transferred to a matrix-valued setting.

Let us elaborate some more on the potential implications of our results. From the theoretical perspective of numerical analysis of a potential PDE-based interpretation the validity of the discrete extremum principle is a fundamental property in classic theory of numerical schemes. This holds in particular for the underlying PDEs of dilation and erosion in the scalar case which are Hamilton-Jacobi equations. It is a cornerstone of numerical analysis of PDEs that a stability notion such as an extremum principle together with consistency could enable to prove convergence of the underlying scheme. Thus our paper may form the first step towards the numerical analysis of a matrix-valued PDE.

Turning to possible implications on the more practical side, our new matrix-valued dilation and erosion provide an interesting basis for building more complex morphological procedures. For instance, they appear well-suited as a building block for morphological levelings. In the near future we aim to explore this possibility and evaluate the performance of the proposed concept in that context.

References

1. Arehart, A.B., Vincent, L., Kimia, B.B.: Mathematical morphology: the Hamilton-Jacobi connection. In: Proceedings Fourth International Conference on Computer Vision, pp. 215–219. IEEE Computer Society Press, Berlin, May 1993
2. Boroujerdi, A.S., Breuß, M., Burgeth, B., Kleefeld, A.: PDE-based color morphology using matrix fields. In: Aujol, J.-F., Nikolova, M., Papadakis, N. (eds.) SSVM 2015. LNCS, vol. 9087, pp. 461–473. Springer, Cham (2015). https://doi.org/10.1007/978-3-319-18461-6_37
3. Breuß, M., Hoeltgen, L., Kleefeld, A.: Matrix-valued levelings for colour images. In: Angulo, J., Velasco-Forero, S., Meyer, F. (eds.) ISMM 2017. LNCS, vol. 10225, pp. 296–308. Springer, Cham (2017). https://doi.org/10.1007/978-3-319-57240-6_24
4. Brockett, R.W., Maragos, P.: Evolution equations for continuous-scale morphology. In: Proceedings IEEE International Conference on Acoustics, Speech and Signal Processing, vol. 3, San Francisco, CA, pp. 125–128, Mar 1992
5. Bruhn, A., Weickert, J., Kohlberger, T., Schnörr, C.: A multigrid platform for real-time motion computation with discontinuity-preserving variational methods. Int. J. Comput. Vis. **70**(3), 257–277 (2006)
6. Burgeth, B., Kleefeld, A.: Morphology for color images via loewner order for matrix fields. In: Hendriks, C.L.L., Borgefors, G., Strand, R. (eds.) ISMM 2013. LNCS, vol. 7883, pp. 243–254. Springer, Heidelberg (2013). https://doi.org/10.1007/978-3-642-38294-9_21
7. Burgeth, B., Kleefeld, A.: An approach to color-morphology based on Einstein addition and Loewner order. Pattern Recogn. Lett. **47**, 29–39 (2014)
8. Burgeth, B., Kleefeld, A.: Order based morphology for color images via matrix fields. In: Westin, C.-F., Vilanova, A., Burgeth, B. (eds.) Visualization and Processing of Tensors and Higher Order Descriptors for Multi-Valued Data. Mathematics and Visualization, pp. 75–95. Springer, Heidelberg (2014). https://doi.org/10.1007/978-3-642-54301-2_4

9. Burgeth, B., Pizarro, L., Breuß, M., Weickert, J.: Adaptive continuous-scale morphology for matrix fields. Int. J. Comput. Vis. **92**(2), 146–161 (2011)

10. Burgeth, B., Welk, M., Feddern, C., Weickert, J.: Morphological operations on matrix-valued images. In: Pajdla, T., Matas, J. (eds.) ECCV 2004, Part IV. LNCS, vol. 3024, pp. 155–167. Springer, Heidelberg (2004). https://doi.org/10.1007/978-3-540-24673-2_13

11. Burgeth, B., Welk, M., Feddern, C., Weickert, J.: Mathematical morphology on tensor data using the Loewner ordering. In: Weickert, J., Hagen, H. (eds.) Visualization and Processing of Tensor Fields, pp. 357–368. Springer, Heidelberg (2006). https://doi.org/10.1007/3-540-31272-2_22

12. Forst, W., Hoffmann, D.: Optimization – Theory and Practice. Springer, New York (2010). https://doi.org/10.1007/978-0-387-78977-4

13. Förstner, W., Gülch, E.: A fast operator for detection and precise location of distinct points, corners and centres of circular features. In: Proceedings ISPRS Intercommission Conference on Fast Processing of Photogrammetric Data, Interlaken, Switzerland, pp. 281–305, June 1987

14. Kleefeld, A., Breuß, M., Welk, M., Burgeth, B.: Adaptive filters for color images: median filtering and its extensions. In: Trémeau, A., Schettini, R., Tominaga, S. (eds.) CCIW 2015. LNCS, vol. 9016, pp. 149–158. Springer, Cham (2015). https://doi.org/10.1007/978-3-319-15979-9_15

15. Laidlaw, D., Weickert, J. (eds.): Visualization and Processing of Tensor Fields: Advances and Perspectives. Springer, Heidelberg (2009). https://doi.org/10.1007/978-3-540-88378-4

16. Löwner, K.: Über monotone Matrixfunktionen. Mathematische Zeitschrift **38**, 177–216 (1934)

17. Pierpaoli, C., Jezzard, P., Basser, P.J., Barnett, A., Di Chiro, G.: Diffusion tensor MR imaging of the human brain. Radiology **201**(3), 637–648 (1996)

18. Rouy, E., Tourin, A.: A viscosity solutions approach to shape-from-shading. SIAM J. Numer. Anal. **29**, 867–884 (1992)

19. Sapiro, G., Kimmel, R., Shaked, D., Kimia, B.B., Bruckstein, A.M.: Implementing continuous-scale morphology via curve evolution. Pattern Recogn. **26**, 1363–1372 (1993)

20. Weickert, J.: Anisotropic Diffusion in Image Processing. Teubner, Stuttgart (1998)

21. Weickert, J., Hagen, H. (eds.): Visualization and Processing of Tensor Fields. Springer, Heidelberg (2006). https://doi.org/10.1007/3-540-31272-2

22. Welk, M., Kleefeld, A., Breuß, M.: Quantile filtering of colour images via symmetric matrices. Math. Morphol. Theor. Appl. **1**, 1–40 (2016)

23. Welk, M., Weickert, J., Becker, F., Schnörr, C., Feddern, C., Burgeth, B.: Median and related local filters for tensor-valued images. Sign. Process. **87**, 291–308 (2007)

24. Westin, C.F., Vilanova, A., Burgeth, B. (eds.): Visualization and Processing of Tensor Fields and Higher Order Descriptors for Multi-Valued Data. Springer, Heidelberg (2014). https://doi.org/10.1007/978-3-642-54301-2

Classification of Hyperspectral Images as Tensors Using Nonnegative CP Decomposition

Mohamad Jouni[1]([✉]) [iD], Mauro Dalla Mura[1,2] [iD], and Pierre Comon[1] [iD]

[1] Univ. Grenoble Alpes, CNRS, Grenoble INP, Gipsa-Lab, 38000 Grenoble, France
{mohamad.jouni,mauro.dalla-mura,pierre.comon}@gipsa-lab.fr
[2] Tokyo Tech World Research Hub Initiative (WRHI), School of Computing, Tokyo Institute of Technology, Tokyo, Japan
http://www.gipsa-lab.grenoble-inp.fr

Abstract. A Hyperspectral Image (HSI) is an image that is acquired by means of spatial and spectral acquisitions, over an almost continuous spectrum. Pixelwise classification is an important application in HSI due to the natural spectral diversity that the latter brings. There are many works where spatial information (e.g., contextual relations in a spatial neighborhood) is exploited performing a so-called spectral-spatial classification. In this paper, the problem of spectral-spatial classification is addressed in a different manner. First a transformation based on morphological operators is used with an example on additive morphological decomposition (AMD), resulting in a 4-way block of data. The resulting model is identified using tensor decomposition. We take advantage of the compact form of the tensor decomposition to represent the data in order to finally perform a pixelwise classification. Experimental results show that the proposed method provides better performance in comparison to other state-of-the-art methods.

Keywords: Hyperspectral imagery · Morphological profiles · Tensor decomposition · Scene classification

1 Introduction

Hyperspectral Imaging is one of the most important tools in remote sensing. A hyperspectral image (HSI) is typically a three-way[1] block of data acquired when many (two-way) images are taken over hundreds of almost-continuous spectral bands and stacked together forming a so called hypercube. HSI is employed in several fields of applications such as astronomy, Earth and planetary observation, monitoring of natural resources, precision agriculture and biomedical imaging.

---

[1] The number of *ways* of an array refers to the number of its indices. A HSI is typically a three-way array of dimensions $I_1 \times I_2 \times J$, where I_1 and I_2 are space dimensions (i.e. pixels) and J denotes the number of spectral bands.

© Springer Nature Switzerland AG 2019
B. Burgeth et al. (Eds.): ISMM 2019, LNCS 11564, pp. 189–201, 2019.
https://doi.org/10.1007/978-3-030-20867-7_15

One of its most common usages is *classification*, i.e., a thematic discrimination of different types of objects present in the image. In other words, the goal of classification is to assign a label to each pixel belonging to the same thematic class according to some of its characteristics (e.g., its spectrum, the homogeneity of its spatial neighborhood or the shape of a region it belongs to). In the case of HSI, the image can be seen as a mapping from a couple of positive integers (x, y) that correspond to horizontal and vertical positioning, to a vector of positive real values that correspond to the radiance measured by the sensor or the reflectance of the scene. Hence, a direct way to classify a HSI is to take the spectral bands as features and the pixels as samples. However, this approach raises some problems. First, in terms of spectral resolution, the acquisition of HSI induces high dimensionality in the spectra, which means that the samples are put in a high dimensional feature space. Second, by only considering the spectral information can be a limiting factor since spatial information (e.g., contextual relations, size and shape of the objects) is not exploited. For this reason, many works aimed at incorporating spatial information based on the pixels' neighborhood, so that each pixel in the HSI has features that include both spectral and spatial information, and this is known in the literature as *spectral-spatial classification*; such previous works could be found in [1–8] with different approaches.

One way to incorporate spatial information in HSI is by using tools defined in the framework of *Mathematical Morphology* (MM). A well-known example is the morphological profile (MP) [9], which is built by stacking the results of a set of openings and closings based on geodesic reconstruction applied on each band of the HSI (or on each component obtained after dimensionality reduction as in [10]) and has been successfully employed in remote sensing [7,11]. Attribute Profiles (AP) [4] are another example, in which a scalar image (e.g., a band of a HSI) is filtered with attribute thinnings and thickenings computed on component trees [4] or self-dual attribute filters based on the the tree-of-shapes [12].

Since HSI is by nature a three-way block of data, applying transformations based on MM results in a four-way block of data, which would be of high rank. Since we care about having *low rank* data, we merge the first two ways that correspond to pixel locations. As a result we obtain a three-way block of data where each dimension could be described by pixels, spectral bands, and spatial features (e.g., by means of morphological transformations), respectively.

In this paper we aim at modeling the 3-way hyperspectral data using tensors, and address the problem of pixelwise classification using tensor tools. Our contribution is inspired by [13], where an Additive Morphological Decomposition (AMD) is applied on HSI, resulting in a 4-way block of data, which is then modeled by tensor means. In [13], the corresponding tensor is dealt with using Tensor-PCA (TPCA) in order to apply dimensionality reduction of the data, which is then passed down to classification. In this paper, our motivation comes from the fact that most tensors enjoy a unique Canonical Polyadic (CP) decomposition [14], and we take advantage of how it represents the data for classification purposes. We sum up our contribution as follows:

- CP decomposition is used as a way to represent the pixel data in a compact form, which is more direct and easy to deal with and works thanks to a low dimensionality of the feature space. Moreover, CP decomposition can be adapted to incorporate the nonnegativity of its entries, which leads to a decomposition that can have a physical meaning.
- The proposed method shows promising results, when comparing it to TPCA approach.

The rest of the paper is organized as follows. In Sect. 2 we briefly talk about adding a new dimension to the data set based on MM. In Sect. 3 we describe CP decomposition, and explain how it can be a suitable way to directly represent the pixel data in a low dimensional feature space ready for classification. Experiments and results are reported in Sect. 4. Finally, a conclusion is drawn in Sect. 5 and some future work is outlined.

2 Mathematical Morphology

Let us consider $\{\bar{\Phi}^i, \underline{\Phi}^i\}_{i=1...m}$, a set of extensive and anti-extensive operators, respectively, where m refers to the size of the filter and the greater m the coarser the filter effect. The MP of a grayscale image (i.e., with scalar pixels value) I using the latter set of operators is defined as:

$$\mathrm{MP}(I) := \{\underline{\Phi}^m(I), \dots, \underline{\Phi}^1(I), I, \bar{\Phi}^1(I), \dots, \bar{\Phi}^m(I)\}, \tag{1}$$

in which the following ordering relation holds $\mathrm{MP}(I)_i \leq \mathrm{MP}(I)_{i+1}$ for $i = 1, \dots, 2m$ and with $\mathrm{MP}(I)_i$ denoting the i-th component of the MP. A MP using Opening and Closing by reconstruction is shown in Fig. 1.

Closings ⟵⟶ **Original** ⟶ **Openings**

Fig. 1. Morphological profile of a portion of the 8th spectral band of the HSI of University of Pavia using 3 different SE; disks with sizes [1,6,11].

In the case of HSI, or more generally in the case of multivariate images, a direct extension to MP is the Extended Morphological Profiles (EMP) [10]. The concept is the same: one simply concatenates the MP of each image along a fourth way and the result is provided as an input to the classifier. Along the same lines, works in the literature extended this concept to other possible rearrangements or derivations, and an example that we adopt in our experiments is the AMD [13].

As explained in [13], first we suppose that we have any image I decomposed through AMD such that:

$$I = \frac{\bar{\Phi}^m(\bar{\Phi}^{m-1}(I)) + \underline{\Phi}^m(\underline{\Phi}^{m-1}(I))}{2} + \sum_{i=1}^{m} \frac{R_i^- - R_i^+}{2} = S + \sum_{i=1}^{m} R_i, \quad (2)$$

The term S is the structure and contains the unfiltered objects of the decomposition, while the terms $\{R_i\}_{i=1...m}$ are the residuals and they contain elements that are filtered out according to the parameters used. Each element of the set $\{S, R_i\}_{i=1...m}$ is a 3-way data set, and at the end they are stacked to form a 4-way data set.

Unlike [13], we prefer to build the morphological way using the set $\{S, R_i^-, R_i^+\}_{i=1...m}$ in order to preserve the nonnegativity of the data, as a result we have a 4-way hyperspectral data set $\mathcal{D} = [S, R_1^-, R_1^+, \ldots, R_m^-, R_m^+]$, which is represented by an $I_1 \times I_2 \times J \times (2m+1)$ array, $I_1 \times I_2$ is the number of pixels, J is the number of spectral bands, and $2m + 1 = K$ is the number of terms in the morphological decomposition.

We note that the case study is not restricted to this kind of application, one could also build a 4th way using attribute thinning and thickenings as done in attribute profiles [4] for example.

3   Tensor Decomposition

In this section, we explain how to process by tensor decomposition the data obtained with AMD. $\mathcal{D}$ is of high rank, this can be explained starting with the fact that usually a 2D image is full rank which is high, and a tensor built upon this image will have a rank that is at least as much as that of the image. So we reshape the first two ways in lexicographic order so that the first dimension becomes $I_1 \times I_2 = I$, and $\mathcal{D}$ becomes an $I \times J \times K$ three-way array.

3.1   CP Decomposition

A third order tensor $\mathcal{D}_r$ is referred to as *decomposable* if it can be written as

$$\mathcal{D}_r \stackrel{\text{def}}{=} a_r \otimes b_r \otimes c_r, \quad (3)$$

where $\otimes$ denotes the tensor product, *e.g.* $[a \otimes b]_{ij} = a_i \, b_j$. A tensor $\mathcal{Y}$ can always be decomposed into a sum of decomposable tensors $\mathcal{D}_r$:

$$\mathcal{Y} = \sum_{r=1}^{R} \lambda_r \mathcal{D}_r, \quad (4)$$

The *tensor rank*, usually denoted by R, is the least number of terms required such that the *CP decomposition* (4) is exact. Suppose we have a 3-way tensor $\mathcal{Y} \in \mathbb{R}^{I \times J \times K}$, each decomposable term $\mathcal{D}_r$ in the CP decomposition can be expressed as in (3), where the set of vectors $\{\boldsymbol{a}_r\}_{r=1...R} \in \mathbb{R}^I$, $\{\boldsymbol{b}_r\}_{r=1...R} \in \mathbb{R}^J$, and $\{\boldsymbol{c}_r\}_{r=1...R} \in \mathbb{R}^K$ form the columns of the so-called factor matrices, $\boldsymbol{A} \in \mathbb{R}^{I \times R}$, $\boldsymbol{B} \in \mathbb{R}^{J \times R}$, and $\boldsymbol{C} \in \mathbb{R}^{K \times R}$. Hence another way to write Eq. 4 is, with an obvious notation:

$$\mathcal{Y} = (\boldsymbol{A}, \boldsymbol{B}, \boldsymbol{C}) \cdot \Lambda \tag{5}$$

The diagonal tensor Λ governs the interaction between the components (columns) of the factor matrices such that components of different r indices do not interact. R can then be seen as directly related to the degrees of freedom in the decomposition.

When $\mathcal{Y}$ is nonnegative, it is often desirable to constrain factor matrices to be also nonnegative. In this case, the rank is called *nonnegative rank* and sometimes denoted R_+. Having nonnegative factors is essential for the sake of physical interpretation of the results where negative values have no meaning. Consequently, the goal is to minimize the cost function:

$$\frac{1}{2}\|\mathcal{Y} - (\boldsymbol{A}, \boldsymbol{B}, \boldsymbol{C}) \cdot \Lambda\|^2 \tag{6}$$
$$\text{s.t. } \boldsymbol{A} \succeq 0, \boldsymbol{B} \succeq 0, \boldsymbol{C} \succeq 0$$

where the symbol $\succeq$ means element-wise nonnegativity.

An interesting aspect is that each of the factor matrices represents one way of the tensor. For example, an element $y_{i,j,k}$ is decomposed into R_+ components and the decomposed information is mapped to the i-th row of $\boldsymbol{A}$, the j-th row of $\boldsymbol{B}$, and the k-th row of $\boldsymbol{C}$.

In the case of HSI, $\boldsymbol{A}$ corresponds to the way of pixels, $\boldsymbol{B}$ corresponds to the way of spectral bands, and $\boldsymbol{C}$ corresponds to the way based on MM. Accordingly each row of $\boldsymbol{A}$ represents a pixel, that of $\boldsymbol{B}$ represents a spectral band, and that of $\boldsymbol{C}$ is defined by a scale or a structure. Thanks to the nonnegativity constraints and especially after normalizing the columns of the factor matrices, one could assume that columns of $\boldsymbol{B}$ represent spectral signatures and those of $\boldsymbol{C}$ represent a combination of scale domination along the set of structuring elements being chosen, and finally each column of $\boldsymbol{A}$ could be seen as an abundance map, that could be reshaped to show a grayscale image, with respect to information received at $\boldsymbol{B}$ and $\boldsymbol{C}$, very similar to the case of spectral unmixing (but for two dimensions only) – except that unmixing is presently not our concern.

In this paper, we take advantage of the low-rank aspect of our CP decomposition, i.e. how it represents the data in a more compact form with almost no loss of information (depending on how exact the decomposition is). First, instead of dealing with a multidimensional data set of large dimensions, we are able to deal with matrices of much smaller size, as explained earlier (since a small rank permits compression). Second, since our approach is pixelwise classification, we mainly think of $\boldsymbol{A}$ as the matrix to deal with. Theoretically speaking, similar pixels should have the same composition of coefficients with respect to the components in $\boldsymbol{A}$, so we take the components (columns) as features and the rows as samples in the classification.

3.2   High-Order Singular Value Decomposition (HOSVD)

Because of the huge dimensions of our tensor, directly applying CP decomposition would be computationally very demanding. One solution to make the decomposition possible is to pass by a pre-processing step by compressing the original tensor into one with smaller dimensions without losing any information. This can be done by providing the minimal compression size for each way such that the information is preserved. In that sense, let's suppose a tensor $\mathcal{Y}$ with dimensions $I \times J \times K$, where $I \gg JK$ (which is our case), the compressed tensor can have a size of $JK \times J \times K$ whenever its rank satisfies $R \leq JK$.

Algorithm 1. COMPRESS

Require: $\mathcal{Y}, way_c, dim_c$
   for $d \in way_c$ **do**
      Unfold $\mathcal{Y}$ into $\boldsymbol{Y}_{(d)}$ such that the mode d takes the second way of the matrix;
      Compute the right singular matrix from the SVD of $\boldsymbol{Y}_{(d)}$, denoted by $\boldsymbol{V}_d$;
      Truncate the columns of $\boldsymbol{V}_d$ by $dim_c(d)$;
   end for
   for $d \notin way_c$ **do**
      $\boldsymbol{V}_d = \boldsymbol{I}_{dim_c(d)}$; (Identity matrix)
   end for
   $\mathcal{G} = \mathcal{Y} \bullet_1 \boldsymbol{V}_1^T \bullet_2 \ldots \bullet_N \boldsymbol{V}_N^T$;
   return $\mathcal{G}$ and $\boldsymbol{V}_1, \ldots, \boldsymbol{V}_N$

This can be explained as follows. First let us denote the d-mode unfolding of $\mathcal{Y}$ by $\boldsymbol{Y}_{(d)}$, which is the matrix flattening of the tensor obtained by reordering the elements with respect to the d-th index. Each of these unfoldings has a matrix rank that essentially appears in its economic SVD. More precisely, after applying the SVD on each unfolding, we get three right singular (orthogonal) matrices that form the basis of the HOSVD factor matrices:

$$\boldsymbol{Y}_{(d)} = \boldsymbol{U}_d \boldsymbol{\Sigma}_d \boldsymbol{V}_d^T \tag{7}$$
$$\forall d = \{1, 2, 3\}$$

These matrices, denoted by $\boldsymbol{V}_1$, $\boldsymbol{V}_2$, and $\boldsymbol{V}_3$ have dimensions $I \times R_1$, $J \times R_2$ and $K \times R_3$ respectively, with $R_1 \leq JK$, $R_2 \leq J$, and $R_3 \leq K$. Hence they can be used to compress the original tensor into smaller dimensions without any loss of information. In fact, this is how HOSVD is computed by finding orthogonal matrices as the basis of the decomposition, and the compressed tensor $\mathcal{G}$ is found by projecting the original one onto the basis of factor matrices. HOSVD of $\mathcal{Y}$ can be written as follows:

$$\mathcal{Y} = \mathcal{G} \underset{1}{\bullet} \boldsymbol{V}_1 \underset{2}{\bullet} \boldsymbol{V}_2 \underset{3}{\bullet} \boldsymbol{V}_3 = (\boldsymbol{V}_1, \boldsymbol{V}_2, \boldsymbol{V}_3) \cdot \mathcal{G} \tag{8}$$

where $\bullet_d$ indicates a summation over the dth tensor index. Subsequently, we shall compress only the first mode, so that $R_1 < JK$, but $R_2 = J$ and $R_3 = K$.

As a consequence, we may choose $\boldsymbol{V}_2$ and $\boldsymbol{V}_3$ both as Identity matrices. The algorithm that we implement to compress an N-way tensor using HOSVD is described in Algorithm 1; $\boldsymbol{way}_c$ is a vector containing the ways that we want to compress and $\boldsymbol{dim}_c$ is a vector containing the compressed dimension size.

3.3 Alternating Optimization-Alternating Direction Method of Multipliers (AO-ADMM)

In order to compute the CP decomposition and cope with the constraints of non-negativity and compression together, the algorithm AO-ADMM [15] is adopted for its efficiency and flexibility especially with the different kinds of constraints. The CP decomposition can be computed via the Alternating Least Squares algorithm (ALS), where one factor matrix is updated alternatingly at a time by fixing the others and minimizing the convex least squares cost function:

$$\boldsymbol{H}_d = \operatorname*{argmin}_{\boldsymbol{H}_d} \frac{1}{2} \| \boldsymbol{Y}_{(d)} - \boldsymbol{W} \boldsymbol{H}_d^T \|^2 \tag{9}$$
$$\forall d = 1 \dots N,$$

where $\boldsymbol{H}_d$ is the factor matrix corresponding to the d-th way of the tensor, and $\boldsymbol{W}$ is the Khatri-Rao product of all the factor matrices excluding $\boldsymbol{H}_d$. The optimization problem (9) needs to be modified when constraints on the factor matrices are imposed, say through regularization functions $r(\boldsymbol{H}_d)$:

$$\boldsymbol{H}_d = \operatorname*{argmin}_{\boldsymbol{H}_d} \frac{1}{2} \| \boldsymbol{Y}_{(d)} - \boldsymbol{W} \boldsymbol{H}_d^T \|^2 + r(\boldsymbol{H}_d) \tag{10}$$
$$\forall d = 1 \dots N,$$

As the name suggests, AO-ADMM works by alternating between the optimization problems of each factor matrix alone while fixing the others. For example in our case, the three factor matrices $(\boldsymbol{H}_1, \boldsymbol{H}_2, \boldsymbol{H}_3)$ are recommended to have nonnegative entries, while only the first one is concerned with compression. AO-ADMM allows to pass different kinds of constraints and parameters for each mode-decomposition and alternates accordingly.

It is important to note that the tensor fed in AO-ADMM is the compressed one that is computed as a result of Sect. 3.2, which allows to have negative entries, and this presents a certain inconsistency with the fact of having nonnegativity constraints on the decomposition. For this reason we present a solution with ADMM inspired by [16]. So our minimization problem goes as follows:

$$\arg\min_{\boldsymbol{H}_1, \boldsymbol{H}_2, \boldsymbol{H}_3, \boldsymbol{\Lambda}} \frac{1}{2} \| \mathcal{G} - (\boldsymbol{H}_1, \boldsymbol{H}_2, \boldsymbol{H}_3) \cdot \boldsymbol{\Lambda} \|^2 \tag{11}$$
$$\text{s.t. } \boldsymbol{V}_1 \boldsymbol{H}_1 \succeq 0, \boldsymbol{H}_2 \succeq 0, \boldsymbol{H}_3 \succeq 0, \boldsymbol{\Lambda} \succeq 0$$

This minimization is executed by ADMM as follows, for $d \in \{2, 3\}$:

$$\left. \begin{aligned} \tilde{\boldsymbol{H}}_1 &\leftarrow (\boldsymbol{W}^T \boldsymbol{W} + \rho \boldsymbol{I})^{-1} (\boldsymbol{W}^T \boldsymbol{G}_{(1)} + \rho(\boldsymbol{U} + \boldsymbol{H}_1)^T) \\ \boldsymbol{H}_1 &\leftarrow \boldsymbol{V}_1^T \max(0, \boldsymbol{V}_1(\tilde{\boldsymbol{H}}_1^T - \boldsymbol{U})) \\ \boldsymbol{U} &\leftarrow \boldsymbol{U} + \boldsymbol{H}_1 - \tilde{\boldsymbol{H}}_1^T \end{aligned} \right\} \tag{12}$$

Algorithm 2. Alternating Optimization (Least Squares loss)

Require: $\mathcal{Y}, \boldsymbol{H}_1, \ldots, \boldsymbol{H}_N, \boldsymbol{U}_1, \ldots, \boldsymbol{U}_N$
  $\boldsymbol{H}_1, \ldots, \boldsymbol{H}_N$ are initialized; $\boldsymbol{U}_1, \ldots, \boldsymbol{U}_N$ are initialized to zero matrices;
  for $d = 1, \ldots, N$ **do**
    Store the different unfoldings $\boldsymbol{Y}_d$;
  end for
  repeat
    for $d = 1, \ldots, N$ **do**
      $\boldsymbol{W} = \odot_{j \neq d} \boldsymbol{H}_j$;
      update $\boldsymbol{H}_d$ and $\boldsymbol{U}_d$ using Algorithm 3;
    end for
    update μ if necessary; (refer to [15] for the update of μ)
  until some termination criterion is reached (number of iterations)
  Normalize the columns of the factor matrices and store the weights in Λ;
  return $\boldsymbol{H}_1, \ldots, \boldsymbol{H}_N, \Lambda$

Algorithm 3. ADMM of a mode d

Require: $\boldsymbol{Y}, \boldsymbol{W}, \boldsymbol{H}, \boldsymbol{U}, k, \mu, \epsilon,$ constraint, $imax, \boldsymbol{V}$
  $\boldsymbol{H}$ and $\boldsymbol{U}$ are already initialized;
  $\boldsymbol{G} = \boldsymbol{W}^T \boldsymbol{W}$;
  $\rho = \text{trace}(\boldsymbol{G})/k$;
  Calculate $\boldsymbol{L}$ from Cholesky decomposition such that $\boldsymbol{G} + (\rho + \mu)\boldsymbol{I}_k = \boldsymbol{L}\boldsymbol{L}^T$;
  $\boldsymbol{F} = \boldsymbol{W}^T \boldsymbol{Y}$;
  $\boldsymbol{H}_f = \boldsymbol{H}$;
  repeat
    $\tilde{\boldsymbol{H}} \leftarrow (\boldsymbol{L}^T)^{-1}\boldsymbol{L}^{-1}(\boldsymbol{F} + \rho(\boldsymbol{H} + \boldsymbol{U})^T + \mu\boldsymbol{H}_f{}^T)$; (See [15] for μ and $\boldsymbol{H}_f$)
    $\boldsymbol{H} \leftarrow$ proximity(constraint,$\tilde{\boldsymbol{H}}^T$,$\boldsymbol{U}$,$\boldsymbol{V}$); (refer to Algorithm 4)
    $\boldsymbol{U} \leftarrow \boldsymbol{U} + \boldsymbol{H} - \tilde{\boldsymbol{H}}^T$;
    Update r and s; (refer to [15] for the updates of r and s)
  until ($r < \epsilon$ and (s is undefined or $s < \epsilon$)) or ($imax > 0$ and $i \geq imax$)
  return $\boldsymbol{H}$ and $\boldsymbol{U}$

Algorithm 4. Proximity Update of $\boldsymbol{H}$ in ADMM

Require: constraint, $\boldsymbol{H}_t, \boldsymbol{U}, \boldsymbol{V}$
  switch (constraint)
  case Nonnegativity:
    $\boldsymbol{H} \leftarrow (\boldsymbol{H}_t - \boldsymbol{U})^+$;
  case Compression and Nonnegativity:
    $\boldsymbol{H}_u \leftarrow \boldsymbol{V}(\boldsymbol{H}_t - \boldsymbol{U})$; $\boldsymbol{H}_u \leftarrow \boldsymbol{H}_u^+$; $\boldsymbol{H} \leftarrow \boldsymbol{V}^T \boldsymbol{H}_u$;
  end switch
  return $\boldsymbol{H}$

$$\left. \begin{array}{l} \tilde{\boldsymbol{H}}_d \leftarrow (\boldsymbol{W}^T\boldsymbol{W} + \rho\boldsymbol{I})^{-1}(\boldsymbol{W}^T\boldsymbol{G}_{(d)} + \rho(\boldsymbol{U} + \boldsymbol{H}_d)^T) \\ \boldsymbol{H}_d \leftarrow \max(0, \tilde{\boldsymbol{H}}_d^T - \boldsymbol{U}) \\ \boldsymbol{U} \;\; \leftarrow \boldsymbol{U} + \boldsymbol{H}_d - \tilde{\boldsymbol{H}}_d^T \end{array} \right\} \qquad (13)$$

Of course, the calculation of $(\boldsymbol{W}^T\boldsymbol{W} + \rho\boldsymbol{I})^{-1}$ should be done once for all outside the loops, as explained in the pseudo-codes Algorithms 2, 3, and 4.

4   Results

4.1   Description of the Dataset

Our dataset is described by a real hyperspectral image, University of Pavia, as shown in Fig. 2 in true colors (by choosing the bands 53, 31, and 8 as Red, Green, and Blue channels respectively). It is an image acquired by the ROSIS sensor with geometric resolution of 1.3 m. The image has 610×340 pixels and 103 spectral bands. The dataset contains a groundtruth image that consists of 9 classes: trees, asphalt, bitumen, gravel, metal sheets, shadows, self-blocking bricks, meadows, and bare soil. There are also 42776 labeled pixels available as test set and 3921 pixels available as training set, the latter is fixed at that.

4.2   Classification

For the classification part we use Support Vector Machines (SVM), which has proved useful in the application of pixelwise classification of HSI. Practically SVM is carried out through the open source machine learning library provided by Libsvm [17]. After decomposing the tensor using CP decomposition, we use the first factor matrix as the set of data to be classified by SVM.

4.3   Results and Discussion

In the following experiments, we fix the morphological transformations on structuring elements whose neighborhoods are defined arbitrarily by disk shapes with varying sizes of radii: 1, 6, 11, 16, 21, and 26 pixels, this accounts to $m = 6$ structuring elements, thus a morphological decomposition of $K = 13$ components. We note that considering the whole set of $\{\boldsymbol{S}, \boldsymbol{\mathcal{R}}_i^-, \boldsymbol{\mathcal{R}}_i^+\}_{i=1...6}$ without dimensionality reduction yielded a classification with better accuracy than that of the set of $\{\boldsymbol{S}, \boldsymbol{\mathcal{R}}_i\}_{i=1...6}$ (96.60% to 93.06% respectively) under the same AMD parameters.

AMD. The HSI of University of Pavia is loaded in its 3rd-order form, $\boldsymbol{\mathcal{I}} \in \mathbb{R}^{610 \times 340 \times 103}$. $\boldsymbol{\mathcal{I}}$ is then decomposed into a morphological structure, going from a dataset of one tensor to a set of tensors ($\boldsymbol{S}$, $\boldsymbol{\mathcal{R}}^-$'s and $\boldsymbol{\mathcal{R}}^+$'s, forming a 4th-order dataset) representing information of the scene with different scales:

$$\boldsymbol{\mathcal{I}} \rightarrow \mathcal{D} = \{\boldsymbol{S}, \boldsymbol{\mathcal{R}}_1^-, \boldsymbol{\mathcal{R}}_1^+, \ldots, \boldsymbol{\mathcal{R}}_6^-, \boldsymbol{\mathcal{R}}_6^+\}$$

such that $\boldsymbol{\mathcal{I}} = \boldsymbol{S} + \sum_{i=1}^6 \frac{\boldsymbol{\mathcal{R}}_i^- - \boldsymbol{\mathcal{R}}_i^+}{2}$. As we mentioned in Sect. 3, $\mathcal{D}$ is vectorized by merging the first two modes resulting in a 3-way tensor of dimensions $207400 \times 103 \times 13$.

Tensors. Now that the data is modeled as a 3rd-order tensor, we seek to decompose it using CP decomposition. The latter is carried out by taking the compressed tensor as input, the factor matrices are initialized randomly (absolute value of i.i.d. standard Gaussian distribution), and only the first factor matrix is compressed after that. The rank is set to different values followed by different initializations of factor matrices.

Classification. The accuracy of the classification is related to the reconstruction error of the CP decomposition since it is important that factor matrix H_1 represents the data in a shape as good as possible. Two factors play major roles in the reconstruction error: the number of iterations and the input rank of the decomposition. Hence we try different values of the two variables and record both the reconstruction error and the result of the overall accuracy as seen in Table 1. Per-class and average accuracy is also recorded for two methods in Table 2. As discussed in Sect. 3.1, H_1 is passed to the classifier since it is considered as the representative of pixel data in compact low-dimensional form.

Table 1. Records of various tests in terms of Reconstruction Error (Rec. Error) (in case of CPD) and Overall Accuracy (OA).

Method	Rec. Error %	OA %
AMD + TPCA + SVM (PC=(10,5))	-	91.09
AMD + CPD + SVM (itr=50,R=10)	18.45	87.87
AMD + CPD + SVM (itr=20,R=20)	11.10	88.85
AMD + CPD + SVM (itr=50,R=20)	10.94	**91.17**
AMD + CPD + SVM (itr=20,R=30)	8.86	**93.76**
AMD + CPD + SVM (itr=50,R=30)	8.72	**94.08**
AMD + CPD + SVM (itr=100,R=30)	8.70	**93.94**
AMD + CPD + SVM (itr=100,R=40)	7.16	**94.03**

Table 2. Records of per-class, overall and average accuracies.

Class	AMD + TPCA (PC=(10,5))	AMD + CPD (itr=50,R=30)
Asphalt	95.54	**98.46**
Meadow	94.08	**96.42**
Gravel	**72.12**	59.83
Tree	97.59	**98.00**
Metal Sheet	**100**	99.64
Bare Soil	81.56	**92.71**
Bitumen	**99.79**	99.69
Brick	76.66	**88.49**
Shadow	**99.87**	86.66
Overall	91.09	**94.08**
Average	90.80	**91.10**

Results. We compare our results to the method of [13]. Looking at Table 1, we can see that the better the reconstruction error in the CP decomposition, the better the overall accuracy of the classification. The accuracy shows promising results for our proposed method. It is worth to stress the influence of rank and/or number of iterations on the reconstruction error. But at some point the error decreases very slowly as more iterations are run, towards what seems to be a limit, which is mainly influenced by the value of the input rank. Figure 2 shows images of the training set, the test set, and the results of various classifications; parameters are mentioned in the caption.

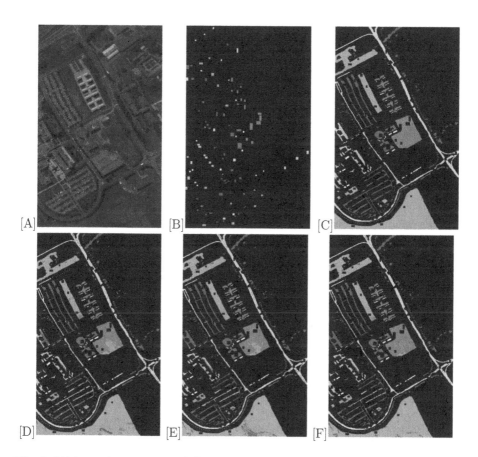

Fig. 2. [A] image in true colors, [B] available training set, [C] test set, [D-F] classification results: [D] AMD+TPCA(15)+SVM (91.09%), [E] AMD+CPD (itr=50,R=20)+SVM (91.17%), [F] AMD+CPD (itr=50,R=30)+SVM (94.08%) (Color figure online)

5 Conclusion

In the framework of pixelwise classification of hyperspectral images, an extraneous diversity is built from hyperspectral data thanks to Mathematical Morphology. Then the use of tensor decomposition allows to take into account both spectral and spatial information of the scene. Experimental results using a real image show that the CP decomposition is a promising way to represent the data in a low dimensional feature space and improve on classification accuracy. In future works, we plan to investigate other potential interests of the CP decomposition, like spectral unmixing of HSI.

References

1. Bruzzone, L., Carlin, L.: A multilevel context-based system for classification of very high spatial resolution images. IEEE Trans. Geosci. Remote Sens. **44**(9), 2587–2600 (2006)
2. Duarte-Carvajalino, J.M., Sapiro, G., Vélez-Reyes, M., Castillo, P.E.: Multiscale representation and segmentation of hyperspectral imagery using geometric partial differential equations and algebraic multigrid methods. IEEE Trans. Geosci. Remote Sens. **46**(8), 2418–2434 (2008)
3. Li, J., Bioucas-Dias, J.M., Plaza, A.: Spectral-spatial hyperspectral image segmentation using subspace multinomial logistic regression and Markov random fields. IEEE Trans. Geosci. Remote Sens. **50**(3), 809–823 (2012)
4. Dalla Mura, M., Benediktsson, J.A., Waske, B., Bruzzone, L.: Morphological attribute profiles for the analysis of very high resolution images. IEEE Trans. Geosci. Remote Sens. **48**(10), 3747–3762 (2010)
5. Fauvel, M., Chanussot, J., Benediktsson, J.A.: A spatial-spectral kernel-based approach for the classification of remote-sensing images. Pattern Recogn. **45**(1), 381–392 (2012)
6. Fauvel, M., Tarabalka, Y., Benediktsson, J.A., Chanussot, J., Tilton, J.C.: Advances in spectral-spatial classification of hyperspectral images. Proc. IEEE **101**(3), 652–675 (2013)
7. Ghamisi, P., Dalla Mura, M., Benediktsson, J.A.: A survey on spectral-spatial classification techniques based on attribute profiles. IEEE Trans. Geosci. Remote Sens. **53**(5), 2335–2353 (2015)
8. Ghamisi, P., et al.: Frontiers in spectral-spatial classification of hyperspectral images. IEEE Geosci. Remote Sens. Mag. (2018)
9. Benediktsson, J.A., Pesaresi, M., Amason, K.: Classification and feature extraction for remote sensing images from urban areas based on morphological transformations. IEEE Trans. Geosci. Remote Sens. **41**(9), 1940–1949 (2003)
10. Benediktsson, J.A., Palmason, J.A., Sveinsson, J.R.: Classification of hyperspectral data from urban areas based on extended morphological profiles. IEEE Trans. Geosci. Remote Sens. **43**(3), 480–491 (2005)
11. Dalla Mura, M., Benediktsson, J.A., Chanussot, J., Bruzzone, L.: The evolution of the morphological profile: from panchromatic to hyperspectral images. In: Prasad, S., Bruce, L., Chanussot, J. (eds.) Optical Remote Sensing, pp. 123–146. Springer, Heidelberg (2011). https://doi.org/10.1007/978-3-642-14212-3_8

12. Cavallaro, G., Dalla Mura, M., Benediktsson, J.A., Plaza, A.: Remote sensing image classification using attribute filters defined over the tree of shapes. IEEE Trans. Geosci. Remote Sens. **54**(7), 3899–3911 (2016)
13. Velasco-Forero, S., Angulo, J.: Classification of hyperspectral images by tensor modeling and additive morphological decomposition. Pattern Recogn. **46**(2), 566–577 (2013)
14. Comon, P.: Tensors: a brief introduction. IEEE Sig. Proc. Mag. **31**(3), 44–53 (2014). hal-00923279
15. Huang, K., Sidiropoulos, N.D., Liavas, A.P.: A flexible and efficient algorithmic framework for constrained matrix and tensor factorization. IEEE Trans. Sign. Process. **64**(19), 5052–5065 (2016)
16. Cohen, J., Farias, R.C., Comon, P.: Fast decomposition of large nonnegative tensors. IEEE Sign. Process. Lett. **22**(7), 862–866 (2015)
17. Chang, C.-C., Lin, C.-J.: LIBSVM: a library for support vector machines. ACM Trans. Intell. Syst. Technol. **2**, 27:1–27:27 (2011). http://www.csie.ntu.edu.tw/~cjlin/libsvm

A Unified Approach to the Processing of Hyperspectral Images

Bernhard Burgeth[1]([✉])[iD], Stephan Didas[2][iD], and Andreas Kleefeld[3][iD]

[1] Saarland University, 66123 Saarbrücken, Germany
burgeth@math.uni-sb.de
[2] Trier University of Applied Sciences, Environmental Campus Birkenfeld,
55761 Birkenfeld, Germany
[3] Forschungszentrum Jülich GmbH, Jülich Supercomputing Centre,
52425 Jülich, Germany

Abstract. Since vector fields, such as RGB-color, multispectral or hyperspectral images, possess only limited algebraic and ordering structures they do not lend themselves easily to image processing methods. However, for fields of symmetric matrices a sufficiently elaborate calculus, that includes, for example, suitable notions of multiplication, supremum/infimum and concatenation with real functions, is available. In this article a vector field is coded as a matrix field, which is then processed by means of the matrix valued counterparts of image processing methods. An approximate decoding step transforms a processed matrix field back into a vector field. Here we focus on proposing suitable notions of a pseudo-supremum/infimum of two vectors/colors and a PDE-based dilation/erosion process of color images as a proof-of-concept. In principle there is no restriction on the dimension of the vectors considered. Experiments, mainly on RGB-images for presentation reasons, will reveal the merits and the shortcomings of the proposed methods.

Keywords: Multispectral image · Hyperspectral image · Matrix field · Mathematical morphology

1 Introduction

The processing of color images has its intricacies, due to the vectorial nature of the data and the fact that a channel-wise treatment is in general insufficient. This already becomes apparent in the fundamental erosion and dilation processes of mathematical morphology for three-channel images where so-called false-color phenomena occur [21]. Countless attempts have been made to overcome these difficulties with a great variety of methods (see [1,27] for excellent surveys and [3,17,20]), especially for three-channel images, each of it with its own merits and drawbacks. Nevertheless, image processing methods are in great demand for multispectral images or for hyperspectral images. The development of multispectral image processing algorithms is of vital importance in a variety

© Springer Nature Switzerland AG 2019
B. Burgeth et al. (Eds.): ISMM 2019, LNCS 11564, pp. 202–214, 2019.
https://doi.org/10.1007/978-3-030-20867-7_16

of applications such as in food safety inspections [26], food quality [18,22,23], and in archaeology [15].

For a typical workflow of the processing of hyperspectral images refer to [16,28] and the references therein. Often these methods are taylormade for images with a certain fixed number of channels or bands restricting their applicability. The aim of this article is to provide a unified approach to multi-channel images, or vectorial data, for that matter, regardless of their dimension. To this end we assume without loss of generality a multi-channel image as a mapping f of the image domain Ω into the d-dimensional hyper-cube Q^d: $f : \Omega \longmapsto Q^d := [0,1]^d$ with $d \in \mathbb{N} \setminus \{0\}$. Typical examples are gray-value images ($d = 1$) and RGB-color images ($d = 3$). This setting can be achieved if the intensities of each of the d recorded frequencies of the multichannel images are normalized to have values in the interval $[0,1]$. However, even RGBα-images are considered to belong to this class ($d = 4$), although the meaning of α is not that of a frequency. Nevertheless, in this article we will refer to such a type of image as a (d-dimensional) multispectral image, and its values in Q^d often as multi-colors. We will concentrate on providing mathematical concepts applicable to multi-spectral images that allow for the construction of the fundamental building blocks for any numerical image processing algorithm: linear combinations, multiplication, concatenation with functions, and fruitful notions of maximum and minimum. Those key-ingredients are available in the case of so called matrix-fields, where we denote any mapping $F : \Omega \longmapsto \mathrm{SYM}(d)$ from a two- or three-dimensional image domain Ω into the real vector space $\mathrm{SYM}(d)$ of $d \times d$, symmetric matrices as a field of symmetric matrices. For the sake of brevity we will refer to them as $\mathrm{SYM}(d)$-valued images, or even shorter, as $\mathrm{SYM}(d)$-fields. Various methods from scalar image processing have been transferred to the matrix-field setting, resulting in diffusion- or transport-type evolutions of matrix fields, e.g. [7,11,12], as well as semi-order-based morphological operations, see [5].

The main idea is to rewrite a multi-spectral image as a $\mathrm{SYM}(d)$-field to take advantage of the image processing concepts available for the later one. Roughly speaking, this "rewriting" of a color vector $c \in Q^d$ amounts to taking its outer or dyadic product with itself $c\,c^\top \in \mathrm{SYM}(d)$. However, one of the challenges will be to reconstruct a multi-spectral image from the processed $\mathrm{SYM}(d)$-field in a reasonable way such that gray-valued mathematical morphology is preserved. The basic structure of the processing strategy is depicted in Fig. 1 in the exemplary case of the pseudo-supremum psup of three-dimensional vectors, resp., their corresponding outer product matrices. Since $\mathrm{SYM}(d)$-fields will play a major role in the sequel, the necessary rudiments of matrix fields and their calculus will

Fig. 1. Proposed processing strategy. Here "$\Longrightarrow$" stands for pre- and post-processing.

be presented in the subsequent section. For further details the reader is referred to [11,12]. We will describe the coding of a multi-spectral image and the corresponding decoding of the matrix field in Sect. 3. Based on [11] we provide in Sect. 4 a short introduction to matrix-valued counterparts of the morphological PDEs of dilation and erosion and their numerical solution schemes. In Sect. 5 we report on experiments performed mainly on three-channel images and on higher dimensional multi-spectral signals as a proof-of-concept, while the last Sect. 6 is devoted to concluding remarks and an outlook.

2   Rudiments of a Calculus for Symmetric Matrices

For the sake of brevity, we present here only the very basic notions from calculus of symmetric matrices: For details see [5] and [7]. Any matrix $S \in \text{SYM}(d)$ can be diagonalized by means of a suitable orthogonal matrix and, furthermore, all the eigenvalues are real: $S = QDQ^\top$. Here, Q is orthogonal, that is, $Q^\top Q = QQ^\top = I$ and $D = \text{diag}(\lambda_1, \ldots, \lambda_n)$ is a diagonal matrix with real entries in decreasing order, $\lambda_1 \geq \ldots \geq \lambda_d$. The matrix S is called positive semidefinite if $\lambda_1 \geq \ldots \geq \lambda_d \geq 0$. We will call a matrix with this property a spd-matrix. If the eigenvalues are strictly positive, the matrix is called positive definite. A matrix S is negative (semi-)definite if $-S$ is positive (semi-)definite. If the matrix S is none of the above, then the symmetric matrix is called indefinite. This gives rise to a partial order "$\geq$" on $\text{SYM}(d)$, often referred to as Loewner order (refer to [2]):

$$A \geq B \text{ if and only if } A - B \text{ is positive semidefinite.}$$

Note that $\text{SYM}(d)$ with this order is not a lattice. Nevertheless, as it is pointed out in [6], and in more detail in [5,8], a rich functional algebraic calculus can be set up for symmetric matrices. This allows to establish numerous filtering and analysis methods for such fields in a rather straight forward manner from their scalar counterparts [11]. We call the matrices $\text{psup}(A, B)$ and $\text{pinf}(A, B)$ as defined in Table 1 for $A, B \in \text{SYM}(d)$ pseudo-supremum resp., pseudo-infimum. They are the upper, resp., lower matrix valued bounds of smallest, resp., largest trace, and as such, acceptable replacements for supremum/infimum in this non-lattice setting, see [10].

3   From Hyperspectral Image to Matrix Field and Back

In this section, we will elaborate on embedding vectorial data of a hyperspectral image into a field of symmetric matrices, and on the restoration of a hyperspectral image from a (processed) field of positive, symmetric matrices. The key is the following simple observation:

Suppose the column vector $0 \neq c = (c_1, \ldots, c_d) \in Q^d$ codes a multicolor, then the outer or dyadic product $cc^\top$ is a matrix which is non-negative,

Table 1. Transferring elements of scalar valued calculus (middle) to the symmetric matrix setting (right).

Setting	Scalar valued	Matrix-valued
Function	$f : \begin{cases} \mathbb{R} \longrightarrow \mathbb{R} \\ x \mapsto f(x) \end{cases}$	$F : \begin{cases} \mathrm{SYM}(d) \longrightarrow \mathrm{SYM}(d) \\ S \mapsto Q\mathrm{diag}(f(\lambda_1), \ldots, f(\lambda_d)) \, Q^\top \end{cases}$
Partial derivatives	$\partial_\omega g,$ $\omega \in \{t, x_1, \ldots, x_d\}$	$\overline{\partial}_\omega S := (\partial_\omega s_{ij})_{ij},$ $\omega \in \{t, x_1, \ldots, x_d\}$
Gradient	$\nabla g(x) := (\partial_{x_1} g(x), \ldots, \partial_{x_d} g(x))^\top,$ $\nabla h(x) \in \mathbb{R}^d$	$\overline{\nabla} G(x) := (\overline{\partial}_{x_1} G(x), \ldots, \overline{\partial}_{x_d} G(x))^\top,$ $\overline{\nabla} H(x) \in (\mathrm{SYM}(d))^d$
Length	$|h|_p := \sqrt[p]{|h_1|^p + \cdots + |h_d|^p},$ $|h|_p \in [0, +\infty[$	$|H|_p := \sqrt[p]{|H_1|^p + \cdots + |H_d|^p},$ $|H|_p \in \{M \in \mathrm{SYM}(d) \mid M \geq 0\}$
Product	$a \cdot b$	$A \bullet B = \frac{1}{2}(AB + BA)$
Supremum	$\sup(a, b)$	$\mathrm{psup}(A, B) = \frac{1}{2}(A + B + |A - B|)$
Infimum	$\inf(a, b)$	$\mathrm{pinf}(A, B) = \frac{1}{2}(A + B - |A - B|)$

symmetric, and has rank one. It can be seen as a very special autocorrelation matrix of the vector c. The only non-zero eigenvalue r satisfies $r = \|c\|_2^2 = \mathrm{trace}(cc^\top)$. Keeping in mind that $\|c\|_\infty \leq 1 \longleftrightarrow c \in Q^d$, it is not difficult to reconstruct the original color vector by performing (theoretically) a spectral decomposition on $cc^\top$, providing us with a unique eigenvector v, which belongs to the largest eigenvalue r, has non-negative entries and euclidean length $\|v\|_2 = \sqrt{r}$. Then $v = c$ holds. However, between coding and decoding the actual processing happens, confronting us with matrices of higher than rank one. Although this effect cannot be avoided completely, some pre- and post-processing steps are in order.

3.1 Vector Data to Matrix Field

The pre-, resp., post-processing of a (multi-)color vector c requires several, notably reversible steps.

1. Renormalization of the color vector c. Let $B_p = \{v \in \mathbb{R}^d \mid \|v\|_p \leq 1\}$ denote the unit ball in $\mathbb{R}^d$ with respect to the p-norm, $p \in [1, \infty]$. Then $Q^d = B_\infty \cap \{v \in \mathbb{R}^d \mid, v_1, \ldots, v_d \geq 0\}$. The function

$$\psi : v \longmapsto \frac{\|v\|_\infty}{\|v\|_2} \cdot v$$

   maps Q^d to $B_2 \cap \{v \in \mathbb{R}^d \mid, v_1, \ldots, v_d \geq 0\}$.
2. This transform is invertible, with

$$\psi^{-1} : v \longmapsto \frac{\|v\|_2}{\|v\|_\infty} \cdot v.$$

The coding of a multispectral image as a matrix field is done by a mapping Φ defined as follows.

Definition 1. *Let* $f : \Omega \longmapsto Q^d$ *be a* d-*dimensional multispectral image. Then define a mapping* Φ *by*

$$\Phi : \begin{cases} Q^d \longrightarrow \mathrm{SYM}(d) \\ c \longmapsto \psi(c) \cdot \psi(c)^\top \end{cases}$$

To each multispectral image f *one associates a matrix field* $F : \Omega \longrightarrow \mathrm{SYM}(d)$ *simply by concatenating* Φ *and* f: $F := \Phi \circ f$.

Writing in short $\psi = \psi(f(x))$ the matrix $F(x) = \Phi(f(x))$ of the color $c = f(x)$ at pixel x has the form

$$F(x) = \Phi(c) = \begin{pmatrix} \psi_1^2 & \cdots & \psi_1\psi_d \\ \vdots & \ddots & \vdots \\ \psi_1\psi_d & \cdots & \psi_d^2 \end{pmatrix}.$$

Note that the normalisation step pays tribute to the fact that the approach relying on the scalar and outer product is closely related to the Euclidean norm rather than the infinity norm the cube Q^d alludes to. Next, we list a few properties of the matrix $\Phi(c)$.

1. The matrix $\Phi(c)$ has rank 1, and is a positive semidefinite matrix.
2. Due to the renormalization of c via ψ, the matrix $\Phi(c) = \psi(c) \cdot \psi(c)^\top$ satisfies trace$(\Phi(c)) \leq 1$ for any color $c \in Q^d$.
3. If $\|c\|_\infty = 1$, then trace$(\Phi(c)) = 1$.
4. A closer look at the construction of ψ and Φ reveals that ψ and Φ are positive-homogeneous of degree 1 and 2, respectively:
   $\psi(t \cdot c) = |t|\psi(c)$ and $\Phi(t \cdot c) = |t|^2\Phi(c)$ for $t \in \mathbb{R}$.
5. The function ψ^{-1} is positive-homogeneous as well.
6. v is a (right-)eigenvector of $\Phi(v)$ with the only non-zero eigenvalue $\|v\|^2$.
7. It is important to note that any positive semidefinite rank-1-matrix can be written as an outer product of a vector v or $-v$ with itself. Hence, we can recover the generating vector uniquely from a positive semidefinite rank-1-matrix via spectral decomposition and requiring that e.g. the last component $v_d \geq 0$. In other words, Φ can be assumed invertible on its range $\Phi(Q^d)$.

Gray-scale images are captured in the multispectral setting by the specification $t \cdot (1, \ldots, 1)$ with $t \in [0, 1]$. Therefore, it is this homogeneity that will ensure the preservation of basic gray-value morphology by the proposed approach. In the next section, we address the decoding, that is, transforming a spd-matrix field back into a proper hyperspectral image.

3.2   Matrix Field to Vector Data: Approximate Inverse of σ

For the decoding of a spd-matrix consisting of rank-1-matrices we simply may use the inverse Φ^{-1} available for rank-1-matrices (only). However, in general, the processing of matrix-fields does not preserve neither the rank-1- nor the spd-property of its matrices. Precisely, it can happen that the pseudo-supremum and pseudo-infimum of two rank-1 matrices is not rank-1. Hence, we have to make do with an approximate inverse mapping $\Phi^{\leftarrow}$. The construction of $\Phi^{\leftarrow}$ boils down to extracting from a general $\mathrm{SYM}(d)$-field a field with symmetric rank-1-matrices. To this end suppose that $F(x)$ is a symmetric matrix at location x. The Eckart-Young-Mirsky theorem (see [13, 14]) provides us with the best rank-1 approximation of $F(x)$ with respect to the Frobenius-Norm for matrices: if $\lambda_{\max}(x)$ is the largest eigenvalue (by absolute value) in the spectrum of $F(x)$ and $v(x)$ is a corresponding eigenvector with $\|v(x)\|_2 = 1$ then

$$F^*(x) := \lambda_{\max}(x)\, v(x)v^\top(x).$$

The transition $F(x)$ to $F^*(x)$ is achieved by terminating the spectral decomposition of $F(x)$ with the first summand:

$$F(x) = \sum_{i=1}^{d} \lambda_i\, v_i(x)v_i(x)^\top \approx \lambda_{\max}(x)\, v(x)v(x)^\top = F^*(x),$$

where $(\lambda_i)_{i=1,\dots,d}$ is arranged in decreasing order. This is a linear projection, hence, positive homogeneous.

From this field F^* we may extract a vector $\Phi^{\leftarrow}(F^*(x)) = \sqrt{\lambda_{\max}(x)}\, v(x)$ (according to property 6). Finally, we may apply the positively homogeneous ψ^{-1} to such a vector hence obtaining a new vector

$$\psi^{-1}\Big(\Phi^{\leftarrow}(F^*(x))\Big) = \psi^{-1}(\sqrt{\lambda_{\max}(x)}\, v(x)) = \sqrt{\lambda_{\max}(x)} \cdot \psi^{-1}(v(x))$$
$$= \sqrt{\lambda_{\max}(x)}\, w(x)$$

as a candidate for the processed color vector at x.

Remark 1. Due to the "geometry" of pseudo-supremum/infimum as a mapping on symmetric matrices, see [9], numerical experiments revealed that

$$1 < \mathrm{trace}(F^*(x)) = \lambda_{\max}(x) \leq 1.05, \quad \text{for } d = 3$$

entailing $\sqrt{\lambda_{\max}(x)}w(x) \notin Q^d$ does not represent a color by a small margin. A fast and "homogeneous" remedy is a rescaling of the rank-1-matrix $F^*(x)$ with a factor $\left(\max(1, \sqrt{\lambda_{\max}(x)})\right)^{-1}$. Since this happens only for nearly "antagonistic" colors $c_1, c_2 \in Q^d$ with $c_1 + c_2 = (1, \dots, 1)$ a simple cut-off is a possibly more convenient choice.

4   Basic PDE-Driven Morphology

The correspondence between real and matrix calculus allows to formulate matrix valued partial differential equations (PDEs), and even matrix valued solution schemes may be gleaned from the real-valued counterparts [5,11].

4.1   Continuous Morphology: Matrix-Valued PDEs

The matrix valued equivalent of the morphological PDEs [4,24] for dilation (+) and erosion (−) proposed for symmetric matrices U in [5] reads for $U(x,t) \in$ SYM(d):

$$\overline{\partial}_t U = \pm |\overline{\nabla} U|_2 \tag{1}$$

with $U(x,t) \in \Omega \times [0,\infty)$. We refer to Table 1 for the bar and norm notation.

4.2   Continuous Morphology: Matrix-Valued Solution Scheme

Due to its simplicity we extend the first-order upwind scheme of Rouy and Tourin [19,25] (RT-scheme) from gray-scale images to the matrix setting to solve (1). We denote by u_{ij}^n the gray value of the image u at the pixel centered at $(ih_x, jh_y) \in \Omega \subset \mathbb{R}^2$ at the time-level $n\tau$ of the evolution with time-step $\tau > 0$. The RT-scheme is expressed in a form that allows directly to extend the coding procedure to the 3D matrix valued setting of SYM(d):

- Instead of gray values u_{ij}^n we employ symmetric matrices $U^n(ih_x, jh_y)$.
- The max-function used below in a scalar-valued setting, is replaced by its matrix-valued generalization psup as given in Table 1, and we proceed likewise with the min-function and pinf.
- The equation can be extended without major difficulties to 3D matrix fields $U^n(ih_x, jh_y, kh_z)$.

For the sake of brevity we restrict ourselves to morphological dilation, the scheme for erosion involves only a simple switch of sign, see (1). The abbreviations we use for forward and backward difference operators are standard, i.e.,

$$D_+^x u_{i,j}^n := u_{i+1,j}^n - u_{i,j}^n \quad \text{and} \quad D_-^x u_{i,j}^n := u_{i,j}^n - u_{i-1,j}^n. \tag{2}$$

These operators can be defined analogously with respect to the y-direction. The Rouy-Tourin scheme we exploit here reads

$$u_{i,j}^{n+1} = u_{i,j}^n + \tau \left(\max\left(\frac{1}{h_x} \max\left(-D_-^x u_{i,j}^n, 0\right), \frac{1}{h_x} \max\left(D_+^x u_{i,j}^n, 0\right) \right)^2 \right.$$

$$\left. + \max\left(\frac{1}{h_y} \max\left(-D_-^y u_{i,j}^n, 0\right), \frac{1}{h_y} \max\left(D_+^y u_{i,j}^n, 0\right) \right)^2 \right)^{1/2} \tag{3}$$

Its performance is very similar to that of the first-order version of a scheme of Osher and Sethian. With this machinery at our disposal we will not process fields of symmetric matrices for their own sake, however. Instead, a color image is coded as a SYM(d)-field with suitable $d = 2, 3, \ldots$ and processed in this "detour space" before being transformed back.

5   Numerical Issues and Experiments

A convenient way to calculate the dominating eigenpair (eigenvalue end eigenvector) is the robust power method. This method is suited to our setting as we may expect a small ratio of second largest to largest eigenvalue for the matrices encountered here. The use of a complete spectral decomposition algorithm is to costly while unnecessary especially with large matrices resulting from high-dimensional multi-spectral images. In a first set of experiments we calculate the $\mathrm{psup}(c_1, c_2)$, resp. $\mathrm{pinf}(c_1, c_2)$ of a pair of colors c_1, c_2. The first experiments are concerned with gray-scales, the results confirm that the proposed approach indeed preserves gray-value morphology as shown in Fig. 2. This property is a direct consequence of the homogeneity of both the transformation of a color vector into a matrix and the pre- resp. post-processing. Black and white are the extreme gray-values, as expected. This remains true when a color, for example blue, is involved, as can be seen in Fig. 3. However, some results for pairs of colors spaced further apart in the RGB-cube are less intuitive, as Fig. 4 reveals. Still, the role of black and white as extreme colors is supported. The outer product of two vectors is a non-linear mapping, hence, it might be instructive to compare the linear combination of two colors (as vectors) with the pseudo-combination of colors via the matrix-valued setting. Some results are depicted in Fig. 5. In Fig. 6 processing of a real-world RGB image with our method is displayed. It is clearly visible that the dilation, internal gradient and external gradient behave as expected for color images. Our second experiment with real-world data displays a section of a printed circuit board with electronic components and labels (cf. Fig. 7). With a white top hat, small bright structures in different colors on darker background, like the conductor paths or the component labelling, are extracted. A morphological shock filter locally choses an dilation or erosion, depending on the sign of the dominant eigenvalue of the morphological Laplacian and clearly sharpens the edges of the image. Of course, the question remains whether the new colors make sense from the perceptional point of view. Proceeding to the multi-channel case, we represent multi-colors as points in a $x - y$-coordinate system, each point representing a channel. The first experiment in Fig. 8 (left) justifies that the pseudo-supremum/pseudo-infimum of two multi-colors deserve their name: if one vector dominates the other componentwise, their pseudo-supremum coincides with the dominating one, while the pseudo-infimum is equal to the dominated vector. However, the proposed method does not act componentwise, hence, some counter-intuitive results might occur, as further experiments reveal, see Fig. 8. We will use the matrix valued version of the Rouy-Tourin-scheme, m-RT-scheme for short, to perform PDE-driven dilation processes in the "detour"-space $\mathrm{SYM}(d)$. A further investigation how the method scales complexity-wise when the number of channels increases to hundreds has to be investigated in the future.

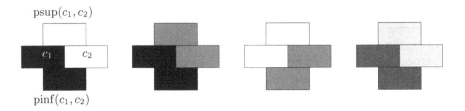

Fig. 2. The p-supremum of two gray values coincides with the maximum of the two. The same holds true for their p-infimum resp. minimum.

Fig. 3. The p-supremum/infimum involving the color blue. The p-supremum of the two shades of blue coincides with their maximum. The same holds true for their p-infimum resp. minimum. (Color figure online)

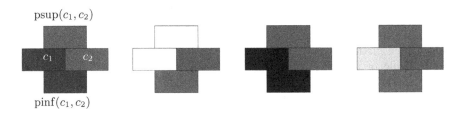

Fig. 4. P-supremum, p-infimum of colors differing significantly in the RGB-cube. (Color figure online)

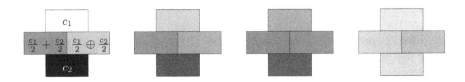

Fig. 5. Averages of pairs of colors c_1, c_2: standard average $\frac{c_1}{2} + \frac{c_2}{2}$ vs. average of the corresponding matrices, indicated by $\frac{c_1}{2} \oplus \frac{c_2}{2}$. (Color figure online)

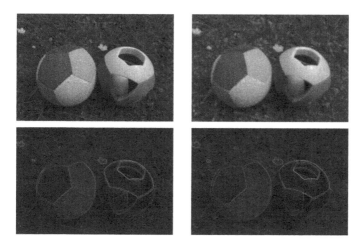

Fig. 6. Morphology on real-world color images. *Top left:* Original image, 512 × 340 pixels, RGB. *Top right:* Dilation with stopping time $t = 2$. *Bottom left:* Internal gradient with stopping time $t = 1$. *Bottom right:* External gradient with stopping time $t = 1$. (Color figure online)

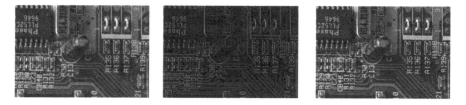

Fig. 7. Morphology on real-world color images. *Left:* Original image, 300 × 200 pixels, RGB. *Middle:* White top hat with stopping time $t = 1$. *Right:* Morphological shock filter with stopping time $t = 1$. (Color figure online)

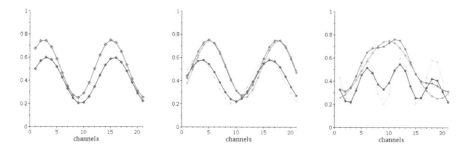

Fig. 8. Images show the performance of the proposed method for some pairs of multicolors, starting from "separated" colors (left) reaching over "weakly separated" (middle) to "interlaced" colors (right). Cyan and gold color indicate the original multicolor, while red and blue represent their pseudo-supremum, resp., pseudo-infimum. In the "separated" case left, p-supremum and p-infimum cover the original multi-colors. (Color figure online)

6   Summary

The proposed unifying approach to the processing of vector valued data, such as color- and multi-spectral images, takes advantage of the methods already available for fields of symmetric matrices. A vector field is coded as a matrix field which is then processed and finally transformed back into a vector field. However, the embedding of a vector field into a matrix field via the outer product of vectors requires pre- and post-processing of the results to allow for a consistent reconstruction of the vectors from the matrices in such a way that gray scale images (as special color-images) are processed according to standard gray-value morphology. Indeed, the methods presented boil down to gray value morphology when applied to gray-scale images. The experiments confirm the applicability of the methods to RGB-images and indicate their conceptual usefulness in the case of multispectral data. Future research will deal with challenging applications to real hyperspectral images, the intricacies of the pre- and post-processing, and, most importantly, the modeling of inter-channel correlations.

References

1. Aptoula, E., Lefèvre, S.: A comparative study on multivariate mathematical morphology. Pattern Recognit. **40**(11), 2914–2929 (2007)
2. Bhatia, R.: Matrix Analysis. Graduate Texts in Mathematics, vol. 169, 1st edn. Springer, Heidelberg (1996). https://doi.org/10.1007/978-1-4612-0653-8
3. Braun, K.M., Balasubramanian, R., Eschbach, R.: Development and evaluation of six gamut-mapping algorithms for pictorial images. In: Color Imaging Conference, pp. 144–148. IS&T - The Society for Imaging Science and Technology (1999)
4. Brockett, R.W., Maragos, P.: Evolution equations for continuous-scale morphology. In: Proceedings of the IEEE International Conference on Acoustics, Speech and Signal Processing, San Francisco, CA, vol. 3, pp. 125–128, March 1992
5. Burgeth, B., Bruhn, A., Didas, S., Weickert, J., Welk, M.: Morphology for tensor data: ordering versus PDE-based approach. Image Vis. Comput. **25**(4), 496–511 (2007)
6. Burgeth, B., Bruhn, A., Papenberg, N., Welk, M., Weickert, J.: Mathematical morphology for tensor data induced by the Loewner ordering in higher dimensions. Signal Process. **87**(2), 277–290 (2007)
7. Burgeth, B., Didas, S., Florack, L., Weickert, J.: A generic approach to diffusion filtering of matrix-fields. Computing **81**, 179–197 (2007)
8. Burgeth, B., Didas, S., Florack, L., Weickert, J.: A generic approach to the filtering of matrix fields with singular PDEs. In: Sgallari, F., Murli, A., Paragios, N. (eds.) SSVM 2007. LNCS, vol. 4485, pp. 556–567. Springer, Heidelberg (2007). https://doi.org/10.1007/978-3-540-72823-8_48
9. Burgeth, B., Kleefeld, A.: An approach to color-morphology based on Einstein addition and Loewner order. Pattern Recognit. Lett. **47**, 29–39 (2014)
10. Burgeth, B., Kleefeld, A.: Towards processing fields of general real-valued square matrices. In: Schultz, T., Özarslan, E., Hotz, I. (eds.) Modeling, Analysis, and Visualization of Anisotropy. MV, pp. 115–144. Springer, Cham (2017). https://doi.org/10.1007/978-3-319-61358-1_6

11. Burgeth, B., Pizarro, L., Breuß, M., Weickert, J.: Adaptive continuous-scale morphology for matrix fields. Int. J. Comput. Vis. **92**(2), 146–161 (2011)
12. Burgeth, B., Pizarro, L., Didas, S., Weickert, J.: 3D-coherence-enhancing diffusion filtering for matrix fields. In: Florack, L., Duits, R., Jongbloed, G., van Lieshout, M.C., Davies, L. (eds.) Mathematical Methods for Signal and Image Analysis and Representation. CIVI, vol. 41, pp. 49–63. Springer, London (2012). https://doi.org/10.1007/978-1-4471-2353-8_3
13. Eckart, C., Young, G.: The approximation of one matrix by another of lower rank. Psychometrika **1**(1), 211–218 (1936)
14. Golub, G.H., Hoffman, A., Stewart, G.W.: A generalization of the Eckart-Young-Mirsky matrix approximation theorem. Linear Algebra Appl. **88–89**, 317–327 (1987)
15. Kamal, O., et al.: Multispectral image processing for detail reconstruction and enhancement of Maya murals from La Pasadita, Guatemala. J. Archaeol. Sci. **26**(11), 1391–1407 (1999)
16. Kleefeld, A., Burgeth, B.: Processing multispectral images via mathematical morphology. In: Hotz, I., Schultz, T. (eds.) Visualization and Processing of Higher Order Descriptors for Multi-Valued Data. MV, pp. 129–148. Springer, Cham (2015). https://doi.org/10.1007/978-3-319-15090-1_7
17. Köppen, M., Nowack, C., Rösel, G.: Pareto-morphology for color image processing. In: Ersbøll, B.K. (ed.) Proceedings of the Eleventh Scandinavian Conference on Image Analysis, vol. 1, pp. 195–202. Pattern Recognition Society of Denmark, Kangerlussuaq, Greenland (1999)
18. Ngadi, M.O., Liu, L.: Chapter 4 - hyperspectral image processing techniques. In: Sun, D.W. (ed.) Hyperspectral Imaging for Food Quality Analysis and Control, pp. 99–127. Academic Press, San Diego (2010)
19. Rouy, E., Tourin, A.: A viscosity solutions approach to shape-from-shading. SIAM J. Numer. Anal. **29**, 867–884 (1992)
20. Serra, J.: Anamorphoses and function lattices (multivalued morphology). In: Dougherty, E.R. (ed.) Mathematical Morphology in Image Processing, pp. 483–523. Marcel Dekker, New York (1993)
21. Serra, J.: The "false colour" problem. In: Wilkinson, M.H.F., Roerdink, J.B.T.M. (eds.) ISMM 2009. LNCS, vol. 5720, pp. 13–23. Springer, Heidelberg (2009). https://doi.org/10.1007/978-3-642-03613-2_2
22. Tsakanikas, P., Pavlidis, D., Nychas, G.J.: High throughput multispectral image processing with applications in food science. PLOS ONE **10**(10), 1–15 (2015)
23. Unay, D.: Multispectral image processing and pattern recognition techniques for quality inspection of apple fruits. Presses univ. de Louvain (2006)
24. van den Boomgaard, R.: Mathematical morphology: extensions towards computer vision. Ph.D. thesis, University of Amsterdam, The Netherlands (1992)
25. Boomgaard, R.: Numerical solution schemes for continuous-scale morphology. In: Nielsen, M., Johansen, P., Olsen, O.F., Weickert, J. (eds.) Scale-Space 1999. LNCS, vol. 1682, pp. 199–210. Springer, Heidelberg (1999). https://doi.org/10.1007/3-540-48236-9_18
26. Yang, C.C., Chao, K., Chen, Y.R.: Development of multispectral image processing algorithms for identification of wholesome, septicemic, and inflammatory process chickens. J. Food Eng. **69**(2), 225–234 (2005)

27. Yeh, C.: Colour morphology and its approaches. Ph.D. thesis, University of Birmingham, UK (2015)
28. Yoon, S.-C., Park, B.: Hyperspectral image processing methods. In: Park, B., Lu, R. (eds.) Hyperspectral Imaging Technology in Food and Agriculture. FES, pp. 81–101. Springer, New York (2015). https://doi.org/10.1007/978-1-4939-2836-1_4

Introducing Multivariate Connected Openings and Closings

Edwin Carlinet[(✉)] and Thierry Géraud

EPITA Research and Development Laboratory (LRDE),
14-16 rue Voltaire, 94270 Le Kremlin-Bicêtre, France
edwin.carlinet@lrde.epita.fr

Abstract. Component trees provide a high-level, hierarchical, and contrast invariant representation of images, suitable for many image processing tasks. Yet their definition is ill-formed on multivariate data, e.g., color images, multi-modality images, multi-band images, and so on. Common workarounds such as marginal processing, or imposing a total order on data are not satisfactory and yield many problems, such as artifacts, loss of invariances, etc. In this paper, inspired by the way the Multivariate Tree of Shapes (MToS) has been defined, we propose a definition for a Multivariate min-tree or max-tree. We do not impose an arbitrary total ordering on values; we only use the inclusion relationship between components. As a straightforward consequence, we thus have a new class of multivariate connected openings and closings.

Keywords: Openings and closings · Component trees ·
Hierarchical representation · Color images · Multivariate data ·
Connected operators

1 Introduction

Mathematical Morphology (MM) offers a large toolbox to design image processing filters that serve as powerful building blocks and enable the user to rapidly build their image processing applications. Operators based on MM have many advantages both from the practitioner and the computer scientist view points: **1.** They are fast to compute, i.e., an erosion/dilation with a structuring element is as fast as a linear filter with a kernel of the same size. **2.** They are numerically stable, i.e., a morphological filter preserves the domain of the values (no rounding is necessary and there is no risk of overflow/underflow during computation). **3.** They are contrast-invariant, i.e., they are suitable for processing low-contrasted images as well as dark/bright images. This property is fundamental to process images with light-varying conditions.

The foundation for MM-based filters relies on complete lattices formed by digital images [20] which are well defined in the binary and grayscale cases for which a natural order exists. On the contrary, there is still no consensus for defining a *natural* order on vectors. The multivariate case is even more subtle

© Springer Nature Switzerland AG 2019
B. Burgeth et al. (Eds.): ISMM 2019, LNCS 11564, pp. 215–227, 2019.
https://doi.org/10.1007/978-3-030-20867-7_17

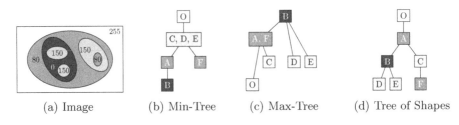

(a) Image   (b) Min-Tree   (c) Max-Tree   (d) Tree of Shapes

Fig. 1. An image (a), and its morphological component trees (b) to (d).

when extending more advanced MM filters such as those requiring a total *ranking* of values. For example, *connected* operators, that only change the values of connected components, have desirable contour-preserving properties and form a widely-used class of filters. Those filters require the values to be totally ordered.

To tackle this problem, many attempts have been done to define a "sensitive" total order which are reviewed in Sect. 2. In this paper, we adopt a different approach, following a *property-based* methodology. After a reminder about connected component-trees in Sect. 2 and how they relate to connected operators, we explain in Sect. 3 our requirements for a multivariate component tree and its construction process and its usage to extend connected filters to multivariate data. Its properties are studied in Sect. 4, and we conclude in Sect. 5.

Due to limited space, this paper only introduces the Multivariate Component Tree (MCT) and the resulting multivariate openings and closings. As a consequence, experiments with these new tools are kept for later.

2   Mathematical Background and State of the Art

2.1   Trees for Morphological Connected Openings and Closings

Connected operators are widely used in Mathematical Morphology for their properties. A connected operator ψ shares with all morphological filters the desirable property to be contrast change invariant. More formally, given a strictly increasing function $\rho : \mathbb{R} \rightarrow \mathbb{R}$ and an image u, ψ must verify $\rho(\psi(u)) = \psi(\rho(u))$. Contrary to structural openings/closings, they may not require prior knowledge about the geometry of what has to be removed. In the binary case, connected operators can only remove some connected sets of pixels. This extends for grayscale images with the threshold-set decomposition principle, connected operators are those that remove (*i.e*, merge) some flat-zones or change their gray level. As a consequence connected operators do not move object boundaries. Opening/closing by reconstruction [22,29] and area openings [30] are some widely used examples of connected filters. They were then extended to *attribute* filters [2] to express more complex forms of filtering, and Salembier et al. [23] proposed a versatile structure, namely the Max-Tree, that brought the potential of MM approaches to a higher level.

With the Max-Tree, filtering an image is as simple as pruning some branches and removing some nodes. Also, it enables much more powerful image filterings with pruning and non-pruning strategies [25], and with second-generation connectivities [17]. In [8], the authors propose a closely related structure, the Tree of Shapes (ToS), to support self-dual connected operators. The *grain filter* [7] is the self-dual counterpart of the *area* opening and removes *extremal* connected components. Most importantly, while morphological trees support fast and advanced filters, they are also hierarchical representations of the image (the reader can refer to Sect. 4.3 of Ronse [21]). As such, they enable a multi-scale image analysis and bring us to an higher level of image understanding. Since then, they have been used for image simplification, segmentation, shape-based image retrieval, compression, image registration and more. To cap it all, those trees are fast to compute [4].

More formally, let an image $u : \Omega \to E$ defined on a domain Ω and taking values on a set E equipped with an ordering relation $\leq$. Let $[u < \lambda]$ (resp. $[u > \lambda]$) with $\lambda \in E$ be a threshold set of u (also called respectively lower cut and upper cut) defined as $[u < \lambda] = \{x \in \Omega, u(x) < \lambda\}$. We denote by $CC(X), X \in \mathcal{P}(\Omega)$ the set of connected components of X. If $\leq$ is a *total* relation, any two connected components $X, Y \in CC([u < \lambda])$ are either disjoint or nested. The set $\mathcal{S} = CC([u < \lambda])$ endowed with the inclusion relation forms a tree called the *Min-Tree*. Its dual tree, defined on the upper cuts $\mathcal{S} = CC([u > \lambda])$, is called the *Max-Tree* (see Figs. 1b and c). The last morphological tree, the *Tree of Shapes (ToS)*, depicted in Fig. 1d, is based on the fusion of the Min- and Max-Trees after having filled the holes of their components. Indeed, given the hole-filling operator $\mathcal{H}$, a *shape* is any element of $\mathcal{S} = \{\mathcal{H}(\Gamma), \Gamma \in CC([u < \lambda])\}_\lambda \cup \{\mathcal{H}(\Gamma), \Gamma \in CC([u > \lambda])\}_\lambda$. Two shapes being also either nested or disjoint, $(\mathcal{S}, \subseteq)$ also forms a tree.

In the rest of this paper, the Min-Tree, Max-Tree and ToS could be used interchangeably, as we will implicitly consider a set of connected components $\mathcal{S}$ endowed with the inclusion relation (*i.e.*, the cover of $(\mathcal{S}, \subseteq)$) to denote the corresponding tree. Also, without loss of generality, we will consider $E = \mathbb{R}^n$ throughout this paper, and we will note $u = \langle u_1, u_2, \cdots, u_n \rangle$ where u is a multiband image and u_k are scalar images.

2.2 Connected Openings and Closings for Multi-band Images

A widely spread solution to extend morphological operators to multi-band images is to process the image channel-wise and finally recombine the results. Marginal processing is subject to the well-known false color problem as it may create new values that were not in the original images. False colors may or may not be a problem by itself (e.g. if the false colors are perceptually close to the original ones) but for image simplification it may produce undesirable artifacts.

Since the problem of defining multivariate connected operators lies in the absence of a total order between values, many attempts have been done to define a total order on vectorial data. Two strategies are mainly used: *conditional (C-) ordering* that gives priorities to some (or all) of the vector components, *reduced*

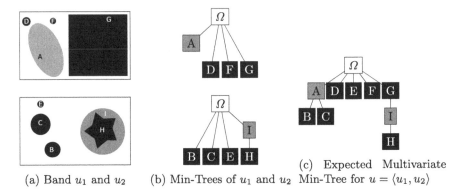

(a) Band u_1 and u_2 (b) Min-Trees of u_1 and u_2 (c) Expected Multivariate Min-Tree for $u = \langle u_1, u_2 \rangle$

Fig. 2. A multi-band image, marginal Min-Tree and the Multivariate Min-Tree that fulfills our requirements.

(R-) ordering that defines a ranking projection function and orders vectors by their rank (a total pre-order).

Commonly used ranking functions are the l_1-norm or the luminance in a given color space. It makes sense if we assume that the geometric information is mainly held by the luminance [7] but it is not that rare to face images where edges are only available in the color space. In other words, this strategy is not sufficient if the geometric information cannot be summed up to a single dimension. In Tushabe and Wilkinson [24], Perret et al. [19], Naegel and Passat [15], authors used another widely used R-ordering, the distance to a reference set of values to extend min- and max- trees to multivariate data for image compression or astronomical object detection. This approach is well-founded whenever the background is uniform or defined as set of values. R-orderings are usually combined with C-orderings (typically a lexicographic cascade) to get a strong ordering (as in [9], to extend grain filters on colors).

More advanced strategies have been designed to build a more "sensitive" total ordering that depends on image content. Velasco-Forero and Angulo [27,28] use machine learning techniques to partition the value space, then a distance to clusters allows to build an ordering. In [13], manifold learning is used to infer a ranking function of values and in [12] a locally-dependent ordering is computed on spatial windows. Lezoray and Elmoataz [11], Lézoray [10] combine both ideas for a manifold learning in a domain-value space capturing small dependencies between a pixel and its neighbors during the construction of the total order. More recently, keeping with content-based ordering approaches, Veganzones et al. [26] proposed an R-ordering based an indexing of the leaves of the image binary partition tree. A review of vector orderings applied to MM can be found in [1].

In [18], the authors propose to deal directly with the partial ordering of values and manipulate the underlying graph structure. While theoretically interesting, a component-graph is algorithmically harder to deal with and the complexity of the structure (in terms of computation time) compels the authors to perform the filtering only locally [16].

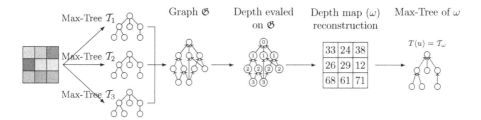

Fig. 3. Scheme of the multivariate component tree construction process.

Finally, in [5], we introduced the Multivariate Tree of Shapes (MToS) as a novel structure to represent color images. While dedicated to extend the Tree of Shapes, the basic idea lies in a sensible strategy to merge several ToS and thus, can be transposed to merge any set of component trees.

3   Extension of Morphological Trees for Multivariate Data

3.1   Requirements for a Multivariate Component Tree

The construction of the Multivariate Component Tree (MCT) is designed from the following premises. First, we have a channel-wise prior about the objects we want to detect and their background (*i.e*, we know channel-wise the contrast "direction"). Second, we have no prior about how objects are spread among the channels, they may be visible in a single, in some or in many channels. Before going further into details of the *what is* a MCT, and *how to get it*, let us start with the *what for* and a description of the properties we want it to have:

Single-band Equivalence. On a single-band image, the Multivariate Component Tree must be the same as the normal component tree.

Preservative Behavior. Given a connected component C of $\mathcal{S}$, if C is either nested or disjoint to every other connected components, then it must appears in the Multivariate Component Tree. This property actually covers the first one. It is illustrated in Fig. 2 with two Min-Trees; since there is no overlap between components from the two channels of u, the Multivariate Min-Tree is intuitive. Note that the "merging" is purely based on the inclusion, there is no *less-than* relation (in terms of values) between the components from u_1 and u_2. Indeed, in this example, I becomes a child of G while its value in u_2 is greater than in u_1. Being agnostic about the value ordering when merging trees is closely related to the marginal contrast change invariance described hereafter.

Invariance to any Marginal Increasing Change of Contrast. For any family of increasing function $\{\rho_1, \rho_2, \cdots \rho_n\}$, $\mathcal{T}(\langle \rho_1(u_1), \rho_2(u_2), \cdots \rho_n(u_n)\rangle) = \mathcal{T}(u)$. This property enforces a fundamental property of morphological representation: well-contrasted and low-contrasted objects are considered equally. This property

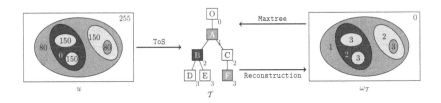

Fig. 4. Equivalence between a hierarchy (here the ToS) and the Max-Tree of its *depth* map.

is twofold. First it enables to consider channels that have different dynamics whereas linear-based approaches generally require a proper prior data normalization. Second, it makes the representation robust when a change of exposure appears only in some channels.

3.2   Multivariate Component Tree Construction

The merging process of the marginal component trees consists in 2 main steps: the construction of a depth map ω and the deduction of a tree from ω. These two steps, designed to get the properties described in Sect. 3.1, are described hereafter.

Getting a Tree Out of a Depth Map. Before going further into details, let us explain and motivate the construction of the *depth* map. The depth conveys the object inclusion level. The lighter an object appears in the depth map, the deeper it stands in the object hierarchy. In [6], we showed that given a hierarchy and its corresponding *depth* map, one can recover the initial hierarchy by computing its max-tree, as illustrated in Fig. 4. The interest of the *depth* map lies in the way it abstracts away the underlying value ordering relation between objects. Indeed, no matter the input hierarchy, whether is it a Min-tree (based on the inclusion of *lower* components), Max-Tree (based on the inclusion of *upper* components) or a ToS (based on the inclusion of *hole-filled lower and upper* components), it is equivalent to the Max-Tree computed on its *depth* map. Figure 5 shows the *depth* maps $\omega_{\mathcal{T}min}$, $\omega_{\mathcal{T}max}$, and $\omega_{\mathcal{T}tos}$ of an image computed from its Min-Tree, its Max-Tree and its ToS. As one can see, dark objects appear bright in (b), bright objects appear bright in (c) and the most inner shapes appear bright in (d). Actually, we can show that for a gray-level image u, $\omega_{\mathcal{T}min} = \rho(255 - u)$ and $\omega_{\mathcal{T}max} = \rho(u)$ with ρ some increasing contrast change. As a consequence, it is straightforward that the Max-Tree of $\omega_{\mathcal{T}}$ is $\mathcal{T}$ itself.

The *depth* map leads to an alternate representation of the image content. Instead of having a representation based on the brightness, we now have a pixel-wise interpretation of the *inclusion level* of objects. In other words, if a pixel appears brighter than its neighbor in the *depth* map, this means that it belongs to a nested sub-object.

(a) Input u (b) $\omega_{\mathcal{T}min}$ (c) $\omega_{\mathcal{T}max}$ (d) $\omega_{\mathcal{T}tos}$

Fig. 5. An input gray-level image (a) and its corresponding *depth* map for the Min-Tree (b), Max-Tree (c) and ToS (d) hierarchies. The dynamic of the depth image has been stretched to $[0-255]$.

From Marginal Component Trees to the Graph of Components. The component trees $\mathcal{T}_1, \mathcal{T}_2, \ldots, \mathcal{T}_n$ are computed on each band $u_1, u_2, \ldots, u_n$ of the input image. Each tree is associated with its set of components $\mathcal{S}_1, \mathcal{S}_2, \ldots, \mathcal{S}_n$. Let $\mathcal{S} = \bigcup \mathcal{S}_i$, we call the Graph of Components (GoC) $\mathfrak{G}$ the cover of $(\mathcal{S}, \subseteq)$. The GoC, depicted in Fig. 3, is actually the inclusion graph of all the connected components computed marginally.

It is worth mentioning even if the GoC is actually a tree, it is not a valid morphological tree. Indeed, in "standard" morphological hierarchies (min-/max-trees) and their extension (the component-graph [18]), for any point x, there exists a *single* smallest component that contains x. As a consequence, a point belongs to a *single* node in the structure. In the GoC, a point may belong to several nodes. For example, in Fig. 6, the points in $(A \cap B)$ belong to both nodes A and B, but $(A \cap B)$ does not exist as a node in any marginal tree. Thus, we cannot just extract a tree (e.g. the minimum spanning tree) from the GoC as it will not be valid.

It is also worth mentioning that the graph of components $\mathfrak{G}$ is different from the component-graph introduced by Passat and Naegel [18]. The latter is the inclusion graph of all connected components based on a partial ordering $\prec$ (*i.e*, the set $\{\mathcal{CC}([u \prec \lambda]), \lambda \in \mathbb{R}^n\}$) while $\mathfrak{G}$ is the inclusion graph of connected components computed marginally (*i.e*, the set $\bigcup_{k=1}^{n}\{\mathcal{CC}([u_k < \lambda]), \lambda \in \mathbb{R}\}$). $\mathfrak{G}$ is thus a sub-graph of the component-graph.

Note that the graph is a "complete" representation of the input image u that can be reconstructed from $\mathfrak{G}$. Indeed, without loss of generality, suppose that $\mathfrak{G}$ is built from $\mathcal{T}_1, \cdots, \mathcal{T}_n$ being Max-Trees. For a component A of $\mathfrak{G}$, let $\lambda_i(A) = \min\{u_i(x), x \in A\}$, then the input can be reconstructed with:

$$u(x) = \left\langle \max_{X:x\in X} \lambda_1(X), \max_{X:x\in X} \lambda_2(X), \cdots, \max_{X:x\in X} \lambda_n(X) \right\rangle. \tag{1}$$

Constructing a *depth* Map from the Graph of Components. As said previously, getting a morphological tree out of the GoC is not as simple as it seems (e.g. with a MST algorithm) as one has to ensure that pixels do not belong to disjoint branches of the tree. We have also seen that the *depth* map is an interesting intermediate representation, which provides pixel-wise description

Fig. 6. A GoC that is an invalid morphological tree. Left: two marginal components from two channels are overlapping. Right: The corresponding Graph of Components (GoC) $\mathfrak{G}$. The points of $A \cap B$ belong to both nodes A and B, so the tree $\mathfrak{G}$ is not valid.

of the organization of the scene (in terms of inclusion) and enables us to extract a tree out of it. As a consequence, our objective is now to build a *depth* map from the GoC. Let ρ be the *depth* of a node A in $\mathfrak{G}$, *i.e*, $\rho(A)$ is the length of the longest path of a component A from the root. Let $\omega : \Omega \to \mathbb{R}$ defined as:

$$\omega(x) = \max_{X \in \mathcal{S}, x \in X} \rho(X). \tag{2}$$

The map ω associates each point x with the depth of the deepest component containing x. Let $\mathbb{C} = \bigcup_{h \in \mathbb{R}} \mathcal{CC}([\omega \geq h])$. $(\mathbb{C}, \subseteq)$ is actually the max-tree of ω. The method is illustrated in Fig. 7 where a two-channel image has overlapping components from red and green components. The trees $\mathcal{T}_1$ and $\mathcal{T}_2$ are computed marginally and merged into the GoC $\mathfrak{G}$ for which we compute the depth of each node (Fig. 7b). Those values are reported in the image space, pixel-wise (Fig. 7c). This is the step which decides which components are going to merge; here B and D are set to the same value. This choice is based on the level of inclusion and no longer on the original pixel values.

While the graph is a complete representation of u, once $\mathfrak{G}$ is flattened to ω, it is not possible to go backward and recover the original connected components. This is the only lossy step part of the process.

Filtering and Reconstruction with the Multivariate Component Tree.
Morphological trees enables to perform openings, closings and extremal filterings in two steps: **1.** a pruning where nodes are removed and pixels are re-assigned; **2.** a restitution where pixels get their final values.

Let $\mathbb{P}$ be a binary predicate that tells if a node is to be removed. For example, $\mathbb{P}(X) = \bar{X} > \alpha$ denotes a predicate that retains all components whose size is above α (grain filter, area opening and closing). Similarly to Eq. (1), the reconstruction after filtering the Max-Tree $\mathcal{T} = (\mathcal{S}, \subseteq)$ of a scalar image u is:

$$\tilde{u}(x) = \max_{X \in \mathcal{S},\ x \in X \,|\, \mathbb{P}(X)} \lambda(X). \tag{3}$$

Equation (3) actually describes the **direct** filtering, the pixels are affected to the level of the *last surviving* node in the hierarchy. This is one of the four filtering strategies described in [14]. The others -namely *min, max, subtractive*- only make sense for a non-increasing predicate, that is a predicate that may keep

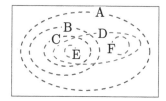

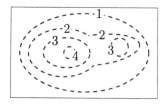

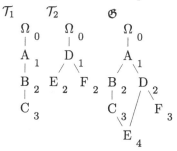

(a) A 2-band image u and its components.

(c) ω image built from $\mathfrak{G}$

(b) The marginal Max-Trees $\mathcal{T}_1$, $\mathcal{T}_2$ and the GoC $\mathfrak{G}$. The depth appears in gray near the nodes.

(d) The Max-Tree $\mathcal{T}_\omega$

Fig. 7. Merging trees with a depth map; in this simple example, we suppose that u verifies: $u_1(\Omega) < u_1(A) < u_1(B) < u_1(C)$ and $u_2(\Omega) < u_2(D) < u_2(E) < u_2(F)$.

a node while its parent is going to be deleted. Interested reader should refer to [14] for more details on filtering strategies.

Last, we need to define the restitution for multivariate data. One could simply define the restitution as in Eq. (1) with a predicate constraint as shown in Eq. (5):

$$\Gamma(x) = \{X \in \mathcal{S}_w \mid x \in X, \mathbb{P}(X)\} \tag{4}$$

$$\tilde{u}(x) = \left\langle \max_{X \in \Gamma(x)} \lambda_1(X), \max_{X \in \Gamma(x)} \lambda_2(X), \cdots, \max_{X \in \Gamma(x)} \lambda_n(X) \right\rangle. \tag{5}$$

Simply stated, each pixel is assigned the infimum of the pixels of its node. The painting *Several Circles*[1] Fig. 8 does not reveal how the filters behave on "real" natural images but enables us to illustrate the troubles we can face when filtering colors. Figure 8a shows that neighboring regions with non-comparable colors (like those of A, B and C) merge at the same level.

The restitution strategy from Eq. (5) (let us call it *Rinf*) leads to several artifacts as shown in Fig. 8b. First, this restitution may affect pixels that belong to components that should be preserved. Second, it creates *false* colors that were not present in the original image (as in region D).

---

[1] *Einige Kreise*, from Vasily Kandinsky, 2016 (Solomon R. Guggenheim Museum, New York, ©2018 Artists Rights Society, New York/ADAGP, Paris).

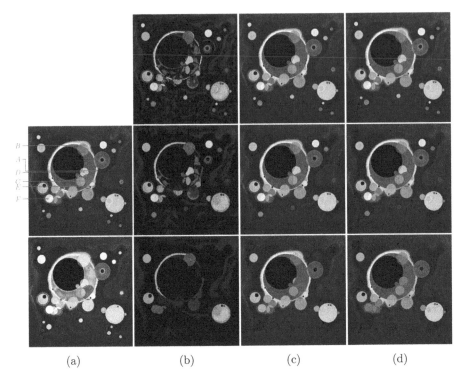

Fig. 8. Comparison of a marginal opening and multivariate openings with several restitution strategies. (a) Original image (top) and its depth map ω_u (bottom). (b) Multivariate opening with restitution strategy *Rinf.* (c) Multivariate opening with restitution strategy *RNC.* (d) Marginal opening. Openings are of sizes 500, 4000 and 16000 (top-down). (Color figure online)

This second problem is shared with the *marginal* processing strategy shown in Fig. 8d. It creates false-colors because it combines marginal filtering results (e.g. in region F). It is also worth noting that, for the same reason, it creates flat zones that are smaller than the filtering grain size (e.g. the region C) and blurs object boundaries. That is why a marginal processing is a legitimate filtering approach when the quality is assessed according to the human perception.

The restitution strategy proposed in [24] tackles those problems by keeping the original pixel values when they belong to a node "living" and using the a color of the contour of the connected component when they belong to a "dead" one. Results of the nearest color strategy are shown in Fig. 8c. It combines the strength of a vectorial approach that produces "real" flat zones with no contour blurring and a quite qualitative color filtering similar to the one obtained by marginal filtering. Yet, there remain some restitution quality troubles for nodes which merge non-comparable regions as we need to choose a single color to reconstruct both regions. It yields color artifacts as in B, and C where orange and magenta pixels have been colorized in cyan from A.

4 Algebraic Properties and Complexity

First, trivially, the Multivariate Min-Tree and the Multivariate Max-Tree are dual by construction. The *preservative behavior* property is also ensured by construction. If a component A is either nested or disjoint to any other component, then this is a "bottleneck" node in $\mathfrak{G}$, in the sense that every path leading to a sub-component of A needs to pass through this node. As a consequence, every pixels in A will have a depth ω greater than $\rho(A)$ (*i.e*, in $\mathcal{CC}[\omega > \rho(A)]$) and there exists a node in $\mathcal{T}_\omega$ corresponding to A.

The openings described previously meet the prerequisites described in Sect. 3.1. Due to the lack of space, we only give the intuition of the proofs. The *single-band equivalence* is straightforward; if there is a single input tree $\mathcal{T}_1$, the graph $\mathfrak{G}$ is thus a tree and $\mathcal{T}_\omega = \mathcal{T}_1$. Whichever restitution strategy is then used, they all are equivalent in this case and produce the same output as the regular attribute opening. The *marginal-contrast-change invariance* is also straightforward; since each tree $\mathcal{T}_i$ is contrast change invariant for the i-th channel (and does not depend on the others), so does the graph $\mathfrak{G}$. Also, since the rest of the process only depends on the inclusion relation between components, the whole process is thus marginal-contrast-change invariant.

Morphological properties for multivariate openings and closings based on $\mathcal{T}_\omega$ depends on the restitution strategy. With the strategy $Rinf$, the openings (resp. closings) are marginally anti-extensive (resp. extensive). However, in the general case they are neither idempotent, nor marginally anti-extensive.

From a computational standpoint, in the classical case of images with low-quantized values (for instance, with values encoded on 8 bit in every channel/band), the marginal trees can be computed in linear time w.r.t. the number N of pixels [4]. The most expensive part is the graph computation which is $O(n^2.H.N)$, where n is the number of channels, H the maximal depth of the trees, and N the number of pixels.

5 Conclusion

This paper introduced preliminary ideas to extend connected filters on multi-band images[2]. This approach relies on the extension of component trees to data for which no natural total order exists. Instead of imposing an arbitrary total ordering, the method tries to combine and merge connected components from several marginal component trees. As such, it can be seen as a local and context-dependent ordering of the pixels based on their level of inclusion in each hierarchy. Beyond multivariate connected filters, our method produces a single hierarchical representation of the image, the MCT, that can be used to study "shapes" in images [3]. As a further work, we plan to figure out a way to prioritize some bands while keeping the idea of "ordering by inclusion level". We also plan to

---

[2] Some extra illustrations can be found on https://publications.lrde.epita.fr/carlinet. 19.ismm.

transform the value space before using the MCT (whether by a principal component analysis, or by projection in a given color-space) to study the effect of the data correlation on the merging process used by the MCT.

References

1. Aptoula, E., Lefèvre, S.: A comparative study on multivariate mathematical morphology. Pattern Recognit. **40**(11), 2914–2929 (2007)
2. Breen, E.J., Jones, R.: Attribute openings, thinnings, and granulometries. Comput. Vis. Image Underst. **64**(3), 377–389 (1996)
3. Cao, F., Lisani, J.-L., Morel, J.-M., Musé, P., Sur, F.: A Theory of Shape Identification. LNM, vol. 1948. Springer, Heidelberg (2008). https://doi.org/10.1007/978-3-540-68481-7
4. Carlinet, E., Géraud, T.: A comparative review of component tree computation algorithms. IEEE Trans. Image Process. **23**(9), 3885–3895 (2014)
5. Carlinet, E., Géraud, T.: MToS: a tree of shapes for multivariate images. IEEE Trans. Image Process. **24**(12), 5330–5342 (2015)
6. Carlinet, E., Géraud, T., Crozet, S.: The tree of shapes turned into a max-tree: a simple and efficient linear algorithm. In: Proceedings of the International Conference on Image Processing (ICIP), pp. 1488–1492, Athens, Greece, October 2018
7. Caselles, V., Monasse, P.: Grain filters. JMIV **17**(3), 249–270 (2002)
8. Caselles, V., Coll, B., Morel, J.-M.: Topographic maps and local contrast changes in natural images. Int. J. Comput. Vis. **33**(1), 5–27 (1999)
9. Coll, B., Froment, J.: Topographic maps of color images. In: Proceedings of the International Conference on Pattern Recognition (ICPR), vol. 3, pp. 609–612 (2000)
10. Lézoray, O.: Complete lattice learning for multivariate mathematical morphology. J. Vis. Commun. Image Represent. **35**, 220–235 (2016)
11. Lezoray, O., Elmoataz, A.: Nonlocal and multivariate mathematical morphology. In: Proceedings of the International Conference on Image Processing (ICIP), pp. 129–132 (2012)
12. Lezoray, O., Meurie, C., Elmoataz, A.: A graph approach to color mathematical morphology. In: Proceedings of the IEEE International Symposium on Signal Processing and Information Technology, pp. 856–861 (2005)
13. Lezoray, O., Charrier, C., Elmoataz, A.: Rank transformation and manifold learning for multivariate mathematical morphology. In: Proceedings of European Signal Processing Conference, vol. 1, pp. 35–39 (2009)
14. Meijster, A., Westenberg, M., Wilkinson, M., Hamza, M.: Interactive shape preserving filtering and visualization of volumetric data. In: Proceedings of the 5th IASTED Conference on Computer Graphics and Imaging (CGIM), pp. 640–643 (2002)
15. Naegel, B., Passat, N.: Component-trees and multi-value images: a comparative study. In: Wilkinson, M.H.F., Roerdink, J.B.T.M. (eds.) ISMM 2009. LNCS, vol. 5720, pp. 261–271. Springer, Heidelberg (2009). https://doi.org/10.1007/978-3-642-03613-2_24
16. Naegel, B., Passat, N.: Towards connected filtering based on component-graphs. In: Hendriks, C.L.L., Borgefors, G., Strand, R. (eds.) ISMM 2013. LNCS, vol. 7883, pp. 353–364. Springer, Heidelberg (2013). https://doi.org/10.1007/978-3-642-38294-9_30

17. Ouzounis, G.K., Wilkinson, M.H.: Mask-based second-generation connectivity and attribute filters. IEEE Trans. PAMI **29**(6), 990–1004 (2007)

18. Passat, N., Naegel, B.: An extension of component-trees to partial orders. In: Proceedings of the International Conference on Image Processing (ICIP), pp. 3933–3936 (2009)

19. Perret, B., Lefèvre, S., Collet, C., Slezak, E.: Connected component trees for multivariate image processing and applications in astronomy. In: Proceedings of the International Conference on Pattern Recognition (ICPR), pp. 4089–4092, August 2010

20. Ronse, C.: Why mathematical morphology needs complete lattices. Signal Process. **21**(2), 129–154 (1990)

21. Ronse, C.: Ordering partial partitions for image segmentation and filtering: merging, creating and inflating blocks. J. Math. Imaging Vis. **49**(1), 202–233 (2014)

22. Salembier, P., Serra, J.: Flat zones filtering, connected operators, and filters by reconstruction. IEEE Trans. Image Process. **4**(8), 1153–1160 (1995)

23. Salembier, P., Oliveras, A., Garrido, L.: Antiextensive connected operators for image and sequence processing. IEEE TIP **7**(4), 555–570 (1998)

24. Tushabe, F., Wilkinson, M. : Color processing using max-trees: a comparison on image compression. In Proceedings of the International Conference on Systems and Informatics (ICSAI), pp. 1374–1380 (2012)

25. Urbach, E.R., Wilkinson, M.H.F.: Shape-only granulometries and grey-scale shape filters. In: Mathematical Morphology and Its Applications to Signal and Image Processing-Proceedings of the International Symposium on Mathematical Morphology (ISMM), vol. 2002, pp. 305–314 (2002)

26. Veganzones, M.Á., Dalla Mura, M., Tochon, G., Chanussot, J.: Binary partition trees-based spectral-spatial permutation ordering. In: Benediktsson, J.A., Chanussot, J., Najman, L., Talbot, H. (eds.) ISMM 2015. LNCS, vol. 9082, pp. 434–445. Springer, Cham (2015). https://doi.org/10.1007/978-3-319-18720-4_37

27. Velasco-Forero, S., Angulo, J.: Supervised ordering in $\mathcal{R}_p$: application to morphological processing of hyperspectral images. IEEE Trans. Image Process. **20**(11), 3301 (2011)

28. Velasco-Forero, S., Angulo, J.: Random projection depth for multivariate mathematical morphology. IEEE J. Sel. Top. Signal Process. **6**(7), 753–763 (2012)

29. Vincent, L.: Morphological grayscale reconstruction in image analysis: applications and efficient algorithms. IEEE Trans. Image Process. **2**(2), 176–201 (1993)

30. Vincent, L.: Morphological area openings and closings for grey-scale images. In: O, Y.L., Toet, A., Foster, D., Heijmans, H.J.A.M., Meer, P. (eds.) Shape in Picture. NATO ASI Series (Series F: Computer and Systems Sciences), vol. 126. Springer, Cham (1994). https://doi.org/10.1007/978-3-662-03039-4_13

Approaches to Multivalued Mathematical Morphology Based on Uncertain Reduced Orderings

Mateus Sangalli and Marcos Eduardo Valle[⊠]

Department of Applied Mathematics,
University of Campinas, Campinas, SP, Brazil
{ra156684,valle}@ime.unicamp.br

Abstract. Mathematical morphology (MM) is a powerful non-linear theory that can be used for signal and image processing and analysis. Although MM can be very well defined on complete lattices, which are partially ordered sets with well defined extrema operations, there is no natural ordering for multivalued images such as hyper-spectral and color images. Thus, a great deal of effort has been devoted to ordering schemes for multivalued MM. In a reduced ordering, in particular, elements are ranked according to the so-called ordering mapping. Despite successful applications, morphological operators based on reduced orderings are usually too reliant on the ordering mapping. In many practical situations, however, the ordering mapping may be subject to uncertainties such as measurement errors or the arbitrariness in the choice of the mapping. In view of this remark, in this paper we present two approaches to multivalued MM based on an uncertain reduced ordering. The new operators are formulated as the solution of an optimization problem which, apart from the uncertainty, can circumvent the false value problem and deal with irregularity issues.

Keywords: Mathematical morphology · Multivalued image ·
Optimization problem · Uncertainty

1 Introduction

Mathematical morphology (MM) is a powerful non-linear theory that uses geometric and topological concepts for signal and image processing and analysis [12,18]. Applications of MM include, for instance, boundary detection, image segmentation and reconstruction, pattern recognition, and signal and image decomposition [4,8,16].

Apart from the geometrical interpretation inherent to many morphological operators, they can be very well defined on an algebraic structure called complete lattices [12,14]. A complete lattice $\mathbb{L}$ is a partially ordered non-empty set in

This work was supported in part by CNPq grant no. 310118/2017-4 and FAPESP grant no. 2019/02278-2.

B. Burgeth et al. (Eds.): ISMM 2019, LNCS 11564, pp. 228–240, 2019.
https://doi.org/10.1007/978-3-030-20867-7_18

which any subset admits both a supremum and an infimum [3,10]. Since the only requirement is a partial order with well-defined extreme operations, complete lattices allowed for the development of morphological operators to multivalued data, including vector-valued images such as color and hyper-spectral images [2,13]. In contrast to real-valued approaches, however, there is no natural ordering for vectors. Therefore, much research on vector-valued MM has been dedicated to finding an appropriate ordering scheme for a given multivalued image processing task. The interested reader can find detailed discussions of multivalued MM in [1,2].

One simple approach to multivalued MM is obtained by applying gray-scale morphological operators in a component-wise manner. From the complete lattice point of view, the component-wise approach is based on the marginal ordering, also known as the product ordering. The marginal ordering is an example of a partial ordering which is not total [11]. A morphological operator based on a non-total ordering scheme may yield *false values*, also called *false colors*. A false value is an element from the set of values that does not belong to the original image and it can be a problem in some vector-valued image processing tasks [7,17]. Furthermore, the information between bands of a vector-valued image (or channels in a color image) are ignored in a component-wise approach.

In order to circumvent the *false values* problem, a great deal of effort has been dedicated to multivalued morphological operators based on a total ordering scheme. Among the total ordering approaches, those obtained by combining a reduced ordering with a look-up table are particularly interesting and computationally cheap [9,21]. The idea behind an approach based on a reduced ordering can be summarized as follows. First, a vector-valued image is transformed into a gray-scale (usually real-valued) image using a surjective mapping h called *ordering mapping*. Then, a flat gray-scale morphological operator is applied on the resulting image and a semi-inverse of h is used to recover a vector-valued image [9]. The semi-inverse mapping can be determined using a look-up table in which the vector-values are ranked accordingly to their h-values [21].

In many practical situations, the h mapping has a continuous range and the probability of two different vectors yield the same h-value is very small. In these situations, the resulting reduced ordering becomes a total ordering when restricted to the range of a vector-valued image [6]. Although total orderings avoid the appearance of *false values*, they are usually irregular in a metric space [6]. Specifically, Chevallier and Angulo showed that under mild conditions there always exist vectors $\mathbf{x}, \mathbf{y}, \mathbf{z}$ such that $\mathbf{x} \leq \mathbf{y} \leq \mathbf{z}$ but $d(\mathbf{x}, \mathbf{z}) < d(\mathbf{x}, \mathbf{y})$, where d denotes a metric and "$\leq$" is a total order. In other words, $\mathbf{z}$ is more similar (or it is closer) to $\mathbf{x}$ than $\mathbf{y}$ in spite of the inequalities $\mathbf{x} \leq \mathbf{y} \leq \mathbf{z}$. Like false values, the irregularity issue may be a problem in some vector-valued image processing tasks.

In this paper, we propose two approaches to multivalued MM that prevent the apparition of false values but may not be defined using a total ordering and, thus, possibly avoiding the irregularity issue. In fact, the novel approaches can deal with the irregularity issue whenever it can be properly measured. The motivation behind our approaches stems from the fact that reduced orderings are also usually too reliant on the mapping h. The mapping h, however, can be subject to uncertainties such as the unavoidable measurement errors under real physical conditions when acquiring a vector-valued image. Broadly speaking, our approaches are derived by relaxing a reduced ordering but including a kind of regularization goal.

The paper is organized as follows: The next section reviews multivalued MM while Sect. 3 briefly presents some approaches based on reduced orderings. We address the uncertainties involved in a reduced ordering and present our approaches in Sect. 4. The paper finishes with some concluding remarks on Sect. 5.

2   Mathematical Morphology for Multivalued Images

First of all, an image is a mapping from a point set $\mathcal{D}$ to a value set $\mathbb{V}$. In particular, a gray-scale image is obtained by considering $\mathbb{V} \subset \bar{\mathbb{R}}$, where $\bar{\mathbb{R}} = \mathbb{R} \cup \{+\infty, -\infty\}$. We speak of a multivalued image when $\mathbb{V} \subset \bar{\mathbb{R}}^k$ for $k \geq 2$. For simplicity, in this paper we shall assume that the domain $\mathcal{D}$ is a finite subset of either $E = \mathbb{R}^2$ or $E = \mathbb{Z}^2$. We denote[1] the set of all images from a domain $\mathcal{D}$ to a value set $\mathbb{V}$ by $\mathcal{V} = \mathbb{V}^{\mathcal{D}}$.

Briefly, morphological operators examine an image by probing it with a small pattern called structuring element [12,18]. The structuring element is used to extract useful information about the geometrical and topological structures on an image. Such as the domain of an image, we assume that a structuring element S corresponds to a finite subset of either $E = \mathbb{R}^2$ or $E = \mathbb{Z}^2$ with $\mathrm{Card}(S) \ll \mathrm{Card}(\mathcal{D})$.

As pointed out in the introduction, complete lattices constitute an appropriate framework for a general theory of MM [12,14]. A partially ordered set $(\mathcal{L}, \leq)$ is a complete lattice if any subset $X \subset \mathcal{L}$ admits a supremum and an infimum, denoted respectively by $\bigvee X$ and $\bigwedge X$. On a complete lattice, the fundamental operators of MM, called erosion and dilation, are defined as follows using an adjunction relationship [12]:

Definition 1. *Let $\mathcal{L}$ be a complete lattice. We say that $\varepsilon : \mathcal{L} \to \mathcal{L}$ and $\delta : \mathcal{L} \to \mathcal{L}$ form an adjunction if*

$$\mathbf{J} \leq \varepsilon(\mathbf{I}) \iff \delta(\mathbf{J}) \leq \mathbf{I}, \quad for\ \mathbf{I}, \mathbf{J} \in \mathcal{L}. \tag{1}$$

If ε and δ form an adjunction, then ε is an erosion and δ is a dilation.

Erosions and dilations are the two elementary operations of MM [18]. Many other morphological operators are obtained by combining erosions and dilations. For example, their compositions yield the so-called opening and closing, which have interesting topological properties and are used as non-linear image filters [18].

---

[1] Throughout this paper, blackboard bold capital letters such as $\mathbb{V}$ and $\mathbb{L}$ are used to denote value sets while calligraphic capital letters like $\mathcal{V}$ and $\mathcal{L}$ are used to denote sets of images.

Let us assume that the value set $\mathbb{V}$, equipped with a certain partial ordering "$\leq$", is a complete lattice. The set $\mathcal{L} = \mathbb{V}^{\mathcal{D}}$ of all images from $\mathcal{D} \subset E$ to a value set $\mathbb{V}$ is also a complete lattice with the pointwise ordering defined by $\mathbf{I} \leq \mathbf{J}$ if and only if $\mathbf{I}(p) \leq \mathbf{J}(p)$ for all $p \in \mathcal{D}$. Furthermore, given a structuring element $S \subset E$, define the operators $\varepsilon_S : \mathcal{L} \to \mathcal{L}$ and $\delta_S : \mathcal{L} \to \mathcal{L}$ as follows:

$$\varepsilon_S(\mathbf{I})(p) = \bigwedge \{\mathbf{I}(p+s) : s \in S, p+s \in \mathcal{D}\}, \quad \forall p \in \mathcal{D}, \tag{2}$$

and

$$\delta_S(\mathbf{I})(p) = \bigvee \{\mathbf{I}(p+s) : s \in S, p+s \in \mathcal{D}\}, \quad \forall p \in \mathcal{D}. \tag{3}$$

It is not hard to show that ε_S and δ_{S^*}, where $S^* = \{-s \mid s \in S\}$ is the reflected structuring element, form an adjunction on $\mathcal{L} = \mathbb{V}^{\mathcal{D}}$ equipped with the pointwise ordering. Thus, ε_S and δ_S defined by (2) and (3) are respectively an erosion and a dilation. In fact, these two operators are the most widely used (flat) elementary operations of either gray-scale and multivalued MM.

3   Reduced Orderings

In a reduced ordering (R-ordering) or h-ordering, the elements are ranked according to a surjective mapping $h : \mathcal{V} \to \mathcal{L}$, where $\mathcal{L}$ is a complete lattice [9]. Given a surjective mapping $h : \mathcal{V} \to \mathcal{L}$, referred to as the ordering mapping, the h-ordering is defined by

$$\mathbf{x} \leq_h \mathbf{y} \iff h(\mathbf{x}) \leq h(\mathbf{y}), \quad \forall \mathbf{x}, \mathbf{y} \in \mathcal{V}. \tag{4}$$

Although being reflexive and transitive, the binary relation "$\leq_h$" may fail to be anti-symmetric. Thus, an h-ordering "$\leq_h$" is in fact a pre-order. Nevertheless, an h-ordering can be used to define elementary morphological operators as follows [9]:

Definition 2. *Consider a surjective mapping h from a value set $\mathcal{V}$ to a complete lattice $\mathcal{L}$. We say that two operators $\varepsilon^h, \delta^h : \mathcal{V} \to \mathcal{V}$ form an h-adjunction if*

$$\mathbf{y} \leq_h \varepsilon^h(\mathbf{x}) \iff \delta^h(\mathbf{y}) \leq_h \mathbf{x}, \quad \text{for } \mathbf{x}, \mathbf{y} \in \mathcal{V}. \tag{5}$$

If ε^h and δ^h form an h-adjunction, we say that ε^h is an h-erosion and δ^h is an h-dilation.

In practice, we obtain an h-adjunction (an h-erosion and an h-dilation) from an adjunction (an erosion and a dilation) on $\mathcal{L}$ [9]. Precisely, if $\varepsilon^h : \mathcal{V} \to \mathcal{V}$ and $\delta^h : \mathcal{V} \to \mathcal{V}$ form an h-adjunction, then there exist adjoint operators $\varepsilon : \mathcal{L} \to \mathcal{L}$ and $\delta : \mathcal{L} \to \mathcal{L}$ such that

$$h\varepsilon^h = \varepsilon h \quad \text{and} \quad h\delta^h = \delta h. \tag{6}$$

In other words, the evaluation of the h-erosion ε^h (or the h-dilation δ^h) by h equals the erosion ε (or the dilation δ) of the evaluation by h. In particular, if we consider $\mathcal{L} = \bar{R}$, then the classical gray-scale elementary operators ε and δ given respectively by (2) and (3) can be used to determine h-morphological operators.

In analogy to the complete lattice case, h-dilations and h-erosions can be combined to yield many h-morphological operators. For example, if ε^h and δ^h form an h-adjunction, then $\gamma^h = \varepsilon^h \delta^h$ is an h-opening and $\phi^h = \delta^h \varepsilon^h$ is an h-closing. Alternatively, in analogy to (6), we can construct a vector-valued operator $\psi^h : \mathcal{V} \to \mathcal{V}$ from a morphological operator $\psi : \mathcal{L} \to \mathcal{L}$ using Proposition 1 below, which is based on the notion of h-increasing operator [9].

Definition 3. *We say that an operator $\psi^h : \mathcal{V} \to \mathcal{V}$ is h-increasing if $\mathbf{x} \leq_h \mathbf{y}$ implies $\psi^h(\mathbf{x}) \leq_h \psi^h(\mathbf{y})$.*

Proposition 1. *An operator $\psi^h : \mathcal{V} \to \mathcal{V}$ is h-increasing if and only if there exists an increasing operator $\psi : \mathcal{L} \to \mathcal{L}$ such that*

$$h\psi^h = \psi h, \qquad (7)$$

In this case, we write $\psi^h \xrightarrow{h} \psi$.

From Proposition 1, any h-increasing operator $\psi^h : \mathcal{V} \to \mathcal{V}$ can be derived from an increasing operator $\psi : \mathcal{L} \to \mathcal{L}$ by means of (7). Needless to say, h-erosions and h-dilations are h-increasing operators. Accordingly, we have $\varepsilon^h \xrightarrow{h} \varepsilon$ and $\delta^h \xrightarrow{h} \delta$ from (6). In a similar fashion, h-openings and h-closings are h-increasing operators. Therefore, an h-opening $\gamma^h : \mathcal{V} \to \mathcal{V}$ and an h-closing $\phi^h : \mathcal{V} \to \mathcal{V}$ can be derived from an opening $\gamma : \mathcal{L} \to \mathcal{L}$ and a closing $\phi : \mathcal{L} \to \mathcal{L}$ such that $h\gamma^h = \gamma h$ and $h\phi^h = \phi h$.

When an h-morphological operator $\psi^h : \mathcal{V} \to \mathcal{V}$, derived from a surjective mapping $h : \mathcal{V} \to \mathcal{L}$ and an increasing operator $\psi : \mathcal{L} \to \mathcal{L}$, is applied to vector-valued images, we consider $\mathcal{V} = \mathbb{V}^{\mathcal{D}}$ and $\mathcal{L} = \mathbb{L}^{\mathcal{D}}$, where $\mathbb{V}$ denotes a vector-valued set and $\mathbb{L}$ is a complete lattice. It turns out that a mapping $h : \mathbb{V} \to \mathbb{L}$ can be extended to images $\mathbf{I} \in \mathcal{V}$ in a pointwise manner as follows:

$$h\mathbf{I}(p) = h(\mathbf{I}(p)), \quad \forall p \in \mathcal{D}. \qquad (8)$$

Although we are concerned with operators for vector-valued images, we shall focus on a mapping $h : \mathbb{V} \to \mathbb{L}$ and assume it is extended to multivalued images using (8). The interested reader is referred to [21] for a review of reduced orderings in vector-valued MM, as well as to [1,5,15,19,20] for some examples of reduced orderings used in vector-valued MM that are found in the literature. Moreover, from Proposition 1, we can compute efficiently a vector-valued h-morphological operator ψ^h from a classical flat gray-scale increasing operator ψ using a look-up table [21]. The interested reader can find in [21] a MATLAB-style pseudo-code for computing $\psi^h(\mathbf{I})$ from an increasing morphological operator ψ that does not introduce false values.

As far as we know, almost all reduced orderings are defined by considering mappings $h : \mathbb{V} \to \mathbb{L} \subset \bar{R}$. Furthermore, we usually have $h(\mathbf{x}) \neq h(\mathbf{y})$ for distinct $\mathbf{x}$ and $\mathbf{y}$ in $\mathbf{I}(\mathcal{D}) = \{\mathbf{I}(p) : p \in \mathcal{D}\}$, the range of the multivalued image $\mathbf{I}$. In this

a) $\mathbf{I}$ b) $h\mathbf{I}$ c) $\gamma_S^h(\mathbf{I})$

Fig. 1. Color image $\mathbf{I}$, its evaluation $h\mathbf{I}$, and the h-opening $\gamma_S^h(\mathbf{I})$ by a disk of radius 10. (Color figure online)

case, the h-ordering yields a total ordering on $\mathbf{I}(\mathcal{D})$. Accordingly, although the h-ordering prevents the occurrence of false values (or colors), it is subject to irregularity issues [6]. A visual interpretation of this remark is illustrated in Fig. 1. Precisely, Fig. 1(a) shows a color image $\mathbf{I}$ while Fig. 1(b) depicts the gray-scale image $h\mathbf{I}$ obtained using the mapping $h : \mathbb{V}_{\text{RGB}} \to [0, 1]$ defined by

$$h(\mathbf{x}) = 0.298936x_R + 0.587043x_G + 0.114021x_B, \forall \mathbf{x} = (x_R, x_G, x_B) \in \mathbb{V}_{\text{RGB}}, \quad (9)$$

where $\mathbb{V}_{\text{RGB}} = [0, 1] \times [0, 1] \times [0, 1]$. Also, Fig. 1(c) shows the h-opening $\gamma_S^h(\mathbf{I})$ by a (flat) disk structuring element S of radius 10 obtained using the look-up table algorithm detailed in [21]. Note some irregularities at the border of the head of the red parrot. Similar irregularities can be observed by considering more complex h-mappings such as the one discussed in [19] and [15].

4 Approximation of h-Increasing Operators

Despite the many successful applications of h-morphological operators, they usually are too sensitive to the h-ordering mapping. Furthermore, there can be many possible sources of uncertainty when dealing with an h-ordering which may affect the outcome of an h-morphological operator. For instance, under real physical conditions, there exist measurement errors when acquiring a vector-valued image. Also, the inevitable presence of rounding errors may cause some distortions on an h-ordering. From the methodological point of view, we may be uncertain about the method, the parameters, or the data used to determine the ordering mapping. In the light of these remarks, let us assume that we only have an approximation $\tilde{h} : \mathcal{V} \to \mathcal{L}$ of the ideal unknown ordering mapping $h : \mathcal{V} \to \mathcal{L}$. Also, let us assume that the complete lattice $\mathbb{L}$ as well as the set of lattice-valued images $\mathcal{L}$ are metric spaces.

According to the previous section, given a vector-valued image $\mathbf{I} \in \mathcal{V}$ and an increasing morphological operator $\psi : \mathcal{L} \to \mathcal{L}$, a $\tilde{h}$-morphological operator $\psi^{\tilde{h}} : \mathcal{V} \to \mathcal{V}$ yields a vector-valued image $\mathbf{J} = \psi^{\tilde{h}}(\mathbf{I}) \in \mathcal{V}$ such that

$$\tilde{h}\mathbf{J} = \psi(\tilde{h}\mathbf{I}). \quad (10)$$

Now, if we are uncertain about the ordering mapping $\tilde{h}$, we propose to replace (10) by the inequality

$$d_{\mathcal{L}}\big(\tilde{h}\mathbf{J}, \psi(\tilde{h}\mathbf{I})\big) \leq \tau, \tag{11}$$

where $\tau \geq 0$ is a prescribed tolerance parameter and $d_{\mathcal{L}} : \mathcal{L} \times \mathcal{L} \to [0, +\infty)$ is a metric. On the one hand, $\tau = 0$ implies the equality (10), which means we have complete confidence in the observed ordering function $\tilde{h} : \mathcal{V} \to \mathcal{L}$. On the other hand, we have no confidence in the ordering function when τ is sufficiently large. In this second case, the vector-valued image $\mathbf{J}$ must be determined using other criteria.

In order to prevent the occurrence of false values, we impose that the range of the output image is contained in the range of the input image. More precisely, in order to preserve local information, we impose that $\mathbf{J}(p)$ is a value in a neighborhood $B_p \subset \mathcal{D}$ of the pixel $p \in \mathcal{D}$. Formally, we have

$$\mathbf{J}(p) \in \mathbf{I}(B_p), \quad \forall p \in \mathcal{D}. \tag{12}$$

We refer to $B_p \subset \mathcal{D}$ as the *local window*. Usually, we define $B_p = \{p + q : q \in B\} \cap \mathcal{D}$ where, like the structuring element, B is a finite subset of either $E = \mathbb{R}^2$ or $E = Z^2$.

Apart from the constraints (11) and (12), we consider an appropriate objective to be minimized such as the irregularities of the output image $\mathbf{J}$. The objective can also circumvent ambiguities when there exists more than one feasible vector-value image $\mathbf{J}$ satisfying both (11) and (12). In mathematical terms, the objective to be minimized is described by a functional $F : \mathcal{V} \to \mathbb{R}$ which associates to a vector-valued image a certain scalar measure. Concluding, we define a new morphological operator as a minimum point of the objective function F subject to the constraints (11) and (12).

4.1 τ-Morphological Operator

Given an increasing morphological operator $\psi : \mathcal{L} \to \mathcal{L}$, an uncertain ordering mapping $\tilde{h} : \mathcal{V} \to \mathcal{L}$, an objective function $F : \mathcal{V} \to \mathbb{R}$, a local window B, and a tolerance $\tau \geq 0$, we define the τ-morphological operator $\psi^{\tau} : \mathcal{V} \to \mathcal{V}$ by setting $\psi^{\tau}(\mathbf{I}) = \mathbf{J}^*$ where $\mathbf{J}^*$ is an optimal solution of the optimization problem:

$$\begin{cases} \text{minimize} & F(\mathbf{J}), \\ \text{subject to} & d_{\mathcal{L}}\big(\tilde{h}\mathbf{J}, \psi(\tilde{h}\mathbf{I})\big) \leq \tau, \\ & \mathbf{J}(p) \in \mathbf{I}(B_p), \quad \forall p \in \mathcal{D}. \end{cases} \tag{13}$$

Note that the optimization problem (13) is feasible if $\psi(\tilde{h}\mathbf{I})(p) \in \tilde{h}\mathbf{I}(B_p)$ for all $p \in \mathcal{D}$. Indeed, if for any $p \in \mathcal{D}$ there exists $q \in B_p$ such that $\psi(\tilde{h}\mathbf{I})(p) = \tilde{h}\mathbf{I}(q)$, then setting $\mathbf{J}(p) = \mathbf{I}(q)$ yields an image $\mathbf{J}$ such that (11) holds true for any $\tau \geq 0$. For example, the optimization problem (13) is feasible if one considers a classical erosion or a classical dilation by a flat structuring element S which is included in the local window B, that is, $S \subset B$. Apart from the feasibility issues, the optimization problem (13) can be computationally intractable. Thus, let us present a simplified version of this problem which scales linearly in the size of the image and the number of bands or channels.

A simplified problem is obtained if the objective function $F : \mathbb{V}^{\mathcal{D}} \to \mathbb{R}$ is formulated in a pointwise manner as follows where $F_p : \mathbb{V} \to \mathbb{R}$, for all $p \in \mathcal{D}$:

$$F(\mathbf{J}) = \sum_{p \in \mathcal{D}} F_p(\mathbf{J}(p)). \tag{14}$$

Although F_p may change for each pixel $p \in \mathcal{D}$, it depends only on the value $\mathbf{x} = \mathbf{J}(p)$. For example, we can define the objective function $F_p : \mathbb{V} \to \mathbb{R}$ as follows for all $p \in \mathcal{D}$ where $\mathbf{R} \in \mathbb{V}^{\mathcal{D}}$ is a reference vector-valued image and $d_{\mathbb{L}}$ denotes a metric on $\mathbb{L}$:

$$F_p(\mathbf{x}) = d_{\mathbb{L}}(\mathbf{x}, \mathbf{R}(p)). \tag{15}$$

More importantly, by considering a pointwise ordering mapping $\tilde{h} : \mathbb{V} \to \mathbb{L}$ and the maximum metric $d_{\mathcal{L}}(\mathbf{I}, \mathbf{J}) = \max\{d_{\mathbb{L}}(\mathbf{I}(p), \mathbf{J}(p)) : p \in \mathcal{D}\}$, the optimization problem (13) can be decomposed into the following optimization problems for all $p \in \mathcal{D}$:

$$\begin{cases} \text{minimize} & F_p(\mathbf{x}), \\ \text{subject to} & d_{\mathbb{L}}(\tilde{h}(\mathbf{x}), \psi(\tilde{h}\mathbf{I})(p)) \leq \tau, \quad \mathbf{x} \in \mathbf{I}(B_p). \end{cases} \tag{16}$$

Then, the pointwise τ-morphological operator $\psi^\tau : \mathbb{V}^{\mathcal{D}} \to \mathbb{V}^{\mathcal{D}}$ is defined by setting, for all $p \in \mathcal{D}$, $\psi^\tau(\mathbf{I})(p) = \mathbf{x}^*$, where $\mathbf{x}^*$ is an optimal solution of (16).

In the worst case, a solution of (16) can be obtained by an exhaustive search on $\mathbf{I}(B_p)$. Moreover, such exhaustive search requires $\mathrm{Card}(B)$ evaluations of F_p and $d_{\mathbb{L}}$, which are usually linear on the number of bands or channels of the multivalued image. Thus, the vector-valued image $\psi^\tau(\mathbf{I})$ can be obtained performing $\mathcal{O}(\mathrm{Card}(\mathcal{D}) \cdot \mathrm{Card}(B))$ operations and evaluations of the objective functions and the metric $d_{\mathbb{L}}$.

Like (13), the optimization problems given by (16) are all feasible if $\psi(\tilde{h}\mathbf{I})(p) \in \tilde{h}\mathbf{I}(B_p), \forall p \in \mathcal{D}$. In our implementations, we alert the user and define $\psi^\tau(\mathbf{I})(p) = \mathbf{I}(p)$ if the optimization problem (16) is unfeasible for some $p \in \mathcal{D}$.

4.2 λ-Morphological Operator

Alternatively, instead of using the restriction $d_{\mathcal{L}}(\tilde{h}\mathbf{J}, \psi(\tilde{h}\mathbf{I})) \leq \tau$, we may add this term multiplied by a scaling factor $\lambda \geq 0$ in the objective function. The resulting optimization problem is:

$$\begin{cases} \text{minimize} & F(\mathbf{J}) + \lambda d_{\mathcal{L}}(\tilde{h}\mathbf{J}, \psi(\tilde{h}\mathbf{I})), \\ \text{subject to} & \mathbf{J}(p) \in \mathbf{I}(B_p). \end{cases} \tag{17}$$

Given an increasing morphological operator $\psi : \mathcal{L} \to \mathcal{L}$, an uncertain mapping $\tilde{h} : \mathcal{V} \to \mathcal{L}$, an objective function $F : \mathcal{V} \to \mathbb{R}$, a local window B, and a parameter

$\lambda \geq 0$, we define the λ-morphological operator $\psi^\lambda : \mathbb{V}^\mathcal{D} \to \mathbb{V}^\mathcal{D}$ by means of the equation $\psi^\lambda(\mathbf{I}) = \mathbf{J}^*$, where $\mathbf{J}^*$ is an optimal solution of (17) for a given $\mathbf{I} \in \mathbb{V}^\mathcal{D}$.

We would like to call the reader's attention to two important issues related to the λ-morphological operator ψ^λ defined by means of (17). First, the parameter $\lambda \geq 0$ in (17) controls the trade-off between the objective function and the ordering mapping $\tilde{h}$. Specifically, the larger the parameter $\lambda \geq 0$ the higher the confidence in the observed mapping $\tilde{h}$. Indeed, if λ is assigned a small value, the ordering mapping $\tilde{h}$ is considered to be uncertain and less emphasis is placed on it. At the other extreme, the similarity between $\tilde{h}\mathbf{J}$ and $\psi(\tilde{h}\mathbf{I})$ dominates the objective function when λ is sufficiently large.

Secondly, but not less important, the optimization problem (17) is always feasible. Thus, in contrast (16), it does not require $\psi(\tilde{h}\mathbf{I})(p) \in \tilde{h}\mathbf{I}(B_p)$, $\forall p \in \mathcal{D}$. As a consequence, $\psi^\lambda : \mathbb{V}^\mathcal{D} \to \mathbb{V}^\mathcal{D}$ is well-defined even when the underlying morphological operator $\psi : \mathbb{L}^\mathcal{D} \to \mathbb{L}^\mathcal{D}$ is based on a non-flat structuring element.

Finally, in analogy to the previous τ-morphological operator, (17) can be simplified if the objective function $F : \mathbb{V}^\mathcal{D} \to \mathbb{R}$ can be written as (15) and the metric $d_\mathcal{L}$ is given by $d_\mathcal{L}(\mathbf{I}, \mathbf{J}) = \sum_{p \in \mathcal{D}} d_\mathbb{L}(\mathbf{I}(p), \mathbf{J}(p))$, where $d_\mathbb{L}$ denotes a metric on $\mathbb{L}$. Using exhaustive search, the λ-morphological operator ψ^λ requires $\mathcal{O}(\mathrm{Card}(\mathcal{D}) \cdot \mathrm{Card}(B))$ operations and evaluations of the objective functions and $d_\mathbb{L}$. GNU Octave source-codes of both τ-morphological and λ-morphological operators are available at https://codeocean.com/capsule/3486771/tree/v1.

4.3   Computational Experiments

This subsection provides some computational experiments to illustrate the new multivalued morphological operators. For simplicity, let us consider the color image $\mathbf{I}$ shown in Fig. 1(a), an opening γ_S by a disk S of radius 10 as the increasing morphological operator ψ, a disk of radius 20 as the local window B, and the functional $F_p : \mathbb{V}_{\mathrm{RGB}} \to \mathbb{R}$ given by (15) with the usual Euclidean distance. Also, let us suppose we only know a noisy version $\tilde{h} : \mathbb{V}_{\mathrm{RGB}} \to [0, 1]$ of the (ideal) ordering mapping $h : \mathbb{V}_{\mathrm{RGB}} \to [0, 1]$ given by (4). In our experiments, $\tilde{h}$ is obtained by adding uncorrelated Gaussian noise with zero mean and variance $\sigma^2 = 0.01$, i.e., $\tilde{h}(\mathbf{x}) = \max\{0, \min\{1, h(\mathbf{x}) + N\}\}$, where $N \sim \mathcal{N}(0, 0.01)$.

First, let us address the effect of the parameters $\tau \geq 0$ and $\lambda \geq 0$ on the operators γ_S^τ and γ_S^λ defined respectively by (16) and (17) with $\psi = \gamma_S$ and the reference image $\mathbf{R} = \mathbf{I}$. Quantitatively, Fig. 2 shows the average *peak signal-to-noise ratio* (PSNR) between the ideal h-opening $\gamma_S^h(\mathbf{I})$ and the operators γ^τ and γ^λ for different values of the parameters $\tau \geq 0$ and $\lambda \geq 0$. The average PSNR have been obtained by performing the same experiment 20 times for each value of the parameters τ and λ. For comparison purposes, we included in Fig. 2 the horizontal lines corresponding to the PSNR values between the ideal h-opening $\gamma_S^h(\mathbf{I})$ and the original image $\mathbf{I}$, the noisy $\tilde{h}$-opening $\gamma_S^{\tilde{h}}(\mathbf{I})$, and the filtered $g\tilde{h}$-opening $\gamma_S^{g\tilde{h}}(\mathbf{I})$. Here, $\gamma_S^{\tilde{h}}(\mathbf{I})$ and $\gamma_S^{g\tilde{h}}(\mathbf{I})$ have been computed using the look-up table algorithm with $\tilde{h}\mathbf{I}$ and $g\tilde{h}\mathbf{I}$, respectively, where g denotes an isotropic Gaussian filter with spread 0.5. In other words, the filtered $g\tilde{h}$-opening

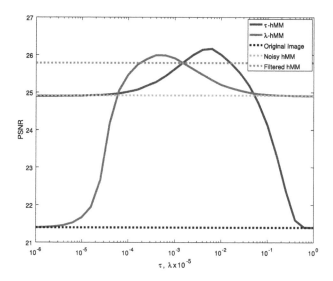

Fig. 2. PSNR between the ideal h-opening $\gamma_S^h(\mathbf{I})$ and the openings $\gamma_S^\tau(\mathbf{I})$, $\gamma_S^\lambda(\mathbf{I})$, $\gamma_S^{\tilde{h}}(\mathbf{I})$, $\gamma_S^{g\tilde{h}}(\mathbf{I})$, or the original image $\mathbf{I}$ by the parameters τ or $\lambda \times 10^{-5}$.

is obtained by trying to remove the Gaussian noise from $\tilde{h}\mathbf{I}$ using a Gaussian filter before computing the h-morphological operator. As expected, we have

$$\lim_{\tau \to 0} \mathrm{PSNR}\big(\gamma_S^\tau(\mathbf{I}), \gamma_S^h(\mathbf{I})\big) = \mathrm{PSNR}\big(\gamma_S^{\tilde{h}}(\mathbf{I}), \gamma_S^h(\mathbf{I})\big) = \lim_{\lambda \to \infty} \mathrm{PSNR}\big(\gamma_S^\lambda(\mathbf{I}), \gamma_S^h(\mathbf{I})\big),$$

and

$$\lim_{\tau \to \infty} \mathrm{PSNR}\big(\gamma_S^\tau(\mathbf{I}), \gamma_S^h(\mathbf{I})\big) = \mathrm{PSNR}\big(\mathbf{I}, \gamma_S^h(\mathbf{I})\big) = \lim_{\lambda \to 0} \mathrm{PSNR}\big(\gamma_S^\lambda(\mathbf{I}), \gamma_S^h(\mathbf{I})\big).$$

Most importantly, the largest PSNR rates are obtained for $\tau = 0.0063$ and $\lambda = 39.81$. For these values, the new operators outperformed the $g\tilde{h}$-opening obtained by filtering the noisy $\tilde{h}\mathbf{I}$ before applying the look-up table algorithm [21]. A visual interpretation of the original image $\mathbf{I}$ and the outcome of the openings $\gamma_S^h(\mathbf{I})$, $\gamma_S^{\tilde{h}}(\mathbf{I})$, $\gamma_S^\tau(\mathbf{I})$, $\gamma_S^\lambda(\mathbf{I})$, and $\gamma_S^{g\tilde{h}}(\mathbf{I})$ is shown in Fig. 3. Comparing Figs. 3(b) and (c), we observe how h-morphological operators are sensitive to the ordering mapping. We can also visualize a significant improvement on the outcome of the new operators $\gamma_S^\tau(\mathbf{I})$ and $\gamma_S^\lambda(\mathbf{I})$, including some kind of regularity in the edge of the head of the red parrot.

Finally, let us turn our attention to the objective function. We previously considered $\mathbf{R} = \mathbf{I}$ but, evidently, another image can be used as the reference image. Figure 4 shows the images $\gamma_S^\tau(\mathbf{I})$ and $\gamma_S^\lambda(\mathbf{I})$ produced by the new operators using $\tau = 0.02$ and $\lambda = 60$ but considering $\mathbf{R} = \gamma_S^M(\mathbf{I})$, where γ_S^M denotes the opening by a disk of radius 10 obtained using the marginal approach. As a consequence, $\gamma_S^\tau(\mathbf{I})$ and $\gamma_S^\lambda(\mathbf{I})$ approximate the marginal opening γ_S^M, constrained by the noisy $\tilde{h}$-mapping, but without introducing *false colors*. In this

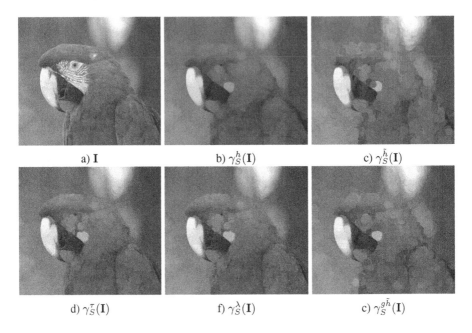

a) $\mathbf{I}$ b) $\gamma_S^h(\mathbf{I})$ c) $\gamma_S^{\bar{h}}(\mathbf{I})$

d) $\gamma_S^\tau(\mathbf{I})$ f) $\gamma_S^\lambda(\mathbf{I})$ c) $\gamma_S^{g\bar{h}}(\mathbf{I})$

Fig. 3. Color image $\mathbf{I}$ and its openings by a disk of radius 10 obtained using different approaches. (Color figure online)

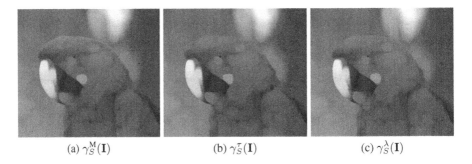

(a) $\gamma_S^M(\mathbf{I})$ (b) $\gamma_S^\tau(\mathbf{I})$ (c) $\gamma_S^\lambda(\mathbf{I})$

Fig. 4. Marginal opening γ_S^M by a disk S of radius 10 and the corresponding γ_S^τ and γ_S^λ operators using γ_S^M as the reference in (15) and the parameters $\tau = 2 \cdot 10^{-2}$ and $\lambda = 60$. (Color figure online)

simple example, the openings shown in Fig. 4 circumvented the irregularity issues present in the openings shown in Figs. 1 and 3. Furthermore, it reveals some of the potential advantages of the new operators ψ^τ and ψ^λ.

5   Concluding Remarks

Despite the rich mathematical background and its successful applications for analysis and processing of color and hyperspectral images [9,20,21], morphological operators based on reduced orderings are usually too sensitive to the ordering mapping. In many practical situations, however, the ordering mapping h is subject to uncertainties such as the unavoidable measurement errors.

In this paper, by assuming that the ordering mapping is uncertain, we proposed two approaches which are formulated as the solution of optimization problems. Precisely, we assume that we only have an approximation $\tilde{h}$ of the ideal ordering mapping h. Given an increasing morphological operator $\psi : \mathcal{L} \to \mathcal{L}$, we define the τ-operator ψ^τ and the λ-operator ψ^λ as the solution of the optimization problems (13) and (17), respectively. Apart from the capability to deal with an uncertain ordering mapping, the new operators can circumvent both the false values problem and the irregularity issues found in many approaches to multivalued mathematical morphology [6,17]. In fact, we avoid false values by imposing $\mathbf{J}(p) \in \mathbf{I}(B_p)$, where B_p is a local window. The irregularity issue can be dealt by considering an adequate objective function. Finally, both optimization problems (13) and (17) scales linearly in the size of the image and the local window if the objective function can be decomposed in a pointwise manner according to (15).

In order to illustrate the new operators and how they are able to deal with the uncertainty present in the ordering mapping, we presented some simple computational experiments with a color image in the RGB color space. By considering an uncertain ordering mapping $\tilde{h}$ obtained by adding Gaussian noise, both τ-opening and λ-opening showed to get closer to the ideal h-opening than the usual approach with or without a filter. The proposed approaches also significantly reduced the visual irregularity present in the images. To illustrate the effects of the choice of the objective function, we used the proposed operators to approximate the marginal opening, which is fairly regular. Visually, both τ-opening and λ-opening are very similar to the marginal opening but they do not introduce false colors.

Future work relating to this paper includes studying how the new operators, in particular the λ-operators given by (17), work with non-flat structuring elements. One can also investigate and compare the performance of the new operators in applications such as image classification or segmentation.

References

1. Angulo, J.: Morphological colour operators in totally ordered lattices based on distances: application to image filtering, enhancement and analysis. Comput. Vis. Image Underst. **107**(1–2), 56–73 (2007). Special issue on color image processing
2. Aptoula, E., Lefèvre, S.: A comparative study on multivariate mathematical morphology. Pattern Recogn. **40**(11), 2914–2929 (2007)
3. Birkhoff, G.: Lattice Theory, 3rd edn. American Mathematical Society, Providence (1993)
4. Braga-Neto, U., Goutsias, J.: Supremal multiscale signal analysis. SIAM J. Math. Anal. **36**(1), 94–120 (2004)
5. Chanussot, J., Lambert, P.: Total ordering based on space filling curves for multivalued morphology. In: Proceedings of the Fourth International Symposium on Mathematical Morphology and Its Applications to Image and Signal Processing, ISMM 1998, pp. 51–58. Kluwer Academic Publishers, Norwell (1998)

6. Chevallier, E., Angulo, J.: The irregularity issue of total orders on metric spaces and its consequences for mathematical morphology. J. Math. Imaging Vis. **54**(3), 344–357 (2016)

7. Comer, M.L., Delp, E.J.: Morphological operations for color image processing. J. Electron. Imaging **8**(3), 279–289 (1999)

8. Gonzalez-Hidalgo, M., Massanet, S., Mir, A., Ruiz-Aguilera, D.: On the choice of the pair conjunction-implication into the fuzzy morphological edge detector. IEEE Trans. Fuzzy Syst. **23**(4), 872–884 (2015)

9. Goutsias, J., Heijmans, H.J.A.M., Sivakumar, K.: Morphological operators for image sequences. Comput. Vis. Image Underst. **62**, 326–346 (1995)

10. Grätzer, G., et al.: General Lattice Theory, 2nd edn. Birkhäuser Verlag, Basel (2003)

11. van de Gronde, J., Roerdink, J.: Group-invariant colour morphology based on frames. IEEE Trans. Image Process. **23**(3), 1276–1288 (2014)

12. Heijmans, H.J.A.M.: Mathematical morphology: a modern approach in image processing based on algebra and geometry. SIAM Rev. **37**(1), 1–36 (1995)

13. Lézoray, O.: Complete lattice learning for multivariate mathematical morphology. J. Vis. Commun. Image Represent. **35**, 220–235 (2016)

14. Ronse, C.: Why mathematical morphology needs complete lattices. Signal Process. **21**(2), 129–154 (1990)

15. Sangalli, M., Valle, M.E.: Color mathematical morphology using a fuzzy color-based supervised ordering. In: Barreto, G.A., Coelho, R. (eds.) NAFIPS 2018. CCIS, vol. 831, pp. 278–289. Springer, Cham (2018). https://doi.org/10.1007/978-3-319-95312-0_24

16. Serra, J.: A lattice approach to image segmentation. J. Math. Imaging Vis. **24**, 83–130 (2006)

17. Serra, J.: The "False Colour" problem. In: Wilkinson, M.H.F., Roerdink, J.B.T.M. (eds.) ISMM 2009. LNCS, vol. 5720, pp. 13–23. Springer, Heidelberg (2009). https://doi.org/10.1007/978-3-642-03613-2_2

18. Soille, P.: Morphological Image Analysis. Springer, Berlin (1999). https://doi.org/10.1007/978-3-662-03939-7

19. Velasco-Forero, S., Angulo, J.: Supervised ordering in $\mathbb{R}^p$: application to morphological processing of hyperspectral images. IEEE Trans. Image Process. **20**(11), 3301–3308 (2011)

20. Velasco-Forero, S., Angulo, J.: Random projection depth for multivariate mathematical morphology. IEEE J. Sel. Top. Signal Process. **6**(7), 753–763 (2012)

21. Velasco-Forero, S., Angulo, J.: Vector ordering and multispectral morphological image processing. In: Celebi, M.E., Smolka, B. (eds.) Advances in Low-Level Color Image Processing. LNCVB, vol. 11, pp. 223–239. Springer, Dordrecht (2014). https://doi.org/10.1007/978-94-007-7584-8_7

Computational Morphology

Efficient 3D Erosion Dilation Analysis by Sub-Pixel EDT

Michael Godehardt[1]([✉]), Dennis Mosbach[1,2], Diego Roldan[1,3],
and Katja Schladitz[1]

[1] Fraunhofer-Institut für Techno- und Wirtschaftsmathematik, Fraunhofer-Platz 1,
67663 Kaiserslautern, Germany
`michael.godehardt@itwm.fraunhofer.de`
[2] Computer Graphics and HCI Group, Technische Universität Kaiserslautern,
67663 Kaiserslautern, Germany
[3] Department of Mathematics, Technische Universität Kaiserslautern,
67663 Kaiserslautern, Germany
`http://www.itwm.fraunhofer.de/en/departments/image-processing/`

Abstract. The micro-structure of materials or natural substrates contributes significantly to macroscopic properties like mechanical strength or filtration performance. The intrinsic volumes or their densities are versatile geometric characteristics of micro-structures and can be estimated efficiently from 3D binary images whose foreground represents the structure of interest. A recent algorithm generalizes this approach to gray value images. That is, the intrinsic volumes are derived for each possible global gray value threshold in the image in a single pass through the image. Here, it is combined with a sub-pixel precise Euclidean distance transform enabling efficient so-called erosion dilation analysis. That means, the densities of the intrinsic volumes are simultaneously computed not only for the structure itself but also for all erosions and dilations by spherical structuring elements of this structure. That way, the algorithm reveals additional information on the local size of critical features of the structure or its complement. The algorithm's power is demonstrated by means of computed tomography image data of rigid foams and stacks of scanning electron microscopies of nano-porous membrane layers sliced by a focused ion beam.

Keywords: Intrinsic volumes · Gray value images ·
Minkowski function · Connectivity function · Spatial analysis

1 Introduction

The possibilities and the demand to spatially image materials micro-structures have grown tremendously during the last decade. Image sizes, complexity of the imaged structures, and detail of the analysis tasks grow at even higher speed, increasing the demand for time and memory efficient algorithms yielding quantitative structural information.

© Springer Nature Switzerland AG 2019
B. Burgeth et al. (Eds.): ISMM 2019, LNCS 11564, pp. 243–255, 2019.
https://doi.org/10.1007/978-3-030-20867-7_19

One very general image analysis tool are the intrinsic volumes. The intrinsic volumes, also known as Minkowski functionals or quermass integrals [1], are in some sense a basic set of geometric structural characteristics [2]. In 3D, they yield information about the volume, surface, mean and Gaussian curvatures of analyzed structures. Various other characteristics describing for instance shape [3] or structure specific features e. g. for open cell foams [4] can be derived. The densities of the intrinsic volumes, combined with erosions and dilations of the structure under consideration have been studied, e. g. to quantify connectivity [5]. Mecke [6] called them Minkowski functions.

For a given segmentation of the gray value image into the component of interest (foreground) and its complement (background), the intrinsic volumes can be efficiently derived from the resulting binary image by Ohser's algorithm [7,8]. In [9], we generalized Ohser's approach to gray value images enabling fast simultaneous calculation of the intrinsic volumes for all possible threshold values in an integer gray value image.

2   Intrinsic Volumes, Euclidean Distance Transform, and Erosion Dilation Analysis

In this section, we summarize the previous work that is combined in the following Sect. 3 to yield a finer erosion dilation analysis than before as demonstrated by the examples in Sect. 4.

2.1   Intrinsic Volumes Estimated Based on 3D Binary Images

The intrinsic volumes (or their densities) are a widely used system of basic geometric characteristics for micro-structures, see e.g. [10–14]. In 3D, the four intrinsic volumes are volume V, surface area S, integral of mean curvature M, and Euler number χ. For macroscopically homogeneous structures, the densities of the respective characteristics are used – volume fraction V_V, specific surface area S_V, and the densities of the integrals of mean curvature M_V and of the Euler number χ_V. From these basic characteristics, further descriptors can be derived. Here, we consider in addition to the densities of the intrinsic volumes the structure model index $\mathrm{SMI} = 12 V_V M_V / S_V^2$ that can be interpreted as a shape factor for a macroscopically homogeneous structure [3,15].

These characteristics can be measured simultaneously and efficiently based on observations restricted to a compact window. The algorithmic idea goes back to [11] and was refined in [8,16]: The local $2 \times 2 \times 2$ pixel configurations in a binary 3D image are coded in an 8bit gray value image of the same size by an appropriate convolution. The intrinsic volumes are then derived as scalar product of a suitable weight vector and the gray value histogram h of this image.

2.2   Intrinsic Volumes as Function of Gray Value Threshold in 3D Images

In [9], Ohser's algorithm was generalized: The local black-or-white pixel configurations induced by thresholding a gray value image are efficiently collected and coded.

For the sake of self-containedness, we repeat the main idea of [9]: Consider the matrix of frequency vectors $h(t)$ for each gray value threshold $t \in \mathbb{R}$. Observe that the contribution of a local $2 \times 2 \times 2$ pixel configuration $p = (p_1, \ldots, p_8)$ can change not more than eight times – each time one of the pixels' gray values coincides with the threshold value. Write $c(p, t)$ for the code of pixel configuration p after thresholding with threshold t. Create $h(c, t)$ for all $c \in \{0, \ldots, 255\}$, $t \in \{0, \ldots, M+1\}$, where M is the image's maximal gray value, by bookkeeping of changing codes for local pixel configurations, only. Thus $h(\cdot, t)$ is the configuration frequency vector for the image thresholded at t.

2.3   Linear Time Euclidean Distance Transform

For a binary image representing a discretized realization of a random closed set X intersected by a cuboidal observation window W as foreground, the Euclidean distance transform (EDT) assigns each pixel x in the background $X^C \cap W$ the Euclidean distance to the foreground. Here, we use the signed version additionally assigning each foreground pixel $x \in X \cap W$ the Euclidean distance to the background.

Commonly used algorithms for calculating distance transforms measure the distances w.r.t. the interface of foreground and background. That is, a surface consisting of lattice points and defined by the chosen discrete adjacency.

Classical algorithms for calculating the exact EDT are based on propagating distances observed locally through several passes through the image, separating the coordinate directions. In 3D, this type of algorithm is due to [17]. Maurer and Raghavan [18] were the first to suggest a dimension reduction Voronoi based algorithm for exact Euclidean distances in 3D, linear in the number of image pixels.

2.4   Sub-pixel Precise Euclidean Distance Transform

EDT algorithms using dimension reduction approaches like [18–20], allow to use a boundary between lattice points as curve or surface of reference In these cases, distances smaller than one lattice spacing can be observed. Lindblad and Sladoje [21] derive this surface by linear interpolation of the neighbor pixel values and applying a threshold.

The general idea behind [18] is to iteratively process sub-images of successively increasing dimensions, first computing shortest distances within individual lines, then 2D slices, before finally obtaining total distances for the entire 3D image. This algorithm assumes that every lattice point outside of the object of interest counts as a Voronoi generator point to which distances get computed.

Lindblad and Sladoje [21] compute the distance to an object's surface using sample points as generators that can deviate from lattice lines in one dimension. If the object is implicitly given as an iso-surface of an image, they propose to use linear interpolation between neighboring pixels to obtain the exact location of the surface point.

2.5   Erosion Dilation Analysis

The Minkowski functions as used e.g. in [6] are the intrinsic volumes of the parallel bodies $X \oplus B_r$ and $X \ominus B_r$, respectively, of the structure X with respect to a ball B_r of radius r. That is, the intrinsic volumes of stepwise erosions/dilations of X with a spherical structuring element are considered as functions of the radius r. From these functions, 3D structural information can be directly derived. In particular the specific Euler number as a function of successive erosions was called connectivity function and used to evaluate bond sizes [5]. In [13], the Euler number combined with successive erosions was used to study the connectivity of firn from the B35 core, see also [8].

Morphological erosions and dilations with a spherical structuring element B can be readily obtained by thresholding the EDT image. More precisely, thresholding the EDT image using r is equivalent to an erosion or a dilation of the foreground X by a ball of radius $r \in \mathbb{R}_+$: $X \oplus B_r = \{x : EDT(X, x) < r\}$ and $X \ominus B_r = \{x : EDT(X, x) < -r\}$. The algorithm from [9] efficiently yields the intrinsic volumes of the parallel bodies $X \oplus B_r$ and $X \ominus B_r$.

3   Intrinsic Volumes on Sub-pixel EDT Images for Refined Erosion Dilation Analysis

Here we combine the approach from [21] with the fast algorithm for the precise EDT by [18] to achieve sub-pixel precision. For that, we modify the implementation of [18] by adding offsets to the intermediate distances computed during each line scan. These offsets represent the distance between the interpolated surface and the next lattice point. As long as they are smaller than the grid spacing, the original algorithm still works as intended.

We apply the sub-pixel precise EDT on an image with a threshold $t \in \mathbb{R}$, denoted by $EDT(X, x)_t$. Note, that t can take values which are in between values of the gray value domain. E.g. for 8 bit gray value images, t lies in $[0, 255)$. For binary images, t lies in $[0, 1)$. Applying $EDT(X, x)_t$ can be interpreted as calculating the distance to the (implicitly given) iso-surface corresponding to t, because the position of the distance generator is computed in each dimension via linear interpolation of the gray values of the two neighboring pixels. Changing this threshold results in virtually dilating/eroding the foreground.

In the following we apply this algorithm to binary images. We derive two different distance maps that can be interpreted as a lower and an upper bound of a central one. For the lower bound we have $EDT(X, x)_0 \leq EDT(X, x)_t$ for each threshold value $t \in [0, 1)$ and each background pixel $x \in X^C$. The upper bound is obtained as $EDT(X, x)_{1-\varepsilon}$ with $\varepsilon = 0.01$. Note that we write $EDT(X, x)_1$ for short in the following. Additionally, we use a distance map w.r.t. a "central" boundary surface $EDT(X, \cdot)_{0.5}$. Note that $EDT(X, x)_0$ coincides exactly with the classical $EDT(X, x)$.

Using this EDT algorithm with sub-pixel precision results in an erosion dilation analysis with improved precision, in particular near the surface, where the observed distance values are rather roughly discretized when applying classical EDT. This is particularly nice when the structure is not very well resolved by the image.

The densities of the intrinsic volumes and derived features for all distances from both the distance map of X as well as the bounding maps $EDT(X)_0$ and $EDT(X)_1$ can be used as an indicator of how reliable the analysis results are. More precisely, we use all distances occurring in the distance map, filter them such that consecutive distances differ by at least 1% of the pixel spacing, and compute for each of those distances the intrinsic volumes.

Note that the structure model index should capture the "structure shape" for both solid structure and pore space. Thus, we switch to the SMI of the complement when eroding the solid structure. We have $S(X \cap W) = S(X^C \cap W)$ and $M(X \cap W) = -M(X^C \cap W)$. Thus, for $r > 0$, we use in the following

$$\mathrm{SMI}^*(X \oplus B_r) = \mathrm{SMI}(X \oplus B_r) \text{ and}$$
$$\mathrm{SMI}^*(X \ominus B_r) = -12 \frac{(V(W) - V((X \ominus B_r) \cap W))M((X \ominus B_r) \cap W)}{S^2((X \ominus B_r) \cap W)}.$$

4 Application Examples

In this section, we apply the algorithm described in the previous Sect. 3 to 3D images of rigid foam samples and of samples from nano-porous membrane layers. Note that the choice is purely motivated by trying to find examples featuring several critical structure thicknesses like strut and wall thickness in the partially closed foams. Although both the considered rigid foams as well as the nano-porous membrane layers are made of ceramic materials, the algorithm is in no way restricted to or tailor-made for ceramic samples.

4.1 Nano-Porous ZrO_2 and Al_2O_3 Ceramics

We compare the structures of nano-porous ceramic samples generated by spin coating. Samples 1 and 2 consist of ZrO_2. In fact, sample 2 is the 1st layer of the nano-porous membrane investigated in [22]. Our sample 3 is the 2nd layer of this nano-porous membrane and consists of Al_2O_3. After spin coating, the samples are sintered at $1\,000$ and $1\,400\,°C$, respectively.

Grain sizes in the range $10–100\,nm$ prevent imaging the structures by computed tomography. Instead, sequential slicing by a focused ion beam (FIB) combined with imaging by scanning electron microscopy (SEM) – so-called FIB-SEM – is applied. Due to the slicing by the focused ion beam, images generated by

FIB-SEM usually feature non-cubic voxels with the resolution in stack direction being considerably coarser than within the SEM images forming the stack. Hence, the sub-pixel precision approach can yield valuable bounds for the true values of characteristics.

For all three samples, the stacks of SEM images are aligned, desheared, and transformed in order to remove distortions introduced by the alignment as described in [22]. Interpolation onto an isotropic lattice results in sample 1 with $986 \times 637 \times 299$ pixels and lattice spacing 3 nm, sample 2 with $980 \times 980 \times 382$ pixels and lattice spacing 3 nm, and sample 3 with $205 \times 205 \times 165$ pixels and lattice spacing 20 nm. This interpolated version of sample 3 is called 3i with "i" for "isotropic" from now on. Additionally, we examine a larger non-interpolated version of sample 3, called sample 3a with "a" for "anisotropic", with $716 \times 168 \times 225$ pixels and lattice spacings 20 nm $\times$ 20 nm $\times$ 16 nm.

Sample 3 is then reconstructed solely based on the energy-selective backscatter electron (EsB) image stack by global gray value thresholding. Samples 1 and 2 are reconstructed by the combination of the secondary electron (SE) and the EsB image stacks by morphological reconstruction by dilation [23]: Both are binarized by global gray value thresholding. Subsequently, the EsB pore image is reconstructed by dilation with the pixelwise maximum of the two pore images as mask image. See [22] for more details (Fig. 1).

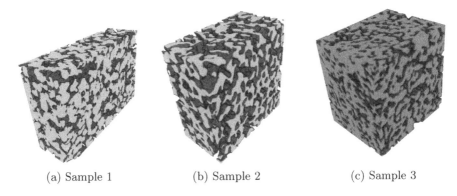

(a) Sample 1   (b) Sample 2   (c) Sample 3

Fig. 1. Renderings of the nano-porous ceramic samples 1–3. Visualized are $986 \times 637 \times 299$ pixels corresponding to a volume of $2.96\,\mu m \times 1.91\,\mu m \times 0.90\,\mu m$ for sample 1, $360 \times 360 \times 190$ pixels corresponding to a volume of $2.97\,\mu m \times 2.97\,\mu m \times 1.14\,\mu m$ for sample 2, and $205 \times 205 \times 165$ pixels corresponding to a volume of $4.1\,\mu m \times 4.1\,\mu m \times 3.3\,\mu m$ for sample 3.

First, we compare the results of the erosion dilation analysis on the anisotropic versus the isotropized samples 3a and 3i to study the influence of the interpolation step. Figure 2 shows the results together with the respective bounds obtained from $EDT(X)_0$ and $EDT(X)_1$. The isotropized 3i is on the one hand less well resolved than 3a and on the other hand smoothed by the interpolation. Both effects reduce the specific surface area slightly, which is visible in Fig. 2(b). Also due to the smoothing, both dilations and erosions yield non-trivial results for a wider distance range. Apart from this, we do not observe significant differences between the interpolated and the non-interpolated version of the sample. Thus, we only report the results for the isotropic version in the following analysis.

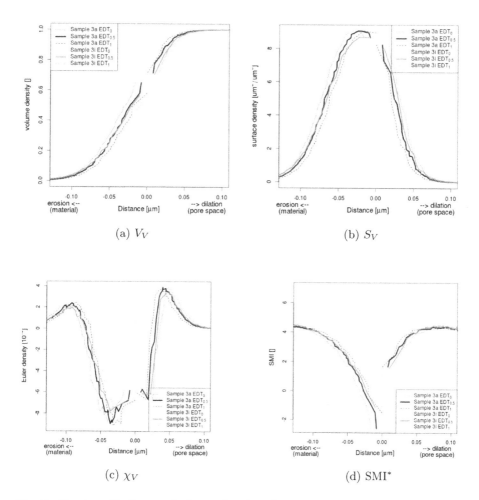

Fig. 2. Erosion dilation analysis results for the nano-porous ceramic sample 3, anisotropic and isotropic versions (3a and 3i) with bounds derived from $EDT(X)_0$ and $EDT(X)_1$.

Results of the erosion dilation analysis for the nano-porous ceramic samples are reported in Fig. 3. The solid structure of sample 2 is clearly finer than the one of samples 1 and 3 while the difference for the pores is not that pronounced, see Fig. 3. Note that the SMI* curves start to fluctuate strongly for larger erosion and dilation radii since there only a few pixels are left to contribute, and the strong erosion/dilation changes the structure of solid structure and pore space, respectively, drastically. In general, the SMI should be trusted only in the distance range where the volume fraction V_V is between 2 and 98 %.

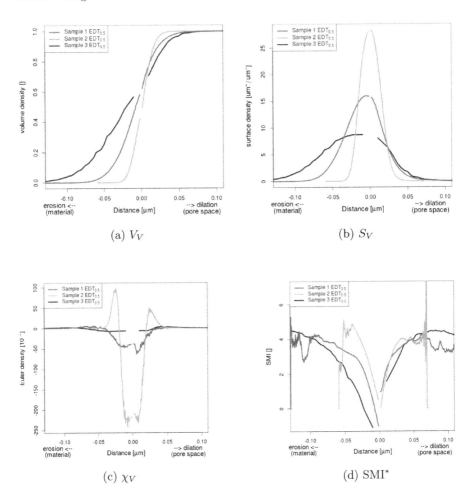

Fig. 3. Erosion dilation analysis results for the nano-porous ceramic samples 1–3.

4.2   Ceramic Foams

We investigate ceramic foams used for filtering metal melts in casting. These filters are produced in large numbers and discarded when loaded. The samples considered here are produced by covering an open cell polymer foam with ceramic slurry, drying, and finally burning. The resulting filters feature a majority of open cells inherited from the polymer foam, some closed cell walls formed by the slurry, and hollow struts where the polymer foam was combusted. The micro-structure of such ceramic foam samples and its relation to the filtration properties has been studied in detail in [24,25]. Here, we consider three samples produced from the same polymer foam but with varying deposition of the ceramic material resulting in varying flow properties. More precisely, samples C01 and C14 weigh the same but the water flow rate differs while it is equal for samples C14 and C20 whose weight differs, see Table 1 (Fig. 4).

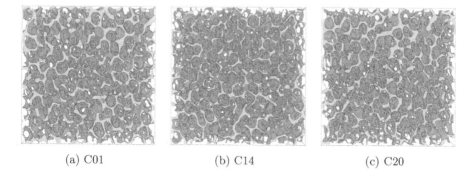

(a) C01   (b) C14   (c) C20

Fig. 4. Renderings of the ceramic foam samples C01, C14, and C20.

Table 1. Ceramic foam samples and their macroscopic properties.

Sample	Weight [g]	Water flow rate [l/s]
C01	17.5	2.105
C14	17.4	1.971
C20	16.2	1.979

The samples of dimensions $50\,mm \times 50\,mm \times 20\,mm$ are imaged by micro-computed tomography at pixel size $34\,\mu m$. Here, we are interested in roughly resolved image data. Thus the images are subsampled to a lattice with spacing $102\,\mu m$. The solid component is segmented by global gray value thresholding using Otsu's algorithm [26]. The struts are closed by connected component labeling and morphological closure. Thus, the Euler number density χ_V yields essentially the negative cell wall density as there is just one connected component and there are no holes.

Results of the erosion dilation analysis for these three samples are reported in Figs. 5, 6 and 7. As expected, we observe all significant local minima and maxima at the same arguments. This is due to the underlying identical polymer foam structure leaving the spatial distribution of the ceramic slurry as the differentiating feature. The solid volume fraction does not reveal significant differences, see Fig. 5(a). The specific surface area of both the eroded and the dilated structures indicates for small radii of the sphere least surface area for C01 and most for C14, see Fig. 6. This is more pronounced on the dilation side and becomes apparent only when looking closer at the dilation range up to 0.6 mm, see Fig. 6(b). The Euler number density singles out C14 when being dilated: χ_V for the dilated C14 is significantly smaller than for the other two samples, see Fig. 7(b). Thus, the dilated C14 contains more tunnels, that is open cell walls, than C01 and C20. This in turn indicates that C01 and C20 include smaller cell walls being closed by the dilation earlier than those in C14.

Nevertheless, none of the curves in Figs. 5, 6 and 7 seems to explain the different weights and water flow rates reported in Table 1. The structure model index reveals distance ranges, where a closer look seems worthwhile, see Fig. 5(b). Indeed, as indicated by the first peak of the dilation curve, C01 has more really large pores than the other two samples rendering a good explanation for the increased water flow rate: Erosion of the pore space with a ball of radius $r = 1.8$ mm, results in 117 connected components for C01, 20 for C14, and 51 for C20. These large pores comprise 0.011% of the pore volume of C01, 0.002% of the pore volume of C14, and 0.008% of the pore volume of C20. If these large pores are reconstructed, they comprise 12% of the pore volume for C01, 2% for C14, and 5.4% for C20.

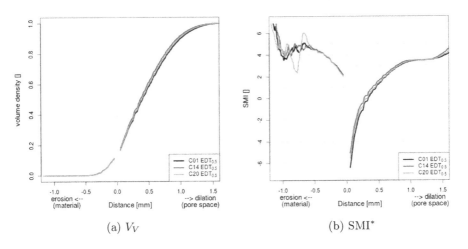

Fig. 5. Erosion dilation analysis results for the ceramic foam samples: solid volume fraction V_V and structure model index.

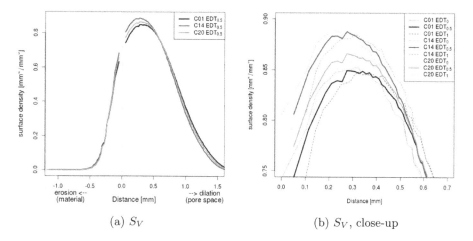

(a) S_V (b) S_V, close-up

Fig. 6. Erosion dilation analysis results for the ceramic foam samples: specific surface area S_V and close-up with envelopes derived from EDT_0 and EDT_1.

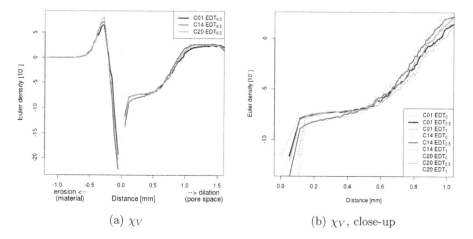

(a) χ_V (b) χ_V, close-up

Fig. 7. Erosion dilation analysis results for the ceramic foam samples: Euler number density χ_V and close-up with envelopes derived from EDT_0 and EDT_1.

5   Discussion

In this paper, we combined the efficient algorithm for calculating the intrinsic volumes for each possible gray value threshold in an integer valued image [9] with the linear time exact Euclidean distance transform from [18] and the sub-pixel precision approach of [21].

We applied it for efficient erosion dilation analysis of nano-porous ceramic membrane layers imaged by FIB-SEM and of ceramic foams imaged by micro-computed tomography. Both examples show, that interesting structural features and differences can be revealed that way at low effort for both preprocessing and the algorithm itself. Thus, it is very well suited for a screening step indicating the direction of further, more detailed analysis by methods dedicated to the specific type of structure of interest. Moreover, the sub-pixel precise EDT can yield bounds for the densities of the intrinsic volumes. This is helpful when studying differences between similar structures. It is particularly valuable in cases, where the resolution is close to being insufficient.

Furthermore, we see potential use in calculating the spherical contact distribution [1] more precisely.

Acknowledgment. This research was supported by the German Federal Ministry of Education and Research [project 03VP00491/5], by the Fraunhofer High Performance Center for Simulation- and Software-Based Innovation, by the German Research Foundation (DFG) within the IRTG 2057 "Physical Modeling for Virtual Manufacturing Systems and Processes", and by the German Academic Exchange Service (DAAD) through PhD program MIC "Mathematics in Industry and Commerce".

We thank Sören Höhn from Fraunhofer IKTS for providing the nano-porous ceramic samples as well as the FIB-SEM images and Claudia Redenbach from Technische Universität Kaiserslautern, Department of Mathematics for sharing the ceramic foam data.

References

1. Stoyan, D., Kendall, W.S., Mecke, J.: Stochastic Geometry and Its Applications, 2nd edn. Wiley, Chichester (1995)
2. Schneider, R., Weil, W.: Stochastic and Integral Geometry. Probability and Its Applications. Springer, Heidelberg (2008). https://doi.org/10.1007/978-3-540-78859-1
3. Ohser, J., Redenbach, C., Schladitz, K.: Mesh free estimation of the structure model index. Image Anal. Stereology **28**(3), 179–186 (2009)
4. Schladitz, K., Redenbach, C., Sych, T., Godehardt, M.: Model based estimation of geometric characteristics of open foams. Methodol. Comput. Appl. Probab. **14**, 1011–1032 (2012)
5. Vogel, H.J.: Morphological determination of pore connectivity as a function of pore size using serial sections. Eur. J. Soil Sci. **48**(3), 365–377 (1997)
6. Mecke, K.: Additivity, convexity, and beyond: application of Minkowski functionals in statistical physics. In: Mecke, K.R., Stoyan, D. (eds.) Statistical Physics and Spatial Statistics. LNP, vol. 554, pp. 111–184. Springer, Heidelberg (2000). https://doi.org/10.1007/3-540-45043-2_6
7. Ohser, J., Nagel, W., Schladitz, K.: Miles formulae for Boolean models observed on lattices. Image Anal. Stereol. **28**(2), 77–92 (2009)
8. Ohser, J., Schladitz, K.: 3D Images of Materials Structures - Processing and Analysis. Wiley VCH, Weinheim (2009)
9. Godehardt, M., Jablonski, A., Wirjadi, O., Schladitz, K.: Fast estimation of intrinsic volumes in 3D gray value images. In: Benediktsson, J.A., Chanussot, J., Najman, L., Talbot, H. (eds.) ISMM 2015. LNCS, vol. 9082, pp. 657–668. Springer, Cham (2015). https://doi.org/10.1007/978-3-319-18720-4_55
10. Vogel, H., Kretzschmar, A.: Topological characterization of pore space in soil–sample preparation and digital image-processing. Geoderma **73**(1), 23–38 (1996)
11. Lang, C., Ohser, J., Hilfer, R.: On the analysis of spatial binary images. J. Microsc. **203**, 303–313 (2001)

12. Arns, C.H., Knackstedt, M.A., Mecke, K.R.: Characterising the morphology of disordered materials. In: Mecke, K.R., Stoyan, D. (eds.) Morphology of Condensed Matter. LNP, vol. 600, pp. 37–74. Springer, Heidelberg (2002). https://doi.org/10.1007/3-540-45782-8_2

13. Freitag, J., Kipfstuhl, S., Faria, S.H.: The connectivity of crystallite agglomerates in low-density firn at Kohnen station, Dronning Maud Land. Antarctica. Ann. Glaciol. 49, 114–120 (2008)

14. Redenbach, C., Rack, A., Schladitz, K., Wirjadi, O., Godehardt, M.: Beyond imaging: on the quantitative analysis of tomographic volume data. Int. J. Mater. Res. 2, 217–227 (2012)

15. Hildebrand, T., Rüegsegger, P.: Quantification of bone microarchitecture with the structure model index. Comput. Methods Biomech. Biomed. Engin. 1(1), 15–23 (1997)

16. Schladitz, K., Ohser, J., Nagel, W.: Measuring intrinsic volumes in digital 3D images. In: Kuba, A., Nyúl, L.G., Palágyi, K. (eds.) DGCI 2006. LNCS, vol. 4245, pp. 247–258. Springer, Heidelberg (2006). https://doi.org/10.1007/11907350_21

17. Saito, T., Toriwaki, J.: New algorithms for Euclidean distance transformations of an n-dimensional digitised picture with applications. Pattern Recogn. 27(11), 1551–1565 (1994)

18. Maurer, C.R., Raghavan, V.: A linear time algorithm for computing exact Euclidean distance transforms of binary images in arbitrary dimensions. IEEE Trans. Pattern Anal. Mach. Intell. 25(2), 265–270 (2003)

19. Breu, H., Gil, J., Kirkpatrick, D., Werman, M.: Linear time Euclidean distance transform algorithms. IEEE Trans. Pattern Anal. Mach. Intell. 17(5), 529–533 (1995)

20. Gavrilova, M.L., Alsuwaiyel, M.H.: Two algorithms for computing the Euclidean distance transform. Int. J. Image Graph. 01(04), 635–645 (2001)

21. Lindblad, J., Sladoje, N.: Exact linear time euclidean distance transforms of grid line sampled shapes. In: Benediktsson, J.A., Chanussot, J., Najman, L., Talbot, H. (eds.) ISMM 2015. LNCS, vol. 9082, pp. 645–656. Springer, Cham (2015). https://doi.org/10.1007/978-3-319-18720-4_54

22. Prill, T., et al.: Simulating permeabilities based on 3D image data of a layered nano-porous membrane. Int. J. Solids Struct. (2019, submitted)

23. Soille, P.: Morphological Image Analysis. Springer, Heidelberg (1999). https://doi.org/10.1007/978-3-662-05088-0

24. Redenbach, C., Wirjadi, O., Rief, S., Wiegmann, A.: Modelling a ceramic foam for filtration simulation. Adv. Eng. Mater. 13(3), 171–177 (2010)

25. Kampf, J., Schlachter, A.L., Redenbach, C., Liebscher, A.: Segmentation, statistical analysis, and modelling of the wall system in ceramic foams. Mater. Charact. 99, 38–46 (2015)

26. Otsu, N.: A threshold selection method from gray level histograms. IEEE Trans. Syst. Man Cybern. 9, 62–66 (1979)

A Fast, Memory-Efficient Alpha-Tree Algorithm Using Flooding and Tree Size Estimation

Jiwoo You[1,2(✉)], Scott C. Trager[1], and Michael H. F. Wilkinson[2]

[1] Kapteyn Astronomical Institute, University of Groningen,
Groningen, The Netherlands
{j.you,sctrager}@astro.rug.nl
[2] Bernoulli Institute of Mathematics, Computer Science and Artificial Intelligence,
University of Groningen, Groningen, The Netherlands
{j.you,m.h.f.wilkinson}@rug.nl

Abstract. The α–tree represents an image as hierarchical set of α-connected components. Computation of α–trees suffers from high computational and memory requirements compared with similar component tree algorithms such as max–tree. Here we introduce a novel α–tree algorithm using (1) a flooding algorithm for computational efficiency and (2) tree size estimation (TSE) for memory efficiency. In TSE, an exponential decay model was fitted to normalized tree sizes as a function of the normalized root mean squared deviation (NRMSD) of edge-dissimilarity distributions, and the model was used to estimate the optimum memory allocation size for α–tree construction. An experiment on 1256 images shows that our algorithm runs 2.27 times faster than Ouzounis and Soille's thanks to the flooding algorithm, and TSE reduced the average memory allocation of the proposed algorithm by 40.4%, eliminating unused allocated memory by 86.0% with a negligible computational cost.

Keywords: Alpha–tree · Mathematical morphology ·
Connected operator · Tree size estimation · Efficient algorithm

1 Introduction

Connected operators are morphological filters that process input images as a set of inter-connected elementary regions called flat-zones [1,2]. Connected operators can modify, merge or remove regions of interest without changing or moving object contours, because they filter pixels based on attributes of connected components that the pixels belong to, instead of some fixed surroundings [3]. The advent of component trees (max– and min–tree) enabled new types of filtering strategies for connected operators by providing efficient data structure to represent hierarchy of image regions [4]. In max–trees (min–trees), connected components with local maxima (minima) become leaf nodes, and those leaf nodes are successively merged to form larger connected components with lower (higher)

© Springer Nature Switzerland AG 2019
B. Burgeth et al. (Eds.): ISMM 2019, LNCS 11564, pp. 256–267, 2019.
https://doi.org/10.1007/978-3-030-20867-7_20

levels, eventually becoming a root node representing the whole image [4]. This creates component-tree hierarchies of level sets, which assumes all connected components are nested around local peaks [6].

The α–tree is a more recent hierarchical image representation, where dissim-ilarities between local pixels define quasi-flat zones [5–8]. The α–tree has shown its usefulness on segmentation of e.g. remote sensing images [6]. Because both the pixels and dissimilarities between neighbouring pixels must be taken into account, the α–tree has multiple times more data elements to process compared to the component tree. Algorithms for component trees have been thoroughly studied and reviewed [9], which showed that flooding algorithms have lower com-putational cost for low-dynamic-range images than algorithms based on Tarjan's union-find algorithm [10]. On the other hand, there have only been a few studies of the α–tree algorithm, and all used algorithms based on the union-find algo-rithm (which will be referred as Ouzounis–Soille's algorithm throughout this paper) [6,7,11].

Memory requirements also play an important role in the α–tree algorithm, because a huge amount of memory is needed to store the α–tree nodes of large images. A large portion of allocated memory is wasted, because the number of α–tree nodes is usually only 40–70% of its maximum (further explained in Sect. 3) [7]. However, previous studies of α– and (even) component tree algorithms usu-ally choose to allocate an oversized amount of memory, because there has been no way to estimate tree sizes before construction. In this paper we introduce a novel α–tree algorithm using (1) the flooding algorithm for computation speed and (2) tree size estimation (TSE) for memory efficiency. We show that TSE accurately estimates the tree size beforehand using the normalized root mean square deviation (NRMSD) of dissimilarity histograms, thus enabling significant reduction in memory usage at small additional computational cost.

2   A Fast, Memory-Efficient α–Tree Algorithm

2.1   Definitions

We represent an image as undirected weighted graph $G = (V, E)$, where V is the set of pixels of an image, and $E = \{\{v_i, v_j\}|i \neq j\}$ is a set of unordered neighbouring pairs (i.e. edges) of V. We define the edge weight $w : E \rightarrow Y$ of an edge $e \in E$ as a symmetric dissimilarity $d(v_i, v_j) = d(v_j, v_i)$ between pixels, where Y is the codomain of w and $W = \{w(e)|e \in E\}$ is the image of w ($W \subseteq Y$). A path $\pi(v \rightsquigarrow u)$ from $v \in V$ to $u \in V$ is a sequence of successively adjacent vertices:

$$\pi(v \rightsquigarrow u) \equiv (v = v_0, v_1, ..., v_{|\pi(v \rightsquigarrow u)|-1} = u). \tag{1}$$

with every pair $\{v_i, v_{i+1}\} \in E$. An α–connected component (α–CC) containing a vertex $v \in V$, or α–CC(v) is defined as

$$\alpha - CC(v) = \{v\} \cup \{u \mid \exists \pi(v \rightsquigarrow u)\forall(v_i \in \pi(v \rightsquigarrow u), v_i \neq u)d(v_i, v_{i+1}) \leq \alpha\}. \tag{2}$$

In words, α–CC(v) is a set of vertices containing v, where for every distant pair (v_i, v_j) of α–CC(v) their exists a path from v_i to v_j in which every successive pair has an incident edge with a weight less than or equal to α. An image can be represented as a partition of α–CCs [6]:

$$\mathbf{P}^\alpha = \{\alpha - \mathrm{CC}(v)|0 \le v \le |V| - 1\}, \cup_v \alpha - \mathrm{CC}(v) = V. \tag{3}$$

For $\alpha = 0$, $\mathbf{P}^0$ forms the finest partition of flat–zones, and as α increases the partition become coarser, eventually becoming a single α–CC that encompasses the whole image ($\mathbf{P}^{\alpha_{max}} = V$). A set of all α–CCs marked by the same vertex (i.e. $\cup_\alpha \alpha$–CC(v)) form a total order:

$$\forall(\alpha_i \le \alpha_j)\ \alpha_i - \mathrm{CC}(v) \subseteq \alpha_j - \mathrm{CC}(v). \tag{4}$$

We label α_j–CC(v) as redundant if there exists α_i–CC(v) such that $\alpha_i < \alpha_j$ and α_i–CC(v) = α_j–CC(v).

2.2 Data Structure

In our algorithm, the α–tree data structure contains the following two arrays:

– *pAry*: Array of indices to leaf nodes for each pixel ($|V|$)
– *pNode*: Array of α–tree nodes ($|V|$)

Note that numbers inside vertical bars indicate array size. A typical representation of an α–tree node is to use *pNode* with size $2|V|$, where the first $|V|$ nodes correspond to leaf nodes representing a single pixel, and the rest correspond to non-leaf nodes. Here we replace single-pixel nodes with indices to parent nodes (*pAry*) to reduce memory usage. *pNode* is an array of structs containing:

– *alpha*: α value of the associated α–CC
– *parent*: An index to the parent node
– *∗attributes*: Attributes of the associated α–CC

Since we used 8-bit images and the L_∞ norm as a dissimilarity metric, *alpha* can be stored in one byte. We used a 32-bit unsigned integer to store *parent*. The *attribute* field represents all attributes stored in a node. In our implementation we used the maximum, the minimum, and the sum of pixels inside the associated α–CC. The size of a single struct including *alpha*, *parent* and all attributes is 19 bytes.

In addition to the α–tree data structure, the α–tree flooding algorithm also requires the following temporary data structures:

– *hqueue*: Hierarchical queue in a one-dimensional array ($|E|$)
– *levelroot*: Array of parent and ancestor nodes ($|W|$)
– *dhist*: Histogram of pixel dissimilarities ($|W|$)
– *dimg*: Array of dissimilarities of all neighbouring pixel pairs ($|E|$)
– *isVisited*: Array that keeps track of visited pixels ($|V|$)

The first two arrays (*hqueue* and *levelroot*) are data structures used in the flooding algorithm which are further explained in the next section. The dissimilarity histogram (*dhist*) is used to allocate queue sizes in each hierarchy of *hqueue* and also to calculate NRMSD in TSE. Dissimilarities of all neighbouring pairs are stored in *dimg* during the computation of *dhist* to avoid computing dissimilarities of the pairs multiple times. The array *isVisited* keeps track of visited pixels, because every pixel in the image needs only a single visit.

2.3   The α–Tree Flooding Algorithm

Flooding is a tree construction algorithm with a top-down design that constructs a tree in depth-first traverse [4]. The α–tree can be interpreted as a min-tree of a hypergraph $V' = (E, V)$ where two edges $e1$ and $e2$ in V' are neighbours if and only if there is a pixel v such that $v \in e1$ and $v \in e2$. Thus building an α–tree of V is equivalent to building a min–tree of V', and therefore any max–tree construction algorithms including the flooding algorithm can be applied to α–tree [11–13].

Our implementation of the flooding algorithm is similar to the max–tree flooding algorithm in [4], but we replace recursive calls with iterative loops as in [3], use a different neighbours-queuing process, and other optimization techniques suitable for low-dynamic-range images. Algorithm 1 shows a pseudocode of our C–implementation of the proposed α–tree algorithm (C–code available at [14, 15]). The overall design of the algorithm is similar to that of a max–tree. The main loop of the algorithm (line 10–42) creates levelroots, connects them to the α–tree, and removes redundant nodes. The inner loop (line 11–32) visits pixels in the hierarchical queue in increasing order of α, queues up neighbouring pixels, and connects the visited pixels to the levelroots. For depth-first traversal of α–tree we used a hierarchical FIFO queue [4], which is a hierarchy of queues, each of which is reserved for queuing pixels with the same α value. During the traversal, *levelroot* stores potential parent and ancestors of the current α–node being processed [3]. Nodes of the α–tree are stored in a one-dimensional array that is pre-allocated with a size estimated by TSE (explained in Sect. 3).

An α–tree is represented as an array of nodes, where a node is a structure storing node information (as defined in Sect. 2.2). An α–tree node is always associated with only a single α–CC, but an α–CC can be represented as a sub-tree of multiple nodes with the same α, where the node that has a parent with higher α is called a levelroot [3] or a canonical element [16]. In Ouzounis–Soille's algorithm a procedure called levelroot-fix or canonicalization is used as a post-processing to merge non-levelroot nodes and its levelroot parents to reduce memory usage and simplify tree structure [9]. However, since the flooding algorithm never creates non-levelroot nodes, our α–tree algorithm does not need an extra procedure to canonicalize the tree [4].

Algorithm 1. The α–tree flooding algorithm

```
void Flood(Pixel *img){
1       Compute dissimilarity histogram and store it in dhist
2       Initialize α-node array using TSE
3       Initialize hqueue using dhist
4       Initialize levelroot[i] as empty for all i
5       alpha_max = the maximum possible α; // 255 for 8-bit images
6       levelroot[alpha_max] = new α–node;
7       current_alpha = alpha_max;
8       curSize = 0; //current size of α-tree
9       hqueue.push(0, alpha_max);

10      while(current_alpha ≤ alpha_max){
11          while(hqueue is not empty for all alpha no higher than current_alpha){
12              p = front of queue;
13              hqueue.pop();
14              if(p is visited){
15                  hqueue.find_min_alpha();
16                  continue;
                }
                // visit all neighbours
17              mark p as visited;
18              for(neighbours q of p){
19                  if(q is not visited){
20                      d = L∞(img[p], img[q]);
21                      hqueue.push(q, d)
22                      if(levelroot[d] is empty)
23                          mark levelroot[d] as node_candidate;
                    }
                }
24              if(current_alpha > hqueue.min_alpha)
25                  current_alpha = hqueue.min_alpha; //go to lower α
26              else
27                  hqueue.find_min_alpha(); //delayed minimum α set
28              //connect a pixel to levelroot
29              if(levelroot[current_alpha] is node_candidate)
30                  levelroot[current_alpha] = &pNode[curSize++] //new α-node
31              connect p to levelroot[current_alpha]
32              pAry[p] = levelroot[current_alpha]
            }
33          if(levelroot[current_alpha] is a redundant node)
34              replace levelroot[current_alpha] with its redundant child node;
35          next_alpha = current_alpha + 1;
36          while(next_alpha ≤ alpha_max && levelroot[next_alpha] is empty)
37              next_alpha = next_alpha + 1;
38          if(levelroot[next_alpha] is node_candidate)
39              levelroot[next_alpha] = &pNode[curSize++] //new α-node
40          connect levelroot[current_alpha] to levelroot[next_alpha]
41          mark levelroot[current_alpha] as empty;
42          current_alpha = next_alpha;
        }
    }
```

3   The Tree Size Estimation

Both α–trees and component trees have high memory requirements. Given an image with $|V|$ pixels, the amount of memory allocated to store α–tree nodes is typically larger than $40|V|$ bytes, and it also needs temporary memory space of up to $6|V| + O(|V|)$ bytes [9]. Most algorithms use a large pre-allocated array to store tree nodes [3,6,7,9], which can store up to $2|V|$ nodes. However this is rarely the case for low-dynamic large images. In [7] only 40–70% of the maximum number of nodes (i.e. $2|V|$ nodes) were created, which means a large memory space is wasted during the construction of α–tree.

There are alternative data structures that could eliminate unused memory space, such as dynamically allocated structures linked by pointers [7]. These data structures allocate each node dynamically on demand and are also more flexible compared with one-dimensional arrays, because it is easier to insert or remove nodes from the tree. However a single node takes more memory space because of the link pointer, and it runs much slower than the array implementation because of the frequent memory allocations. In [7] the dynamically allocated data structure took 48% more time than the array on average in building α–tree from images with 0.78 megapixels, and 109% more time on images with 431.3 megapixels. Memory allocation is also locking, preventing use in parallel.

Another way to deal with the problem of memory usage is to use one-dimensional arrays with a smaller array size and increase it on demand using the *realloc*() function whenever the array is full [17]. In this way the average memory usage can be reduced, but several problems arise from the memory reallocation. If there is no memory space to extend the array in place, *realloc*() relocates the array by allocating a new array with a new size and copying the entire data from the old to the new array. Such a cumbersome reallocation procedure significantly increases computational cost, and the memory usage might even exceed the maximum memory space needed to store $2|V|$ nodes, because of the use of temporary memory space in reallocation.

The best way to reduce the excessive memory usage is to accurately estimate the tree size before building the tree. The size of the tree is generally proportional to the size of the image, but it is also heavily dependent on the image content. We have found a simple measure that can be used to accurately estimate the tree size, thus significantly reducing the memory usage at almost zero computational cost. Let the normalized tree size (NTS) be defined as follows:

$$Normalized\ tree\ size\ (NTS) = \frac{number\ of\ nodes\ in\ tree}{2|V|}. \tag{5}$$

If the tree size is at its maximum, $NTS \approx 1$ for large $|V|$. If all dissimilarities of neighbouring pixel pairs in the image have distinct values, the α–tree built from that image will become a complete binary tree with $2|V|$ nodes, and thus NTS of that tree becomes its maximum. In such a case the pixel dissimilarity histogram should be flat, that is, $dhist[w] = 1$ for all $w \in W$. On the other extreme, the smallest tree size can be observed when all the pixels have the same intensity, and thus the entire image becomes a single α–CC. In such a case, NTS $= 1/(2|V|)$ and $dhist[w_0] = |Y|$ for some w_0 and $dhist[w] = 0$ for $w \neq w_0$, where Y is the codomain of the edge weighting as defined in Sect. 2.1 ($|Y| = |E|$). Based on these observations, we hypothesize that the NTS of an image can be estimated by the root mean square deviation (RMSD) between $dhist$ and a flat distribution as follows:

$$\text{RMSD} = \frac{\sqrt{\sum_{i \in Y} (dhist[i] - 1)^2}}{|Y|}. \tag{6}$$

The RMSD is zero when the $dhist$ is a flat histogram, which means all neighboring pairs have unique edge weights ($|W| = |Y|$). For low-dynamic-range images with high-resolution, $|W| << |Y|$ ($|W|$ is typically bounded by image bit-depth), thus the RMSD of those images cannot be zero. We define the normalized RMSD (NRMSD) as the RMSD divided by its maximum:

$$\text{NRMSD} = \sqrt{\frac{\sum_{i \in Y} (dhist[i] - 1)^2}{(|Y| - 1)|Y|}}, \tag{7}$$

and some rearranging yields more easily calculable form,

$$\text{NRMSD} = \sqrt{\frac{\sum_{i \in W} dhist[i]^2 - |Y|}{(|Y| - 1)|Y|}}. \tag{8}$$

which gives a normalized value between 0 and 1. The NRMSD requires little more than $|W|$ multiplications, where $|W| \leq 256$ for 8-bit greyscale images. We find that if the NTS is plotted as a function of the NRMSD, it is very close to the exponential decay function shown in Fig. 1(a). Thus, we define the estimate of the NTS as an exponential decaying model as follows:

$$TSE(x) = Ae^{-\sigma x} + B \tag{9}$$

The parameters A, B, and σ are optimized using the test datasets in the next section.

4   Experiments

We use 1256 pictures including 639 greyscale aerial pictures selected from the database of the Netherlands Institute of Military History [18], with sizes ranging from 2430×3500 to 3289×3500 (*Holland* dataset). The other pictures include 617 more diverse colour images (*General* dataset), including pictures of natural scenery, artificial structures, maps, arts, and synthetic images with size ranging from 366×550 to 13394×10119. Images from the *General* dataset were also converted to greyscale images, and thus we have 1256 greyscale images and 617 colour images overall. The experiment was conducted on an Intel i7–6700HQ processor with 16 GB memory.

Table 1 shows processing speed and memory usage of the proposed and the conventional α–tree algorithm by Ouzounis and Soille [6]. The experiments were conducted on the full 1256 greyscale dataset, which includes the 639 *Holland* dataset images and 617 *General* dataset (greyscale–converted), and the 617 colour dataset. The proposed algorithm runs 2.27 (2.84) times faster than Ouzounis–Soille's for greyscale (colour) images. There is more recent study on the fast single-thread α–tree algorithm by Havel *et al.* which achieved 5.56Mpixel/s on colour images using different hardware (i7–2670QM with 8GB memory) and a different dataset [7]. Full comparison of the processing speed and the memory usage between the proposed algorithm and the Havel's algorithm should be conducted in the future study.

We applied TSE to both flooding and Ouzounis–Soille's algorithm, and TSE reduced the memory usage of flooding and Ouzounis–Soille's algorithm by 40.4% and 14.1%, while decreasing the processing speed by only 1.7% and 0.3%, respectively. We have not implemented TSE on colour α–tree algorithms here because dissimilarities between colour pixels have higher dynamic range than their greylevel counterparts, which is out of the scope of this paper. Here we used L_∞ dissimilarity for colour images to reduce the dynamic range of dissimilarities. We found that the exponential decay modeling of tree sizes does not work on colour image α–trees with L_∞ dissimilarities, possibly because of the loss of information in the calculation of dissimilarity.

Table 1. The processing speed and the memory usage of the proposed and the previous α–tree algorithm. The proposed algorithm runs 2.27 (2.84) times faster than Ouzounis–Soille's in greyscale (colour) images. Tree size estimation (TSE) reduced memory usage of flooding and Ouzounis–Soille's algorithm by 40.4% and 14.1%, respectively.

Image dataset	α–tree algorithm	Processing speed (Mpixel/s)	Memory usage (byte/pixel)
1256 greyscale images (*Holland + General*)	Flooding	14.00	62.12
	Flooding + TSE[a]	13.76	37.00
	Ouzounis-Soille	6.18	90.00
	Ouzounis-Soille + TSE	6.16	77.31
617 colour images (*General*)	Flooding[a]	6.72	110.12
	Ouzounis-Soille	2.37	122.00

[a]proposed

4.1 TSE Performance

We compare the time and memory usage of TSE for different memory management methods. We model the NTS as an exponential decay function of the NRMSD as in (8), and optimize parameters using the full dataset. The optimal values were $A = 1.3901$, $B = -0.1906$ and $\sigma = 2.1989$. The upper bound of tree sizes is computed as

$$\text{TS}_{max}(\text{NRMSD}) = 2|V|(\text{TSE}(\text{NRMSD}) + M), \tag{10}$$

with the estimate of TSE as in (9), and M a constant determining the upper bound.

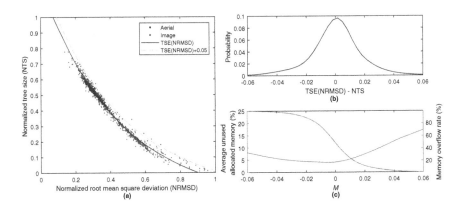

Fig. 1. (a) The exponential decay modeling of the normalized tree size (NTS) as a function of the normalized root mean square deviation (NRMSD). The solid line shows the least square model fit, and the dashed line shows the upper bound of NTS used as memory allocation sizes. (b) The probability mass function of TSE estimate error from leave-p-out cross validation with $p = 1244$. (c) The rate of unused allocated memory and memory overflow as a function of M.

Figure 1(a) shows the NTS as a function of the NRMSD using the test images. The solid line shows the least square fit of the exponential decay model, and the dashed line shows the upper bound of the NTS, which is used as the memory allocation size of TSE for a given NRMSD (which should be multiplied by $2|V|$ as in Eq. 10). The *General* dataset is more widely distributed than the *Holland* dataset, especially near the tail of the model, because the former contains more images with large (quasi-)flat zones, which results in high NRMSD. Figure 1(b) shows the probability mass function of TSE estimate error from leave-p-out cross validation (CV) with $p = 1244$. In the CV 12 randomly sampled images from the dataset are used to fit the TSE model, and the remaining 1244 images are used to validate the model. The training and validating are repeated 1000 times. Despite the small size of training set, we find that approximately 80% of TSE estimates differs from the true NTS by no more than 0.02. Figure 1(c) shows the rate of unused allocated memory and memory overflow, as a function of M. As M increases, the memory overflow rate decreases, but too-high M will lead to a high unused allocated memory rate. We choose $M = 0.05$ to minimize unused allocated memory rate with zero memory overflow.

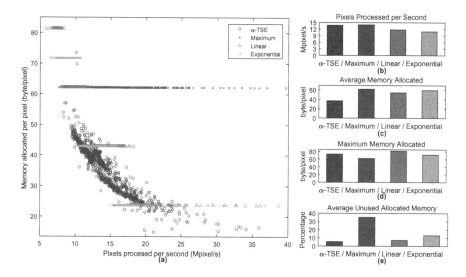

Fig. 2. (a) Execution speed and memory usage of TSE, *Maximum*, *Linear*, and *Exponential*. TSE clearly allocates adaptive memory sizes for different images. Bar plots on the right side show **(b)** pixels processed per second, **(c)** the average memory allocated, **(d)** the maximum memory allocated, and **(e)** the average unused allocated memory.

We analyze the computational and memory costs of TSE and other conventional memory management methods to verify the performance of TSE. Conventional methods typically allocate the maximum amount of memory for $2|V|$ nodes (*Maximum*). Alternatively we could either first allocate a small initial memory space of nodes and increase it linearly every time the memory overflows (*Linear*), or exponentially by multiplying it (*Exponential*). In our implementation, both *Linear* and *Exponential* use initial space of $0.4|V|$; increment in *Linear* is $0.4|V|$ and the multiplication factor in *Exponential* is 2. All methods use clipping to prevent them from allocating memory for more than $2|V|$ nodes.

Figure 2(a) shows the execution speed and memory usage of TSE, *Maximum*, *Linear*, and *Exponential*. To compute memory allocated per pixel, we count all of the memory space of data structures used to build the α–tree defined in Sect. 2.2. This figure shows that TSE allocates memory sizes adapted to the image sizes. The right column of the figure shows pixels processed per second (Fig. 2(b)), the average memory allocated (Fig. 2(c)), the maximum memory allocated (Fig. 2(d)), and the average unused allocated memory (Fig. 2(e)). For TSE, these measures are 14.00 Mpixel/s, 37.00 bytes/pixel, 73.36 bytes/pixel and 5.4%, respectively. Compared to *Maximum*, TSE reduces the average memory (−40.4%) and average unused allocated memory (−84.6%) with a negligible decrease in the execution speed (−1.7%). The maximum memory is increased by 18.1%, but this is due to a couple of outlying images (on the top side of Fig. 2(a)). *Linear* and *Exponential* also show decreases in the average memory (−12.5% and −4.1%) and the unused allocated memory (−79.9% and −62.9%), but they also show a significant increase in the maximum memory allocated ($+30.9\%$ and $+15.5\%$) and a decrease in the execution speed (-12.7% and -12.9%).

We find a high correlation (0.943) between NRMSD and the processing speed; this is because, as NRMSD increases, the number of nodes to be processed decreases, which leads to an increase in the processing speed. An accurate estimation of the processing speed can be useful in the parallel construction of an α–tree in future works.

5   Conclusion

We have introduced a novel α–tree algorithm using a flooding algorithm and TSE. We have modified and optimized a flooding algorithm for α–tree construction and find that our α–tree flooding algorithm runs 2.27 times faster than the conventional Ouzounis–Soille algorithm. Our algorithm also significantly reduces the memory usage using TSE: TSE reduced the average allocated memory by 40.4% and reduced unused allocated memory from 35.2% to 5.4%, with a negligible execution speed reduction (-1.7%). In future work, the new α–tree algorithm can be applied to existing parallel algorithms to further increase the speed or in a hybrid parallel α–tree algorithm for high-dynamic-range images using pilot tree as in [19], where TSE can be used to estimate computation and memory requirement of pilot tree nodes to maximize load balance. We can also apply TSE to component trees and investigate if we can obtain similar results as in α–trees.

Acknowledgements. This paper is based on research developed in the DSSC Doctoral Training Programme, co-funded through a Marie Skłodowska-Curie COFUND (DSSC 754315).

References

1. Salembier, P., Serra, J.: Flat zones filtering, connected operators, and filters by reconstruction. IEEE Trans. Image Process. **7**(4), 1153–1160 (1995)
2. Salembier, P., Wilkinson, M.H.F.: Connected operators: a review of region-based morphological image processing techniques. IEEE Signal Process. Mag. **26**(6), 136–157 (2009)
3. Wilkinson, M.H.F.: A fast component-tree algorithm for high dynamic-range images and second generation connectivity. In: 2011 18th IEEE International Conference on Image Processing (ICIP), Brussels, Belgium, pp. 1021–1024 (2011)
4. Salembier, P., Oliveras, A., Garido, L.: Antiextensive connected operators for image and sequence processing. IEEE Trans. Image Process. **7**(4), 555–570 (1998)
5. Soille, P.: Constrained connectivity for hierarchical image partitioning and simplification. IEEE Trans. Pattern Anal. Mach. Intell. **30**(7), 1132–1145 (2008)
6. Ouzounis, G.K., Soille, P.: The alpha-tree algorithm, theory, algorithms, and applications. JRC Technical Reports, Joint Research Centre, European Commission (2012)
7. Havel, J., Merciol, F., Lefèvre, S.: Efficient tree construction for multiscale image representation and processing. J. Real-Time Image Proc. **2016**, 1–18 (2016)

8. Merciol, F., Lefèvre, S.: Fast image and video segmentation based on α-tree multiscale representation. In: International Conference on Signal Image Technology Internet Systems, Naples, Italy, pp. 336–342, November 2012

9. Carlinet, E., Géraud, T.: A comparative review of component tree computation algorithms. IEEE Trans. Image Process. **23**(9), 3885–3895 (2014)

10. Berger, C., Géraud, T., Levillain, R., Widynski, N., Baillard, A., Bertin, E.: Effective component tree computation with application to pattern recognition in astronomical imaging. In: IEEE International Conference on Image Processing (ICIP), San Antonio, TX, USA, vol. 4, pp. 41–44, September 2007

11. Najman, L., Cousty, J., Perret, B.: Playing with Kruskal: algorithms for morphological trees in edge-weighted graphs. In: Hendriks, C.L.L., Borgefors, G., Strand, R. (eds.) ISMM 2013. LNCS, vol. 7883, pp. 135–146. Springer, Heidelberg (2013). https://doi.org/10.1007/978-3-642-38294-9_12

12. Najman, L.: On the equivalence between hierarchical segmentations and ultrametric watersheds. J. Math. Imaging Vis. **40**(3), 231–247 (2011)

13. Soille, P., Najman, L.: On morphological hierarchical representations for image processing and spatial data clustering. In: Köthe, U., Montanvert, A., Soille, P. (eds.) WADGMM 2010. LNCS, vol. 7346, pp. 43–67. Springer, Heidelberg (2012). https://doi.org/10.1007/978-3-642-32313-3_4

14. You, J.: Alpha-tree algorithm for greyscale images. GitHub repository (2019). https://github.com/jwRyu/AlphaTreeGrey

15. You, J.: Alpha-tree algorithm for 3-channel colour images. GitHub repository (2019). https://github.com/jwRyu/AlphaTreeRGB

16. Najman, L., Couprie, M.: Building the component tree in quasi-linear time. IEEE Trans. Image Process. **15**(11), 3531–3539 (2006)

17. International Organization for Standardization, ISO/IEC 9899:TC3: Programming Languages – C, September 2007

18. Nederlands Instituut voor Militaire Historie. https://www.flickr.com/people/nimhimages. Accessed 29 Mar 2019

19. Moschini, U., Meijster, A., Wilkinson, M.H.F.: A hybrid shared-memory parallel max-tree algorithm for extreme dynamic-range images. IEEE Trans. Pattern Anal. Mach. Intell. **40**(3), 513–526 (2018)

Preventing Chaining in Alpha-Trees Using Gabor Filters

Xiaoxuan Zhang(iD) and Michael H. F. Wilkinson(✉)(iD)

University of Groningen, 9747 AG Groningen, The Netherlands
{zhang.xiaoxuan,m.h.f.wilkinson}@rug.nl

Abstract. Hierarchical segmentation using α-trees can suffer from unwanted leakage or chaining effects, which lower segmentation quality by reducing the depth of the hierarchy. In this paper we introduce a new way to prevent the chaining effect of α-trees. It relies on the odd 2-D Gabor filter. A series of clean, noisy, and blurred synthetic images was used to test the ability of improved α-tree to stop the chaining effect obtaining a 99.8% segmentation accuracy, and we compared it with the contrast-based α-tree proposed previously by Soille. Two remote sensing images were also used to test the performance of the methods on natural images. The results showed that both 4-CN and 8-CN odd Gabor filter based α-trees can prevent the chaining effect efficiently.

Keywords: Image segmentation · Mathematical morphology · Alpha-tree · Gabor filter

1 Introduction

Hierarchical segmentation is an important branch of mathematical morphological image segmentation. As one kind of hierarchical image representation, the α-tree algorithm [9] is used for image segmentation and pattern recognition, especially for very high resolution remote sensing images [5]. The α-tree is created by hierarchically linking regions based on dissimilarities between neighbouring pixels. Thus the leaves of the α-tree are formed by flat zones [13], i.e., connected regions of constant grey level or colour of maximal extent. The flat zones are merged by linking them to their most similar neighbours, which process is iterated until the root of the tree, which represents the entire image, is reached. The computation of α-trees is efficient because dissimilarities between pixels are computed just once, and all merges of a specific dissimilarity value α are performed in a single step. By contrast, in binary partition trees [12] the initial formation of flat zones might be the same. However, if there are multiple pairs of flat zones with the same dissimilarity values, only a single merge is performed, after which the dissimilarities are recomputed, leading to a worst case complexity of $O(N^2 \log N)$, as opposed to quasi-linear complexity in the case of the α-tree.

This increased efficiency comes at a price, however. By performing multiple merges at a given dissimilarity value α simultaneously, long chains of pixels

B. Burgeth et al. (Eds.): ISMM 2019, LNCS 11564, pp. 268–280, 2019.
https://doi.org/10.1007/978-3-030-20867-7_21

with mutual dissimilarities of e.g. one, could lead to regions of very different grey levels being merged. This *chaining through transition zones* [15] leads to very low-depth trees containing little information. Attempts to prevent regions of high pixel diversity occurring at low levels in the α-tree through constrained connectivity [14] do not help as the constraints only select a subset of the α-tree's nodes, so if the desired nodes do not exist in the tree, no constraint can conjure them up. Thus, the problem with the chaining effect of the α-tree which is caused by the transition areas is not solved by constraints. Therefore, it is necessary to find an effective way to protect α-trees from the chaining effect.

Chaining effects are usually caused by transition areas, in particular edges that are either blurred or not exactly axis aligned [15]. Usually, a transition is between two adjacent objects and the width of such a zone is close to 1 pixel. Its length depends on the common edge of these two objects. The orientation of transition is decided by the direction of the common edge. Dissimilarities along the major axis of the transition zone are typically much smaller than at right angles to it. This fact was used by Soille to change the dissimilarity measures to take into account the presence of a transition region [15], and mitigate the chaining effect. Here we propose the use of odd Gabor filters [1–3] to detect transition regions, and alter the dissimilarities based on a models of the human visual system. The goal is to preserve the depth of the α-tree as much as possible, to provide a deeper hierarchy of segmentation, which should make it easier to find an optimal one.

After the proposition of Gabor function by Gabor [3] and the improvements by Daugman [1,2], plenty of research has shown that the 2-dimensional (2-D) Gabor filter (GF) is a very useful tool for texture and pattern analysis [4,10,11]. The even and odd 2-D GF has excellent performance for orientation and edge detection [8], respectively. The even GF has been used to detect narrow objects with the improved α-tree [19]. In this paper, the odd GF is introduced to enhance the edge recognition ability of α-tree, by incorporating an edge-weighted dissimilarity measure.

We review the α-tree and the chaining effect in Sect. 2, and the contrast-based α-tree in Sect. 3. The methodology of the odd Gabor filter based α-ω-tree method is introduced in Sect. 4. In Sect. 5, the performances of original, contrast-based, and odd GF-based α-ω-tree methods are compared. The conclusions are in Sect. 6.

2   The Representation of α-Tree and Chaining Effect

Let E be the image domain of the grey-scale image I. A *path* between two different elements in E is an arbitrary composition of a chain of adjacent pixels. The dissimilarity along the path is the maximum of all the pairwise dissimilarities between adjacent pixels in the path. For a pixel x in E the α-*connected component* (α-CC), to which x belongs is the union of all the pixels connected to x by a path of dissimilarity less than or equal to α. Given a totally ordered set of values of α, all the α-CCs of E can construct the hierarchy $\{\Psi_\alpha^E\}$ indexed by α, and all the levels of hierarchy can be represented as an α-*tree* [9].

Since the pixel connectivity in 2-D images is usually defined in two ways, i.e. 4-connected neighbourhood (4-CN) and 8-connected neighbourhood (8-CN), there are correspondingly two kinds of α-CC, 4-CN based α-CC and 8-CN based α-CC. Therefore, the α-tree also has two types, the 4-CN α-tree (Fig. 1(a)) and the 8-CN α-tree (Fig. 1(b)).

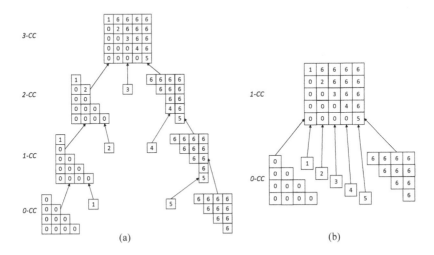

Fig. 1. The 4-CN (a) and 8-CN (b) α-trees for a simple image with the hierarchy of various α-CCs. The objects could not be detected with 8-CN α-tree (b) because of the chaining effect.

Because the α-tree is based on a *local* dissimilarity measure, the chaining effect might result in unexpected, highly heterogeneous partitions (α-CCs), even at low values of α. One example is shown in Fig. 1. The 5×5 image has two flat areas with the values 0 and 6, respectively. However, there is a transition area between them with an increasing step of 1. When we build a 4-CN α-tree for this image as shown in Fig. 1(a), the objects could be obtained easily with a relatively large $\alpha = 2$. On the other hand, with the 8-CN α-tree (Fig. 1(b)), the objects cannot be detected from the image. When $\alpha = 1$, the entire image has already merged into one partition as the root of the tree due to the chaining effect.

3   The Contrast-Based α-Tree

In order to solve the problem of the chaining effect of the α-tree, several ways were proposed by Soille [14,16]. One type of improved α-tree is the α-ω-tree, which is based on the notion of α-ω-CCs [14]. The α-ω-CCs use a range threshold ω to limit the maximum dissimilarity between any pair of pixels within the α-CCs. Only nodes in the α-tree that meet the range threshold criterion set by

ω are considered valid nodes of the α-ω-tree. It should be noted that the α-ω-tree cannot completely deal with the chaining effect caused by transition areas as seen in Fig. 1(b). This is because the α-ω-tree contains only a subset of the nodes of the α-tree from which it is derived. Therefore, Soille introduced another kind of improved α-tree, the contrast-based α-tree, in 2011 [16].

The contrast-based α-tree makes full use of the transition features described in Sect. 1. It is based on the minimum α-dissimilarity along the most contrasted increasing paths, which is "the absolute difference of the intensity values of its end points" [16]. Since the contrast of the pixel on the most contrasted increasing path can be calculated with the morphological gradient ρ, the minimum α-dissimilarity along the most contrasted increasing paths of a pixel can be obtained by the minimum of the gradients by erosion ρ^ε and dilation ρ^δ at the pixel. This operation is called *inf-increment*, and when it is restricted in a specific direction, the inf-increment is defined as

$$p^\wedge = \vee(\rho_\theta^\varepsilon \wedge \rho_\theta^\delta), \ \theta \in \begin{cases} \{0, \pi/2\} & \text{4-CN} \\ \{0, \pi/4, \pi/2, 3\pi/4\} & \text{8-CN} \end{cases} \tag{1}$$

where $\rho_\theta^\varepsilon = id - \varepsilon_\theta$, $\rho_\theta^\delta = \delta_\theta - id$. ρ_θ^ε and ρ_θ^δ are the elementary directional erosion and dilation in the direction θ. Then the α-tree can be built using the contrast-based dissimilarity defined as

$$d^*(p, q) = max\{|f(p) - f(q)|, p^\wedge(p), p^\wedge(q)\} \tag{2}$$

Note that using this dissimilarity measure can also be combined with constrained connectivity to create contrast-based α-ω-trees.

Basically, the contrast-based α-tree uses the fact that transition zones leading to chaining can be detected using an edge detector, and that modifying the pixel dissimilarities based on maximum edge strength can reduce the chaining effect. In the following we explore a different edge detector, based on models of the visual cortex.

4   The Odd Gabor Filter Based α-Tree

Based on the spatial features of transition areas, the edge enhanced dissimilarity measure is introduced in this paper. As a useful tool for edge detection, odd 2-D GF was used to detect the strength of edge signals in various directions [8] and the edge signal bank of objects can be obtained. The odd 2-D GF kernel $(g_{odd}(x, y))$ is shown in Fig. 2(a) and is given by [8,20]

$$g_{odd}(x, y) = \exp(-\frac{x'^2 + \gamma^2 y'^2}{2\sigma^2}) \sin(2\pi\frac{x'}{\lambda} + \phi) \quad where \begin{cases} x' = x \cos \theta + y \sin \theta \\ y' = -x \sin \theta + y \cos \theta \end{cases} \tag{3}$$

where γ is the spatial aspect ratio; λ is the wavelength and its inverse proportion $1/\lambda$ is the spatial-frequency; σ is the standard deviation of the Gaussian factor;

σ/λ is the spacial frequency bandwidth; ϕ ($\phi \in (-\pi, \pi]$) is the phase of kernel and θ ($\theta \in [0, \pi)$) is the detection direction. In this study, we fix the values of γ, σ/λ, ϕ to constant values equal with 0.5, 0.56, and 0, respectively following [7]. We choose λ set to less than or equal to 5, because the transition usually is 1 pixel wide. The value of σ can be calculated by $\sigma/\lambda \times \lambda$. The value of θ is in the set $\{0, \pi/4, \pi/2, 3\pi/4\}$ because the maximum connected neighbourhood in this study is 8.

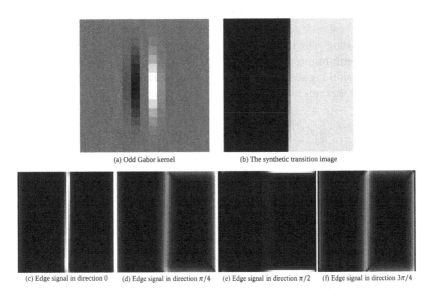

(a) Odd Gabor kernel   (b) The synthetic transition image

(c) Edge signal in direction 0   (d) Edge signal in direction $\pi/4$ (e) Edge signal in direction $\pi/2$ (f) Edge signal in direction $3\pi/4$

Fig. 2. The odd Gabor kernel (a), the synthetic transition image (b) in direction 0, and the edge signal bank of the image ((c)–(d)).

With the odd GF kernel, the edge signal ($e_\theta(x, y)$) in various directions in image I can be computed by absolute value of the convolution

$$e_\theta(x, y) = |(I * g_{odd})(x, y)| \tag{4}$$

This yields the edge signal bank e_θ in 4 directions $\{0, \frac{\pi}{4}, \frac{\pi}{2}, \frac{3\pi}{4}\}$, see Fig. 2(c)–(f). From the figure it is easy to see that, when the direction of the odd GF kernel is nearly parallel to the transition area, the edge signal is the strongest. Conversely, at right angles to the direction of the zone, the edge signal is very weak, and in the two intermediate directions, the strength of the signal lies between that of the parallel and vertical directions. Based on these characteristics obtained through odd GF, the direction of the strongest edge signal θ_{max} of each pixel was computed as follows:

$$\theta_{max} = \begin{cases} \arg\max(\forall e_\theta, \theta \in \{0, \frac{\pi}{2}\}) & \text{4-CN} \\ \arg\max(\forall e_\theta, \theta \in \{0, \frac{\pi}{4}, \frac{\pi}{2}, \frac{3\pi}{4}\}) & \text{8-CN} \end{cases} \tag{5}$$

In this way, the edge signal of pixels in their main edge direction $\theta_{\max}$ can be obtained, which is denoted as e_Θ. Due to the existence of the texture and the noise in natural images, not all the e_Θ of pixels are necessary. Therefore, the threshold (T) of e_Θ is necessary to distinguish the edges of objects and the other details. In this study, T was experimentally set to be twice the average of all non-zero maximum edge strengths e_Θ in the image. Other methods for edge thresholding could also be used, such as the method used in [17], and developed further in [18].

Combining the notion of the α-tree and the edge signal, an edge signal weight w_e of pixels can be used to prevent the chaining effect in the α-tree. Let the dissimilarity measure δ between two pixels (p and q) be the absolute difference of them. Using the $\theta_{\max}$ and T as described above, we can introduce an edge signal based dissimilarity measure $\delta_e(p, q)$ as

$$
\delta_e(p, q) = \begin{cases} \delta(p, q) \times w_e & \text{if the edge } (p, q) \text{ is along the main edge direction and} \\ & e_\Theta(p) \geq T \\ \delta(p, q) & \text{otherwise.} \end{cases} \tag{6}
$$

In this way only the dissimilarity measure of pixels on edges along their main edge direction will be enhanced, and the odd GF based α-tree can be built in the same way as the original α-tree, once all dissimilarities have been computed.

5   Results and Discussion

5.1   Segmentation Results on Synthetic Images

In this initial study, a series of 51×51 synthetic transition images were used (Fig. 2(b)). One transition image is composed of three areas, the dark, light flat areas, and a one-pixel-wide transition area between them. The intensities of the dark and light flat areas are 0 and 208 respectively. In the transition area, the value of pixels increases from 4 to 204 in steps of 4. Based on the image described above, a series of images in which the transition is in 17 different evenly distributed directions could be obtained.

Both noisy and blurred synthetic images were used to test the segmentation ability of the methods. Gaussian noise is based on the standard Box–Muller transform and a standard deviation of 10, and mean of 0. The scale of Gaussian filter for the blurred images is 3×3 with standard deviation of 0.8.

Because the composition of these images is simple, it is easy to obtain the ground truth images and determine the accuracy of the segmentation results. The partitions of dark and light areas are very clear, and the only issue is the transition areas. In this study, whatever the partition of the pixels in the transition belongs to, the dark areas, the light areas, or the areas composed of pixels in the transition, they are all acceptable and treated as correct, because the main purpose of our study was to prevent merging of the dark and light areas. Based on this, we compared the segmentation performance of the original, contrast-based, and the odd GF based α-trees as shown in Table 1. In order to get better

Table 1. The α_{max} which could detect both dark and light partitions from the images with different methods. The w_e is set to 100 in this case. "N" means that the dark and light partitions could not be detected with a $\alpha > 4$.

Images	Methods	Transition images in various directions								
		0, π/2	π/32, 15π/32	π/16, 7π/16	3π/32, 13π/32	π/8, 3π/8	5π/32, 11π/32	3π/16, 5π/16	7π/32, 9π/32	π/4
Clean synthetic images	4-CN α-tree	N	83	95	99	99	103	103	103	103
	8-CN α-tree	N	N	N	N	N	N	N	N	N
	4-CN contrast α-tree	51	55	61	67	71	73	81	85	N
	8-CN contrast α-tree	51	55	61	67	71	75	83	91	99
	4-CN odd GF α-tree	103	103	103	103	103	103	103	103	103
	8-CN odd GF α-tree	103	103	103	103	103	103	103	103	103
Noisy images	4-CN α-tree	35.4	78.1	88.4	87.8	89.3	94.4	94.1	93.3	97.4
	8-CN α-tree	31.6	32.7	33.3	33.4	33.0	32.5	32.9	31.6	33.0
	4-CN contrast α-tree	39.7	44.1	51.4	53.1	56.2	64.3	70.4	86.6	35.8
	8-CN contrast α-tree	41.9	44.8	46.2	50.5	61.7	65.9	71.3	80.0	84.6
	4-CN odd GF α-tree	99.3	98.5	99.7	97.5	97.8	98.2	98.4	99.1	97.2
	8-CN odd GF α-tree	94.6	92.9	92.3	92.2	89.4	90.2	92.1	93.7	94.0
Blurred images	4-CN α-tree	53	42	48	37	48	42	46	47	64
	8-CN α-tree	52	39	38	37	37	38	38	38	37
	4-CN contrast α-tree	38	35	39	43	44	38	42	46	50
	8-CN contrast α-tree	38	38	41	44	46	52	57	64	61
	4-CN odd GF α-tree	77	75	74	74	74	65	65	65	64
	8-CN odd GF α-tree	76	62	60	47	37	38	45	46	64

statistics, we tested five times for noisy images in each method. All the segmentation results have an accuracy greater than 99.8%, which means there are only five pixels or fewer segmented incorrectly. The chief difference in performance lies in the depth of the α-tree, and we consider deeper trees to be better, because this indicates better resistance to the chaining effect.

The performance of the original 4-/8-CN α-trees is shown in the first two rows for each type of images as examples to evaluate the results of the improved α-tree methods. For the clean synthetic image, the chaining effect of the 4-CN α-tree is not very obvious except when the transition is in the direction of 0 and $\pi/2$. Especially when the direction is between $5\pi/32$–$11\pi/32$, the maximum α (α_{max}), with which the dark and light partitions can be detected independently, could reach a peak value of 103. Moving away from these maxima, the value of α_{max} decreases gradually. However, the 8-CN α-tree completely fails to distinguish the light and dark areas with an α larger than the transition step. The performance of the original 4-CN α-tree on noisy images is similar to the clean images in most of cases. In the cases of direction 0 and $\pi/2$ for the 4-CN α-tree and all the directions for the 8-CN α-tree, the α_{max} could reach the value of 8–9 times the step size because the Gaussian noise breaks up the regular "step-ladder" of the transition zone. There is no clear differences to process transitions in various directions with 8-CN α-tree.

For blurred images, the original 4-/8-CN α-trees have very different α_{max} compare to the cases described above. Both 4-CN and 8-CN α-trees have a relatively large α_{max} when the transition is in the direction of 0 or $\pi/2$. For the images in other directions, the α_{max}s for 4-/8-CN α-trees have small changes and the peak of the α_{max} appears in the direction of $\pi/4$ for the 4-CN α-tree as usual.

From the table, it is obvious that the contrast-based α-trees can solve the chaining effect problem in most of the clean images. The improvement is obvious for the images which have transition in the directions of 0 and $\pi/2$ with the 4-CN method, and all the images in various directions processed by the 8-CN method. With the 8-CN contrast-based α-tree, the α_{max} increases from 51 when the direction of transition is 0, reaches a peak 99 when the direction is $\pi/4$, and decreases to 51 again with the direction of $\pi/2$. Although the 4-CN contrast-based α-tree has a similar trend, it has some drawbacks compared to the original method. Except for the $0, \pi/2$ case, all the α_{max}s are less than the original method, and it even aggravates the chaining effect in the $\pi/4$ case, unlike the original 4-CN α-tree.

However, this problem is not obvious for the noisy and blurred images. Although the α_{max}s of the 4-CN contrast-based α-tree are less than the original 4-CN α-tree in most cases, the chaining effect is not obvious so much, and there is an improvement for the $0, \pi/2$ images. The 8-CN contrast-based α-tree works much better than the original 8-CN α-tree with great improvements.

For the odd GF based α-tree, the α_{max} depends on the w_e, and in the clean images cases, the relationship between them is as Fig. 3(a). In Table 1, only the α_{max} obtained by a large w_e is shown. The ability to prevent the chaining effect of the 4-CN and 8-CN odd GF based α-trees is very steady and it can be controlled by the w_e. The maximum of the α_{max} in this case is 103 in all the directions for both 4-CN and 8-CN methods. The performance of the 4-CN and 8-CN odd GF α-trees is also perfect for noisy images comparing with the other two kinds of α-trees. The average level of α for 4-CN method is a little bit higher than the 8-CN method but there is no obvious differences.

However, when the odd GF based α-trees are used to process the blurred images, the prevention of chaining effect is not as easy as the two cases above. Although the 4-CN method has the largest α_{max} among the 6 kinds of methods, the ability to stop the chaining effect has a valley in the direction of $\pi/4$, and the

α_{max} of the 4-CN odd GF based α-tree and the original method are completely same. The valley of the 8-CN odd GF based α-tree is in the direction of $\pi/8$ and $3\pi/8$. This is because there should be two directions of strongest edge signal, but the methods only chooses one direction as the main edge signal direction, and that leads to the α_{max} of the odd GF based α-tree is decided by the unprocessed δ.

Both the odd GF based and the contrast-based α-trees have the ability of stopping chaining effect. However, if we change the principle of segmentation accuracy described before, let the pixels on the transition areas must belong to the dark or light areas. Then, a larger α_{max} usually means a better segmentation result. One example is shown in Fig. 3(b) and (c). The average α_{max} of the contrast-based method for the noise images in direction 0 and $\pi/2$ is 41.9 but it is 94.6 for the odd GF based method. Therefore, it is harder for the contrast-based method to merge the pixels of the transition into the dark or light partitions under the premise of the dark and light partitions can be detected independently. More generally, a deeper α-tree means that there are more nodes to select as valid nodes when using constrained connectivity, so more likelihood that the resulting α-ω-tree contains valid image structures.

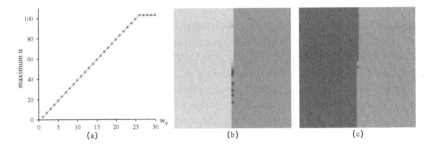

Fig. 3. The relationship between w_e and the α_{max} for the clean synthetic image (a). The segmentation result of the noisy image with the α_{max} using 8-CN contrast-based (b) and odd GF based (c) α-tree.

5.2   Segmentation Results on Remote Sensing Images

In this subsection, two remote sensing images are used to show the performances of the α-tree methods. The images are the Airborne Visible/Infrared Imaging Spectrometer (AVIRIS) Indian pines image (Fig. 4(a)), and the Reflective Optics System Imaging Spectrometer (ROSIS) Pavia university image (Fig. 5(a)).

Figures 4(b)–(g) and 5(b)–(g) show the segmentation results of three kinds of α-trees. With the certain α, the amount of partitions of the original α-tree is obviously less than the contrast and odd GF based α-tree, which means both improved α-tree methods are effective for chaining effect preventing. Comparing the contrast-based and odd GF based α-trees, they show their segmentation advantages in different situations. The contrast-based α-trees works better in 8-CN cases, and the odd GF based α-trees have better results in 4-CN cases. One reason for these results is that the 8-CN odd GF α-tree is not good at the

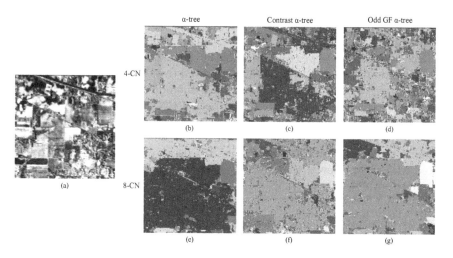

Fig. 4. The segmentation results of AVIRIS Indian pines image based on the 4-/8-CN α-trees, contrast α-trees, and odd GF α-trees respectively from left to right when $\alpha = 30$. The w_e for odd GF α-trees is 5.

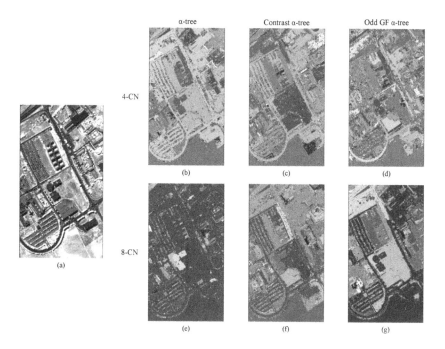

Fig. 5. The segmentation results of ROSIS Pavia university image based on the 4-/8-CN α-trees, contrast α-trees, and odd GF α-trees respectively from left to right when $\alpha = 20$. The w_e for odd GF α-trees is 5.

blurred edges when the transitions are around the direction of $\pi/8$ and $3\pi/8$. Another reason is that the size of GF kernel is not sensitive enough for very detailed objects.

Since we consider deeper trees to be better, the depths of different α-trees are also analysed in Table 2. Obviously, the improved trees have more levels than the original α-trees in most cases. The depths of the odd GF based α-trees can be deeper than the contrast-based α-tree in both 4-CN and 8-CN cases. However, this is depends on the value of w_e and the complexity of the images. In the case of the Pavia university, the depths of contrast-based α-trees are similar to the original α-trees. The table also shows that the depth of odd GF based α-trees is related with w_e. However, the effect of w_e of 8-CN odd GF based α-tree is not as clear in both cases.

Table 2. The depth of different kinds of α-trees based on the AVIRIS Indian pines image and the ROSIS Pavia university image.

Images	4-CN methods					8-CN methods				
	α-tree	Contrast α-tree	Odd GF α-tree $w_e = 2$	Odd GF α-tree $w_e = 5$	Odd GF α-tree $w_e = 10$	α-tree	Contrast α-tree	Odd GF α-tree $w_e = 2$	Odd GF α-tree $w_e = 5$	Odd GF α-tree $w_e = 10$
Indian pines	91	113	110	139	151	80	114	87	100	103
Pavia university	137	136	157	179	200	122	127	129	130	130

6 Conclusions

The chaining effect caused by the transition areas is especially obvious for the 8-CN α-tree and for the 4-CN α-tree when the transition is in horizontal or vertical direction. The 4-/8-CN contrast-based α-trees were proved to be very effective algorithms to stop the chaining effect. However, the 4-CN contrast-based α-tree might be invalid in some special cases, such as when the transition is in $\pi/4$ direction. In this study, the odd Gabor filter based α-trees are introduced to prevent the chaining effect and they were compared with the original and the contrast-based α-trees. The tests based on the series of clean, noisy, and blurred synthetic images showed that both 4-CN and 8-CN odd GF based α-trees reduce the chaining effect, and the performance is better than the contrast-based α-tree in most of the cases, except the 8-CN method for blurred image in $\pi/8$, $3\pi/8$ directions. This ability could even be controlled over a certain range by the w_e depending on the requirement. The performance of odd GF based α-trees on two remote sensing images showed reduced chaining as well. The odd GF based α-trees showed better prevention of chaining in the 4-CN cases, but showed only modest gains in the 8-CN cases, where the contrast-based method works better.

Future Work. This work complements work on using even Gabor filters for better preservation of thin line-like structures in α-trees [19]. In future work we will study how best to combine these two approaches. Furthermore, these approaches need extension to colour and other multi-band images, or even tensor images. We might also consider replacing Gabor filters with other visual cortex models, such as derivatives of Gaussians [6]. The optimal range of the w_e still needs more tests. We will also explore other ways of implementing edge weights, perhaps in a more continuous fashion.

References

1. Daugman, J.G.: Two-dimensional spectral analysis of cortical receptive field profiles. Vis. Res. **20**(10), 847–856 (1980). https://doi.org/10.1016/0042-6989(80)90065-6

2. Daugman, J.G.: Uncertainty relation for resolution in space, spatial frequency, and orientation optimized by two-dimensional visual cortical filters. J. Opt. Soc. Am. A **2**(7), 1160 (1985). https://doi.org/10.1364/JOSAA.2.001160, https://www.osapublishing.org/abstract.cfm?URI=josaa-2-7-1160

3. Gabor, D.: Theory of communication. J. Inst. Electr. Eng. Part III Radio Commun. Eng. **93**(26), 429–457 (1946). citeulike-article-id: 4452465. https://doi.org/10.1049/ji-3-2.1946.0074

4. Grigorescu, S.E., Petkov, N., Kruizinga, P.: Comparison of texture features based on Gabor filters. IEEE Trans. Image Process. **11**(10), 1160–1167 (2002). https://doi.org/10.1109/TIP.2002.804262

5. Gueguen, L., Ouzounis, G., Pesaresi, M., Soille, P.: Tree based representations for fast information mining from VHR images. In: Datcu, M. (ed.) Proceedings of the ESA-EUSCJRC Eight Conference on Image Information Mining, pp. 15–20 (2012)

6. Koenderink, J.J.: The structure of images. Biol. Cybern. **50**(5), 363–370 (1984)

7. Kruizinga, P., Petkov, N.: Nonlinear operator for oriented texture. IEEE Trans. Image Process. **8**, 1395–1407 (1999)

8. Mehrotra, R., Namuduri, K., Ranganathan, N.: Gabor filter-based edge detection. Pattern Recognit. **25**(12), 1479–1494 (1992). https://doi.org/10.1016/0031-3203(92)90121-X. http://www.sciencedirect.com/science/article/pii/003132039290121X

9. Ouzounis, G.K., Soille, P.: The Alpha-tree algorithm. Publications Office of the European Union, Luxembourg, December 2012. https://doi.org/10.2788/48773

10. Petkov, N.: Biologically motivated computationally intensive approaches to image pattern recognition. Future Gener. Comput. Syst. **11**(4–5), 451–465 (1995). https://doi.org/10.1016/0167-739X(95)00015-K

11. Riaz, F., Hassan, A., Rehman, S.: Texture classification framework using Gabor filters and local binary patterns. In: Arai, K., Kapoor, S., Bhatia, R. (eds.) SAI 2018. AISC, vol. 858, pp. 569–580. Springer, Cham (2019). https://doi.org/10.1007/978-3-030-01174-1_43

12. Salembier, P., Garrido, L.: Binary partition tree as an efficient representation for image processing, segmentation and information retrieval. IEEE Trans. Image Process. **9**(4), 561–576 (2000)

13. Salembier, P., Serra, J.: Flat zones filtering, connected operators, and filters by reconstruction. IEEE Trans. Image Process. **4**, 1153–1160 (1995)

14. Soille, P.: Constrained connectivity and connected filters. IEEE Trans. Pattern Anal. Mach. Intell. **30**(7), 1132–1145 (2008)

15. Soille, P., Grazzini, J.: Constrained connectivity and transition regions. In: Wilkinson, M.H.F., Roerdink, J.B.T.M. (eds.) ISMM 2009. LNCS, vol. 5720, pp. 59–69. Springer, Heidelberg (2009). https://doi.org/10.1007/978-3-642-03613-2_6

16. Soille, P.: Preventing chaining through transitions while favouring it within homogeneous regions. In: Soille, P., Pesaresi, M., Ouzounis, G.K. (eds.) ISMM 2011. LNCS, vol. 6671, pp. 96–107. Springer, Heidelberg (2011). https://doi.org/10.1007/978-3-642-21569-8_9

17. Wilkinson, M.H.F.: Optimizing edge detectors for robust automatic threshold selection: coping with edge curvature and noise. Graph. Mod. Image Process. **60**, 385–401 (1998)

18. Wilkinson, M.H.F.: Gaussian-weighted moving-window robust automatic threshold selection. In: Petkov, N., Westenberg, M.A. (eds.) CAIP 2003. LNCS, vol. 2756, pp. 369–376. Springer, Heidelberg (2003). https://doi.org/10.1007/978-3-540-45179-2_46

19. Zhang, X., Wilkinson, M.H.F.: Improving narrow object detection in alpha trees. Mathematical Morphology - Theory and Application (2019, Submitted)

20. Zhu, Z., Lu, H., Zhao, Y.: Scale multiplication in odd Gabor transform domain for edge detection. J. Vis. Commun. Image Represent. **18**(1), 68–80 (2007). https://doi.org/10.1016/j.jvcir.2006.10001

A Graph Formalism for Time and Memory Efficient Morphological Attribute-Space Connected Filters

Mohammad Babai[1(✉)], Ananda S. Chowdhury[2], and Michael H. F. Wilkinson[1]

[1] University of Groningen, Groningen, The Netherlands
{m.babai,m.h.f.wilkinson}@rug.nl
[2] Jadavpur University, Kolkata, India
as.chowdhury@jadavpuruniversity.in

Abstract. Attribute-space connectivity has been put forward as a means of improving image segmentation in the case of overlapping structures. Its main drawback is the huge memory load incurred by mapping a N-dimensional image to an $(N + \dim(A))$-dimensional volume, with $\dim(A)$ the dimensionality of the attribute vectors used. In this theoretical paper we introduce a more space and time efficient scheme, by representing attribute spaces for analysis of binary images as a graph rather than a volume. Introducing a graph formalism for attribute-space connectivity opens up the possibility of using attribute-space connectivity on 3D volumes or using more than one attribute dimension, without incurring huge memory costs. Furthermore, the graph formalism does not require quantization of the attribute values, as is the case when representing attribute spaces in terms of $(N + \dim(A))$-dimensional discrete volumes. Efficient processing of high dimensional data produced by multi-sensor detection systems is another advantage of application of our formalism.

Keywords: Image segmentation · Connectivity · Attribute-spaces · Graph representations · Graph morphology · Morphological filtering

1 Introduction

In many image-processing applications, one needs to segment the image into disjoint connected components. This task may fail when the structures overlap, because segmentation aims to provide a partition rather than a cover of the image domain. A solution to this problem was proposed in [24], by introducing the concept of connectivity in higher dimensional spaces, known as *attribute spaces*. In this scheme, which was only defined for binary images, each foreground pixel (x, y) in the original image is mapped to a non-empty set $\{(x, y, a_i)\}$, with a_i indicating the ith attribute value assigned to the pixel, with i from some index set I. Transforming the image into a higher dimensional space could theoretically be a solution to segment overlapping structures into disjoint segments, as shown

© Springer Nature Switzerland AG 2019
B. Burgeth et al. (Eds.): ISMM 2019, LNCS 11564, pp. 281–294, 2019.
https://doi.org/10.1007/978-3-030-20867-7_22

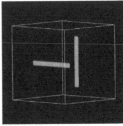

Fig. 1. An example of attribute-space connectivity: (Left) Original image containing a single connected component; (right) mapping to a 3D orientation space separates the horizontal from the vertical bar.

in Fig. 1, where all foreground pixels in a 2D binary image are mapped to a 3D orientation space, by assigning one or more orientations to each pixel, e.g. using a scheme similar to that in [8].

The theory put forward in [24] can readily be extended to deal with 3D volumes as input, but expansion to 4D would be very costly in terms of memory use. Likewise, the theory can also be extended to more than one attribute dimension, e.g. orientation and curvature, which again would lead to 4D data hypercubes, even if we restrict ourselves to 2D inputs.

There is a further drawback to representation of the images or volumes transformed into the attribute space. List of attribute values with a pre-defined length for all pixels of an image (multi-dimensional arrays) requires prior discretization of the attribute values. On binary images, many attributes, such as width might be by nature discrete, so do not pose a problem. However, other attributes, such as orientation and curvature are not necessarily discrete, which means we need to quantize the data for representation in an N-dimensional array.

We therefore turn to mathematical morphology on graphs [23], which has recently come under renewed attention [5,6,13]. An image can be interpreted as a graph where vertices are represented by the pixels and the edges are defined by the adjacency relations on those pixels. In a recently submitted paper [1] we made a first attempt at using the theory of attribute-space connectivity in the domain of graphs. In the current work, we improve upon the previous attempt, by assigning attribute values not just to vertices, but also to the edges themselves, which allows a far simpler formalism, and obviates the need of using tunable parameters such as thresholds, which was required in the previous method.

The rest of the paper is organised as follows: The basics of connectivity and filtering are briefly described in Sect. 2. In Sect. 3 we explain in short the basics of attribute-spaces on graphs, translating original definition of attribute-space transform pairs to the graph domain. We describe our method to compute connected components in Sect. 4 and finally, in Sect. 5 we discuss our method and future work.

2   Basic Concepts of Connectivity and Filtering

In mathematical morphology images are usually represented as subsets X of some image domain M [19]. Members of X are the foreground pixels, and connectivity between pixels is defined by 4 or 8 neighbourhood relationships. Connected components are usually defined using path connectivity, but a more general framework is that of *connectivity classes* or *connections*. A connectivity class $\mathcal{C}$ is the family of all connected subsets of M, with the following properties:

- $\emptyset \in \mathcal{C}$.
- Singletons $\{x\} \in \mathcal{C}$ for all $x \in M$.
- For any family $\{X_i\} \subseteq \mathcal{C}$, $\bigcap_i X_i \neq \emptyset$ implies $\bigcup_i X_i \in \mathcal{C}$.

as defined in [20]. Each connectivity class is associated with a family of connectivity opening Γ_x, with one associated with each $x \in M$.

$$\Gamma_x(X) = \bigcup \{C \in \mathcal{C} \mid x \in C \wedge C \subseteq X\}. \tag{1}$$

The function $\Gamma_x(X)$ returns the connected component of X to which x belongs, if $x \in X$ and the empty set otherwise.

These notions extend readily to complete lattices [20], and therefore also to a graph formalism, in which a binary image X is represented as a subgraph $G = (V, E)$ of some maximal graph $G_M = (V_M, E_M)$ with $V \subseteq V_M$, $E \subseteq E_M$ and the set of vertices $V_M \subseteq M$ and a set of edges $E_M \subseteq (M \times M)$. The lattice $\mathcal{L}_G$ associated with G_M in this way is simply the set of all subgraphs of G_M, with the empty graph $\mathbf{0}$ as global infimum and G_M as supremum, and the partial order relation $a \preccurlyeq b$ mean a is a subgraph of b. All that has really changed is making the neighbourhood relationship explicit in the form of the set of edges E. The vertices of G represent the pixels and the edges are determined by the neighbourhood relations. Each vertex $v \in V$ has a maximum of 8 edges, in a 2D image with 8-connectivity scheme. The spatial coordinates, grey or colour value or any other local pixel attributes are stored in the nodes, depending on the graph embedding or the properties of the input image; an example method to generate a graph from an image is illustrated in Algorithm 2.1. The graph G can be constructed in $O(N)$ time, where N is the total number of pixels in the image. Note that in general graph nodes could be of an arbitrary dimension and can have arbitrary and non-uniform neighbouring relations.

Algorithm 2.1 Create a graph G from an image.

Input: (Image img)
Output:(Graph G)

```
1: for all p (pixels) ∈ img do
2:      Create node g and store all attributes of p in g.
3:      Add indices of all neighbours of p to neighbour-list of g.
4:      Insert g into G.
5: end for
```

The connectivity classes described in [17,20,24] can be used to define connectivity. Every connectivity class $\mathcal{C} \subseteq \mathcal{L}_G$ can be considered as the family of all connected elements of a lattice; in this case the family of all connected subgraphs of some maximal graph $G_M = (V_M, E_M)$. Usually a graph $G = (V, E) \preccurlyeq G_M$ is considered connected if $\forall a, b \in V$, there is a sequence $(x_1, ..., x_n) \subseteq V$ with $x_1 = a$ and $x_n = b$ and all edges $e_i = (x_i, x_{i+1}) \in E$. We can define connected components $\{C_i\}$ of any graph $G = (V, E)$ in the usual way, as connected subgraphs of G of maximal extent, i.e. $\{ \forall C_i, \nexists(C \succ C_i) : C \preccurlyeq G \wedge C \in \mathcal{C}\}$. Connectivity openings can be defined in a similar way, by allowing connectivity openings marked more generally by any member of the connectivity class $\mathcal{C}_G$, not just the family of singleton graphs $\mathcal{S}_G$ which contains all graphs $g \in \mathcal{L}_G$ of the form $(\{v\}, \emptyset)$ or $(\emptyset, \{e\})$, i.e. both singleton vertices and singleton edges. We consider all elements of $\mathcal{S}_G$ connected. A binary connected opening Γ_x on $G \in \mathcal{L}(G)$ for any connected subgraph x of G_M can be defined as:

$$\Gamma_x(G) = \bigvee \{C \in \mathcal{C} \mid x \preccurlyeq C \wedge C \preccurlyeq G\}. \tag{2}$$

$\Gamma_x(G)$ is the connected component in which $x \preccurlyeq G$ is contained and $\mathbf{0}$ otherwise. Generalisations such as second-generation connectivities as developed on images [3,15,20] should also extend quite naturally to the graph framework.

Attribute filters form a class of the connected filters. Analogous to binary images, we can define connected filters [18], and in particular attribute filters [4] which remove or select connected components that fulfil a certain criterion T as:

$$\Gamma^T(G) = \bigvee_{x \in \mathcal{S}_G} \Gamma_T(\Gamma_x(G)). \tag{3}$$

with Γ_T the *trivial filter*, which when applied to some graph C, returns C if it meets some criterion T, and $\mathbf{0}$ otherwise. Usually, T is of the form $T(C) = Attr(C) \geq \tau$, with the property of being increasing; τ is a threshold value [4] and $Attr(C)$ is an attribute of C.

As can be seen, by casting graph-based connected filtering into a lattice framework, connectivity and connected filtering readily extends to the graph case. Extensions such as hyperconnections [16,27], which have also been cast into lattice formalism should also be straightforward. The same does not hold true for attribute-space connections, for which only a set-based formalism exists. In the following section we will extend the formalism to the lattice of graphs $\mathcal{L}_G$.

3   Attribute-Space Connectivity

In [24] attribute-space connectivity and filters were introduced in the case of binary images. The core idea is to transform the image into a higher-dimensional domain in which overlapping structures are readily separated. An image X is transformed into a higher dimensional "attribute space" $(M \times A)$, with A the attribute dimension. The latter represents some local measure of the image, such as width or orientation. This is done by mapping every pixel $x \in X$ to one or more tuples $(x, a_i) \in (X \times A)$ using some operator $\Omega : \mathcal{P}(M) \to \mathcal{P}(M \times A)$; $\mathcal{P}$ denotes the power set and i is from an index set I. Note that $\Omega(X)$ yields a binary image in $(M \times A)$ and A is typically a subset of $\mathbb{Z}$ or $\mathbb{R}$. These separated structures can then be filtered in the $(M \times A)$ space using an appropriate attribute filter on the connected components. After application of the selected filters the results could be projected back onto M by the inverse operator $\Omega^{-1} : \mathcal{P}(M \times A) \to \mathcal{P}(M)$. In summary, the required properties [25] of the pair (Ω, Ω^{-1}) are:

1. $\Omega(\emptyset) = \emptyset$,
2. $\forall x \in M : \Omega(\{x\}) \in \mathcal{C}_{M \times A}$,
3. The mapping $\Omega : \mathcal{P}(M) \to \mathcal{P}(M \times A)$ satisfies the condition that for all $X \in \mathcal{P}(M)$, each $x \in X$ is mapped to at least one point $(x, a) \in \Omega(X)$, with $a \in A$,
4. The inverse mapping $\Omega^{-1} : \mathcal{P}(M \times A) \to \mathcal{P}(M)$ satisfies the condition that for all $Y \in \mathcal{P}(M \times A)$, each $(x, a) \in Y$ is projected to $x \in \Omega^{-1}(Y)$. In other words, every tuple is mapped to the actual pixel in the original image domain (M),
5. $\forall X \in \mathcal{P}(M) : \Omega^{-1}(\Omega(X)) = X$,
6. Ω^{-1} is increasing; $\{\forall X, Y \in \mathcal{P}(M \times A) : X \leq Y \implies \Omega^{-1}(X) \leq \Omega^{-1}(Y)\}$.

Since axiom 4 defines Ω^{-1} to be a projection and thus being increasing, the axiom 6 is redundant [25], and we will therefore leave it out in the following discussion. As we have discussed earlier, a naive application of this idea incurs a huge memory overhead, which in turn increases computational cost.

Alternatively, we simply only duplicate the required vertices and edges of G, to form a new graph $G^A = (V^A, E^A)$ with $V^A = (V \times A)$, and $E^A \subseteq (V^A \times V^A)$. As our recently submitted paper [1], we denote each element of V^A as $v_{ij}^A = (v_i, a_j)$, with i and j from index sets over V and A respectively. For every v_{ij}^A, a_j is referred to as the node's *attribute value*. In our previous work edges in G^A are defined as

$$E^A = \{((v_{ij}^A, v_{k,l}^A) \in V^A \times V^A \mid (v_i, v_k) \in E) \wedge (||a_j - a_l|| < D)\}, \qquad (4)$$

with D some threshold. Thus nodes in G^A are connected by an edge, if their counterparts in graph G are connected by an edge and their difference in attribute value is small enough. This has a slight problem in that some edges might be removed entirely by Ω, and unless certain slightly artificial measure were taken, back-projection by Ω^{-1} would not restore them. This would imply that $\Omega^{-1}(\Omega(G)) \neq G$ for some graphs G, which is clearly undesirable.

Here, we choose a more general approach, in that transformation operator Ω assigns not only one or more attributes to the vertices, but also one or more *pairs* of attributes to each edge $e \in E$. Thus, we can write

$$\Omega(G) = (\Omega_V(G), \Omega_E(G)) \qquad (5)$$

Algorithm 3.1 Compute Attribute Space for a graph G.

Input:(Graph G)
Output:(Graph G^A)

1: **for all** x ∈ G **do**
2:     **if** (x is foreground) **then**
3:         Compute (x, a_i) and (e, a_{1j}, a_{2j}).
4:         Store (a_i) in x and (a_{1j}) and (a_{2j}) in e.
5:     **end if**
6: **end for**

---

with $\Omega_V : \mathcal{L}_G \rightarrow \mathcal{P}(M \times A)$ essentially the same operator on vertices as in [1], which yields V^A, and $\Omega_E : \mathcal{L}_G \rightarrow \mathcal{P}((M \times A) \times (M \times A))$ a mapping from a single edge in the original graph to one or more edges in $G^A = \Omega(G)$. As before, $\mathcal{P}$ denotes the power set. We assume the result G^A is subgraph of some maximal graph G^A_M embedded in the attribute space, with corresponding lattice of subgraphs $\mathcal{L}^A_G$.

Edges e^A in these latter graphs can be denoted either as pairs of vertices in the attribute domain (v^A_1, v^A_2) or equivalently, as triples (e, a_1, a_2), with $e \in M \times M$, and $a_1, a_2 \in A$. Therefore Ω_E may assign different attribute values to each endpoint of the edges in G^A; the determined value depends on the local properties of the attached structures. This method is illustrated in Algorithm 3.1; note that adding an attribute a_i internally to a nodes' attributes list is equivalent to splitting the node. One can construct G^A from G in $O(|V|)$ time.

We can readily incorporate the mapping of (4) into this mechanism, but at the same time allow disconnected edges to appear in the transformed graph G^A. Attribute values are assigned to each end-point of every edge connecting active regions of the graph G; these values may differ per end-point of the edges. This is equivalent to creating a copy of each edge for each attribute value. This operation allows for an edge to be active in one attribute-space while not connected

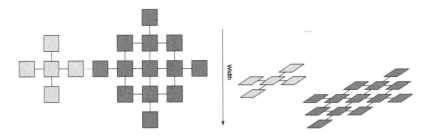

Fig. 2. Loose edges in attribute space: (left) graph G showing two diamond-shaped structures linked in M by the red edge; (right) width-space projection G^W in which all edges and vertices are assigned the value of the diameter of the largest inscribed city-block disk in which they fit. The red edge does not fit into any disk, and is assigned a zero value, disconnecting the structures in width space. Without filtering, back projection of G^W onto the original space yields the original graph. Note that we use L_1 norm to determine the width-based attribute for each node. (Color figure online)

(active) in another attribute-space as a consequence of not fulfilling a certain criterion; this simplifies the inverse operator Ω^{-1}. Ω^{-1} converts G^A to G by merely projecting the attribute dimensions for all vertices and edges in G^A to the original domain of G. This effect is illustrated in Fig. 2, which uses a width space transform Ω_W (see Sect. 3.1). The centre edge linking the two diamond shapes in the original graph G is assigned a low width at either side, resulting in it being disconnected in $G^W = \Omega_W(G)$. Without filtering G^W, the back projection operator Ω_W^{-1} will simply result in the original graph G, as desired.

The operator $\Omega(G) = (\Omega_V(G), \Omega_E(G))$, in Eq. 5, is basically the same as $\Omega(X)$ in the case of binary images. It is merely split in two operators, one to determine the mapping for the vertices and one for the mapping of the edges, respectively. Thus, the previously discussed set of required properties apply to $\Omega(G)$ as well. In summary $\Omega(G)$ has the following properties:

1. $\Omega(0) = 0$,
2. $\forall G \in \mathcal{S}_G : \Omega(G) \in \mathcal{C}_{M \times A}$,
3. $\Omega : \mathcal{L}_G \rightarrow \mathcal{L}_G^A$ is a mapping such that for any $G = (V, E) \in \mathcal{L}_G$ each $v \in V$ is mapped to at least one corresponding element $(v, a) \in \Omega_V(G)$, and each $e \in E$ is mapped to at least one corresponding element $(e, a_1, a_2) \in \Omega_E(G)$,
4. $\Omega^{-1} : \mathcal{L}_G^A \rightarrow \mathcal{L}_G$ is a mapping such that for every $G^A = (V^A, E^A) \in \mathcal{L}_G^A$, each $(v, a) \in V^A$ is projected onto $v \in \Omega_V^{-1}(G^A)$, and each $(e, a_1, a_2) \in E^A$ is projected onto $v \in \Omega_E^{-1}(G^A)$,
5. $\Omega^{-1}(\Omega(G)) = G, \forall G \in \mathcal{L}_G$

Using these properties, one can define an attribute-space connection $\mathcal{A} \subseteq \mathcal{L}_G$ generated by the transformation pair (Ω, Ω^{-1}) and connection $\mathcal{C}_A$ on $\mathcal{L}_G^A$ as:

$$\mathcal{A}_\Omega = \{C \in \mathcal{L}_G \mid \Omega(C) \in \mathcal{C}_A\}. \tag{6}$$

Using the definitions above one can define attribute-space connected filters as:

$$\Phi^A = \Omega_A^{-1}(\Phi(\Omega_A(G))). \tag{7}$$

With $G \in \mathcal{L}_G$ and $\Phi : \mathcal{L}_G^A \rightarrow \mathcal{L}_G^A$ a connected filter and $(\Omega_A, \Omega_A^{-1})$ a transformation pair for the attribute A.

3.1   Width-Based Attributes

Local width is an attribute that could be used to transform a binary image $X \in \mathcal{P}(M)$ to $Y \in \mathcal{P}(M \times A)$, with $A \in \mathbb{Z}^+$. The corresponding mapping Ω_W could be implemented using an opening transform defined by granulometry $\{\theta_r\}$ [11]. Here, each operator $\theta_r : E \rightarrow E$ is an opening with the structuring element Θ_r. The opening transform Ω_W on the binary image X could be defined as follows:

$$\Omega_W(X) = \max\{r \in A \mid x \in \theta_r(X)\}. \tag{8}$$

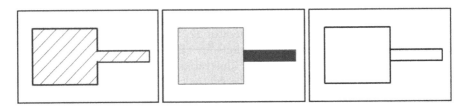

Fig. 3. An example of assigned width-based attributes to the pixels of a binary image. (left) Original image. (middle) transformed image to the width-based attribute space. Each pixel is assigned the radius of the largest ball. Darker colour indicates smaller radius. (right) partitioning of X using a threshold τ. Because of abrupt changes in r, X is split in two components.

If $\theta_r(X) = X \circ \Theta_r$ with $\circ$ denoting a structural opening with a ball-shaped structuring element Θ_r of radius r, an opening transform assigns the radius of the largest circle to each $x \in X$ such that $x \in X \circ \Theta_r$. We can now write the mapping $\Omega_W : \mathcal{P}(E) \to \mathcal{P}(E \times \mathbb{Z}^+)$ as:

$$\Omega_W(X) = \{(x, Y) | x \in X\} \; with \; Y = \Omega_W(x) \tag{9}$$

and Ω_W^{-1} as:

$$\Omega_W^{-1}(Y) = \{(x \in E) | (x, y) \in Y\}. \tag{10}$$

An example of a width space transformation is shown in Fig. 3. In the case of graph representation of an image, the original operator Ω_W consists of a pair $(\Omega_{W,V}, \Omega_{W,E})$. $(\Omega_{W,V})$ assigns width values to each node and $(\Omega_{W,E})$ determines a width-based attribute value for each end-point of each edge. In the example shown in Fig. 3 we used L_∞ to determine the value of r for each pixel, whereas in Fig. 2 the city-block distance (L_1) is used to compute the radius of the largest disk; note that the edges of the left diamond receive a value of $r = 3$ and the right diamond $r = 4$. The red edge receives a smaller attribute value at both ends; thus it is not connected in some width-based spaces.

3.2   Orientation Attributes

Using the binary-image based attribute-space representation, an operator $\Omega_\beta :$ $\mathcal{P}(M) \to \mathcal{P}(M \times A)$, with $A = \{\beta_1, \beta_2, \ldots, \beta_n\}$, representing N different orientations, can be defined which assigns one or more orientation-attributes β_i to each active node (pixel) $x \in X$ [24]. The result is a binary image $f : (M \times A) \to \{0, 1\}$, which would be the indicator function of the set $\Omega_\beta(X) \subseteq M \times A$.

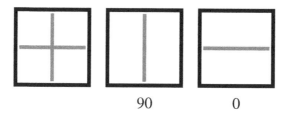

90   0

Fig. 4. An example of assigned orientation attribute to pixels of an image. Each attribute-space corresponds to a attribute-value and contains the entire image including the background pixels. (left) original image; (middle) pixels with $\beta = 90°$; (right) pixels with $\beta = 0°$. The central pixel is active in all orientations while the other pixels have only one orientation-attribute.

A naive implementation of this idea would simply create n copies of the original image X (a 2D array storing the pixels), and set the pixels $x \in \Omega_\beta(X)$ to active. Figure 4 illustrates this idea by showing a few slices of such a volume. This straightforward approach has a significant impact on the required memory size and computational intensity of the algorithm. Alternatively we represent the binary image as a subgraph G of G_M, in which we only include the active nodes and relevant edges, and omit all others. As discussed before, the original operator Ω_β now consists of a pair $(\Omega_{\beta,V}, \Omega_{\beta,E})$, where $\Omega_{\beta,V}$ creates one or more copies of all the vertices of G to the new graph G^β. However, rather than assigning discrete values $\beta_1, \beta_2, \ldots, \beta_n$, we can now assign any value in the domain $[0, 180)$. A similar mapping is applied to the edges by $\Omega_{\beta,E}$. In the example from Fig. 1, this means only the centre pixel of the cross needs to be copied more than once, so only a tiny memory penalty is incurred. A similar scheme is applied to the edges connecting the pixels; each edge receives an orientation value for each end-point; one could use Algorithm 3.1 to compute $G^\beta = (V^\beta, E^\beta)$. By way of comparison, the original approach would require a memory cost of $O(|G_M| n)$, with n the number of values β might take, and $|G_M|$ the storage size of G_M, whereas the new approach requires $O(|G| n_i)$, with n_i the number of intersections of components. This also means that all subsequent processing is on a far smaller graph, which reduces computational cost.

4   Connected Components Analysis, Higher-Dimensional Attributes and Filtering

Application of the discussed operator Ω^A on G constructs G^A; one can then apply connected component analysis and attribute filtering on G^A and mapping the results to the original domain using Ω^{-1}. Connected component analysis on graphs can be achieved efficiently, for example, by applying Tarjan's Union-Find algorithm [21,22], in pseudo-linear time complexity. This represents disjoint sets as a forest of spanning trees over G^A. Initially, all vertices in G^A are considered singleton trees. The algorithm then processes all edges in G^A. The algorithm merges the trees to which the vertices are linked by each edge inspected. This is done by letting the root pointer of one tree point to the root node of the other. Because not all edges link two vertices in our case, we only perform this merge if the edge actually links to existing vertices of G^A. During merger, path compression and union-by-rank are used to keep root paths as short as possible. For details see [21, 22], and for more recent papers on its application to connected filtering check [9, 10, 12].

In the preceding discussion we have focused on regular 4- or 8-connectivity. However, using graph representation allows for processing patterns with non-uniform neighbouring definitions in high dimensional data. For example consider the observable patterns in point clouds produced by multi-sensor detector networks. In those structures, different nodes have a varying number of neighbours and connectivity structures based on the sensor densities in different locations. Furthermore, the notion of connectivity has been generalised [3, 15, 17, 20], allowing objects separated by small gaps (caused e.g. by noise) to be considered connected. It has been shown that these forms of connectivity can also be implemented in a graph formalism [14, 26]. The memory and compute loads do tend to increase, but only by a constant factor.

During the union-find process, we can compute various higher-order derived attributes as the components grow in size at each merge. One of the easiest attribute is the area (or volume in 3D), which simply involves counting the number of vertices involved [10]; at each node $g \in G^A$, we store a record to hold the value of the largest area to which g belongs. When handling thin and elongated structures, one of the obvious attributes to compute is the curvature of the processed patterns. One can determine the curvature for a set of at least three pixels. These attributes are updated while the connected components grow. The higher order attributes could serve not only for filtering or segmentation but also as criterion for merging connected components, specially when small gaps are allowed between the components. Because we only copy nodes and edges to an attribute-space if they are active in that particular space, the size of the required memory remains smaller than number of nodes and edges times number of attributes. Using this scheme, some nodes may contain a vector of attributes and the others may have only a few. For example, the central node in Fig. 1 left panel may contain width and multiple orientation attributes while all other pixels contains at most two attribute values namely the width and a single value for the local orientation. Once we have transformed G to G^A by determining all attributes $a \in A$ for vertices and edges of G, connected filters could be defined to select or discard connected components based on the attribute values. A possible method for connected component analysis is shown in Algorithm 4.1.

We have applied this formalism to process the data produced by the Straw Tube Tracker (STT) of the PANDA experiment [7], to reconstruct sub-atomic particle tracks [1]. This detector (STT) is an irregular-graph containing more than 4000 tubes with a varying length between 150 cm down to a few centimetres. The tubes are arranged in layers around the beam pipe (z-axis). A number of these layers are tilted with $+3$ or $-3°$ with respect to the z- axis. Because of

---

Algorithm 4.1 Connected component analysis.

Input: (Graph G^A, a set of criteria T.)
Output: (Graph $G' \in G$. The output is the maximum sub-graph G' of G fulfilling the criterion T.)

```
1: for all (v' ∈ V^A) and (e' ∈ E^A) do
2:     if v' and e' fulfil t_i then
3:         Add v' and e' to G'
4:     end if
5: end for
6: for all (v', e') ∈ G' do
7:     Compute Ω^{-1}(v', e')
8: end for
```

---

this tilt structure, the number of neighbours of the tubes varies between 2 and 22 tubes. When a charged particle passes through the STT, it turns a subset of the tubes on and produces a cloud of points containing a binary image in $3D$. Reconstructing paths using these points needs to be performed at run-time (at 20 MHz). Figure 5 shows an example of a pattern produced with such a multi-sensor detection system. We have computed the orientation attribute for all active tubes in the graph. One can observe that each node has a few local orientation attributes and the value of these attributes vary strongly between different nodes. Our method limits the number of copies of each node to the number of actual computed attributes for it. During the later stages the constructed connected components were merged on the basis of their distances where the length of the connected components attached to each edge was used to weight the preferred path growth-directions. Using the curvature of the line connecting the points in the space as an attribute for merging the larger components, is an obvious choice; but this requires fitting a curve through the points which is a computational intensive operation. One can evaluate the Mahalanobis distance of a point cloud as a measure for the strongest elongation direction. During the later merging operations, we have used Mahalanobis distance, as a measure for the orientation compatibility between different components; these higher order attributes were updated as the components grew. The results of this hierarchical merging based on heterogeneous conditions is shown in the panels in the second row of Fig. 5.

5 Discussion

We have introduced a new and generic graph-based formalism to construct memory-efficient methods for computing attribute-spaces and attribute-space connected filters for binary images. This formalism allows for processing larger images in higher dimensional attribute-spaces. We have studied the transformation pair $\Omega_A = (\Omega, \Omega^{-1})$ and their properties and relations, to transform graphs G to an attribute-space graph G^A and vice versa. By representing an image X as a graph, we store the determined attribute values computed using Ω_A for all active nodes in the vertices and assign attribute values to each endpoint of the edges connecting those nodes. In this formalism, the nodes and the corresponding edges are copied only for actual computed attribute values;

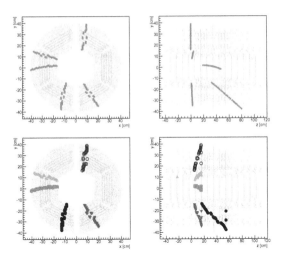

Fig. 5. First row: (left): an example event shown in the xy-plane. The grey dots represent the detector tubes centres and the red dots the read-out positions of activated sensors. (right): the same pattern shown in the yz-plane. Second row: (left): reconstructed particle paths using an implementation of this formalism in the xy-plane. (right) reconstructed paths in the yz-plane. In the images in the second row, each colour and marker combination represents a different connected component (particle path). The differences between the reconstructed (second row) and the true (first row) particle paths in the yz-plane is partially because of the detector read-out imperfections. (Color figure online)

this mechanism allows for computing non-discrete attribute values. In general, the nodes representing the data points (pixels) at intersection points of structures in an image receive a larger number of attributes whereas all the other vertices receive only a few. We utilise this feature in our method which leads to a significant reduction of the size of the required memory for representing the image together with its copies in different attribute-spaces. This reduction makes it possible to handle larger graphs or to compute and store a larger set of distinct attributes which could be used to design heterogeneous conditions based attribute-filters for segmentation or connected component analysis. The graph-based interpretation allows for efficient and straightforward embedding of higher order attributes, such as width and curvature for each constructed connected component, which could be used for hierarchical filtering based on these attribute-values. For example, one can perform width-based merging followed by orientation-based merge operations. We have applied this method to the point-clouds produced by a multi-sensor detector to reconstruct subatomic particle trajectories. Because in this application each connected component is defined by the list of its member vertices, one can query the structure for selective processing of the available components and patterns.

Future work will consist of extending our formalism to handle grey-scale images using node-weighted and edge-weighted graphs. Further, we want to explore the option of assigning a *range* of attribute-values to the edges and vertices in G^A rather than a single value or a pair of values. This step would enable us to process hyper-graphs [2] and design filtering and pattern analysis on hypergraph-based structures and networks. Applying this formalism to realistic and real-life graph-based data is another option to explore; this will also lead to quantitative results on comparative computation time and memory requirement.

Acknowledgement. We want to thank the Centre for Information Technology (CIT) of the Rijksuniversiteit Groningen for their support and facilitating this work.

References

1. Babai, M., Kalantar-Nayestanaki, N., Messchendorp, J.G., Wilkinson, M.H.F.: An efficient attribute-space connected filter on graphs to reconstruct paths in point-clouds. Pattern Recognition (submitted)
2. Bloch, I., Bretto, A.: Mathematical morphology on hypergraphs, application to similarity and positive kernel. Comput. Vis. Image Underst. **117**(4), 342–354 (2013)
3. Braga-Neto, U., Goutsias, J.: A theoretical tour of connectivity in image processing and analysis. J. Math. Imag. Vis. **19**, 5–31 (2003)
4. Breen, E.J., Jones, R.: Attribute openings, thinnings and granulometries. Comp. Vis. Image Understand. **64**(3), 377–389 (1996)
5. Cousty, J., Najman, L., Dias, F., Serra, J.: Morphological filtering on graphs. Comput. Vis. Image Underst. **117**(4), 370–385 (2013). https://doi.org/10.1016/j.cviu.2012.08.016. http://www.sciencedirect.com/science/article/pii/S107731421 200183X, special issue on Discrete Geometry for Computer Imagery
6. Cousty, J., Najman, L., Serra, J.: Some morphological operators in graph spaces. In: Wilkinson, M.H.F., Roerdink, J.B.T.M. (eds.) ISMM 2009. LNCS, vol. 5720, pp. 149–160. Springer, Heidelberg (2009). https://doi.org/10.1007/978-3-642-03613-2_14
7. Erni, P.C.W., Keshelashvili, I., Krusche, B., et al.: Technical design report for the panda straw tube tracker. Eur. Phys. J. A **49**, 25 (2013). Special Article - Tools for Experiment and Theory
8. Franken, E., Duits, R.: Crossing-preserving coherence-enhancing diffusion on invertible orientation scores. Int. J. Comput. Vis. **85**, 253–278 (2009)
9. Géraud, T.: Ruminations on Tarjan's union-find algorithm and connected operators. In: Ronse, C., Najman, L., Decenciére, E. (eds.) Mathematical Morphology: 40 Years On, vol. 30. Springer, Dordrecht (2005)
10. Meijster, A., Wilkinson, M.H.F.: A comparison of algorithms for connected set openings and closings. IEEE Trans. Pattern Anal. Mach. Intell. **24**(4), 484–494 (2002)
11. Nacken, P.F.M.: Image analysis methods based on hierarchies of graphs and multi-scale mathematical morphology. Ph.D. thesis, University of Amsterdam, Amsterdam, The Netherlands (1994)
12. Najman, L., Couprie, M.: Building the component tree in quasi-linear time. IEEE Trans. Image Proc. **15**, 3531–3539 (2006)
13. Najman, L., Cousty, J.: A graph-based mathematical morphology reader. Pattern Recogn. Lett. **47**, 3–17 (2014)

14. Oosterbroek, J., Wilkinson, M.H.F.: Mask-edge connectivity: theory, computation, and application to historical document analysis. In: Proceedings of 21st International Conference on Pattern Recognition, pp. 3112–3115 (2012)

15. Ouzounis, G.K., Wilkinson, M.H.F.: Mask-based second generation connectivity and attribute filters. IEEE Trans. Pattern Anal. Mach. Intell. **29**, 990–1004 (2007)

16. Perret, B., Lefevre, S., Collet, C., Slezak, E.: Hyperconnections and hierarchical representations for grayscale and multiband image processing. IEEE Trans. Image Process. (2011). https://doi.org/10.1109/TIP.2011.2161322

17. Ronse, C.: Set-theoretical algebraic approaches to connectivity in continuous or digital spaces. J. Math. Imag. Vis. **8**, 41–58 (1998)

18. Salembier, P., Serra, J.: Flat zones filtering, connected operators, and filters by reconstruction. IEEE Trans. Image Proc. **4**, 1153–1160 (1995)

19. Serra, J.: Mathematical morphology for Boolean lattices. In: Serra, J. (ed.) Image Analysis and Mathematical Morphology, II: Theoretical Advances, pp. 37–58. Academic Press, London (1988)

20. Serra, J.: Connectivity on complete lattices. J. Math. Imag. Vis. **9**(3), 231–251 (1998)

21. Tarjan, R.E.: Efficiency of a good but not linear set union algorithm. J. ACM **22**, 215–225 (1975)

22. Tarjan, R.E.: Data Structures and Network Algorithms. SIAM, Philadelphia (1983)

23. Vincent, L.: Graphs and mathematical morphology. Signal Processing **16**(4), 365–388 (1989). https://doi.org/10.1016/0165-1684(89)90031-5. http://www.sciencedirect.com/science/article/pii/0165168489900315, special Issue on Advances in Mathematical Morphology

24. Wilkinson, M.H.F.: Attribute-space connectivity and connected filters. Image Vis. Comput. **25**, 426–435 (2007)

25. Wilkinson, M.H.F.: Hyperconnectivity, attribute-space connectivity and path openings: theoretical relationships. In: Wilkinson, M.H.F., Roerdink, J.B.T.M. (eds.) ISMM 2009. LNCS, vol. 5720, pp. 47–58. Springer, Heidelberg (2009). https://doi.org/10.1007/978-3-642-03613-2_5

26. Wilkinson, M.H.F.: A fast component-tree algorithm for high dynamic-range images and second generation connectivity. In: 2011 Proceedings of International Conference Image Processing, pp. 1041–1044 (2011)

27. Wilkinson, M.H.F.: Hyperconnections and openings on complete lattices. In: Soille, P., Pesaresi, M., Ouzounis, G.K. (eds.) ISMM 2011. LNCS, vol. 6671, pp. 73–84. Springer, Heidelberg (2011). https://doi.org/10.1007/978-3-642-21569-8_7

Machine Learning

Ultimate Levelings with Strategy for Filtering Undesirable Residues Based on Machine Learning

Wonder Alexandre Luz Alves[1(✉)], Charles Ferreira Gobber[1(✉)],
Dennis José da Silva[2], Alexandre Morimitsu[2], Ronaldo Fumio Hashimoto[2],
and Beatriz Marcotegui[3]

[1] Informatics and Knowledge Management Graduate Program,
Universidade Nove de Julho, São Paulo, Brazil
wonder@uni9.pro.br, gobber@uninove.edu.br
[2] Department of Computer Science, Institute of Mathematics and Statistics,
Universidade de São Paulo, São Paulo, Brazil
[3] Centre for Mathematical Morphology, Mines-ParisTech, Fontainebleau, France

Abstract. Ultimate levelings are operators that extract important image contrast information from a scale-space based on levelings. During the residual extraction process, it is very common that some residues are selected from undesirable regions, but they should be filtered out. In order to avoid this problem some strategies can be used to filter residues extracted by ultimate levelings. In this paper, we introduce a novel strategy to filter undesirable residues from ultimate levelings based on a regression model that predicts the correspondence between objects of interest and residual regions. In order to evaluate our new approach, some experiments were carried out with a plant dataset and the results show the robustness of our method.

1 Introduction

Residual operators are transformations that involve combinations of morphological operators with differences. Morphological gradient, top-hat transforms, skeleton by maximal balls and ultimate opening are some examples of residual operators widely used in image processing applications.

There is a class of important residual operators called *ultimate levelings* [5]. They are residual operators that analyze the evolution of the residual values between two consecutive operators on a scale-space of levelings and keep the maximum residues for each pixel. The residual value of these operators can reveal important contrast information in images. This class of operators includes maximum difference of openings (resp., closings) by reconstruction [15], differential morphological profiles [21], ultimate attribute openings (resp., closings) [22], differential attribute profiles [9], shape ultimate attribute openings (resp., closings) [12], differential area profiles [19] and ultimate grain filters [3]. They have successfully been used as a preprocessing step in various applications

© Springer Nature Switzerland AG 2019
B. Burgeth et al. (Eds.): ISMM 2019, LNCS 11564, pp. 297–309, 2019.
https://doi.org/10.1007/978-3-030-20867-7_23

such as texture features extraction [15], segmentation of high-resolution satellite imagery [9,20], text location [1,23] and segmentation of building façades [12].

Due to the design of the ultimate levelings, some residues extracted by them can be from undesirable regions. In this sense, several researches have been proposed in recent years, introducing strategies to filter undesirable residues during the residual extraction process [2,3,5,12]. They are based on attributes extracted from residual regions. However, when the number of attributes increases the strategy construction becomes difficult, because you need to find thresholds to apply the strategy. In order to deal with this problem we can made use of machine learning. In addition, the combination of machine learning and morphological operators is successfully applied in a large range of problems, such that: morphological operators learning [16], convolutional nets and watershed cuts [8], morphological profiles [9], and others.

Given the above considerations, the main contribution of this paper is a novel strategy to filter undesirable residues based on machine learning. Since ultimate levelings can be efficiently computed from morphological trees, information can be extracted from branches of the trees to obtain features for a regression or a classification model. This idea is inspired by maximally stable extremal regions [17] and morphological profiles [9], but we made it considering contrast information extracted by the residues. In order to apply and evaluate our new approach, we chose the plant bounding box detection problem [18].

The remainder of this paper is structured as follows. Sections 2 and 3 briefly recall some definitions and properties about the morphological trees and the ultimate levelings. The original contribution of this paper is given in Sect. 4, where we introduce a novel strategy to filter undesirable residues from the ultimate levelings based on a regression model that predicts the correspondence between objects of interest and residual regions. Experimental results are shown in Sect. 5. Finally, Sect. 6 concludes this work and presents some future research directions.

2   Theoretical Background

For decades, image representations through trees have been proposed to carry out tasks of image processing and analysis, such as: filtering, segmentation, pattern recognition, contrast extraction, compression and others. In this scenario, the image is represented by means of a tree, then all tasks are performed through information extraction or modifications in the tree itself, and finally an image is reconstructed from the modified tree.

In order to build the trees considered in this paper, we need the following definitions. First, we consider images as mappings from a Cartesian grid $\mathcal{D} \subset \mathbb{Z}^2$ to a discrete set of $k \geq 1$ integers $\mathbb{K} = \{0, 1, \ldots, k-1\}$. These mappings can be decomposed into *lower (strict)* and *upper (large)* level sets, i.e., for any $\lambda \in \mathbb{K}$, $\mathcal{X}_\downarrow^\lambda(f) = \{p \in \mathcal{D} : f(p) < \lambda\}$ and $\mathcal{X}_\lambda^\uparrow(f) = \{p \in \mathcal{D} : f(p) \geq \lambda\}$. From these sets, we define two other sets $\mathcal{L}(f)$ and $\mathcal{U}(f)$ composed by the connected components (CCs) of the lower and upper level sets of f, i.e., $\mathcal{L}(f) = \{\tau \in \mathcal{CC}(\mathcal{X}_\downarrow^\lambda(f)) : \lambda \in$

$\mathbb{K}\}$ and $\mathcal{U}(f) = \{\tau \in \mathcal{CC}(\mathcal{X}_\lambda^\uparrow(f)) : \lambda \in \mathbb{K}\}$, where $\mathcal{CC}(\mathcal{X})$ denotes the sets of either 4 or 8-CCs of $\mathcal{X}$, respectively.

The ordered pairs consisting of the CCs of the lower and upper level sets and the usual inclusion set relation, i.e., $(\mathcal{L}(f), \subseteq)$ and $(\mathcal{U}(f), \subseteq)$, induce two dual trees [7,13,24] called *component trees*. It is possible to combine them into a single tree in order to obtain the so-called *tree of shapes*. Then, let $\mathcal{P}(\mathcal{D})$ denote the *powerset* of $\mathcal{D}$ and let $sat : \mathcal{P}(\mathcal{D}) \to \mathcal{P}(\mathcal{D})$ be the operator of saturation [7] (or filling holes). Then, $\mathcal{SAT}(f) = \{sat(\tau) : \tau \in \mathcal{L}(f) \cup \mathcal{U}(f)\}$ be the family of CCs of the upper and lower level sets with holes filled. The elements of $\mathcal{SAT}(f)$, called *shapes*, are nested by the inclusion relation and thus the pair $(\mathcal{SAT}(f), \subseteq)$ induces a tree which is called *tree of shapes* [7].

In tree of shapes, and also component trees (max-tree and min-tree), each pixel $p \in \mathcal{D}$ is associated only to the smallest Connected Component (CC) of the tree containing it; and through parenthood relationship, it is also associated to all its ancestor nodes. Then, we denote by $\mathcal{SC}(\mathcal{T}, p)$ the smallest CC containing p in tree $\mathcal{T}$. Similarly, we say $p \in \mathcal{D}$ is a *compact node pixel* (CNP) of a given CC $\tau \in \mathcal{T}$ if and only if τ is the smallest CC containing p, i.e., $\tau = \mathcal{SC}(\mathcal{T}, p)$. Figure 1 shows examples of min-tree, max-tree, and tree of shapes, where CNPs are highlighted in red.

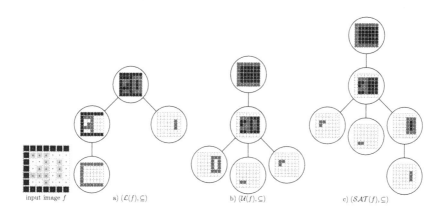

Fig. 1. Min-tree (a), max-tree (b) and tree of shapes (c) as compact representations of $(\mathcal{L}(f), \subseteq)$ and $(\mathcal{U}(f), \subseteq)$, and $(\mathcal{SAT}(f), \subseteq)$, respectively, of the image f. Only Compact Node Pixels (CNPs) are stored and they are highlighted in red. (Color figure online)

3   Ultimate Levelings

Ultimate levelings constitute a wider class of residual operators defined from a scale-space of levelings $\{\psi_i : i \in \mathcal{I}\}$ [2,3,5]. An ultimate leveling analyzes the evolution of residual values from a family of consecutive primitives, i.e. $r_i^+(f) = [\psi_i(f) - \psi_{i+1}(f) \vee 0]$ and $r_i^-(f) = [\psi_{i+1}(f) - \psi_i(f) \vee 0]$, keeping the maximum positive and negative residues for each pixel. Thus, contrasted objects can be detected if a relevant residue is generated when they are filtered out by one of these levelings.

More precisely, the ultimate leveling $\mathcal{R}$ is defined for any image f as follows:

$$\mathcal{R}(f) = \mathcal{R}^+(f) \vee \mathcal{R}^-(f), \tag{1}$$

where $\mathcal{R}^+(f) = \sup_{i \in \mathcal{I}} \{r_i^+(f)\}$ and $\mathcal{R}^-(f) = \sup_{i \in \mathcal{I}} \{r_i^-(f)\}$.

Residual values of these operators can reveal important contrasted structures in the image. In addition to these residues, other associated information can be obtained such as properties of the operators that produced the maximum residual value. For example, W. Li [15] introduced a function $q_{\mathcal{I}_{max}} : \mathcal{D} \to \mathcal{I}$ that associates to each pixel the major index that produces the maximum non-null residue, i.e.,

$$p \in \mathcal{D}, q_{\mathcal{I}_{max}}(p) = \begin{cases} q_{\mathcal{I}_{max}}^+(p), \text{ if } [\mathcal{R}(f)](p) > [\mathcal{R}^+(f)](p), \\ q_{\mathcal{I}_{max}}^-(p), \text{ otherwise.} \end{cases} \tag{2}$$

where $q_{\mathcal{I}_{max}}^+(p) = \max\{i + 1 : [r_i^+(f)](p) = [\mathcal{R}^+(f)](p) > 0\}$ and $q_{\mathcal{I}_{max}}^-(p) = \max\{i + 1 : [r_i^-(f)](p) = [\mathcal{R}^-(f)](p) > 0\}$.

The ultimate levelings can be efficiently implemented thanks to the theorem proposed in W. A. L. Alves et. al [4], which shows that an increasing family of levelings $\{\psi_i : i \in \mathcal{I}\}$ can be obtained through a sequence of pruned trees $(\mathcal{T}_f^0, \mathcal{T}_f^1, \dots, \mathcal{T}_f^{\mathcal{I}MAX})$ from the structure of the max-tree, min-tree or tree of shapes $\mathcal{T}_f$ constructed from the image f. Then, the i-th positive (resp. negative) residue $r_i^+(f)$ can be obtained from the set of nodes $\mathcal{N}r(i) = \mathcal{T}_f^i \setminus \mathcal{T}_f^{i+1}$, i.e., $\forall \tau \in \mathcal{N}r(i)$,

$$r_{\mathcal{T}_f^i}^+(\tau) = \begin{cases} \texttt{level}(\tau) - \texttt{level}(\texttt{Parent}(\tau)), & \text{if } \texttt{Parent}(\tau) \notin \mathcal{N}r(i), \\ \texttt{level}(\tau) - \texttt{level}(\texttt{Parent}(\tau)) \\ \quad + \ r_{\mathcal{T}_f^i}^+(\texttt{Parent}(\tau)), & \text{otherwise,} \end{cases} \tag{3}$$

where $\texttt{level}(\tau)$ and $\texttt{Parent}(\tau)$ are functions that represent the gray level and the parent node of τ in $\mathcal{T}_f$, respectively. Thus, the i-th positive (resp. negative) residue $r_i^+(f)$ is given as follows:

$$\forall p \in \mathcal{D}, [r_i^+(f)](p) = \begin{cases} r_{\mathcal{T}_f^i}^+(SC(\mathcal{T}_f^i, p)), & \text{if } SC(\mathcal{T}_f^i, p) \in \mathcal{N}r(i), \\ 0, & \text{otherwise.} \end{cases} \tag{4}$$

Those facts lead to efficient algorithms for computing ultimate levelings and its variations [3, 10].

Ultimate levelings are operators that extract residual information from primitive families. During the residual extraction process, it is very common that undesirable regions of the input image contain residual information that should be filtered out. These undesirable residual regions often include desirable residual regions due to the design of the ultimate levelings which consider maximum residues. Thus, residual information can be improved by filtering of residues extracted from undesirable regions [2,3,5,12]. We can decide whether a residue $r_i^+(f)$ (resp., $r_i^-(f)$) is filtered out or not, just checking nodes $\tau \in \mathcal{N}r(i)$ that satisfy a given filtering criterion $\Omega : \mathcal{P}(\mathcal{D}) \to \{\text{desirable, undesirable}\}$. Thus, we just calculate the ultimate leveling $\mathcal{R}$ for residues r_i^+ (resp., r_i^-) such that satisfy the criterion Ω. So, positive (resp. negative) residues are redefined as follows:

$$\forall \tau \in \mathcal{N}r(i), r_{T_f^i}^{\Omega+}(\tau) = \begin{cases} r_{T_f^i}^+(\tau), & \text{if } \exists C \in \mathcal{N}r(i) \\ & \quad \text{such that } \Omega(C) \text{ is desirable;} \\ 0, & \text{otherwise,} \end{cases} \tag{5}$$

and thus redefined the ultimate levelings with strategy for filtering from undesirable residues as follows: $\mathcal{R}_\Omega(f) = \mathcal{R}_\Omega^+(f) \vee \mathcal{R}_\Omega^-(f)$, where, $\mathcal{R}_\Omega^+(f) = \sup\{r_i^{\Omega+}(f) : i \in \mathcal{I}\}$ and $\mathcal{R}_\Omega^-(f) = \sup\{r_i^{\Omega-}(f) : i \in \mathcal{I}\}$.

Given the above considerations, in this paper we are interested in the maximum desirable residues provided by Ultimate Levelings. In this sense, is presented in the Sect. 4 a new approach to construct a strategy based on machine learning techniques.

4   Strategy to Filter Undesirable Residues Based on Machine Learning

In this section, we present the construction of a strategy to filter undesirable residues from the ultimate levelings based on a regression that predicts the correspondence between objects of interest and residual regions.

For this, we recall that an increasing family of levelings $\{\psi_i : i \in \mathcal{I}\}$ can be obtained through a sequence of pruned trees $(T_f^0, T_f^1, \ldots, T_f^{\mathcal{I}_{MAX}})$ from the structure of the max-tree, min-tree or tree of shapes T_f constructed from the image f. In addition, thanks to Eq. 4, a sequence of residues $(r_0(f), r_1(f), \ldots, r_{|\mathcal{I}-1|}(f) : r_i(f) = \psi_i(f) - \psi_{i+1}(f))$ can be obtained by the sequence of residual nodes $(\mathcal{N}r(0), \mathcal{N}r(1), \ldots, \mathcal{N}r(|\mathcal{I}|-1) : \mathcal{N}r(i) = T_f^i \setminus T_f^{i+1})$. Thus, from a sequence of residual nodes, we can use the tree structure to define strategies for filtering undesirable residues of the ultimate levelings.

Given this consideration, our idea is to represent each residue through an attribute vector extracted from overlapping residual regions with low contrast. Therefore, we define a partial order relation $\preceq$ on the collection of residual nodes $\mathcal{B}r = \{\mathcal{N}r(i) : i = 0, 1, \ldots, |\mathcal{I}| - 1\}$ such that for any $\mathcal{N}r(i), \mathcal{N}r(j) \in \mathcal{B}r$, we write $\mathcal{N}r(i) \preceq \mathcal{N}r(j)$ if and only if the nodes belonging to $\mathcal{N}r(i)$ are descendants of some node belonging to $\mathcal{N}r(j)$ in T_f.

This is equivalent to $\bigcup\{p \in \tau : \tau \in \mathcal{N}r(i)\} \subseteq \bigcup\{p \in \tau : \tau \in \mathcal{N}r(j)\}$. Now, we use this poset $(\mathcal{B}r, \preceq)$ in order to extract an attribute vector on a sequence of residual nodes, that is defined as follows:

$$\mathcal{B}_\Delta(i) = (\mathcal{N}r(i-k) \in \mathcal{B}r : -\Delta \le k \le \Delta), \tag{6}$$

where $\mathcal{N}r(i - \Delta) \preceq \mathcal{N}r(i - \Delta + 1) \preceq \ldots \preceq \mathcal{N}r(i) \preceq \mathcal{N}r(i+1) \preceq \ldots \preceq \mathcal{N}r(i+\Delta)$ and $\Delta \in \mathbb{N}$ is a parameter that defines the amount of residual regions.

In addition, if there are bifurcations in the path between the nodes $\mathcal{N}r(i - \Delta)$ to $\mathcal{N}r(i + \Delta)$, we choose the path with biggest accumulated area to define the sequence. This is based on the fact that all descendent nodes of a given node are subset of it, then the path of descendant with biggest accumulated area contains more region in common with the node, it means that the nodes of the path are more stable [14]. Note that the length of $\mathcal{B}_\Delta(i)$ is $2\Delta + 1$. An example of poset $(\mathcal{B}r, \preceq)$ and the sequence $\mathcal{B}_\Delta(i)$ is showed in Fig. 2. Then, we define an attribute vector $\Lambda(i) \in \mathbb{R}^{n(2\Delta+1)}$ on a sequence $\mathcal{B}_\Delta(i)$ for any $i \in \mathcal{I}$, as follows:

$$\Lambda(i) = \big(\kappa_j(\mathcal{N}r) : \mathcal{N}r \in \mathcal{B}_\Delta(i) \text{ and } j = 1, 2, \dots, n\big), \tag{7}$$

where $\kappa_j : \mathbb{P}(\mathcal{D}) \to \mathbb{R}$ is an attribute (feature) extracted from the largest residual region belonging to $\mathcal{N}r \in \mathcal{B}_\Delta(i)$ and n is the number of attributes.

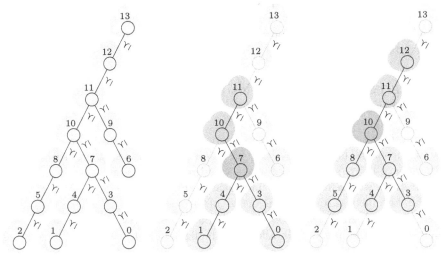

a) The labels $0, 1, \dots, 13$ are related to the sets $\mathcal{N}r(0), \mathcal{N}r(1), \dots, \mathcal{N}r(13)$.

b) Ascendant (in green) and descendant paths (in orange) of $\mathcal{N}r(7)$ (in red).

c) Ascendant (in green) and descendant paths (in orange) of $\mathcal{N}r(10)$ (in red).

Fig. 2. (a) An example of feature extraction using our approach for $\Delta = 2$. In (b) the path formed by $\mathcal{N}r(10) \preceq \mathcal{N}r(11)$ is the ascendant path of $\mathcal{N}r(7)$ and there are two descendant paths of $\mathcal{N}r(7)$ formed by $\mathcal{N}r(0) \preceq \mathcal{N}r(3)$ and by $\mathcal{N}r(1) \preceq \mathcal{N}r(4)$. In (c) there is a similar situation. (Color figure online)

To construct the regression, we define a training set (X, Y) with $m \in \mathbb{N}$ labeled samples such that $X \subseteq \mathbb{R}^m \times \mathbb{R}^{n(2\Delta+1)}$ is a set of attribute vectors of the residual regions and $Y \subseteq \mathbb{R}^m$ is a set of measure of similarities. Thus, we have that $\Lambda(i) \in X$ is an attribute vector that represents the residual region of r_i and $\mathtt{Match}(\mathcal{N}r(i), \mathtt{Label}(f)) \in Y$ is the value of measure of similarity between the residual region of r_i and a labeled region $R \in \mathtt{Label}(f)$ of an image f. This measure of similarity is defined as follows:

$$\texttt{Match}(\mathcal{N}r(i), \texttt{Label}(f)) = \max_{R \in \texttt{Label}(f)} \frac{|R \cap \tau_{\text{region}}|}{|R \cup \tau_{\text{region}}|}, \tag{8}$$

where $\tau_{\text{region}} = \bigcup\{p \in \tau : \tau \in \mathcal{N}r(i)\}$ is residual region of r_i. An example of training set extraction is shown in Fig. 3. We remember that, a machine learning model usually tries to find a function $h : \mathbb{R}^{n(2\Delta+1)} \to \mathbb{R}$, also called the hypothesis, that best fits the training set (X, Y).

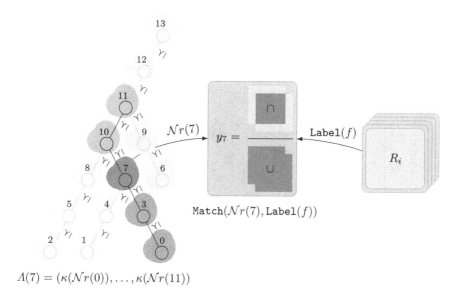

$$\Lambda(7) = (\kappa(\mathcal{N}r(0)), \dots, \kappa(\mathcal{N}r(11)))$$

Fig. 3. An example of training set computation. For each residual node set $\mathcal{N}r(i)$ we extract an attribute vector $\Lambda(i)$ and the matching value y_i. Thus, the pair $(\Lambda(i), y_i)$ is a sample.

Once h has been trained, we can decide when a residue $r_i^+(f)$ (resp., $r_i^-(f)$) is desirable, if the prediction of $\Lambda(i)$ is greater than a threshold ε, that is:

$$\forall \tau \in \mathcal{N}r(i), r_{T_f^i}^{h+}(\tau) = \begin{cases} r_{T_f^i}^+(\tau), & \text{if } h(\Lambda(i)) > \varepsilon, \\ 0, & \text{otherwise,} \end{cases} \tag{9}$$

and thus redefined the ultimate levelings with strategy for filtering from undesirable residues as follows: $\mathcal{R}_h(f) = \mathcal{R}_h^+(f) \vee \mathcal{R}_h^-(f)$, where, $\mathcal{R}_h^+(f) = \sup\{r_i^{h+}(f) : i \in \mathcal{I}\}$ and $\mathcal{R}_h^-(f) = \sup\{r_i^{h-}(f) : i \in \mathcal{I}\}$.

4.1 Selecting Disjoint Residual Regions

As mentioned in [5], the ultimate levelings can produce nesting of residual regions. Selecting disjoint residual regions (DRR) is a good way to deal with this.

Thus, let π_F be the set of $\mathcal{N}r(i)$ that contains nodes in the path started from a leaf F to the root of $\mathcal{T}_f$. During this path, we say that the best residual region is the one with greatest prediction, that is:

$$\text{DRR} = \Big\{ S : S = \underset{\mathcal{N}r(i) \in \pi_F}{\arg\max} \{ h(\Lambda(i)) : \nexists \mathcal{N}r(j), \mathcal{N}r(j) \preceq \mathcal{N}r(i) \\ \text{with } h(\Lambda(j)) > h(\Lambda(i)) \} \Big\} \quad (10)$$

Thus, we select all best residual node sets applying Eq. 10 for each leaf F in $\mathcal{T}_f$ and then we redefined Eq. 9 only with the residues $\mathcal{N}r(i)$ belonging to the set DRR. An illustration of the disjoint residual node sets selection that apply the Eq. 10 is shown in Fig. 4.

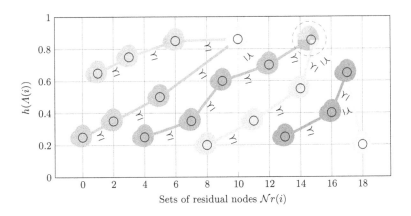

Fig. 4. An example of DRR selection. We start exploring paths from leaves to root of the tree according to Eq. 10. Residual node sets of more than one color represent residual node sets that are the best for two or more paths. For example the residual node set $\mathcal{N}r(15)$ (marked by a dashed red circle) is the best of the paths $\mathcal{N}r(1) \preceq \cdots \preceq \mathcal{N}r(15) \preceq \cdots \preceq \mathcal{N}r(18)$, $\mathcal{N}r(0) \preceq \cdots \preceq \mathcal{N}r(15) \preceq \cdots \preceq \mathcal{N}r(18)$, $\mathcal{N}r(4) \preceq \cdots \preceq \mathcal{N}r(15) \preceq \cdots \preceq \mathcal{N}r(18)$ and $\mathcal{N}r(8) \preceq \cdots \preceq \mathcal{N}r(15) \preceq \cdots \preceq \mathcal{N}r(18)$. (Color figure online)

5   Results and Experiments

In this section we present the results of experiments conducted with our new approach. To this purpose, we choose an important vision task problem called plant bounding box detection. Plants have a very complex morphology, so the detection of their bounding box is considered a difficult task. A good plant dataset in literature is provided by Minervine et al. [18]. In respect of plant bounding box detection task, the dataset is composed by three subsets: Ara2012, Ara2013-Canon and Ara2013-Rpi, totalizing 70 images with size between 3108×2324 and 2592×1944 pixels.

An overview of our approach applied to plant bounding box detection is shown in Fig. 5. First, we constructed regression model using features based on contrasts, colors and shapes. After, this regression model is used to select the desirable residues (see Eq. 9). Finally we define the bounding boxes using the DRR extracted of the desirable residues (see Eq. 10). The main parameters are: (1) the primitives obtained by the area attribute with min area of 50 and max area of 95,000 and the bounding box area attribute with min area of 150 and max area of 400,000; (2) the attributes chose to extract the features of the residual regions: 3 contrast attributes (residue, maximum residue and altitude), 3 color attributes (green level, distance of RGB foreground and distance of RBG background) and 3 shape attributes (TBMR, compactness, eccentricity). Totalizing 9 type of feautures, i.e., $k = 9$; (3) the minimal value ε to avoid false bounding boxes in DRR (see Eq. 10).

Thus we trained h with a Multilayer Neural Network using 3 hidden layers with $(50, 30, 10)$ units and a learning rate of 0.005 found by a grid search. The model was trained on the 1.59 million of training samples and evaluated on the 32 thousand of validation and test samples. We notice overfitting when the number of neurons and hidden layers were set to high values.

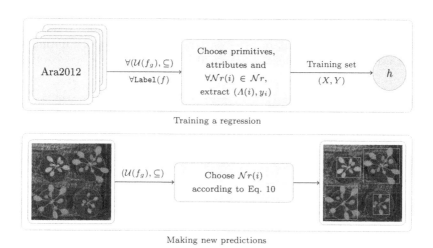

Fig. 5. Overview of our approach. First we train a regression model h. Then, we can make new predictions to detect bounding boxes according to Eq. 10.

In order to study the influence of Δ in plant bounding box detection, we conducted some experiments varying this parameter. We remember that the choice of the type of features and the number of them are usually challenges in machine learning. Thus, to select the best combination of both, we performed a grid search tuned by cross-validation as shown in Fig. 6.

The results are summarized in Fig. 6. They reveal that Δ parameter and the number of features have a big influence in the model, because when we increase Δ and the number of features the MSE of all models tends to decrease. The best h was obtained with $\Delta = 3$ and 63 features.

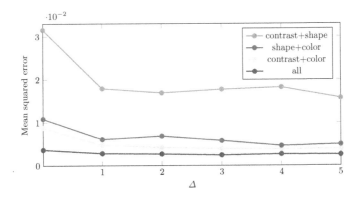

Fig. 6. Evolution of Δ during model validation. Note that, the number of features is $k(2\Delta + 1)$, where k is the number of attributes. For example, consider $\Delta = 3$, the number of features for all attributes is $9 \times (2 \times 3 + 1) = 63$.

Another parameter that influences the result of plant bounding box detection is ε. This parameter determines the minimum prediction value for a residual node set to be considered desirable. In Fig. 7 is shown the evolution of ε during tests. The best accuracy was obtained using $\varepsilon = 0.7$.

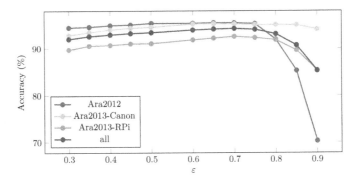

Fig. 7. Evolution of ε during tests.

Previously, we presented results in plant bounding box detection using non supervised strategies [6,11]. In order to compare them with our new approach, we chose the metrics provided by Minervine et al [18]. Those measures are: SBD that is a kind of accuracy metric; DiC that is the difference between predicted bounding box and the annotated bounding box, and $|DiC|$ that is the absolute value of DiC. In a general way, the results obtained by our new approach are the best results of all our works in plant bounding box detection. Those results are summarized in Table 1.

Table 1. Results obtained with our new approach compared to our previously works. In a general way the results of our new approach is the best of our works.

Dataset	Approach	$SBD[\%]$	DiC	$|DiC|$
Ara2012	Mumford Shah Energy [11]	92.00	0.20	0.20
	MSER [6]	91.90	0.20	0.20
	TBMR [6]	93.60	0.00	0.00
	Our	**95.43**	−0.18	0.43
Ara2013-Canon	Mumford Shah Energy [11]	87.50	0.10	0.30
	MSER [6]	89.40	0.10	0.10
	TBMR [6]	92.00	0.20	0.20
	Our	**95.20**	0.14	0.37
Ara2013-RPi	Mumford Shah Energy [11]	80.30	0.10	0.40
	MSER [6]	83.20	0.20	0.40
	TBMR [6]	84.00	0.30	0.30
	Our	**92.55**	−0.33	0.48
Mean (All)	Mumford Shah Energy [11]	85.80	0.00	0.30
	MSER [6]	87.60	0.10	0.20
	TBMR [6]	89.30	0.10	0.10
	Our	**94.23**	−0.11	0.42

6   Conclusion

This paper presented a novel approach to construct a strategy to filter undesirable residues from the ultimate levelings based on a regression that predicts the correspondence between objects of interest and residual regions. As it is known, during the residual extraction process, it is very common that undesirable regions of the input image contain residual information that should be filtered out. In this sense, we designed a filter based on the regression to filter out residues extracted from undesirable regions. The results obtained applying the filter in plant bounding box detection reveal the robustness of our approach. Further studies should investigate the following areas: (*i*) apply regression prediction as a weight of Ultimate Levelings residues, (*ii*) explore classifiers models instead of regressions and (*iii*) methods to determine Δ parameter.

Acknowledgements. This study was financed in part by the CAPES - Coordenação de Aperfeiçoamento de Pessoal de Nível Superior (Finance Code 001); FAPESP - Fundação de Amparo a Pesquisa do Estado de São Paulo (Proc. 2018/15652-7); CNPq - Conselho Nacional de Desenvolvimento Científico e Tecnológico (Proc. 428720/2018-8).

References

1. Alves, W., Hashimoto, R.: Text regions extracted from scene images by ultimate attribute opening and decision tree classification. In: Proceedings of the 23rd SIBGRAPI Conference on Graphics, Patterns and Images, pp. 360–367 (2010)
2. Alves, W., Morimitsu, A., Castro, J., Hashimoto, R.: Extraction of numerical residues in families of levelings. In: 2013 26th SIBGRAPI - Conference on Graphics, Patterns and Images (SIBGRAPI), pp. 349–356 (2013)
3. Alves, W.A.L., Hashimoto, R.F.: Ultimate grain filter. In: IEEE International Conference on Image Processing, pp. 2953–2957 (2014)
4. Alves, W.A.L., Morimitsu, A., Hashimoto, R.F.: Scale-space representation based on levelings through hierarchies of level sets. In: Benediktsson, J.A., Chanussot, J., Najman, L., Talbot, H. (eds.) ISMM 2015. LNCS, vol. 9082, pp. 265–276. Springer, Cham (2015). https://doi.org/10.1007/978-3-319-18720-4_23
5. Alves, W.A., Hashimoto, R.F., Marcotegui, B.: Ultimate levelings. Comput. Vis. Image Underst. **165**, 60–74 (2017)
6. Alves, W.A.L., Gobber, C.F., Hashimoto, R.F.: Plant bounding box detection from desirable residues of the ultimate levelings. In: Campilho, A., Karray, F., ter Haar Romeny, B. (eds.) ICIAR 2018. LNCS, vol. 10882, pp. 474–481. Springer, Cham (2018). https://doi.org/10.1007/978-3-319-93000-8_53
7. Caselles, V., Monasse, P.: Geometric Description of Images as Topographic Maps. Springer, Heidelberg (2010). https://doi.org/10.1007/978-3-642-04611-7
8. Couprie, C., Farabet, C., Najman, L., LeCun, Y.: Convolutional nets and watershed cuts for real-time semantic labeling of RGBD videos. J. Mach. Learn. Res. **15**, 3489–3511 (2014)
9. Dalla Mura, M., Benediktsson, J.A., Waske, B., Bruzzone, L.: Morphological attribute profiles for the analysis of very high resolution images. IEEE Trans. Geosci. Remote Sens. **48**, 3747–3762 (2010)
10. Fabrizio, J., Marcotegui, B.: Fast implementation of the ultimate opening. In: Wilkinson, M.H.F., Roerdink, J.B.T.M. (eds.) ISMM 2009. LNCS, vol. 5720, pp. 272–281. Springer, Heidelberg (2009). https://doi.org/10.1007/978-3-642-03613-2_25
11. Gobber, C.F., Alves, W.A., Hashimoto, R.F.: Ultimate leveling based on Mumford-Shah energy functional applied to plant detection. In: Progress in Pattern Recognition, Image Analysis, Computer Vision, and Applications, pp. 220–228 (2017)
12. Hernández, J., Marcotegui, B.: Shape ultimate attribute opening. Image Vis. Comput. **29**, 533–545 (2011)
13. Jones, R.: Connected filtering and segmentation using component trees. Comput. Vis. Image Underst. **75**(3), 215–228 (1999). https://doi.org/10.1006/cviu.1999.0777
14. Kimmel, R., Zhang, C., Bronstein, A., Bronstein, M.: Are MSER features really interesting? IEEE Trans. Pattern Anal. Mach. Intell. **33**(11), 2316–2320 (2011)
15. Li, W., Haese-Coat, V., Ronsin, J.: Residues of morphological filtering by reconstruction for texture classification. Pattern Recogn. **30**, 1081–1093 (1997)
16. Masci, J., Angulo, J., Schmidhuber, J.: A learning framework for morphological operators using counter-harmonic mean. In: Hendriks, C.L.L., Borgefors, G., Strand, R. (eds.) Mathematical Morphology and Its Applications to Signal and Image Processing, pp. 329–340 (2013)
17. Matas, J., Chum, O., Urban, M., Pajdla, T.: Robust wide-baseline stereo from maximally stable extremal regions. Image Vis. Comput. **22**, 761–767 (2004)

18. Minervini, M., Fischbach, A., Scharr, H., Tsaftaris, S.A.: Finely-grained annotated datasets for image-based plant phenotyping. Pattern Recogn. Lett. **81**, 80–89 (2016)
19. Ouzounis, G.K., Pesaresi, M., Soille, P.: Differential area profiles: decomposition properties and efficient computation. IEEE Trans. Pattern Anal. Mach. Intell. **34**(8), 1533–1548 (2012)
20. Pesaresi, M., Benediktsson, J.: A new approach for the morphological segmentation of high-resolution satellite imagery. IEEE Trans. Geosci. Remote Sens. **39**, 309–320 (2001)
21. Pesaresi, M., Benediktsson, J.: Image segmentation based on the derivative of the morphological profile. In: Goutsias, J., Vincent, L., Bloomberg, D. (eds.) Mathematical Morphology and its Applications to Image and Signal Processing. Computational Imaging and Vision, vol. 18, pp. 179–188. Springer, Boston (2000). https://doi.org/10.1007/0-306-47025-X_20
22. Retornaz, T., Marcotegui, B.: Scene text localization based on the ultimate opening. In: Proceedings of the 8th International Symposium on Mathematical Morphology and its Applications to Image and Signal Processing, pp. 177–188 (2007)
23. Retornaz, T., Marcotegui, B.: Ultimate opening implementation based on a flooding process. In: The 12th International Congress for Stereology (2007)
24. Salembier, P., Oliveras, A., Garrido, L.: Anti-extensive connected operators for image and sequence processing. IEEE Trans. Image Process. **7**(4), 555–570 (1998)

Max-Plus Operators Applied to Filter Selection and Model Pruning in Neural Networks

Yunxiang Zhang[1,2,3(✉)], Samy Blusseau[3], Santiago Velasco-Forero[3],
Isabelle Bloch[2], and Jesús Angulo[3]

[1] Ecole Polytechnique, 91128 Palaiseau, France
`yunxiang.zhang@polytechnique.edu`
[2] LTCI, Télécom ParisTech, Université Paris-Saclay, Paris, France
[3] Centre for Mathematical Morphology, Mines ParisTech, PSL Research University,
Fontainebleau, France
`samy.blusseau@mines-paristech.fr`

Abstract. Following recent advances in morphological neural networks, we propose to study in more depth how Max-plus operators can be exploited to define morphological units and how they behave when incorporated in layers of conventional neural networks. Besides showing that they can be easily implemented with modern machine learning frameworks, we confirm and extend the observation that a Max-plus layer can be used to select important filters and reduce redundancy in its previous layer, without incurring performance loss. Experimental results demonstrate that the filter selection strategy enabled by a Max-plus layer is highly efficient and robust, through which we successfully performed model pruning on two neural network architectures. We also point out that there is a close connection between Maxout networks and our pruned Max-plus networks by comparing their respective characteristics. The code for reproducing our experiments is available online (for code release, please visit https://github.com/yunxiangzhang.).

Keywords: Mathematical morphology ·
Morphological neural networks · Max-plus operator · Deep learning ·
Filter selection · Model pruning

1 Introduction

During the previous era of high interest in artificial neural networks, in the late 1980s, morphological networks [4,27] were introduced merely as an alternative to their linear counterparts. In a nutshell, they consist in changing the elementary neural operation from the usual scalar product to dilations, erosions and more general rank filters. Besides the practical achievements in this research line, which reached state-of-the-art results at its time [19], it questioned the main practices in the field of artificial neural networks on crucial topics such as architectures and optimization methods, improving their understanding.

B. Burgeth et al. (Eds.): ISMM 2019, LNCS 11564, pp. 310–322, 2019.
https://doi.org/10.1007/978-3-030-20867-7_24

After the huge technological leap taken by deep neural networks in the last decade, understanding them remains challenging and insights can still be brought by testing alternatives to the most popular practices. However, since the emergence of highly efficient deep learning frameworks (Caffe, Torch, TensorFlow, *etc.*), morphological neural networks have been very little re-investigated [2,17]. Motivated by the promising results in [2], this paper proposes to extend and deepen the study on morphological neural networks with Max-plus layers. More precisely, we aim to validate and exploit a specific property, namely that Max-plus layer (*i.e.* a dilation layer) following a conventional linear layer tends to select a reduced number of filters from the latter, making the others useless.

Our findings and contributions are three-fold. First, we propose an efficient training framework for morphological neural networks with Max-plus blocks, and theoretically demonstrate that they are universal function approximators under mild conditions. Secondly, we perform extensive experiments and visualization analysis to validate the robustness and effectiveness of the filter selection property provided by Max-plus blocks. Thirdly, we successfully applied the resulting Max-plus blocks to the task of model pruning on different neural network architectures. Related work is briefly summarized in Sect. 2. Then we show how to introduce Max-plus operators in neural networks in Sect. 3. Experimental results are discussed in Sect. 4.

2   Related Work

Morphological neural networks were defined almost simultaneously in two different ways at the end of the 1980s [4,27]. Davidson [4] introduced neural units that can be seen as pure dilations or erosions, whereas Wilson [27] focused on a more general formulation based on rank filters, in which max and min operators are two particular cases. Davidson's definition interprets morphological neurons as bounding boxes in the feature space [20,21,24]. In the latter studies, networks were trained to perform perfectly on training sets after few iterations, but little attention was drawn to generalization. Only recently, a backpropagation-based algorithm was adopted and improved constructive ones [28]. Still, the "bounding-box" approach does not seem to generalize well to test set when faced with high-dimensional problems like image analysis.

Wilson's idea, on the other hand, inspired hybrid linear/rank filter architectures, which were trained by gradient descent [18] and backpropagation [19]. In this case, the geometrical interpretation of decision surfaces becomes much richer, and the resulting framework was successfully applied to an image classification problem. The previously mentioned study [2] is one of the latest in this area, introducing a hybrid architecture that experimentally shows an interesting property on network pruning. In this classification experiment [2], each Max-plus unit shows, after training, one or two large weight values compared to the others. At the inference stage, this non-uniform distribution of weight values induces a selection of important filters in the previous layer, whereas the other filters (the majority) are no longer used in the subsequent classification task.

Therefore, after removal of the redundant filters, the network behaves exactly as it did before pruning. In this paper, we focus on the exploration of this property and show that it becomes more stable and effective provided that a proper regularization is applied.

Model pruning stands for a family of algorithms that explore the redundancy in model parameters and remove unimportant ones while not compromising the performance. Rapidly increased computational and storage requirements for deep neural networks in recent years have largely motivated research interests in this domain. Early works on model pruning dates back to [9,13], which prune model parameters based on the Hessian of the loss function. More recently, the model pruning problem was addressed by first dropping the neuronal connections with insignificant weight value and then fine-tuning the resulting sparse network [8]. This technique was later integrated into the Deep Compression framework [7] to achieve even higher compression rate. The HashNets model [3] randomly groups parameters into hash buckets using a low-cost hash function and performs model compression *via* parameter sharing. While previous methods achieved impressive performance in terms of compression rate, one notable disadvantage of these non-structured pruning algorithms lies in the sparsity of the resulting weight matrices, which cannot lead to speedup without dedicated hardwares or libraries.

Various structured pruning methods were proposed to overcome this subtlety in practical applications by pruning at the level of channels or even layers. Filters with smaller L_1 norm were pruned based on a predefined pruning ratio for each layer in [14]. Model pruning was transformed into an optimization problem in [16] and the channels to remove were determined by minimizing next layer reconstruction error. A branch of algorithms in this category employs L_p regularization to induce shrinkage and sparsity in model parameters. Sparsity constraints were imposed in [15] on channel-wise scaling factors and pruning was based on their magnitude, while in [26] group-sparsity was leveraged to learn compact CNNs via a combination of L_1 and L_2 regularization. One minor drawback of these regularization-based methods is that the training phase generally requires more iterations to converge. Our approach also falls into the structured pruning category and thus no dedicated hardware or libraries are required to achieve speedup, yet no regularization is imposed during model training. Moreover, in contrast to most existing pruning algorithms, our method does not need fine-tuning to regain performance.

3   Max-Plus Operator as a Morphological Unit

3.1   Morphological Perceptron

In traditional literature on machine learning and neural networks, a perceptron [22] is defined as a linear computational unit, possibly followed by a non-linear activation function. Among all popular choices of activation functions, such as logistic function, hyperbolic tangent function and rectified linear unit

(ReLU) function, ReLU [5] generally achieves better performance due to its simple formulation and non-saturating property. Instead of multiplication and addition, the morphological perceptron employs addition and maximum, which results in a non-linear computational unit. A simplified version [2] of the initial formulation [4,20] is defined as follows.

Definition 1 (Morphological Perceptron). *Given an input vector* $\mathbf{x} \in \mathbb{R}^n_{max}$ *(with* $\mathbb{R}_{max} = \mathbb{R} \cup \{-\infty\}$*), a weight vector* $\mathbf{w} \in \mathbb{R}^n_{max}$*, and a bias* $\mathbf{b} \in \mathbb{R}_{max}$*, the morphological perceptron computes its activation as:*

$$a(\mathbf{x}) = \max \left\{ \mathbf{b}, \max_{i \in \{1,...,n\}} \{\mathbf{x}_i + \mathbf{w}_i\} \right\} \tag{1}$$

where $\mathbf{x}_i$ *(resp.* $\mathbf{w}_i$*) denotes the i-th component of* $\mathbf{x}$ *(resp.* $\mathbf{w}$*).*

This model may also be referred to as $(\max, +)$ perceptron since it relies on the $(\max, +)$ semi-ring with underlying set $\mathbb{R}_{max}$. It is a dilation on the complete lattice $((\mathbb{R} \cup \{\pm\infty\})^n, \leq_n)$ with $\leq_n$ the Pareto ordering.

3.2   Max-Plus Block

Based on the formulation of the morphological perceptron, we define the Max-plus block as a standalone module that combines a fully-connected layer (or convolutional layer) with a Max-plus layer [2]. Let us denote the input vector of the fully-connected layer[1], the input and output vectors of the Max-plus layer respectively by $\mathbf{x}$, $\mathbf{y}$ and $\mathbf{z}$, whose components are indexed by $i \in \{1, ..., I\}$, $j \in \{1, ..., J\}$ and $k \in \{1, ..., K\}$, respectively. The corresponding weight matrices are denoted by $\mathbf{w}^f \in \mathbb{R}^{I \times J}_{max}$ and $\mathbf{w}^m \in \mathbb{R}^{J \times K}_{max}$. Then the operation performed in this Max-plus block is (see Fig. 1):

$$\begin{aligned} \mathbf{y}_j &= \sum_{i \in \{1,...,I\}} \mathbf{x}_i \cdot \mathbf{w}^f_{ij} \\ \mathbf{z}_k &= \max_{j \in \{1,...,J\}} \left\{ \mathbf{y}_j + \mathbf{w}^m_{jk} \right\} \end{aligned} \tag{2}$$

Note that the bias vector of the fully-connected layer (convolutional layer) is removed in our formulation, since its effect overlaps with that of the weight matrix $\mathbf{w}^m$. In addition, the bias vector of the Max-plus layer is shown to be ineffective in practice and is therefore not used here.

3.3   Universal Function Approximator Property

The result presented here is very similar to the approximation theorem on Max-out networks[2] [6], based on Wang's work [25]. As shown in [6], Maxout networks with enough affine components in each Maxout unit are universal function approximators. Recall that a model is called a universal function approximator if it can approximate arbitrarily well any continuous function provided

---

[1] This formulation can be easily generalized to the case of convolutional layers.
[2] Note that the classical universal approximation theorems for neural networks (see for example [10]) do not hold for networks containing max-plus units.

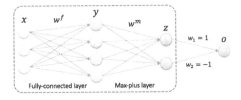

Fig. 1. A Max-plus block with two output units, applied to an input vector x.

enough capacity. Similarly, provided that the input vector (or input feature maps) $\mathbf{y} \in \mathbb{R}_{max}^{J}$ of the Max-plus layer may have arbitrarily many affine components (or affine feature maps), we show that a Max-plus model with just two output units in its Max-plus block can approximate arbitrarily well any continuous function of the input vector (or input feature maps) $\mathbf{x} \in \mathbb{R}^{I}$ of the block on a compact domain. A diagram illustrating the basic idea of the proof is shown in Fig. 1.

Theorem 1 (Universal function approximator). *A Max-plus model with two output units in its Max-plus block can approximate arbitrarily well any continuous function of the input of the block on a compact domain.*

Proof. We provide here a sketch of the proof, which follows very closely the one of Theorem 4.3 in [6]. By Proposition 4.2 in [6], any continuous function defined on a compact domain $C \subset \mathbb{R}^{I}$ can be approximated arbitrarily well by a piecewise linear (PWL) continuous function g, composed of k affine regions. By Proposition 4.1 in [6], there exist two matrices $W_1, W_2 \in \mathbb{R}^{k \times I}$ and two vectors $b_1, b_2 \in \mathbb{R}^k$ such that

$$\forall x \in C, \; g(x) = \max_{1 \le j \le k} \{W_{1j}x + b_{1j}\} - \max_{1 \le j \le k} \{W_{2j}x + b_{2j}\}, \tag{3}$$

where W_{ij} is the j-th row of matrix W_i and b_{ij} the j-th coefficient of b_i, $i = 1, 2$. Now, Eq. 3 is the output of the Max-plus block of Fig. 1, provided $\mathbf{w}^f = [W_1; W_2] \in \mathbb{R}^{2k \times I}$ is the matrix of the fully connected layer, $\mathbf{w}_1^m = [b_1^T, -\infty, \ldots, -\infty] \in \mathbb{R}_{max}^{2k}$ and $\mathbf{w}_2^m = [-\infty, \ldots, -\infty, b_2^T] \in \mathbb{R}_{max}^{2k}$ the two rows of $\mathbf{w}^m$. This concludes the proof. $\qed$

4   Experiments

In this section, we present experimental results for our Max-plus blocks by integrating them in different types of neural networks. All neural network models shown in this section are implemented with the open-source machine learning library TensorFlow [1] and trained on benchmark datasets MNIST [12] or CIFAR-10 [11] depending on the model complexity and capability.

4.1   Filter Selection Property

In an attempt to reproduce and confirm the experimental results reported in [2], we first implemented a simple Max-plus model composed of a fully-connected layer with J units followed by a Max-plus layer with ten units, namely a Max-plus block in our terminology, to perform image classification on MNIST dataset. In contrast to the original formulation in [2], the two bias vectors are removed for practical concerns explained in Sect. 3.2.

Table 1 summarizes the classification accuracy achieved by this simple model on the validation set of MNIST dataset for different values of J in columns 3 to 5. Note that all the experiments contained in these 3 columns are conducted under the same training setting (initial learning rate, learning rate decay steps, batch size, optimizer, *etc.*) except for parameter initialization (each column corresponds to a different random seed). The performance of the original model in [2] is included in column 2 for comparison. As shown in Table 1, provided a proper initialization, our simple Max-plus model generally achieves an improved performance compared to the original model. More interestingly, through horizontal comparison across different runs, we find that the performance of this naive Max-plus model is highly sensitive to parameter initialization.

Table 1. Classification accuracy on the validation set of MNIST dataset with three different random seeds.

J units	Model [2]	Max-plus			Max-plus + dropout		
24	84.3%	76.7%	**84.4%**	76.0%	93.7%	93.7%	**94.2%**
32	84.8%	84.5%	76.5%	**94.0%**	**94.6%**	94.2%	93.7%
48	84.6%	**94.0%**	93.8%	75.8%	**94.6%**	94.5%	94.5%
64	92.1%	85.4%	94.7%	**94.9%**	94.8%	**95.1%**	94.8%
100	–	**94.8%**	85.3%	93.7%	94.8%	**95.5%**	95.2%
144	–	**95.1%**	85.8%	85.4%	95.3%	95.7%	**95.9%**

In order to gain more insight into this instability problem, we follow the approach of [2] and visualize the weight matrix of the Max-plus layer $\mathbf{w}^m \in \mathbb{R}_{max}^{J \times 10}$ by splitting and reshaping it into 10 gray-scale images $\mathbf{w}_{\cdot 1}^m \ldots \mathbf{w}_{\cdot i}^m \ldots \mathbf{w}_{\cdot 10}^m$, each corresponding to a specific class (Fig. 2, left). In addition we also visualized, as on the right hand side of Fig. 2, the ten linear filters from the fully-connected layer that correspond to the maximum value of each weight vector $\mathbf{w}_{\cdot i}^m$, defined as:

$$\text{Filter}_i = \mathbf{w}_{\cdot j_{\max}^{(i)}}^f \in \mathbb{R}_{max}^I \quad \text{where} \quad j_{\max}^{(i)} = \underset{j \in \{1,\ldots,J\}}{\arg \max} \left\{ \mathbf{w}_{ji}^m \right\}. \tag{4}$$

Figure 2 shows specifically the weights of the Max-plus model that achieves an accuracy of 85.8% with $J = 144$ units (fifth column and last row in Table 1). We notice that there exists a severe filter-collision problem in the Max-plus block, namely different output units select the same linear filter in the fully-connected layer to compute their outputs. More specifically, class 3 and class 8 share the same *argmax* (highlighted by red circles) in their weight vectors and consequently select the same linear filter (visualized filters for class 3 and class 8 are the same). This collision between output units directly leads to classification confusion (because the Max-plus layer is the last layer in this simple model) since the classes that employ the same linear filter are completely indistinguishable. Furthermore, we only observed this filter-collision problem in the experiments that achieve relatively poor performance compared to the others (same model with different initialization). This observation consistently verifies our hypothesis that filter-collision is at the root of lower performance.

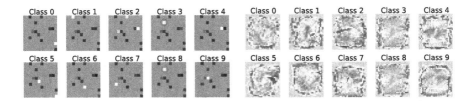

Fig. 2. Visualization of the weight matrix $\mathbf{w}^m$ (left) and the 10 linear filters that correspond to the maximum value of each weight vector $\mathbf{w}^m_{.i}$ (right). Visualization of the weight matrix $\mathbf{w}^m$ (left) and the 10 linear filters that correspond to the maximum value of each weight vector $\mathbf{w}^m_{.i}$ (right). (Color figure online)

In order to separate the output units that got stuck with the same linear filter, we applied a dropout regularization [23] to randomly switch off the neuronal connections between the fully-connected layer and the Max-plus layer during training. Empirically, we found this approach highly effective, as the performance of the dropout-regularized Max-plus model becomes much less sensitive to parameter initialization. Table 1 shows that the Max-plus model with dropout regularization achieves both a better performance and more stable results. We further validate the effect provided by dropout regularization through an additional experiment. With $J = 144$, we carried out 25 runs with different initializations for different dropout ratios. As shown in Fig. 3, dropout reduces dramatically the variability across experiments. The interpretation of this improvement is that dropout forces each class to use more than one linear filter to represent its corresponding digit. This allows for a more general representation and limits the risk of collisions between classes. Interestingly, we observe a slight accuracy drop when dropout ratio surpasses a certain level. This indicates that a trade-off between stability and performance needs to be found, although a large range of dropout values (between 25% and 75%) show robustness to random seed values without penalizing the effectiveness on classification accuracy.

Whereas in general (with or without dropout) linear filters $\mathbf{w}^f_{.j(i)}$ that correspond to large values of $\mathbf{w}^m_{.i}$ are similar to the images in Fig. 2 (right), *i.e.* digit-like shape with high contrast, those corresponding to smaller values in

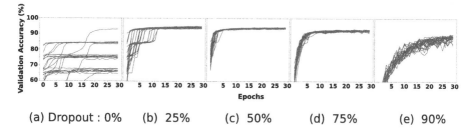

(a) Dropout : 0%   (b) 25%   (c) 50%   (d) 75%   (e) 90%

Fig. 3. Classification accuracy per epochs for 25 runs with different random initialization and dropout ratios on MNIST validation set. Dropout values between 25% and 75% in (b–d) show robustness to random seed values without penalizing the effectiveness on classification accuracy.

the weight matrix $\mathbf{w}^m$ are noisy or low-contrast images showing the shape of a specific digit. This means that these filters were activated by few training examples. Figure 4 (left) shows several linear filters of this kind. The visualization above suggests that large values in $\mathbf{w}^m$ correspond to linear filters of the fully-connected layer that contain rich information for the subsequent classification task, and hence that strongly respond to the samples of a specific class. If that holds, then it is likely that only these filters shall achieve the maximum in Eq. (2), and contribute to the classification output. Therefore, we might be able to reduce the complexity of a model while not degrading its performance by exploiting this filter selection property of Max-plus layers.

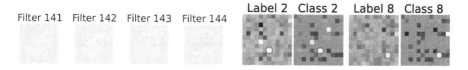

Fig. 4. Visualization of several linear filters corresponding to small values in $\mathbf{w}^m$ (left) and comparison between the activations of the fully-connected layer for two training examples and their corresponding weight vectors in $\mathbf{w}^m$ (right).

In order to further validate the stability of this connection before taking advantage of it, we also visualized the activations of the fully-connected layer as gray-scale images for several training examples and compare[3] them to the visualization of $\mathbf{w}^m$. As shown in Fig. 4 (right), there is a clear correspondence between the maximum activation value of a training example with label i and the largest two or three values of the weight vector $\mathbf{w}^m_{\cdot(i+1)}$, which indicates that the linear filters corresponding to large values in the weight matrix $\mathbf{w}^m$ are effectively used for the subsequent classification task.

---

[3] For example, if a training example has label i, then we compare its activation vector $\mathbf{y} \in \mathbb{R}^J_{max}$ with the weight vector $\mathbf{w}^m_{\cdot(i+1)} \in \mathbb{R}^J_{max}$.

4.2 Application to Model Pruning

Now that our approach to filter selection via Max-plus layers is proved to be quite effective and stable, we formalize our model pruning strategy as follows: given a fixed threshold $s \in [0,1]$, for each weight vector $\mathbf{w}^m_{\cdot(i+1)}$, we only keep the values that are larger than $s \times \max_{j \in \{1,...,J\}} \left\{ \mathbf{w}^m_{j(i+1)} \right\} + (1-s) \times$ $\min_{j \in \{1,...,J\}} \left\{ \mathbf{w}^m_{j(i+1)} \right\}$ and the linear filters that correspond to these retained values. Therefore, if in total J_r linear filters are kept in the pruned model, then the remaining parameters in the Max-plus layer no longer form a weight matrix but a weight vector of size J_r, where each entry corresponds to a linear filter in the fully-connected layer. The pruned fully-connected layer and the pruned Max-plus layer combined together perform a standard linear transformation followed by a maximum operation over uneven groups. Note that the pruning process is conducted independently for each output unit $\mathbf{z}_k$ of the Max-plus block $(1 \leq k \leq 10)$, thus the number of retained linear filters in the pruned model may vary from one class to another. From now on, we shall call these selected linear filters by *active filters* and the others by *non-active filters*. Figure 5 shows a graphical illustration of the comparison between the original model and the pruned model.

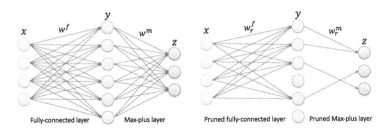

Fig. 5. Illustration of the comparison between the original Max-plus model (left) and the pruned Max-plus model (right). Here we have $\mathbf{w}^f_r \in \mathbb{R}^{I \times J_r}_{max}$ and $\mathbf{w}^m_r \in \mathbb{R}^{J_r}_{max}$.

We tried different pruning levels on this simple Max-plus model by varying the threshold s and tested the pruned models on the validation set and test set of MNIST dataset. We plotted the resulting classification accuracy in function of the number of active filters in Fig. 6. The performance of a single-layer softmax model and a single-layer Maxout model (number of affine components in each Maxout unit is two) is also provided for comparison.

As we can see on the diagram, the performance of the pruned Max-plus model is quite inferior to that of a single-layer softmax model when only one active filter is allowed to be selected for each class, *i.e.* the threshold is fixed to 1.0. However, as we relaxed the constraint on the number of total active filters by decreasing the threshold, the accuracy recovers rapidly and approaches that of the unpruned Max-plus model in a monotonic way. With exactly 24 active

J_r	Accuracy	J_r	Accuracy
10	82.7%	18	90.4%
11	86.4%	19	94.6%
12	86.4%	20	94.7%
13	89.9%	21	94.7%
14	90.0%	22	94.8%
15	90.2%	23	94.8%
16	90.2%	24	95.7%
17	90.3%	25	95.7%

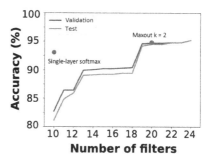

Fig. 6. Classification accuracy of pruned Max-plus model in function of the number of retained active filters.

filters retained in the pruned model, we achieve a full-recovery of the original Max-plus model performance, which means that the other 120 linear filters do not contribute to the classification task. Moreover, we can achieve comparable performance as the 2-degree Maxout model with roughly the same amount of parameters, which again validates the effectiveness of our Max-plus models.

With the same method, we successfully performed model pruning on a much more challenging CNN model by replacing the last fully-connected layer with a Max-plus layer. In order to facilitate the training of deep Max-plus model, we resort to transfer learning by initializing the two convolutional layers with pre-trained weights. The pruned Max-plus model achieves slightly better performance than the CNN model while reducing 94.8% of parameters of the second last fully-connected layer and eliminating the last fully-connected layer compared to the CNN model. Note that we could achieve a full-recovery of the unpruned Max-plus model performance with only ten active filters in this case, namely one linear filter for each output unit. Table 2 summarizes the architectures of the CNN model, the unpruned Max-plus model and the pruned Max-plus model, along with their classification accuracy on the test set of CIFAR-10 dataset.

Table 2. The architectural specifications of the CNN model, unpruned and pruned Max-plus model, along with their performance on the test set of CIFAR-10 dataset.

CNN	Max-plus	Pruned Max-plus
conv(5*5)	conv(5*5)	conv(5*5)
maxpool(2*2)	maxpool(2*2)	maxpool(2*2)
conv(5*5)	conv(5*5)	conv(5*5)
maxpool(2*2)	maxpool(2*2)	maxpool(2*2)
fc(384)	fc(384)	fc(384)
fc(192)	fc(192)	fc(10)
fc(10)	maxplus(10)	maxplus(10)
83.5%	83.9%	83.9%

4.3   Comparison to Maxout Networks

It is noticeable that the pruned Max-plus network differs from Maxout networks only in their grouping strategy for maximum operations. Maxout networks impose this rigid constraint in a way that each Maxout unit has an equal number of affine components while Max-plus networks are more tolerant in this respect. Figure 7 shows a graphical illustration of this comparison.

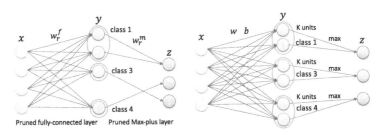

Fig. 7. Illustration of the comparison between the pruned Max-plus model (left) and Maxout model (right).

This higher flexibility hence endows Max-plus blocks with the capability of adapting the number of active filters used for each output unit accordingly. If a latent concept (say digit 1, which can be easily confused with digit 7) is considerably tougher to capture than some other concepts (say digits 3, 4 and 8, each of which has a relatively unique shape among the ten Arabic numbers), then the Max-plus layer will select more linear filters to abstract it than for the others. For example, the partition of the 24 active filters for the ten digit classes in Sect. 4.2 is $[2, 3, 3, 1, 2, 3, 3, 3, 2, 2]$, which is consistent with our point. This adaptive behavior of the filter selection property of Max-plus blocks makes the pruned Max-plus network more computationally efficient (fewer model parameters, smaller run-time memory footprint and faster inference) and is highly desirable in real-life applications.

5   Conclusions and Future Work

In this work we went a step further on a very new and promising topic, namely the reduction of deep neural networks with Max-plus blocks. Our experiments show strong evidence that model pruning via this method is compatible with high performance when a proper dropout regularization is applied during training. This was tested on data and architectures of variable complexity. Just as interesting as the obtained results are the many questions raised by these new insights. In particular, training these architectures is a challenging task which requires a better understanding on them, both theoretical and practical. We observed that training a deep model containing a Max-plus block is not straightforward, as we needed to resort to transfer learning. New optimization tricks will be needed to train deep architectures with *several* Max-plus blocks. The extension to a convolutional version of Max-plus blocks is also an open question, which we hope can be addressed based on the elements provided by this work.

Acknowledgements. This work was partially funded by a grant from Institut Mines Telecom.

References

1. Abadi, M., Barham, P., Chen, J., Chen, Z., et al.: TensorFlow: a system for large-scale machine learning. In: 12th USENIX Symposium on Operating Systems Design and Implementation, vol. 16, pp. 265–283 (2016)
2. Charisopoulos, V., Maragos, P.: Morphological perceptrons: geometry and training algorithms. In: Angulo, J., Velasco-Forero, S., Meyer, F. (eds.) ISMM 2017. LNCS, vol. 10225, pp. 3–15. Springer, Cham (2017). https://doi.org/10.1007/978-3-319-57240-6_1
3. Chen, W., Wilson, J., Tyree, S., Weinberger, K., Chen, Y.: Compressing neural networks with the hashing trick. In: International Conference on Machine Learning, pp. 2285–2294 (2015)
4. Davidson, J.L., Ritter, G.X.: Theory of morphological neural networks. In: Digital Optical Computing II, vol. 1215, pp. 378–389 (1990)
5. Glorot, X., Bordes, A., Bengio, Y.: Deep sparse rectifier neural networks. In: 14th International Conference on Artificial Intelligence and Statistics, pp. 315–323 (2011)
6. Goodfellow, I.J., Warde-Farley, D., Mirza, M., Courville, A., Bengio, Y.: Maxout networks. arXiv preprint arXiv:1302.4389 (2013)
7. Han, S., Mao, H., Dally, W.J.: Deep compression: compressing deep neural networks with pruning, trained quantization and Huffman coding. In: International Conference on Learning Representations (2015)
8. Han, S., Pool, J., Tran, J., Dally, W.: Learning both weights and connections for efficient neural network. In: Advances in Neural Information Processing Systems, pp. 1135–1143 (2015)
9. Hassibi, B., Stork, D.G.: Second order derivatives for network pruning: optimal brain surgeon. In: Advances in Neural Information Processing Systems, pp. 164–171 (1993)
10. Hornik, K., Stinchcombe, M., White, H.: Multilayer feedforward networks are universal approximators. Neural Netw. **2**(5), 359–366 (1989)
11. Krizhevsky, A., Hinton, G.: Learning multiple layers of features from tiny images. University of Toronto, Technical report (2009)
12. LeCun, Y., Bottou, L., Bengio, Y., Haffner, P.: Gradient-based learning applied to document recognition. Proc. IEEE **86**(11), 2278–2324 (1998)
13. LeCun, Y., Denker, J.S., Solla, S.A.: Optimal brain damage. In: Advances in Neural Information Processing Systems, pp. 598–605 (1990)
14. Li, H., Kadav, A., Durdanovic, I., Samet, H., Graf, H.P.: Pruning filters for efficient ConvNets. In: International Conference on Learning Representations (2016)
15. Liu, Z., Li, J., Shen, Z., Huang, G., Yan, S., Zhang, C.: Learning efficient convolutional networks through network slimming. In: ICCV, pp. 2755–2763 (2017)

16. Luo, J., Wu, J., Lin, W.: ThiNet: a filter level pruning method for deep neural network compression. In: ICCV, pp. 5068–5076 (2017)
17. Mondal, R., Santra, S., Chanda, B.: Dense morphological network: an universal function approximator. arXiv preprint arXiv:1901.00109 (2019)
18. Pessoa, L.F., Maragos, P.: MRL-filters: a general class of nonlinear systems and their optimal design for image processing. IEEE Trans. Image Process. **7**(7), 966–978 (1998)
19. Pessoa, L.F., Maragos, P.: Neural networks with hybrid morphological/rank/linear nodes: a unifying framework with applications to handwritten character recognition. Pattern Recogn. **33**(6), 945–960 (2000)
20. Ritter, G.X., Sussner, P.: An introduction to morphological neural networks. In: 13th International Conference on Pattern Recognition, vol. 4, pp. 709–717 (1996)
21. Ritter, G., Urcid, G.: Lattice algebra approach to single-neuron computation. IEEE Trans. Neural Netw. **14**(2), 282–295 (2003)
22. Rosenblatt, F.: Principles of neurodynamics: perceptrons and the theory of brain mechanisms. Technical report, Cornell Aeronautical Lab Inc., Buffalo, NY (1961)
23. Srivastava, N., Hinton, G., Krizhevsky, A., Sutskever, I., Salakhutdinov, R.: Dropout: a simple way to prevent neural networks from overfitting. J. Mach. Learn. Res. **15**(1), 1929–1958 (2014)
24. Sussner, P., Esmi, E.L.: Morphological perceptrons with competitive learning: lattice-theoretical framework and constructive learning algorithm. Inf. Sci. **181**(10), 1929–1950 (2011)
25. Wang, S.: General constructive representations for continuous piecewise-linear functions. IEEE Trans. Circ. Syst. I Regul. Pap. **51**(9), 1889–1896 (2004)
26. Wen, W., Wu, C., et al.: Learning structured sparsity in deep neural networks. In: Advances in Neural Information Processing Systems, pp. 2074–2082 (2016)
27. Wilson, S.S.: Morphological networks. In: Visual Communications and Image Processing IV, vol. 1199, pp. 483–496 (1989)
28. Zamora, E., Sossa, H.: Dendrite morphological neurons trained by stochastic gradient descent. Neurocomputing **260**, 420–431 (2017)

Part-Based Approximations for Morphological Operators Using Asymmetric Auto-encoders

Bastien Ponchon[1,2(✉)], Santiago Velasco-Forero[1], Samy Blusseau[1],
Jesús Angulo[1], and Isabelle Bloch[2]

[1] Centre for Mathematical Morphology, Mines ParisTech, PSL Research University,
Fontainebleau, France
`bastien.ponchon@gmail.com`, `samy.blusseau@mines-paristech.fr`
[2] LTCI, Télécom ParisTech, Université Paris-Saclay, Paris, France

Abstract. This paper addresses the issue of building a part-based representation of a dataset of images. More precisely, we look for a non-negative, sparse decomposition of the images on a reduced set of atoms, in order to unveil a morphological and interpretable structure of the data. Additionally, we want this decomposition to be computed online for any new sample that is not part of the initial dataset. Therefore, our solution relies on a sparse, non-negative auto-encoder where the encoder is deep (for accuracy) and the decoder shallow (for interpretability). This method compares favorably to the state-of-the-art online methods on two datasets (MNIST and Fashion MNIST), according to classical metrics and to a new one we introduce, based on the invariance of the representation to morphological dilation.

Keywords: Non-negative sparse coding · Auto-encoders ·
Mathematical morphology · Morphological invariance ·
Representation learning

1 Introduction

Mathematical morphology is strongly related to the problem of data representation. Applying a morphological filter can be seen as a test on how well the analyzed element is represented by the set of invariants of the filter. For example, applying an opening by a structuring element B tells how well a shape can be represented by the supremum of translations of B. The morphological skeleton [14,17] is a typical example of description of shapes by a family of building blocks, classically homothetic spheres. It provides a disjunctive decomposition where components - for example, the spheres - can only contribute positively as they are combined by supremum. A natural question is the optimality of this additive decomposition according to a given criterion, for example its sparsity - the number of components needed to represent an object. Finding a sparse

© Springer Nature Switzerland AG 2019
B. Burgeth et al. (Eds.): ISMM 2019, LNCS 11564, pp. 323–334, 2019.
https://doi.org/10.1007/978-3-030-20867-7_25

disjunctive (or part-based) representation has at least two important features: first, it allows *saving resources* such as memory and computation time in the processing of the represented object; secondly, it provides a *better understanding* of this object, as it reveals its most elementary components, hence operating a dimensionality reduction that can alleviate the issue of model over-fitting. Such representations are also believed to be the ones at stake in human object recognition [18].

Similarly, the question of finding a sparse disjunctive representation of a whole database is also of great interest and will be the main focus of the present paper. More precisely, we will approximate such a representation by a non-negative, sparse linear combination of non-negative components, and we will call *additive* this representation. Given a large set of images, our concern is then to find a smaller set of non-negative image components, called dictionary, such that any image of the database can be expressed as an additive combination of the dictionary components. As we will review in the next section, this question lies at the crossroad of two broader topics known as sparse coding and dictionary learning [13].

Besides a better understanding of the data structure, our approach is also more specifically linked to mathematical morphology applications. Inspired by recent work [1,20], we look for image representations that can be used to efficiently calculate approximations to morphological operators. The main goal is to be able to apply morphological operators to massive sets of images by applying them only to the reduced set of dictionary images. This is especially relevant in the analysis of remote sensing hyperspectral images where different kinds of morphological decomposition, such as morphological profiles [15] are widely used. For reasons that will be explained later, sparsity and non-negativity are sound requirements to achieve this goal. What is more, whereas the representation process can be learned offline on a training dataset, we need to compute the decomposition of any new sample *online*. Hence, we take advantage of the recent advances in deep, sparse and non-negative auto-encoders to design a new framework able to learn part-based representations of an image database, compatible with morphological processing.

The existing work on non-negative sparse representations of images are reviewed in Sect. 2, that stands as a baseline and motivation of the present study. Then we present in Sect. 3 our method before showing results on two image datasets (MNIST [9] and Fashion MNIST [21]) in Sect. 4, and show how it compares to other deep part-based representations. We finally draw conclusions and suggest several tracks for future work in Sect. 5. The code for reproducing our experiments is available online[1].

---

[1] For code release, visit https://gitlab.telecom-paristech.fr/images-public/asymae_morpho.

2   Related Work

2.1   Non-negative Sparse Mathematical Morphology

The present work finds its original motivation in [20], where the authors set the problem of learning a representation of a large image dataset to quickly compute approximations of morphological operators on the images. They find a good representation in the sparse variant of Non-negative Matrix Factorization (sparse NMF) [7], that we present hereafter.

Consider a family of M images (binary or gray-scale) $\mathbf{x}^{(1)}, \mathbf{x}^{(2)}, ..., \mathbf{x}^{(M)}$ of N pixels each, aggregated into a $M \times N$ data matrix $\mathbf{X} = (\mathbf{x}^{(1)}, \mathbf{x}^{(2)}, ..., \mathbf{x}^{(M)})^T$ (the ith row of $\mathbf{X}$ is the transpose of $\mathbf{x}^{(i)}$ seen as a vector). Given a feature dimension $k \in \mathbb{N}^*$ and two numbers s_H and s_W in $[0,1]$, a sparse NMF of $\mathbf{X}$ with dimension k, as defined in [7], is any solution of the problem

$$\mathbf{HW} = \arg\min \|\mathbf{X} - \mathbf{HW}\|_2^2 \quad \text{s.t.} \quad \begin{cases} \mathbf{H} \in \mathbb{R}^{M \times k}, \mathbf{W} \in \mathbb{R}^{k \times N} \\ \mathbf{H} \geq 0, \ \mathbf{W} \geq 0 \\ \sigma(\mathbf{H}_{:,j}) = s_H, \sigma(\mathbf{W}_{j,:}) = s_W, \ 1 \leq j \leq k \end{cases}$$

$$(1)$$

where the second constraint means that both $\mathbf{H}$ and $\mathbf{W}$ have non-negative coefficients, and the third constraint imposes the degree of sparsity of the columns of $\mathbf{H}$ and lines of $\mathbf{W}$ respectively, with σ the function defined by

$$\forall \mathbf{v} \in \mathbb{R}^p, \quad \sigma(\mathbf{v}) = \frac{\sqrt{p} - \|\mathbf{v}\|_1 / \|\mathbf{v}\|_2}{\sqrt{p} - 1}. \tag{2}$$

Note that σ takes values in $[0,1]$. The value $\sigma(\mathbf{v}) = 1$ characterizes vectors $\mathbf{v}$ having a unique non-zero coefficient, therefore the sparsest ones, and $\sigma(\mathbf{v}) = 0$ the vectors whose coefficients all have the same absolute value. Hoyer [7] designed an algorithm to find at least a local minimizer for the problem (1), and it was shown that under fairly general conditions (and provided the L_2 norms of $\mathbf{H}$ and $\mathbf{W}$ are fixed) the solution is unique [19].

In the terminology of representation learning, each row $\mathbf{h}^{(i)}$ of $\mathbf{H}$ contains the *encoding* or *latent features* of the input image $\mathbf{x}^{(i)}$, and $\mathbf{W}$ holds in its rows a set of k images called the dictionary. In the following, we will use the term *atom images* or *atoms* to refer to the images $\mathbf{w}_j = \mathbf{W}_{j,:}$ of the dictionary. As stated by Eq. (1), the atoms are combined to approximate each image $\mathbf{x}^{(i)} := \mathbf{X}_{i,:}$ of the dataset. This combination also writes as follows:

$$\forall i \in \{1, ..., M\}, \quad \mathbf{x}^{(i)} \approx \hat{\mathbf{x}}^{(i)} = \mathbf{H}_{i,:}\mathbf{W} = \mathbf{h}^{(i)}\mathbf{W} = \sum_{j=1}^k h_{i,j}\mathbf{w}_j. \tag{3}$$

The assumption behind this decomposition is that the more similar the images of the set, the smaller the required dimension to accurately approximate it. Note that only $k(N + M)$ values need to be stored or handled when using the previous approximation to represent the data, against the NM values composing the original data.

By choosing the sparse NMF representation, the authors of [20] aim at approximating a morphological operator ϕ on the data $\mathbf{X}$ by applying it to the atom images $\mathbf{W}$ only, before projecting back into the input image space. That is, they want $\phi(\mathbf{x}^{(i)}) \approx \Phi(\mathbf{x}^{(i)})$, with $\Phi(\mathbf{x}^{(i)})$ defined by

$$\Phi(\mathbf{x}^{(i)}) := \sum_{j=1}^{k} h_{i,j}\phi(\mathbf{w}_j). \tag{4}$$

The operator Φ in Eq. (4) is called a **part-based approximation** to ϕ. To understand why non-negativity and sparsity allow hoping for this approximation to be a good one, we can point out a few key arguments. First, sparsity favors the support of the atom images to have little pairwise overlap. Secondly, a sum of images with disjoint supports is equal to their (pixel-wise) supremum. Finally, dilations commute with the supremum and, under certain conditions that are favored by sparsity it also holds for the erosions. To precise this, let us consider a flat, extensive dilation δ_B and its adjoint anti-extensive erosion ε_B, B being a flat structuring element. Assume furthermore that for any $i \in [1, M], (j, l) \in [1, k]^2$ with $j \neq l$, $\delta_B(h_{i,j}\mathbf{w}_j) \bigwedge \delta_B(h_{i,l}\mathbf{w}_l) = 0$. Then on the dataset $\mathbf{X}$, δ_B and ε_B are equal to their approximations as defined by Eq. (4), that is to say:

$$\delta_B(\mathbf{x}^{(i)}) = \delta_B \left(\sum_{j=1}^{k} h_{i,j}\mathbf{w}_j \right) = \delta_B \left(\bigvee_{j\in[1,k]} h_{i,j}\mathbf{w}_j \right) = \bigvee_{[1,k]} \delta_B(h_{i,j}\mathbf{w}_j)$$
$$= \sum_{j=1}^{k} \delta_B(h_{i,j}\mathbf{w}_j) = \sum_{j=1}^{k} h_{i,j}\delta_B(\mathbf{w}_j) := D_B(\mathbf{x}^{(i)})$$

and similarly, since $\delta_B(x) \wedge \delta_B(y) = 0 \Rightarrow \varepsilon_B(x \vee y) = \varepsilon_B(x) \vee \varepsilon_B(y)$ for δ_B extensive, we also get $\varepsilon_B(\mathbf{x}^{(i)}) = \sum_{j=1}^{k} h_{i,j}\varepsilon_B(\mathbf{w}_j) := E_B(\mathbf{x}^{(i)})$. It follows that the same holds for the opening $\delta_B\varepsilon_B$. The assumption we just made is obviously too strong and unlikely to be verified, but this example helps realize that the sparser the non-negative decomposition, the more disjoint the supports of the atom images and the better the approximation of a flat morphological operator.

As a particular case, in this paper we will focus on part-based approximations of the dilation by a structuring element B, expressed as:

$$D_B(\mathbf{x}^{(i)}) := \sum_{j=1}^{k} h_{i,j}\delta_B(\mathbf{w}_j), \tag{5}$$

that we will compare with the actual dilation of our input images to evaluate our model, as shown in Fig. 1.

2.2   Deep Auto-encoders Approaches

The main drawback of the NMF algorithm is that it is an *offline* process, the encoding of any new sample with regards to the previously learned basis $\mathbf{W}$ requires either to solve a computationally extensive constrained optimization problem, or to release the Non-Negativity constraint by using the pseudo-inverse $\mathbf{W}^+$ of the basis. The various approaches proposed to overcome this shortcoming rely on Deep Learning, and especially on deep auto-encoders, which are widely used in the representation learning field, and offer an *online* representation process.

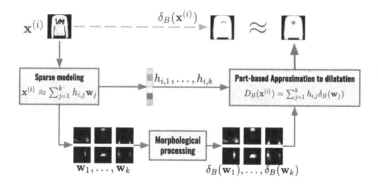

Fig. 1. Process for computing the part-based-approximation of dilation.

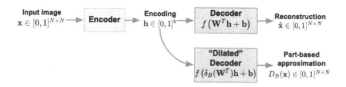

Fig. 2. The auto-encoding process and the definition of part-based approximation to dilation by a structuring element B in this framework.

An auto-encoder, as represented in Fig. 2, is a model composed of two stacked neural networks, an encoder and a decoder whose parameters are trained by minimizing a loss function. A common example of loss function is the mean square error (MSE) between the input images $\mathbf{x}^{(i)}$ and their reconstructions by the decoder $\hat{\mathbf{x}}^{(i)}$:

$$L_{AE} = \frac{1}{M} \sum_{i=1}^{M} L(\mathbf{x}^{(i)}, \hat{\mathbf{x}}^{(i)}) = \frac{1}{M} \sum_{i=1}^{M} \frac{1}{N} ||\hat{\mathbf{x}}^{(i)} - \mathbf{x}^{(i)}||_2^2. \tag{6}$$

In this framework, and when the decoder is composed of a single linear layer (possibly followed by a non-linear activation), the model approximates the input images as:

$$\hat{\mathbf{x}}^{(i)} = f\left(\mathbf{b} + \mathbf{h}^{(i)}\mathbf{W}\right) = f\left(\mathbf{b} + \sum_{j=1}^{k} h_{i,j}\mathbf{w}_j\right) \tag{7}$$

where $\mathbf{h}^{(i)}$ is the encoding of the input image by the encoder network, $\mathbf{b}$ and $\mathbf{W}$ respectively the bias and weights of the linear layer of the decoder, and f the (possibly non-linear) activation function, that is applied pixel-wise to the output of the linear layer. The output $\hat{\mathbf{x}}^{(i)}$ is called the *reconstruction* of the input image $\mathbf{x}^{(i)}$ by the auto-encoder. It can be considered as a linear combination of atom images, up to the addition of an offset image $\mathbf{b}$ and to the application of the activation function f. The images of our learned dictionary are hence the columns of the weight matrix $\mathbf{W}$ of the decoder. We can extend the definition of part-based approximation, described in Sect. 2.1, to our deep-learning architectures, by applying the morphological operator to these atoms $\mathbf{w}_1, \dots, \mathbf{w}_k$, as pictured by the *"dilated decoder"* in Fig. 2. Note that a central question lies in how to set the size k of the latent space. This question is beyond the scope of this study and the value of k will be arbitrarily fixed (we take $k = 100$) in the following.

The NNSAE architecture, from Lemme *et al.* [11], proposes a very simple and shallow architecture for online part-based representation using linear encoder and decoder with tied weights (the weight matrix of the decoder is the transpose of the weight matrix of the encoder). Both the NCAE architectures, from Hosseini-Asl *et al.* [6] and the work from Ayinde *et al.* [2] that aims at extending it, drop this transpose relationship between the weights of the encoder and of the decoder, increasing the capacity of the model. Those three networks enforce the non-negativity of the elements of the representation, as well as the sparsity of the image encodings using various techniques.

Enforcing Sparsity of the Encoding. The most prevalent idea to enforce sparsity of the encoding in a neural network can be traced back to the work of Lee *et al.* [10]. This variant penalizes, through the loss function, a deviation S of the expected activation of each hidden unit (*i.e.* the output units of the encoder) from a low fixed level p. Intuitively, this should ensure that each of the units of the encoding is activated only for a limited number of images. The resulting loss function of the sparse auto-encoder is then:

$$L_{AE} = \frac{1}{M} \sum_{i=1}^{M} L(\mathbf{x}^{(i)}, \hat{\mathbf{x}}^{(i)}) + \beta \sum_{j=1}^{k} S(p, \sum_{i=1}^{M} h_j^{(i)}), \qquad (8)$$

where the parameter p sets the expected activation objective of each of the hidden neurons, and the parameter β controls the strength of the regularization. The function S can be of various forms, which were empirically surveyed in [22]. The approach adopted by the NCAE [6] and its extension [2] rely on a penalty function based on the KL-divergence between two Bernoulli distributions, whose parameters are the expected activation and p respectively, as used in [6]:

$$S(p, t_j) = KL(p, t_j) = p \log \frac{p}{t_j} + (1 - p) \log \frac{1 - p}{1 - t_j} \quad \text{with } t_j = \sum_{i=1}^{M} h_j^{(i)} \qquad (9)$$

The NNSAE architecture [11] introduces a slightly different way of enforcing the sparsity of the encoding, based on a parametric logistic activation function at the output of the encoder, whose parameters are trained along with the other parameters of the network.

Enforcing Non-negativity of the Decoder Weights. For the NMF (Sect. 2.1) and for the decoder, non-negativity results in a part-based representation of the input images. In the case of neural networks, enforcing the non-negativity of the weights of a layer eliminates cancellations of input signals. In all the aforementioned works, the encoding is non-negative since the activation function at the output of the encoder is a sigmoid. In the literature, various approaches have been designed to enforce weight positivity. A popular approach is to use an asymmetric weight decay, added to the loss function of the network, to enact more decay on the negative weights that on the positive ones. However this approach, used in both the NNSAE [11] and NCAE [6] architectures, does not ensure that all weights will be non-negative. This issue motivated the variant of the NCAE architecture [2,11], which uses either the L_1 rather than the L_2 norm, or a smoothed version of the decay using both the L_1 and the L_2 norms. The source code of that method being unavailable, we did not use this more recent version as a baseline for our study.

3   Proposed Model

We propose an online part-based representation learning model, using an asymmetric auto-encoder with sparsity and non-negativity constraints. As pictured in Fig. 3, our architecture is composed of two networks: a deep encoder and a shallow decoder (hence the asymmetry and the name of AsymAE we chose for our architecture). The encoder network is based on the discriminator of the info-GAN architecture introduced in [4], which was chosen for its average depth, its use of widely adopted deep learning components such as batch-normalization [8], 2D-convolutional layers [5] and leaky-RELU activation function [12]. It has been designed specifically to perform interpretable representation learning on datasets such as MNIST and Fashion-MNIST. The network can be adapted to fit larger images. The decoder network is similar to the one presented in Fig. 2. A Leaky-ReLU activation has been chosen after the linear layer. Its behavior is the same as the identity for positive entries, while it multiplies the negative ones by a fixed coefficient $\alpha_{lReLU} = 0.1$. This activation function has shown better performances in similar architectures [12]. The sparsity of the encoding is achieved using the same approach as in [2,6] that consists in adding to the previous loss function the regularization term described in Eqs. (8) and (9).

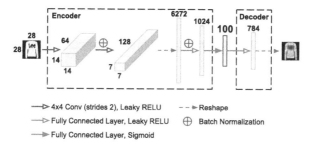

Fig. 3. Our proposed auto-encoder architecture.

We only enforced the non-negativity of the weights of the decoder, as they define the dictionary of images of our learned representation and as enforcing the non-negativity of the encoder weights would bring nothing but more constraints to the network and lower its capacity. We enforced this non-negativity constraint explicitly by projecting our weights on the nearest points of the positive orthant after each update of the optimization algorithm (such as the stochastic gradient descent). The main asset of this other method that does not use any additional penalty function, and which is quite similar to the way the NMF enforces non-negativity, is that it ensures positivity of all weights without the cumbersome search for good values of the parameters of the various regularization terms in the loss function.

4   Experiments

To demonstrate the goodness and drawbacks of our method, we have conducted experiments on two well-known datasets MNIST [9] and Fashion MNIST [21]. These two datasets share common features, such as the size of the images (28×28), the number of classes represented (10), and the total number of images (70000), divided in a training set of 60000 images and a test set of 10000 images. We compared our method to three baselines: the sparse-NMF [7], the NNSAE [11], the NCAE [6]. The three deep-learning models (AsymAE (ours), NNSAE and NCAE) were trained until convergence on the training set, and evaluated on the test set. The sparse-NMF algorithm was ran and evaluated on the test set. Note that all models but the NCAE may produce reconstructions that do not fully belong to the interval $[0, 1]$. In order to compare the reconstructions and the part-based approximation produced by the various algorithms, their outputs will be clipped between 0 and 1. There is no need to apply this operation to the output of NCAE as a sigmoid activation enforces the output of its decoder to belong to $[0, 1]$. We used three measures to conduct this comparison:

- the reconstruction error, that is the pixel-wise mean squared error between the input images $\mathbf{x}^{(i)}$ of the test dataset and their reconstruction/approximation $\hat{\mathbf{x}}^{(i)}$: $\frac{1}{MN} \sum_{i=1}^{M} \sum_{j=1}^{N} (\mathbf{x}_j^{(i)} - \hat{\mathbf{x}}_j^{(i)})^2$;
- the sparsity of the encoding, measured using the mean on all test images of the sparsity measure σ (Eq. 2): $\frac{1}{M} \sum_{i=1}^{M} \sigma(\mathbf{h}^{(i)})$;
- the approximation error to dilation by a disk of radius 1, obtained by computing the pixel-wise mean squared error between the dilation δ_B by a disk of radius 1 of the original image and the part-based approximation D_B to the same dilation, using the learned representation: $\frac{1}{MN} \sum_{i=1}^{M} \sum_{j=1}^{N} (D_B(\mathbf{x}^{(i)})_j - \delta_B(\mathbf{x}^{(i)})_j)^2$.

The parameter settings used for NCAE and the NNSAE algorithms are the ones provided in [6,11]. For the sparse-NMF, a sparsity constraint of $S_h = 0.6$ was applied to the encodings and no sparsity constraint was applied on the atoms of the representation. For our AsymAE algorithm, $p = 0.05$ was fixed for the sparsity objective of the regularizer of Eq. (9), and the weight of the sparsity regularizer in the loss function in Eq. (8) was set to $\beta = 0.001$ for MNIST and $\beta = 0.0005$ for Fashion-MNIST. Various other values have been tested for each algorithm, but the improvement of one of the evaluation measures usually came at the expense of the two others. Quantitative results are summarized in Table 1. Reconstructions by the various approaches of some sample images from both datasets are shown in Fig. 4.

Table 1. Comparison of the reconstruction error, sparsity of encoding and part-based approximation error to dilation produced by the sparse-NMF, the NNSAE, the NCAE and the AsymAE, for both MNIST and Fashion-MNIST datasets.

Model	Reconstruction error	Sparsity of code	Part-based approximation error to dilation
MNIST			
Sparse-NMF	0.011	**0.66**	**0.012**
NNSAE	0.015	0.31	0.028
NCAE	0.010	0.35	0.18
AsymAE	**0.007**	0.54	0.069
Fashion MNIST			
Sparse-NMF	0.011	**0.65**	**0.022**
NNSAE	0.029	0.22	0.058
NCAE	0.017	0.60	0.030
AsymAE	**0.010**	0.52	0.066

Both the quantitative results and the reconstruction images attest the capacity of our model to reach a better trade-off between the accuracy of the reconstruction and the sparsity of the encoding (that usually comes at the expense of the former criteria), than the other neural architectures. Indeed, in all conducted experiments, varying the parameters of the NCAE and the NNSAE as an attempt to increase the sparsity of the encoding came with a dramatic increase of the reconstruction error of the model. We failed however to reach a trade-off as good as the sparse-NMF algorithm that manages to match a high sparsity of the encoding with a low reconstruction error, especially on the Fashion-MNIST dataset. The major difference between the algorithms can be seen in Fig. 5 that pictures 16 of the 100 atoms of each of the four learned representations. While sparse-NMF manages, for both datasets, to build highly interpretable and clean part-based representations, the two deep baselines build representations that picture either too local shapes, in the case of the NNSAE, or too global ones, in the case of the NCAE. Our method suffers from quite the same issues as the NCAE, as almost full shapes are recognizable in the atoms. We noticed through experiments that increasing the sparsity of the encoding leads to less and less local features in the atoms. It has to be noted that the L_2 Asymmetric Weight Decay regularization used by the NCAE and NNSAE models allows for a certain proportion of negative weights. As an example, up to 32.2% of the pixels of the atoms of the NCAE model trained on the Fashion-MNIST dataset are non-negative, although their amplitude is lower than the average amplitude of the positive weights. The amount of negative weights can be reduced by increasing the corresponding regularization, which comes at the price of an increased reconstruction error and less sparse encodings. Finally Fig. 6 pictures the part-based approximation to dilation by a structuring element of size one, computed using the four different approaches on ten images from the test set. Although the quantitative results state otherwise, we can note that our approach yields a quite interesting part-based approximation, thanks to a good balance between a low overlapping of atoms (and dilated atoms) and a good reconstruction capability.

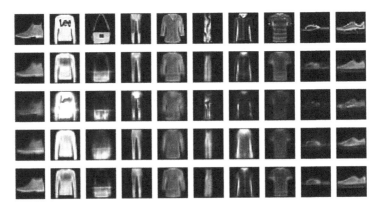

Fig. 4. Reconstruction of the Fashion-MNIST dataset (first row) by the sparse-NMF, the NNSAE, the NCAE and the AsymAE.

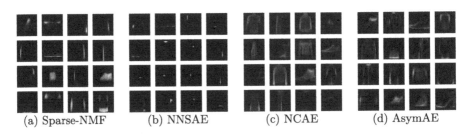

(a) Sparse-NMF   (b) NNSAE   (c) NCAE   (d) AsymAE

Fig. 5. 16 of the 100 atom images of the four compared representations of Fashion-MNIST dataset.

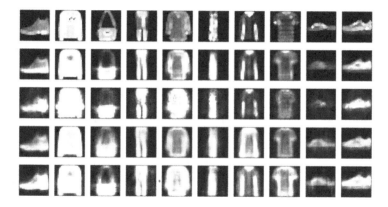

Fig. 6. Part-based approximation of the dilation by a structuring element of size 1 (first row), computed using the sparse-NMF, the NNSAE, the NCAE and the AsymAE.

5   Conclusions and Future Works

We have presented an online method to learn a part-based dictionary representation of an image dataset, designed for accurate and efficient approximations of morphological operators. This method relies on auto-encoder networks, with a deep encoder for a higher reconstruction capability and a shallow linear decoder for a better interpretation of the representation. Among the online part-based methods using auto-encoders, it achieves the state-of-the-art trade-off between the accuracy of reconstructions and the sparsity of image encodings. Moreover, it ensures a strict (that is, non approximated) non-negativity of the learned representation. These results would need to be confirmed on larger and more complex images (e.g. color images), as the proposed model is scalable. We especially evaluated the learned representation on an additional criterion, that is the commutation of the representation with a morphological dilation, and noted that all online methods perform worse than the offline sparse-NMF algorithm. A possible improvement would be to impose a major sparsity to the dictionary images an appropriate regularization. Additionally, using a morphological layer [3,16] as a decoder may be more consistent with our definition of part-based approximation, since a representation in the (max, +) algebra would commute with the morphological dilation by essence.

Acknowledgments. This work was partially funded by a grant from Institut Mines-Telecom and MINES ParisTech.

References

1. Angulo, J., Velasco-Forero, S.: Sparse mathematical morphology using non-negative matrix factorization. In: Soille, P., Pesaresi, M., Ouzounis, G.K. (eds.) ISMM 2011. LNCS, vol. 6671, pp. 1–12. Springer, Heidelberg (2011). https://doi.org/10.1007/978-3-642-21569-8_1
2. Ayinde, B.O., Zurada, J.M.: Deep learning of constrained autoencoders for enhanced understanding of data. CoRR abs/1802.00003 (2018)

3. Charisopoulos, V., Maragos, P.: Morphological perceptrons: geometry and training algorithms. In: International Symposium on Mathematical Morphology and Its Applications to Signal and Image Processing, pp. 3–15, April 2017. https://doi.org/10.1007/978-3-319-57240-6_1

4. Chen, X., Duan, Y., Houthooft, R., Schulman, J., Sutskever, I., Abbeel, P.: Info-GAN: interpretable representation learning by information maximizing generative adversarial nets. CoRR abs/1606.03657 (2016)

5. Dosovitskiy, A., Springenberg, J.T., Brox, T.: Learning to generate chairs with convolutional neural networks. CoRR abs/1411.5928 (2014)

6. Hosseini-Asl, E., Zurada, J.M., Nasraoui, O.: Deep learning of part-based representation of data using sparse autoencoders with nonnegativity constraints. IEEE Trans. Neural Networks Learn. Syst. 27(12), 2486–2498 (2016)

7. Hoyer, P.O.: Non-negative matrix factorization with sparseness constraints. CoRR cs.LG/0408058 (2004)

8. Ioffe, S., Szegedy, C.: Batch normalization: accelerating deep network training by reducing internal covariate shift. CoRR abs/1502.03167 (2015)

9. LeCun, Y., Cortes, C.: MNIST handwritten digit database (2010). http://yann.lecun.com/exdb/mnist/

10. Lee, H., Ekanadham, C., Ng, A.Y.: Sparse deep belief net model for visual area V2. In: Platt, J.C., Koller, D., Singer, Y., Roweis, S.T. (eds.) Advances in Neural Information Processing Systems 20, pp. 873–880 (2008)

11. Lemme, A., Reinhart, R.F., Steil, J.J.: Online learning and generalization of parts-based image representations by non-negative sparse autoencoders. Neural Networks 33, 194–203 (2012)

12. Maas, A.L.: Rectifier nonlinearities improve neural network acoustic models. In: International Conference on Machine Learning (2013)

13. Mairal, J., Bach, F.R., Ponce, J.: Sparse modeling for image and vision processing. CoRR abs/1411.3230 (2014)

14. Maragos, P., Schafer, R.: Morphological skeleton representation and coding of binary images. IEEE Trans. Acoust. Speech Sig. Process. 34(5), 1228–1244 (1986)

15. Pesaresi, M., Benediktsson, J.A.: A new approach for the morphological segmentation of high-resolution satellite imagery. IEEE Trans. Geosci. Remote Sens. 39(2), 309–320 (2001)

16. Ritter, G., Sussner, P.: An introduction to morphological neural networks. In: 13th International Conference on Pattern Recognition, vol. 4, pp. 709–717, September 1996. https://doi.org/10.1109/ICPR.1996.547657

17. Soille, P.: Morphological Image Analysis: Principles and Applications. Springer Science & Business Media, Heidelberg (2013)

18. Tanaka, K.: Columns for complex visual object features in the inferotemporal cortex: clustering of cells with similar but slightly different stimulus selectivities. Cereb. Cortex 13(1), 90–9 (2003)

19. Theis, F.J., Stadlthanner, K., Tanaka, T.: First results on uniqueness of sparse non-negative matrix factorization. In: 13th IEEE European Signal Processing Conference, pp. 1–4 (2005)

20. Velasco-Forero, S., Angulo, J.: Non-Negative Sparse Mathematical Morphology, chap. 1. In: Advances in Imaging and Electron Physics, vol. 202. Elsevier Inc., Academic Press (2017)

21. Xiao, H., Rasul, K., Vollgraf, R.: Fashion-MNIST: a novel image dataset for benchmarking machine learning algorithms. arXiv:1708.07747 (2017)

22. Zhang, L., Lu, Y.: Comparison of auto-encoders with different sparsity regularizers. In: International Joint Conference on Neural Networks (IJCNN), pp. 1–5 (2015)

Segmentation

Marker Based Segmentation Revisited

Fernand Meyer[⊠]

MINES ParisTech, PSL-Research University, CMM-Centre de Morphologie
Mathématique, Paris, France
fernand.meyer@mines-paristech.fr

Abstract. Watershed segmentation generally produces a severe over-segmentation. Marker based segmentation creates watershed partitions in which each region contains a marker, avoiding the oversegmentation. We obtain a coherent picture of marker based segmentation by modelling it on node or edge weighted graphs. Links are established between the methods published in the literature and some misconceptions are corrected.

Keywords: Weighted graphs · Watershed ·
Marker based segmentation · Flooding

1 Introduction

The watershed is a powerful tool for segmenting images when it is used in conjunction with markers. Without the introduction of markers, the images get severely oversegmented by the watershed transform. With marker based segmentation, a marker is chosen within each object to segment and within the background. The most significant contours of the image between the markers are then detected, creating a partition in which each region contains a marker [12]. The method is robust as the shape or exact position of the markers have no importance. It is powerful as the contours are found automatically as soon as the markers have been placed. The methods can be categorized as follows.

Intuitively, a grey tone image may be considered as a topographic surface. The regional minima serve as sources, pouring water which produces a uniform flooding over the domain, whose level steadily increases. All points flooded by the same source will be assigned to the same region. This intuitive definition has inspired various algorithms, differing by the way the flood progresses [1,4,14]. With this formulation, it is easy to introduce markers: instead of placing a source at each regional minimum, one places sources only at the position of the markers [5].

An alternative solution consists in applying the plain watershed transform to a modified topographic surface whose regional minima are all placed at the position of the markers [7,12].

If several marker based segmentations have to be performed, it is advantageous to proceed in two steps. A first watershed partition associated to all regional minima is produced and modelled as a region adjacency graph (RAG).

© Springer Nature Switzerland AG 2019
B. Burgeth et al. (Eds.): ISMM 2019, LNCS 11564, pp. 337–349, 2019.
https://doi.org/10.1007/978-3-030-20867-7_26

Each catchment basin is represented by a node. Two nodes are connected by an edge if the corresponding basins are neighbors; this edge is weighted by the altitude of the pass point between both basins. If a set of markers is chosen, one then constructs a minimum spanning forest (MSF) of the RAG, in which each tree is rooted in a marker. Each tree spans a region of the desired partition. This formulation permits an easy adjustment of the segmentation, when the set of markers is modified [6].

These methods are heterogeneous: the two first apply to images (which may be modelled as node weighted graphs), the last one to the RAG, an edge weighted graph. Expressing them in the abstract framework of node or edge weighted graphs establishes many bridges between them but also highlight some differences which lead to frequent misconceptions in the past.

2   Watershed Partitions on Node Weighted Graphs

We will establish below that each marker based segmentation of a weighted graph associated with a family of markers boils down to constructing a watershed partition of a node weighted graph associated both to the initial graph and to the markers, in which the only regional minima are the markers.

The watershed on a node weighted graph G may be defined by observing how a drop of water evolves from node to node: the drop of water may move from a node p to a node q, if e_{pq} is a flowing edge of p, i.e. it verifies $\nu_p \geq \nu_q$.

A flowing path is then a succession of flowing nodes $p, q, s...$, each edge connecting two successive nodes being a flowing edge of the first one. A regional minimum X is a maximal subgraph of G, in which all nodes have the same weight and without neighboring nodes holding a lower weight.

Each node of the graph is linked by one or several flowing paths with one or several regional minima. All nodes linked with the same regional minimum constitute the catchment zone of this regional minimum. Catchment zones may overlap. They overlap less if the graph is pruned so as to only keep the steepest flowing paths. In a watershed partition, the catchment basins do not overlap. If a node belongs to the catchment zone of distinct regional minima, it will be attributed to one and only one of them. For each node p attributed to a catchment basin of a regional minimum there exists a flowing path linking p and the minimum entirely included in the basin [9,10].

3   Flooding Distances on Weighted Graphs

We first establish a number of bridges between node and edge weighted graphs, as we will have to cross these bridges in both directions all along the paper.

3.1   Weighted Graphs

A *graph* $G = [\mathcal{N}, \mathcal{E}]$ contains a set $\mathcal{N}$ of *nodes* (designated with small letters: p, q, r) and a set $\mathcal{E}$ of *edges* (e_{pq} being the edge connecting p and q). Edges and/or nodes may be weighted with weights taken in a completely ordered set $\mathcal{T}$ with a smallest value 0 and a largest value O.

In the node weighted graph $G(\nu, nil)$, the node p takes the weight ν_p. In the edge weighted graph $G(nil, \eta)$, the edge e_{pq} takes the weight η_{pq}. In $G(\nu, \eta)$ the nodes take the weights ν and the edges the weights η.

Both operators $(\delta_{en}, \varepsilon_{ne})$ form an adjunction between node and edge weights:

- $(\delta_{en}\eta)_{pq} = \nu_p \vee \nu_q$ is a *dilation* from node weights to edge weights.
- $(\varepsilon_{ne}\nu)_p = \bigwedge\limits_{s \text{ neighbors of } p} \eta_{ps}$ is an *erosion* from edge weights to node weights.

We associate to the adjunction the opening on edge weights $\gamma_e = \delta_{en}\varepsilon_{ne}$ and the closing on node weights $\varphi_n = \varepsilon_{ne}\delta_{en}$.

3.2   The Flooding Distance

A path $\pi_{pq} = p - p_1 - p_2 \cdots q$ is a sequence of nodes, in which two successive nodes are linked by an edge.

If π_{pq} is a path linking p and q in $G(nil, \eta)$, we define its sup-section $|\pi_{pq}|_e$ (or simply $|\pi_{pq}|$ if there is no ambiguity) as the highest weight of the edges belonging to π_{pq}. If $p = q$ we define it as the highest weight of the edges belonging to $\varnothing$, i.e. $|\pi_{pp}|_e = o$.

If π_{pq} is a path linking p and q in $G(\nu, nil)$, we define its sup-section $|\pi_{pq}|_n$ (or simply $|\pi_{pq}|$ if there is no ambiguity) as the highest weight of the nodes belonging to π_{pq}. If $p = q$ it takes the value $|\pi_{pp}|_n = \nu_p$.

Definition 1. *If* Π_{pq} *is the family of paths linking* p *and* q *in the graph* G, *we define the flooding distance between* p *and* q *by* $\chi(p, q) = \bigwedge\limits_{\pi \in \Pi_{pq}} |\pi|$.

Consider now a node weighted graph $G(\nu, nil)$ and the edge weighted graph $G(nil, \delta_{en}\nu)$, in which each edge is assigned the weight of its highest extremity: $(\delta_{en}\nu)_{pq} = \nu_p \vee \nu_q$. The weights of the highest node and the highest edge within π_{pq} are identical: $|\pi_{pq}|_n = |\pi_{pq}|_e$. This establishes the following lemma and explains why, as we establish later, both graphs have identical floodings.

Lemma 1. *The flooding distances between distinct nodes in* $G(\nu, nil)$ *and* $G(nil, \delta_{en}\nu)$ *are identical:* $\chi(p, q) = \bigwedge\limits_{\pi \in \Pi_{pq}} |\pi|_n = \bigwedge\limits_{\pi \in \Pi_{pq}} |\pi|_e = \chi_e(p, q)$.

Let T be a minimum spanning tree (MST) T of $G(nil, \eta)$. The unique path of the tree connecting two nodes is a shortest path for the flooding distance between both nodes [3].

We will base the marker based segmentation of a graph associated to a set of markers on the flooding distance.

Definition 2. *A partition* (Ξ_i) *of the nodes is a marker based segmentation (MBS) of the graph* G *associated to a family of markers* (m_i), *if* (Ξ_i) *constitutes a Voronoï partition (VP) of the markers for the flooding distance:* $p \in \Xi_k \Leftrightarrow \forall m_i \neq m_k : \chi(p, m_k) \leq \chi(p, m_i)$.

All classical algorithms for constructing a shortest distance to a node on a graph are easily adapted to shortest flooding distances [11,13]. Gondran and Minoux have unified many shortest path algorithms under the name of "path algebra". Flooding shortest path algorithms are expressed in the algebra (min,max), whereas the classical shortest path algorithms are expressed in the algebra (min,+) [2].

3.3   Floodings on Graphs

Fig. 1A presents a flooded topographic surface covered by a number of lakes. We represent it as a node weighted graph $G(\nu, nil)$ in which each node represents a pixel of the image. Two nodes are linked by an edge if the corresponding pixels are neighbors. The nodes hold two weights, the ground level ν equal to the grey tone of the corresponding pixel and the flooding level $\omega \geq \nu$ representing the altitude of the lake covering the pixel. If a node p is dry, then $\omega_p = \nu_p$.

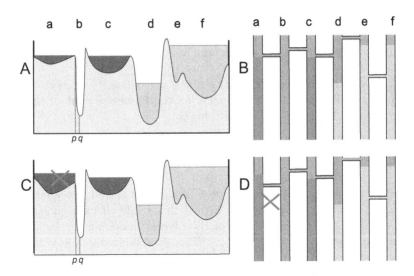

Fig. 1. A and B: a physically valid flood distribution; C and D: an impossible flood distribution

The same flood distribution is considered in Fig. 1B at the scale of the catchment basins. Modelled as an edge weighted graph $G(nil, \eta)$, it is represented as a tank network: each catchment basin is represented by a node in the graph and a tank in the tank network. The nodes representing two neighboring catchment basins are linked by an edge in the graph and the corresponding tanks by a pipe. The weight of the edge and the level of the pipe are equal to the altitude of the pass point between the catchment basins. If a catchment basins is covered by a lake, the corresponding node gets a weight equal to this level and the tank is filled up to this level. If the catchment basin is dry, the node weight is 0 and the tank remains empty, i.e. takes the weight o.

Figure 1A and B thus represent the same flood distribution observed at the scale of the pixels and at the scale of the catchment basins. On the contrary, Fig. 1C and D represent a flood distribution which is physically impossible.

The function τ on the nodes of the node weighted graph $G(\nu, nil)$ is a n-*flooding* of this graph, i.e. is a stable distribution of fluid if it verifies: $\tau \geq \nu$ and for any couple of neighboring nodes (p, q) we have $\tau_p > \tau_q \Rightarrow \tau_p = \nu_p$ (criterion $n1$). The criterion $n2$ is equivalent: $\tau_p \leq \tau_q \vee \nu_p$.

The function τ on the nodes of the edge weighted graph $G(nil, \eta)$ is an e-*flooding* of this graph, i.e. is a stable distribution of fluid if it verifies the criterion: for any couple of neighboring nodes (p, q) we have: $\tau_p > \tau_q \Rightarrow \eta_{pq} \geq \tau_p$ (criterion $e1$). The criterion $e2$ is equivalent [8, 11]: $\tau_p \leq \tau_q \vee \eta_{pq}$.

Criterion $n1$ is verified for the pairs of nodes p and q in Fig. 1A but not in Fig. 1C. The equivalent flooding observed at the scale of the catchment basins has a similar behavior: criterion $e1$ is verified for the tanks (a, b) and (e, d) in Fig. 1B but is not verified for the tanks (a, b) in Fig. 1D.

Equivalent Floodings in $G(\nu, nil)$ and $G(nil, \delta_{en}\nu)$. Since $\tau \geq \nu$, we have $\tau_q = \tau_q \vee \nu_q$, which permits to write the following equivalences for a pair (p, q) of neighboring nodes: $\{\tau$ n-flooding of $G(\nu, nil)\} \Leftrightarrow$

$$\left\{\tau_p \leq \tau_q \vee \nu_p = \tau_q \vee \nu_q \vee \nu_p = \tau_q \vee (\delta_{en}\nu)_{pq}\right\} \Leftrightarrow \{\tau \text{ e-flooding of } G(nil, \delta_{en}\nu)\}.$$

Lemma 2. *If a function $\tau \geq \nu$ is a valid flooding on one of the graphs $G(\nu, nil)$ or $G(nil, \delta_{en}\nu)$, it also is a valid flooding on the other.*

The supremum of all floodings of a weighted graph $G(nil, \eta)$ below a ceiling function ω is itself a flooding, called dominated flooding of G under ω. Following the preceding lemma, we may state:

Lemma 3. *The dominated flooding of the graph $G(\nu, nil)$ under the ceiling function $\omega \geq \nu$ is also the dominated flooding under ω of the graph $G(nil, \delta_{en}\nu)$.*

Flooding Distances Are Floodings. Let τ be the flooding distance to a node Ω in a weighted graph and p, q two neighboring nodes. Let $\pi_{\Omega q}$ be the lowest path between Ω and q. The concatenation of $\pi_{\Omega q} \rhd e_{qp}$ of $\pi_{\Omega q}$ and the edge e_{qp} is a path between Ω and p but not necessarily the lowest path: $\tau_p \leq |\pi_{\Omega q} \rhd e_{qp}|$.

- for $G(nil, \eta)$: $\tau_p \leq |\pi_{\Omega q} \rhd e_{qp}| = |\pi_{\Omega q}|_e \vee \eta_{qp} = \tau_q \vee \eta_{qp}$, characterizing the valid floodings of $G(nil, \eta)$. If q belongs to the shortest path between Ω and p, e_{pq} is called flooding edge and $\tau_p = \tau_q \vee \eta_{qp}$.
- for $G(\nu, nil)$: $\tau_p \leq |\pi_{\Omega q} \rhd e_{qp}| = |\pi_{\Omega q}|_e \vee \nu_p = \tau_q \vee \nu_p$, characterizing the valid floodings of $G(\nu, nil)$. If q belongs to the shortest path between Ω and p, e_{pq} is called flooding edge and $\tau_p = \tau_q \vee \nu_p$.

4   Marker Based Segmentation on an Edge Weighted Graph

4.1   A Marker Based Segmentation Associated to the Minimum Spanning Forests of the Graph

We aim to construct a marker based segmentation of an edge weighted graph $G(nil, \eta)$, verifying $\eta > 0$, associated to a family (m_i) of markers (in Fig. 2A the green and violet nodes). We construct an augmented graph G_Ω (in Fig. 2B) by adding to the graph G a dummy node Ω linked with each marker by an edge with weight 0; T_Ω (in Fig. 2C) is a minimum spanning tree of G_Ω.

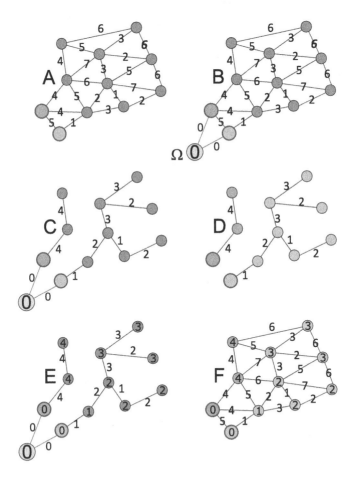

Fig. 2. A: A graph $G(nil, \eta)$ with two markers nodes; B: The augmented $G_\Omega(nil, \eta)$; C: The minimum spanning tree T_Ω of G_Ω; D: Suppressing dummy node Ω and dummy edges in T_Ω produces the minimum spanning forest (T_i); E: The flooding distance to the node Ω inside the tree T_Ω; F: The final marker based segmentation of G. (Color figure online)

Suppressing the dummy node and cutting the dummy edges in the minimum spanning tree T_Ω (in Fig. 2C) produces a minimum spanning forest $F = (T_i)$ (in Fig. 2D) in which each tree T_k is rooted in the marker m_k [6]. We call N_i the nodes spanned by the tree T_i. The family (N_i) forms a partition of the nodes of G.

Each node $p \in T_k$ is connected by a unique path $\pi_{m_k p}$ in T_Ω and T_k; belonging to the MST of G, this path is a geodesic of the flooding distance. The path $e_{\Omega m_k} \triangleright \pi_{m_k p}$ is then one of the lowest paths connecting Ω and p in G_Ω, since the edge $\eta_{\Omega m_k} = 0$. Now, if $\pi_{m_i p}$ is one of the lowest paths between another marker m_i and p, then $e_{\Omega m_i} \triangleright \pi_{m_i p}$ is a path between Ω and p, which cannot be lower than the lowest path $e_{\Omega m_k} \triangleright \pi_{m_k p}$ between Ω and p : $|e_{\Omega m_i} \triangleright \pi_{m_i p}| \geq |e_{\Omega m_k} \triangleright \pi_{m_k p}|$. But as $\eta_{\Omega m_i} = \eta_{\Omega m_k} = 0$, $|e_{\Omega m_i} \triangleright \pi_{m_i p}| = |\pi_{m_i p}| = \chi(p, m_i)$ and $|e_{\Omega m_k} \triangleright \pi_{m_k p}| = |\pi_{m_k p}| = \chi(p, m_k) = \tau_p$, where $\chi(s, t)$ is the flooding distance between s and t in the graph G. This shows that for each node $p \in N_k : \chi(p, m_k) \leq \chi(p, m_i)$ for $m_i \neq m_k$. The family (N_i) thus forms a Voronoï tessellation for the flooding distance, associated to the markers m_i, i.e. a MBS. In Fig. 2D, all nodes of a tree have been assigned the color of the marker.

Lemma 4. *Given an edge weighted graph $G(nil, \eta)$ and a family of markers, any minimum spanning forest $G(nil, \eta)$, in which each tree is rooted in a marker spans a marker based segmentation of the graph.*

Any algorithm for constructing a minimum spanning tree is easily adapted for constructing a minimum spanning forest.

Discussion

Minimum spanning forests do not constitute the only solutions to marker based segmentation. They are not even necessarily the best ones as shows Fig. 3. The nodes a and b serve as markers of the edge weighted graph in Fig. 3A. There exist two symmetrical minimum spanning forests rooted in the markers; one of them is illustrated by Fig. 3D. It breaks the symmetry of the graph and markers. On the contrary, the spanning forest in Fig. 3E respects the symmetry but is not a minimum spanning forest. It constitutes a better marker based segmentation of the graph. Below we explain how to find such additional MBS.

4.2   The Topography of the Graph $G(\tau, nil)$

We now study the topography of the graph $G(\tau, nil)$, in which τ is equal to 0 on the markers and equal to the flooding distance to the markers for all other nodes. If $F = (T_i)$ is a minimum spanning forest (MSF) spanned in the markers, τ may be computed on each tree separately, since the unique path connecting a node $p \in T_k$ with the root m_k in the tree F_k is a geodesic of the flooding distance. Since $\eta > 0$, we have $\tau_p > 0$, whereas $\tau_{m_k} = 0$. As the node p is thus connected by a non increasing path with a lower node, it cannot be a regional minimum of the graph $G(\tau, nil)$. In other terms, $G(\tau, nil)$ has no regional minima outside the marker nodes and for each marker node $\tau = 0$. The marker nodes constitute the only regional minima of $G(\tau, nil)$, but they may coalesce: two neighboring

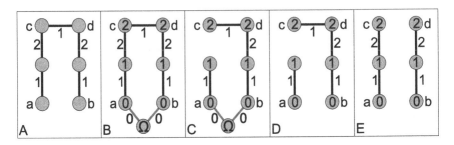

Fig. 3. Image A presents an edge weighted graph in which the nodes a and b are chosen as markers. The minimum spanning forest in image D breaks the symmetry of the figure. The spanning forest in image E respects the symmetry but is not minimal.

marker nodes in the graph G, each with a weight $\tau = 0$, belong to a unique regional minimum of the graph $G(\tau, nil)$. Some catchment basins of the graph $G(\tau, nil)$ will thus contain several markers. For this reason does a watershed partition of the graph $G(\tau, nil)$ not constitute a marker based segmentation of $G(nil, \eta)$ associated to the markers.

In Fig. 2F precisely, the two marker nodes are neighbors and form a unique regional minimum of the graph $G(\tau, nil)$. They are separated by an edge with weight 5, which is higher than 0, the weight of its extremities. Such edges verifying $\tau_p \vee \tau_q < \eta_{pq}$ are not flooding edges, and should be cut in the graph $G(\tau, \eta)$ before constructing a watershed partition, which will then constitute a valid MBS of the graph $G(nil, \eta)$. The following example shows another harmful effect of non flooding edges.

Figure 4A presents an edge weighted graph $G(nil, \eta)$ with 3 markers, respectively in red, blue and green color. The minimum spanning forest of Fig. 4B spans a valid marker based segmentation.

We now examine whether a watershed partition of the graph $G(\tau, nil)$ also constitutes a valid MBS. The flooding distance τ to the markers and the edge weights of the graph $G(\tau, \eta)$ is represented in Fig. 4C. Contrarily to the previous example, the markers are isolated from each other and thus form isolated regional minima of the graph $G(\tau, nil)$. A watershed partition of $G(\tau, nil)$ has been constructed in Fig. 4D. Each region of the watershed partition of the graph $G(\tau, nil)$ contains one marker and has the same color. However the contours of the partition are not satisfactory. For instance the nodes a and b have been assigned to the red marker as a is the lowest neighbor of the node b. The edge e_{ba} is thus a flowing edge of the graph $G(\tau, nil)$ but is not a flooding edge. The node b has not been flooded by the node a through the edge e_{ab} since $\tau_b(=6) < \tau_a(=3) \vee \eta_{ab}(=8)$.

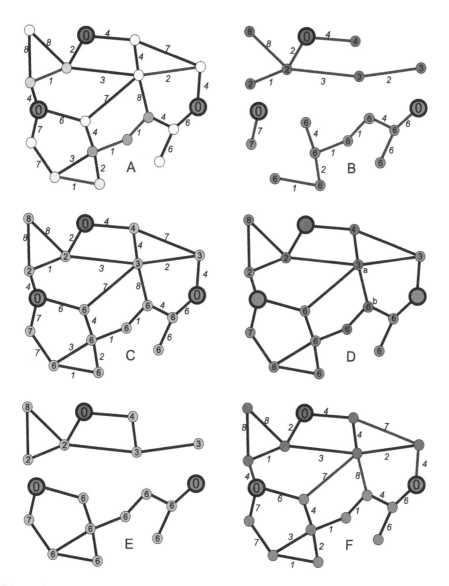

Fig. 4. A: an edge weighted graph $G(nil, \eta)$ with three markers; B: a minimum spanning forest of G, with each tree rooted in a marker; C: the graph $G(\tau, \eta)$ where η represents the shortest flooding distance to the markers; D: a watershed partition of the graph $G(\tau, nil)$ and poor segmentation; E: the pruned graph $\widetilde{G}(\tau, nil)$ in which the non flooding edges have been suppressed; F: A watershed partition of the graph $\widetilde{G}(\tau, nil)$ and correct segmentation. The edges in red are the non flooding edges. (Color figure online)

4.3   A Pruning Suppressing the Non Flooding Edges

A node p is flooded by a neighboring node q through the edge e_{pq} if $\tau_p = \tau_q \vee \eta_{pq}$. Similarly q is flooded by p if $\tau_q = \tau_p \vee \eta_{pq}$. If none of these two relations are verified, which happens when $\tau_p < \tau_q \vee \eta_{pq}$ and $\tau_q < \tau_p \vee \eta_{pq}$, then no flooding path crosses the edge e_{pq}. Both preceding inequalities are equivalent with $\tau_p \vee \tau_q < \eta_{pq}$. Edges verifying $\tau_p \vee \tau_q < \eta_{pq}$ are "non-flooding edges" of the graph $G(\tau, nil)$. Suppressing these non flooding edges produces the pruned graph $\widetilde{G}(\tau, nil)$. Fig. 4E presents the pruned graph $\widetilde{G}(\tau, nil)$ and Fig. 4F one of its watershed partitions, which also constitutes a valid MBS of $G(nil, \eta)$ as states the following theorem.

Theorem 1. *The marker based segmentations of the graph $G(nil, \eta)$ associated to a family of markers (m_i) are identical with the watershed partitions of the graph $\widetilde{G}(\tau, nil)$, where τ is the flooding distance to the markers in the graph $G(nil, \eta)$.*

Proof. (1) Let (X_i) be a watershed partition of the graph $\widetilde{G}(\tau, nil)$. We will show that it is a MBS of $G(nil, \eta)$. Each node of X_k is linked with the marker m_k by a flowing path contained in X_k. If e_{pq} is a flowing edge of this path, we have $\tau_p \geq \tau_q$. Thus $\tau_p = \tau_p \vee \tau_q \geq \eta_{pq}$ as the edge e_{pq} has been preserved by the pruning. As $\tau_p \geq \tau_q$ and $\tau_p \geq \eta_{pq}$, we have $\tau_p \geq \tau_q \vee \eta_{pq}$.
On the other hand τ results from a flooding of G: $\tau_p \leq \tau_q \vee \eta_{pq}$. Finally $\tau_p = \tau_q \vee \eta_{pq}$, showing that the flowing edge e_{pq} is also a flooding edge of $G(\tau, nil)$. Each flowing path of $\widetilde{G}(\tau, nil)$ is also a flooding path. Thus any watershed partition (X_i) of the graph $\widetilde{G}(\tau, nil)$ also constitutes a marker based segmentation of the graph $G(nil, \eta)$.
(2) Inversely, if (N_i) is a MBS of the graph $G(nil, \eta)$, each node p is linked with a marker m_k by a flooding path π. If s and t are two successive nodes on this path if one follows π in the direction from p to m_k, then $\tau_s = \tau_t \vee \eta_{st}$. Thus $\tau_t \vee \tau_s \geq \eta_{st}$, the edge e_{st} is a flooding edge, is not pruned and is still present in $\widetilde{G}(\tau, nil)$. On the other hand $\tau_s = \tau_t \vee \eta_{st} \geq \tau_t$, showing that the edge e_{st} is also a flowing edge. The path π is thus a flowing path linking p with m_k in the graph $\widetilde{G}(\tau, nil)$. It results that the MBS (N_i) is also a watershed partition of the graph $\widetilde{G}(\tau, nil)$.

5   Marker Based Segmentation of a Node Weighted Graph

We now aim at constructing a marker based segmentation of a node weighted graph $G(\nu, nil)$ associated to a family of markers. We suppose $\nu > 0$ (if $\nu \geq 0$ we add a constant value to all nodes). We set $\nu = 0$ on all marker nodes. As $G(\nu, nil)$ and $G(nil, \delta_{en}\nu)$ have identical floodings, the methods of the previous section are applicable to the graph $G(nil, \delta_{en}\nu)$ but get simpler.

Any minimum spanning forest of the graph $G(nil, \delta_{en}\nu)$ rooted in the markers spans a particular marker based segmentation. This furnishes a first series of solutions.

The function τ equal to 0 on the markers and equal to the flooding distance of each node to the markers on the other nodes may be computed on the graph $G(nil, \delta_{en}\nu)$ as previously, or, with simpler and faster algorithm, on the graph $G(\nu, nil)$ [11]. All edges in the graph $G(\tau, \delta_{en}\nu)$ are then flooding edges. Indeed, as τ is a flooding of $G(\nu, nil)$, it verifies $\tau \geq \nu$. If p and q are two neighboring nodes, we have $\tau_p \geq \nu_p$ and $\tau_q \geq \nu_q$, hence $\tau_p \vee \tau_q \geq \nu_p \vee \nu_q = (\delta_{en}\nu)_{pq} = \eta_{pq}$. As no pruning is necessary, any watershed partition of the graph $G(\tau, nil)$ constitutes a valid marker based segmentation.

We have thus established the validity of frequently used method on images [12], modelled as node weighted graphs: compute the highest flooding of the graph $G(\nu, nil)$ under a ceiling function equal to 0 on the markers and equal to O on all other nodes. This highest flooding is precisely the function τ. Any watershed partition of this function constitutes a valid marker based segmentation, and vice versa.

6 Conclusion

The time has come to assess the results of this study. By jointly studying the watershed and floodings on node and edge weighted graphs, we have been able to thoroughly revisit marker based segmentation.

We have adopted a common definition of marker based segmentation of a weighted graph as a Voronoï tessellation of the markers for the flooding distance on the graphs. All shortest path algorithms on graphs may be adapted to the flooding distance and produce a MBS.

We have proved that any minimum spanning forest of a graph $G(nil, \eta)$ with trees rooted in the markers spans a MBS, but not all MBS are of this type. This construction is particularly advantageous in case of interactive segmentation. Indeed, after a MSF (F_i) has been constructed associated to a first set of markers, it is easy to add or suppress markers, and produce the associated segmentation, simply by a local processing of the graph [6]. If an added marker m_n belongs to an existing tree F_k, one simply cuts one of the highest edges of the unique path linking m_n and m_k. If the marker m_k is suppressed one adds to F_k one of the edges with the lowest weight connecting F_k with a neighboring tree of the forest, merging both trees.

As the floodings are identical on a node weighted graph $G(\nu, nil)$ and on the derived graph $G(nil, \delta_{en}\nu)$, we are able to apply to node weighted graphs all methods developed for edge weighted graphs. This is the case of MBS based on minimum forests. A minimum forest of $G(nil, \delta_{en}\nu)$ rooted in the markers is constructed and may be easily adjusted if one adds or suppresses a marker.

We have proved the validity of the popular method for marker based segmentation, sometimes called swamping: compute the highest flooding of the graph $G(\nu, nil)$ under a ceiling function equal to 0 on the markers and equal to O on all other nodes [12]. Any watershed partition of this function constitutes a valid marker based segmentation and vice versa. However, this method is not directly applicable to edge weighted graphs. If τ is the flooding distance to the markers in the graph $G(nil, \eta)$, one has first to prune the graph $G(\tau, \eta)$ by suppressing the non flooding edges. The MBS of $G(nil, \eta)$ are then identical with the watershed partitions of $\widetilde{G}(\tau, nil)$. We have thus shown that marker based segmentation for both node or edge weighted graph boils down to the construction of a watershed partition of a node weighted graph, for which fast and unbiased algorithms exist (such as [5]); they permit to divide and share among the markers the large lakes created by the flooding.

My first significant contribution to the field of mathematical morphology was to regularize the watershed segmentation by introducing markers. After some 40 years, this paper closes a long loop...

References

1. Beucher, S., Lantuéjoul, C.: Use of watersheds in contour detection. In: Proceedings of the International Workshop on Image Processing, Real-Time Edge and Motion Detection/Estimation (1979)
2. Gondran, M.: Path algebra and algorithms. In: Roy, B. (ed.) Combinatorial Programming: Methods and Applications. ASIC, vol. 19, pp. 137–148. Springer, Dordrecht (1975). https://doi.org/10.1007/978-94-011-7557-9_6
3. Hu, T.C.: The maximum capacity route problem. Oper. Res. **9**(6), 898–900 (1961)
4. Lantuéjoul, C., Beucher, S.: On the use of the geodesic metric in image analysis. J. Microsc. **121**, 39 (1981)
5. Meyer, F.: Un algorithme optimal de ligne de partage des eaux. In: Proceedings 8ème Congrès AFCET, Lyon-Villeurbanne, pp. 847–857 (1991)
6. Meyer, F.: Minimum spanning forests for morphological segmentation. In: Serra, J., Soille, P. (eds.) Mathematical Morphology and Its Applications to Image Processing. CIVI, vol. 2, pp. 77–84. Springer, Dordrecht (1994). https://doi.org/10.1007/978-94-011-1040-2_11
7. Meyer, F.: Flooding and segmentation. In: Goutsias, J., Vincent, L., Bloomberg, D.S. (eds.) Mathematical Morphology and its Applications to Image and Signal Processing. CIVI, vol. 18, pp. 189–198. Springer, Boston (2002). https://doi.org/10.1007/0-306-47025-X_21
8. Meyer, F.: Flooding edge or node weighted graphs. In: Hendriks, C.L.L., Borgefors, G., Strand, R. (eds.) ISMM 2013. LNCS, vol. 7883, pp. 341–352. Springer, Heidelberg (2013). https://doi.org/10.1007/978-3-642-38294-9_29
9. Meyer, F.: Watersheds on weighted graphs. Pattern Recognit. Lett. **47**, 72–79 (2014)
10. Meyer, F.: Topographical Tools for Filtering and Segmentation 1: Watersheds on Node- or Edge-Weighted Graphs. ISTE/Wiley, London/Hoboken (2019)
11. Meyer, F.: Topographical Tools for Filtering and Segmentation 2: Floodings and Marker-Based Segmentation on Node- or Edge-Weighted Graphs. ISTE/Wiley, Etats-Unis/Newark (2019)
12. Meyer, F., Beucher, S.: Morphological segmentation. J. Vis. Commun. Image Represent. **1**(1), 21–46 (1990)

13. Moore, E.F.: The shortest path through a maze. In: Proceedings of the International Symposium on Theory of Switching, vol. 30, pp. 285–292 (1957)
14. Vincent, L., Soille, P.: Watersheds in digitalspaces: an efficient algorithm based on immersion simulations. IEEE Trans. Pattern Anal. Mach. Intell. **1**, 583 (1991)

Fast Marching Based Superpixels Generation

Kaiwen Chang and Bruno Figliuzzi[✉]

Centre for Mathematical Morphology, Mines ParisTech, PSL Research University,
35 rue Saint-Honoré, 77300 Fontainebleau, France
bruno.figliuzzi@mines-paristech.fr

Abstract. In this article, we present a fast-marching based algorithm for generating superpixel (FMS) partitions of images. The idea behind the algorithm is to draw an analogy between waves propagating in an heterogeneous medium and regions growing on an image at a rate depending on the local color and texture. The FMS algorithm is evaluated on the Berkeley Segmentation Dataset 500. It yields results in term of boundary adherence that are comparable to the ones obtained with state of the art algorithms including SLIC or ERGC, while improving on these algorithms in terms of compactness and density.

Keywords: Superpixel segmentation · Eikonal equation ·
Fast marching algorithm

1 Introduction

Superpixel algorithms refer to a class of techniques designed to partition an image into small regions that group perceptually similar pixels. Superpixel representations are generally more meaningful than raw pixel representations and easier to analyze. One of the advantages of using superpixels is notably that the number of superpixels is much smaller than the number of pixels in the image, which leads to a reduction in the amount of calculations required for further processing. In addition, thanks to their homogeneity, superpixels constitute subregions of the image on which it is particularly relevant to compute features. As a consequence, superpixel algorithms are commonly employed as a pre-processing step in a number of applications, including depth estimation [1], object classification or image segmentation [2].

Defining requirements for a "good" superpixel segmentation algorithm is an ambiguous task [3]. Several criteria are nevertheless commonly retained in the literature to evaluate the quality of a superpixel partition. First, it is generally admitted that superpixels should preserve the boundaries in the image [3,4]. This property, referred to as boundary adherence, is for instance of paramount importance for segmentation applications. Several metrics have been defined in the literature to quantify boundary adherence, including *boundary recall*, which

© Springer Nature Switzerland AG 2019
B. Burgeth et al. (Eds.): ISMM 2019, LNCS 11564, pp. 350–361, 2019.
https://doi.org/10.1007/978-3-030-20867-7_27

characterizes the proximity between the superpixels contours and the boundaries in the image, and *undersegmentation error*, which quantifies the "leakage" of the superpixels crossing an actual boundary of the image. The shape of the superpixels returned by a segmentation algorithm also constitutes an important characteristic of the algorithm. Obviously, algorithms returning superpixels with highly tortuous contours are more likely to exhibit high adherence to boundaries [3,5], but the resulting partition can hardly be considered to be satisfactory. Criteria including *average compactness* or *superpixels contours density* are classically used to characterize the shape of the superpixels [3,5]. Finally, for real-time applications, in addition to these requirements, superpixels should also be fast to compute and memory efficient.

An extensive amount of literature in computer vision and image analysis is dedicated to the topic of superpixels. We refer the readers interested by a review of the field to the articles [3,4]. Several approaches have been considered in the literature to construct superpixel partitions. Graph based methods build upon a graph representation of the image. In this representation, each pixel constitutes a node of the graph and all couples of adjacent pixels are linked by an edge. Classical graph-based algorithms for generating superpixels include the normalized cut algorithm (NC) of Shi and Malik [6,7] or the efficient graph-based segmentation algorithm (GS) of Felzenswalb and Huttenlocher [8]. By contrast, clustering based methods for superpixel segmentation proceed by iteratively refining clusters of pixels until some convergence criterion is met. Clustering based methods notably include mean shift [9], watershed [10–12], turbopixel [13] or waterpixel [5] algorithms, respectively.

The Simple Linear Iterative Clustering (SLIC) algorithm proposed by Achanta *et al.* [4,14] is an archetypal example of clustering based algorithm. Due to its ease of use and its good performance, SLIC is ranked among the most used algorithms for computing a superpixel segmentation. In addition, the SLIC algorithm has a linear complexity, its implementation is available [4] and it offers the possibility to weight the trade-off between boundary adherence and shape or size similarity. SLIC constructs a superpixel partition by applying a K-means clustering algorithm on local patches of the image. During initialization, K cluster centers locations are selected on the image using a grid with uniform spacing S. Each pixel is then associated to the closest cluster center in the image according to a distance involving the color proximity and the physical distance between the pixel and the seed, respectively. The search for similar pixels is restricted to a neighborhood of size $2S$ by $2S$ around each cluster center. After the assignment step, the K cluster centers are updated. The color and the location associated to the cluster centers are set equal to the average color and location of the pixels of the cluster. The L^2 distance between the previous and the new locations is used to compute a residual error E. The aforementioned procedure is iterated until the error E converges. The clusters of pixels obtained after this procedure are usually not connected. A post-processing step is therefore applied to reassign the disjoint pixels to nearby superpixels.

Most superpixel algorithms make use of color and spatial information. Recently, several studies have demonstrated that using texture information could improve the superpixel performance. In 2017, Xiao et al. [15] proposed to introduce both texture and gradient distance terms in SLIC's distance formula. In their work, the weight of each term is adapted to the discriminability of the features in the image. Their algorithm achieves better performances than state of art algorithms including SLIC and LSC [16]. Giraud et al. [17] proposed to measure the texture distance between a pixel and a superpixel seed by considering the average distance between a square patch around that pixel and similar patches inside the superpixel found by a nearest neighbor method.

In this article, we introduce a novel clustering-based algorithm for generating a superpixel partition of a given image termed fast-marching based superpixels (FMS). Following an idea originally introduced for the Eikonal-based region growing for efficient clustering algorithm (ERGC) of Buyssens et al. [18–20], we rely upon the Eikonal equation and the fast marching algorithm to assign the pixels of the images to the relevant clusters. The advantage of FMS when compared to SLIC is that no post-processing is required to obtain connected superpixels. The outline of the article is as follows. In Sect. 2, we present the FMS algorithm as well as its practical implementation. In Sect. 3, we use the Berkeley segmentation dataset 500 to illustrate the algorithm and we compare its results to the ones obtained with SLIC and ERGC. Conclusions are drawn in the last section.

2   Algorithm

The idea behind the FMS algorithm is to draw an analogy between waves propagating in an heterogeneous medium and regions growing on an image at a rate depending on the local color and texture. A similar idea was proposed by Buyssens et al. in 2014 [18–20]. However, both approaches differ in several aspects, including the expression for the local velocity as a function of the image content, the use of texture features or the region update during propagation.

Eikonal equation is a non-linear partial differential equation describing the propagation of waves in a domain $\Omega \subset \mathbb{R}^2$. At each point x of the domain, the local velocity is denoted $F(x)$. For all x in Ω, Eikonal equation reads:

$$\begin{cases} ||\nabla U(x)|| = \dfrac{1}{F(x)} & \forall x \in \Omega \\ U(x) = 0 & \forall x \in \partial\Omega, \end{cases} \quad (1)$$

where $\partial\Omega$ denotes the boundary of the domain Ω. The solution $U(x)$ to Eikonal equation can conveniently be interpreted as the minimal travelling time required for the wave to travel from $\partial\Omega$ to point x. Fast computational algorithms are available to approximate the solution of Eikonal equation, including the fast marching algorithm [21, 22].

In what follows, we adopt the following notations. A pixel in image $\mathcal{I}$ is denoted by p and its coordinates by (x, y). We denote by $\mathbf{C}(p)$ the color at pixel

p in the CIELAB color space. Similarly, we denote by $\mathbf{T}(p)$ a vector of features characterizing the local texture at pixel p. We initialize the algorithm by selecting N *seeds* or *cluster centers* locations $\{s_1, s_2, ..., s_N\}$. Then, we associate a velocity field $F_i(p)$ to each seed s_i depending on the color and the texture of both the seed and the pixel location $p := (x, y)$. The labels of the seeds are then gradually propagated from the labeled pixels to the unlabeled pixels according to the local velocities $\{F_i(x)\}_{i=1,...,N}$. On the domain defined by the image, Eikonal equation therefore reads

$$\begin{cases} ||\nabla U(p)|| = \dfrac{1}{F(p)} & \forall p \in I \\ U(p) = 0 & \forall p \in \Gamma. \end{cases} \tag{2}$$

In this expression, $F(p)$ is the local velocity at pixel p, Γ corresponds to the subset of the image $\mathcal{I}$ constituted by the seeds $\{s_1, s_2, ..., s_N\}$ and $U(p)$ is the minimal traveling time from Γ to p.

2.1 Region Growing

In this section, we describe in greater details the implementation of the algorithm. Let us denote by K the desired number of superpixels. We initialize the algorithm by selecting N ($N < K$) seeds on a regular grid. To avoid placing a seed on a boundary, we place the seed at the local minimum of the gradient in a 3×3 neighborhood of the nodes of the grid. A similar strategy is used for selecting the initial seeds in the SLIC algorithm [4]. The propagation algorithm is initialized as follows:

1. The traveling time map U is initialized:

$$U(p) = \begin{cases} 0 \text{ if } p \in \Gamma, \\ +\infty \text{ otherwise.} \end{cases} \tag{3}$$

2. The label map is initialized:

$$L(p) = \begin{cases} i \text{ if } p = s_i, \\ 0 \text{ otherwise.} \end{cases} \tag{4}$$

3. The pixels $\{s_1, .., s_N\}$ are grouped in a set referred to as the *narrow band*. All other pixels are labeled as *far*.

At each iteration, we extract the pixel $\tilde{p} := (X, Y)$ in the narrow band with the smallest arrival time and we label it as *frozen*. Then, we compute the arrival time for all neighbor pixels that are not labelled as frozen. To that end, we use the velocity field corresponding to the label $L := L(\tilde{p})$ associated to pixel $\tilde{p}$. Note that frozen pixels are used to compute the arrival times in other pixels, but their arrival times are never recomputed. Hence, the arrival time at a neighbor $p := (x, y)$ of $\tilde{p}$ is computed by solving Eikonal equation

$$||\nabla U(x, y)|| F_L(x, y) = 1. \tag{5}$$

To approximate the gradient term, we rely on the formula

$$
\begin{aligned}
||\nabla U(x,y)||^2 &= \max(U_{x,y} - U_{x,y+1}, U_{x,y} - U_{x,y-1}, 0)^2 \\
&+ \max(U_{x,y} - U_{x+1,y}, U_{x,y} - U_{x-1,y}, 0)^2
\end{aligned}
\tag{6}
$$

Note that in this expression, we only consider the neighbors with label L. If a neighbor of p has a label distinct than L, then we consider its associated arrival time to be equal to $+\infty$. We introduce this slight modification of the original fast marching algorithm to efficiently keep track of the labels propagation. Using the gradient discretization (6), Eikonal equation becomes

$$
\begin{aligned}
&\max(U_{x,y} - U_{x,y+1}, U_{x,y} - U_{x,y-1}, 0)^2 \\
&+ \max(U_{x,y} - U_{x+1,y}, U_{x,y} - U_{x-1,y}, 0)^2 = \frac{1}{F_L(x,y)^2}.
\end{aligned}
\tag{7}
$$

In this equation, the only unknown is the arrival time $U_{x,y}$ at pixel $p := (x,y)$. We have

$$
\begin{aligned}
&\max(U_{x,y} - U_{x,y+1}, U_{x,y} - U_{x,y-1}, 0)^2 = \\
&(U_{x,y} - \min(U_{x,y+1}, U_{x,y-1}))^2.
\end{aligned}
\tag{8}
$$

Hence, $U_{x,y}$ is solution of the quadratic equation

$$
\begin{aligned}
&(U_{x,y} - \min(U_{x,y+1}, U_{x,y-1}))^2 + \\
&(U_{x,y} - \min(U_{x+1,y}, U_{x-1,y}))^2 = \frac{1}{F_L(x,y)^2}.
\end{aligned}
\tag{9}
$$

Equation (9) has two distinct solutions. To respect the consistency of the scheme, the arrival time must be higher than the arrival time $U(\tilde{p})$ associated to the point selected in the narrow band. Hence, the arrival time $U_{x,y}$ is necessarily the largest solution of (9). At this point, we can encounter two distinct situations:

- When the neighbor point (x,y) is already in the narrow band, it has necessarily been associated an arrival time $U_{x,y}^{old}$. If $U_{x,y} < U_{x,y}^{old}$, then we affect the arrival time $U_{x,y}$ to (x,y) as well as the label $L := L(\tilde{p})$ of point $\tilde{p}$. On the contrary, if $U_{x,y} > U_{x,y}^{old}$, then the label and the arrival time at (x,y) remain unchanged.
- When the neighbor point (x,y) is not in the narrow band, we affect to it the arrival time $U_{x,y}$, the label $L := L(\tilde{p})$ and we add it to the narrow band.

At each iteration of the algorithm, it is necessary to extract the element of the narrow band with the smallest arrival time. To reduce the complexity of the algorithm, the elements of the narrow band are stored in a binary heap. We refer the reader interested by more details on the fast marching algorithm implementation to the original articles [21,22]. The algorithm stops when the narrow band is empty. At that point, each pixel of the image has necessarily been labeled, yielding a superpixel partition of $\mathcal{I}$.

2.2   Local Velocity Model

The algorithm described in the previous section can be used with numbers of non-negative velocity models $F(p)$. We describe in this section the velocity model used in this article, which relies upon both color and texture information. Natural images usually contain textured regions composed of repeated texels with strong color changes. A pixel in a textured region can therefore exhibit a strong color difference with the cluster center. To account for texturedness, we consider a velocity model incorporating both local color and texture information. Hence, we consider the distance between a pixel p and a seed s_i to be

$$D(p, s_i) = w_0 \|\boldsymbol{C}_p - \boldsymbol{C}_i\|_2 + w_1 \|\boldsymbol{T}_p - \boldsymbol{T}_i\|_2, \tag{10}$$

where $\boldsymbol{T}_p$ is a vector of features characterizing the local texture at pixel p, $\boldsymbol{T}_i$ is the corresponding vector of features for pixel s_i, and w_0 and w_1 are positive coefficients weighting color and texture contributions, respectively.

To obtain the texture features, we compute the local response of the image with a bank of Gabor filters [23]. The kernel of the Gabor filters is given by the expression

$$g(x, y; f_0, \theta, \phi, \sigma_{x'}, \sigma_{y'}) = \exp\left\{-\frac{1}{2}\left[\frac{x'^2}{\sigma_{x'}^2} + \frac{y'^2}{\sigma_{y'}^2}\right]\right\} \cos(2\pi f_0 x') \tag{11}$$

where $x' = x\cos\theta + y\sin\theta$, $y' = -x\sin\theta + y\cos\theta$, f_0 is the frequency of the sinusoidal wave along x' axis, θ is the angle between the x' and x axes, and σ_x, σ_y are the standard deviation of the gaussian filter in the space domain along x' and y' axes. The relation between the frequency resolution and the space resolution is $\sigma_u = 1/(2\pi\sigma_{x'})$, $\sigma_v = 1/(2\pi\sigma_{y'})$. The vector $\boldsymbol{T}_p$ stacks the 8 responses of the Gabor filtering at pixel p.

The local velocity $F_i(p)$ associated with the propagation of region i is finally obtained by relying on the exponential kernel

$$F(p, i) = \exp(-w_0 \|\boldsymbol{C}_p - \boldsymbol{C}_i\|_2 - w_1 \|\boldsymbol{T}_p - \boldsymbol{T}_i\|_2). \tag{12}$$

The velocity is maximal and equal to 1 when the pixel and the seed share the same color and texture characteristics.

2.3   Refinement

Images usually contain regions of interest over a very large range of scales. Hence, selecting the seeds according to a grid with uniform spacing might not be the optimal strategy to obtain a superpixel segmentation with good boundary adherence. In this section, we present a strategy built on the map of arrival time to refine the superpixel segmentation.

We recall that the number of desired superpixels is denoted by K. The algorithm presented in the previous sections yields a partition of the image into N connected segments corresponding to the initial seeds $\{s_1, s_2, ..., s_N\}$, as well as a map $U(p)$ storing the arrival time at each pixel p. The map $U(p)$ contains a significant amount of information that can be used to refine the superpixel partition. When the arrival time is high at some pixel p in a region $\mathcal{R}_i$ associated to seed s_i, it means that the region boundary propagating from s_i has traveled through pixels that are highly dissimilar to s_i. The arrival time found in each region $\mathcal{R}_i$ therefore constitute an interesting criterion to assess whether or not this region should be subsequently splitted. A similar refinement strategy is used in the article [19], where the distance map is used at the end of the over segmentation to generate new superpixels for refinement.

The refinement is conducted as follows. Let us denote by $\mathcal{B}$ the set of pixels belonging to the superpixel boundaries and by $\delta(\mathcal{B})$ the dilated set of $\mathcal{B}$ by a disk of radius 2 pixels. Then, the maximal arrival time in region $\mathcal{R}_i$ is defined to be

$$t_i = \max_{p \in \mathcal{R}_i \cap \delta(\mathcal{B})^c} U(p), \qquad (13)$$

where $\delta(\mathcal{B})^c$ is the complementary set of $\delta(\mathcal{B})$. To further refine the superpixel segmentation, we select the k regions with highest t_i and we add seeds at the corresponding locations before re-propagating. We exclude the pixels that belong to the region $\delta(\mathcal{B})$ to avoid implanting a seed directly on a boundary. This procedure is repeated iteratively until the desired number K of superpixels is obtained and enables to significantly improve the boundary adherence of the resulting superpixel partition.

3   Experiments and Discussion

In this section, we evaluate the performance of the FMS algorithm and compare it with two state of the-art superpixel algorithms, SLIC [4] and ERGC [18–20]. To that end, we use the Berkeley Segmentation Dataset 500 (BSDS500) [24]. This dataset contains 500 images and provides several ground truth manual segmentations for each image.

3.1   Metrics

To evaluate the performance of the algorithms, we adopt four metrics: boundary recall, undersegmentation error, compactness and contour density.

- *Boundary recall* is the ratio of the number of true positive contour pixels with regard to the number of contour pixels in the ground truth segmentation. To tolerate small localization errors, contour pixels that lie within 2 pixels from a real contour pixel are considered as true positives. High boundary recall means good adherence to boundaries.
- *Undersegmentation error* measures the leakage of a superpixel overlapping with a gound truth segment. We adopt the formulation proposed by Neubert and Protzel in [25].

$$\mathrm{UE}_{NP}(G, S) = \frac{1}{N} \sum_{G_i} \sum_{S_j \cap G_i \neq \emptyset} \min \left\{ \mid S_j \cap G_i \mid, \mid S_j - S_j \cap G_i \mid \right\} \qquad (14)$$

where S_j is a superpixel and G_i is a gound truth segment. We take the minimum of the number of overlapping pixels, and the number of non overlapping pixels within S_j as the leakage.

- *Compactness* is defined as the ratio of the region area with respect to a circle with the same perimeter as the superpixel, weighted by the ratio of pixel numbers inside the region [26].

$$COMP(S_i) = \sum_{S_i} \frac{|S_i|}{N} \frac{4\pi A(S_i)}{P(S_i)^2} \tag{15}$$

High compactness indicates a high regularity of the superpixel shapes, which is desirable for some applications.

- *Contour density* was first proposed in [5]. It is defined as

$$CD = \frac{|C|}{N} \tag{16}$$

where $|C|$ is the total number of pixels on superpixel contours and N is the number of pixels in the image. A low contour density tends to indicate that the tortuosity of the superpixels contours remains small. This is a desirable effect, since tortuous contours tend to artificially increase boundary recall.

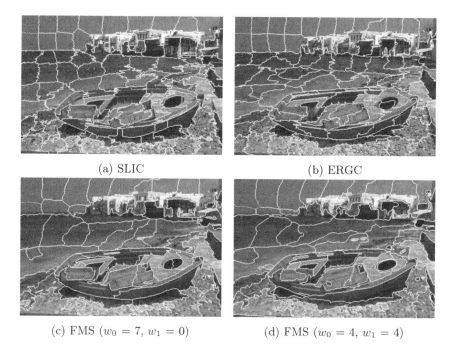

(a) SLIC   (b) ERGC

(c) FMS ($w_0 = 7$, $w_1 = 0$)   (d) FMS ($w_0 = 4$, $w_1 = 4$)

Fig. 1. Illustration of the superpixel segmentation on an image of the BSDS500

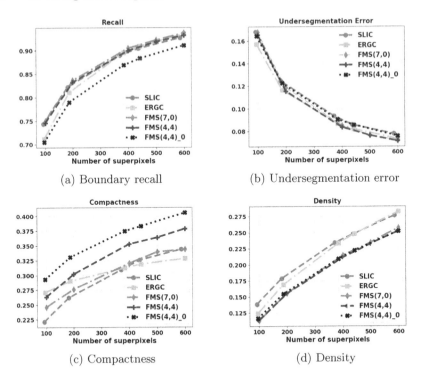

(a) Boundary recall

(b) Undersegmentation error

(c) Compactness

(d) Density

Fig. 2. Comparison between SLIC, ERGC and FMS algorithms on the BSDS500

3.2 Result and Discussion

Figure 2 displays the recall, undersegmentation error, compactness and density of FMS, SLIC and ERGC averaged over the 500 images of BSDS500. For a fair comparison, the parameters of SLIC and ERGC are optimized by the code of Stutz in [3]. To run the FMS algorithm, we selected four directions ($0°$, $45°$, $90°$ and $135°$) for the Gabor kernels of the filter banks. The frequency was set equal to $f_0 = 0.1$. Two kernels with standard deviation σ set equal to 2 and 4 are used to perform the filtering. Overall, this results in a bank of filters containing 8 kernels. The number of initial seeds N is set to $0.6K$ through experiment. The superpixels constructed with these algorithms are illustrated in Fig. 1.

The parameters w_0 and w_1 chosen for weighting the color and texture distances must be selected with caution. Overall, increasing w_0 and w_1 lead to an increase of the boundary recall. This is due to the fact that the propagation velocity becomes more sensitive to strong color and texture variations. However, for high values of the weights e.g. $w_0, w_1 > 8$, the local velocity can fall to very low values. This has the unwanted effect to deteriorate the compactness of the obtained superpixels and leads to the obtention of many small isolated super-pixels corresponding to small contrasted part of the image. Hence, the choice of parameters results from a trade-off between boundary adherence performance and topological considerations. In our experiments, we selected the parameters $w_0 = 4$ and $w_1 = 4$ by relying on a grid search to obtain an optimal boundary adherence. We also computed the superpixel partition using the parameters $w_0 = 7$ and $w_1 = 0$ to evaluate the influence of texture.

In term of recall, we can note that the FMS algorithm yields slightly better results than SLIC and ERGC algorithms. The refinement strategy also significantly improves the results. We can note that the results are better with the set of parameters $w_0 = 7$ and $w_1 = 0$ than with $w_0 = 4$ and $w_1 = 4$. The undersegmentation error is comparable to the one obtained with SLIC and ERGC. ERGC is the best performer when the number of superpixels N remains below 400, but FMS exhibits a slightly lower undersegmentation error when $N > 400$. The main advantage of the FMS algorithm is that it provides good topological metrics: the compactness of FMS is higher than the one of SLIC and ERGC, and the density of contours is significantly lower. In particular, the version of FMS with parameters $w_0 = 4$ and $w_1 = 4$ yields a boundary recall and an undersegmentation error similar to ERGC and SLIC, but with higher compactness. In addition, due to the refinement procedure, the size of the superpixels constructed with the FMS algorithm is locally adapted depending on the content of the image. One should nevertheless keep in mind that the refinement is performed at the expense of the superpixels uniformity: the refinement yields superpixels are heterogeneous in scale and size.

In terms of complexity, if N is the number of pixels in the image, then the complexity is $O(N \log N)$ at each propagation step. Hence, if there are L refinement steps, the complexity of the algorithm is in $O(LN \log N)$.

Finally, it is interesting to point out the main differences between the FMS and the ERGC algorithms [18–20]. ERGC is a clustering-based algorithm that computes superpixel partitions by relying upon the Eikonal equation. A significant difference between both algorithms is that their local velocity models differ. For ERGC, the local velocity field is simply given by:

$$F(p, i) = \frac{1}{\|C_p - C_i\|_2}. \tag{17}$$

In particular, the velocity is very high when the pixel and the seed have similar color characteristics. Another important difference is that ERGC only considers local color information to compute the local velocity, while FMS also relies on texture information. Finally, an additional difference between both approaches is that the cluster centers are iteratively updated according to the new pixels entering the clusters. By contrast, in the FMS algorithm, the cluster centers remain fixed.

4   Conclusion

In this article, we presented a fast-marching based algorithm for generating superpixel partitions of images. The FMS algorithm was evaluated on the Berkeley Segmentation Database 500. It yields results in term of boundary recall and undersegmentation error that are comparable to the ones obtained with state of the art algorithms including SLIC or ERGC, while improving on these algorithms in terms of compactness and density. A possible extension of this work could be to incorporate gradient information in the local velocity model to better account for the presence of contours and discontinuity in the image. It would also be of interest to try to automatically estimate optimal parameters w_0 and w_1 depending on the processed image.

References

1. Zitnick, C.L., Kang, S.B.: Stereo for image-based rendering using image oversegmentation. Int. J. Comput. Vis. **75**(1), 49–65 (2007)
2. Fulkerson, B., Vedaldi, A., Soatto, S.: Class segmentation and object localization with superpixel neighborhoods. In: 2009 IEEE 12th International Conference on Computer Vision, pp. 670–677. IEEE (2009)
3. Stutz, D., Hermans, A., Leibe, B.: Superpixels: an evaluation of the state-of-the-art. Comput. Vis. Image Underst. (2017)
4. Achanta, R., Shaji, A., Smith, K., Lucchi, A., Fua, P., Süsstrunk, S.: SLIC superpixels compared to state-of-the-art superpixel methods. IEEE Trans. Pattern Anal. Mach. Intell. **34**(11), 2274–2282 (2012)
5. Machairas, V., Decenciere, E., Walter, T.: Waterpixels: superpixels based on the watershed transformation. In: 2014 IEEE International Conference on Image Processing (ICIP), October, pp. 4343–4347 (2014)
6. Shi, J., Malik, J.: Normalized cuts and image segmentation. IEEE Trans. Pattern Anal. Mach. Intell. **22**(8), 888–905 (2000)
7. Malik, J., Belongie, S., Leung, T., Shi, J.: Contour and texture analysis for image segmentation. Int. J. Comput. Vis. **43**(1), 7–27 (2001)
8. Felzenszwalb, P.F., Huttenlocher, D.P.: Efficient graph-based image segmentation. Int. J. Comput. Vis. **59**(2), 167–181 (2004)
9. Comaniciu, D., Meer, P.: Mean shift: a robust approach toward feature space analysis. IEEE Trans. Pattern Anal. Mach. Intell. **24**(5), 603–619 (2002)
10. Vincent, L., Soille, P.: Watersheds in digital spaces: an efficient algorithm based on immersion simulations. IEEE Trans. Pattern Anal. Mach. Intell. **6**, 583–598 (1991)
11. Beucher, S., Meyer, F.: The morphological approach to segmentation: the watershed transformation. Opt. Eng. N.Y. Marcel Dekker Inc. **34**, 433–433 (1992)
12. Figliuzzi, B., Chang, K., Faessel, M.: Hierarchical segmentation based upon multi-resolution approximations and the watershed transform. In: Angulo, J., Velasco-Forero, S., Meyer, F. (eds.) ISMM 2017. LNCS, vol. 10225, pp. 185–195. Springer, Cham (2017). https://doi.org/10.1007/978-3-319-57240-6_15
13. Levinshtein, A., Stere, A., Kutulakos, K.N., Fleet, D.J., Dickinson, S.J., Siddiqi, K.: TurboPixels: fast superpixels using geometric flows. IEEE Trans. Pattern Anal. Mach. Intell. **31**(12), 2290–2297 (2009)
14. Achanta, R., Susstrunk, S.: Superpixels and polygons using simple non-iterative clustering. In: 2017 IEEE Conference on Computer Vision and Pattern Recognition (CVPR), Honolulu, HI, July, pp. 4895–4904. IEEE (2017)
15. Xiao, X., Gong, Y., Zhou, Y.: Adaptive superpixel segmentation aggregating local contour and texture features. In: 2017 IEEE International Conference on Acoustics, Speech and Signal Processing (ICASSP), March, pp. 1902–1906 (2017)

16. Li, Z., Chen, J.: Superpixel segmentation using linear spectral clustering. In: 2015 IEEE Conference on Computer Vision and Pattern Recognition (CVPR), June, pp. 1356–1363 (2015)
17. Giraud, R., Ta, V.-T., Papadakis, N., Berthoumieu, Y.: Texture-aware superpixel segmentation. arXiv:1901.11111 [cs], January 2019
18. Buyssens, P., Toutain, M., Elmoataz, A., Lézoray, O.: Eikonal-based vertices growing and iterative seeding for efficient graph-based segmentation. In: IEEE International Conference on Image Processing (ICIP 2014), Paris, France, October, p. 5 (2014)
19. Buyssens, P., Gardin, I., Ruan, S., Elmoataz, A.: Eikonal-based region growing for efficient clustering. Image Vis. Comput. **32**(12), 1045–1054 (2014)
20. Buyssens, P., Gardin, I., Ruan, S.: Eikonal based region growing for superpixels generation: application to semi-supervised real time organ segmentation in CT images. IRBM **35**(1), 20–26 (2014)
21. Sethian, J.A.: A fast marching level set method for monotonically advancing fronts. Proc. Nat. Acad. Sci. **93**(4), 1591–1595 (1996)
22. Sethian, J.A.: Fast marching methods. SIAM Rev. **41**(2), 199–235 (1999)
23. Jain, A.K., Farrokhnia, F.: Unsupervised texture segmentation using Gabor filters. Pattern Recogn. **24**(12), 1167–1186 (1991)
24. Arbelaez, P., Maire, M., Fowlkes, C., Malik, J.: Contour detection and hierarchical image segmentation. IEEE Trans. Pattern Anal. Mach. Intell. **33**(5), 898–916 (2011)
25. Neubert, P., Protzel, P.: Superpixel benchmark and comparison. In: Forum Bildverarbeitung 2010, pp. 205–218 (2012)
26. Schick, A., Fischer, M., Stiefelhagen, R.: Measuring and evaluating the compactness of superpixels. In: Proceedings of the 21st International Conference on Pattern Recognition (ICPR2012), November, pp. 930–934 (2012)

Characterization and Statistics of Distance-Based Elementary Morphological Operators

Arlyson Alves do Nascimento[1] and Marcos Eduardo Valle[2(✉)]

[1] Federal Institute of Education, Science and Technology of Alagoas,
Maceió, AL, Brazil
[2] Department of Applied Mathematics, University of Campinas,
Campinas, SP, Brazil
valle@ime.unicamp.br

Abstract. Mathematical morphology (MM) is a powerful theory widely used for image processing and analysis. Distance-based morphological operators are parametrized by a reference value so that a dilation enlarges regions of an image similar to the reference while an erosion shrinks them. In this paper, we first characterize distance-based erosions, dilations, and gradients as a function of the reference. Then, assuming the reference is a sample from a random variable, we use a descriptive statistic to obtain relevant information on a certain distance-based morphological operator. We validate our approach by presenting a successful application of some statistics of distance-based morphological operators for edge detection in color images.

Keywords: Mathematical morphology · Multivalued images ·
Distance-based order · Descriptive statistics · Edge detection

1 Introduction

Mathematical morphology (MM) is a powerful non-linear theory widely used for image processing and analysis [11,22]. The first morphological operators have been conceived by Matheron and Serra in the early 1960s for binary images [11]. Although binary MM has been effectively extended to gray-scale images using level-sets, umbra, and fuzzy set theory [24], the extension of MM to multivalued images is not general because there is no natural ordering for vectors [2]. Indeed, much research on multivalued MM focused on finding an appropriate ordering for a given image processing task [2,7,9,15,25,28]. A detailed account on multivalued MM can be found in [2,3,28].

In this paper, we focus on a distance-based approach which have been investigated and generalized by many researchers including [1,2,8,14,20,26–28]. Precisely, we focus on the distance-based approach to multivalued MM obtained by

This work was supported in part by CNPq under grant no. 310118/2017-4 and FAPESP under grant no. 2019/02278-2.

B. Burgeth et al. (Eds.): ISMM 2019, LNCS 11564, pp. 362–374, 2019.
https://doi.org/10.1007/978-3-030-20867-7_28

first comparing the distance between the value of a pixel and a reference value r followed by a lexicographical cascade to resolve ambiguities [2]. Although it has been extended using support-vector machines, krigging, and fuzzy set theory [20,27], we shall focus our attention in the simplest distance-based approach because it depends on a single meaningful parameter – the reference r. Precisely, we first characterize the outcome of an elementary distance-based morphological operator as a function of the reference. Then, we address the case in which we are uncertain about the appropriate reference value r. Using probability theory, we assume that the reference is modeled by a random variable. Furthermore, to obtain relevant information on the distance-based morphological operators, we use descriptive statistics such as the expectation, the mode, or some measures of dispersion. For illustrative purpose, we apply some statistics of distance-based morphological operators for edge detection in color image and, despite the lack of interpretation, we obtain some promising results.

The paper is organized as follows: Next section reviews the distance-based approach and presents a characterization of distance-based erosions, dilations, and gradients. Statistics of the distance-based morphological operators are addressed in Sect. 4 while the computational experiment for edge detection is provided in Sect. 5. The paper finishes with some concluding remarks in Sect. 6.

2 Distance-Based Elementary Morphological Operators

A multivalued image can be modeled as a mapping $f : \mathcal{D} \to \mathbb{V}$, from a point set $\mathcal{D}$ into a vector-valued space $\mathbb{V}$. In this paper, we assume that the point set is $\mathcal{D} \subseteq \mathbb{Z}^2$. Also, let us assume that the value set $\mathbb{V} \subset \mathbb{R}^n$. Additional requirements on $\mathbb{V}$ will be included accordingly throughout the paper.

Although MM can be defined in more general mathematical structures such as sponges [10] and inf-lattices [12], in this paper we focus on a particular but very useful approach to multivalued MM based on complete lattices [11,19]. Recall that a partially ordered set $(\mathbb{L}, \leq)$ is a complete lattice if every subset $X \subseteq \mathbb{L}$ has an infimum and a supremum on $\mathbb{L}$, denoted respectively by $\bigwedge X$ and $\bigvee X$ [4].

Like the classical approaches to MM, erosions and dilations of a lattice-valued image $f : \mathcal{D} \to \mathbb{V}$ can be defined in terms of a structuring element B. The structuring element B is used to extract relevant information about the shape and form of objects in the probed image f [23]. The two elementary non-linear, translation invariant morphological operators for lattice-valued images can defined as follows using a structuring element: Let the value set $\mathbb{V}$, equipped with a partial order "$\leq$", be a complete lattice. The erosion and the dilation of a lattice-valued image $f : \mathcal{D} \to \mathbb{V}$ by a structuring element B, denoted respectively by $\varepsilon_B(f)$ and $\delta_B(f)$, are the lattice-valued images given by

$$\varepsilon_B(f)(x) = \bigwedge f(B_x) \quad \text{and} \quad \delta_B(f)(x) = \bigvee f(B_x), \quad \forall x \in \mathcal{D}, \qquad (1)$$

where

$$f(B_x) = \{f(x+b) : b \in B, x+b \in \mathcal{D}\}, \qquad (2)$$

is the set of the values of the image $\boldsymbol{f}$ in the pixels selected by the structuring element B translated by $x \in \mathcal{D}$. We would like to recall that the elementary morphological operators given by (1) satisfies the following fundamental relationship from algebraic mathematical morphology called adjunction [11]: Given two images $\boldsymbol{f}$ and $\boldsymbol{g}$, we have $\delta_B(\boldsymbol{f}) \leq \boldsymbol{g}$ if and only if $\boldsymbol{f} \leq \varepsilon_{\check{B}}(\boldsymbol{g})$, where $\check{B} = \{-b : b \in B\}$ is the reflected structuring element.

The fundamental requirement in (1) is a partial order with well defined extreme operations. It turns out, however, that there is no natural ordering for vectors. Therefore, much research on multivalued MM have been focused on determining appropriate vector ordering schemes [2,7,9,15,25,28]. A detailed account on the various approaches to multivalued MM can be found in [2,3]. In the following, we focus on the distance-based approach examined by Angulo [2].

Briefly, a distance-based order is a reduced order in which the elements are ranked by comparing the distance to a certain reference value $\boldsymbol{r} \in \mathbb{V}$. In order to avoid ambiguities, the usual lexicographical cascade is used in the subsequent comparisons. Formally, suppose that $\mathbb{V} \subseteq \mathbb{R}$ is equipped with a metric $d : \mathbb{V} \times \mathbb{V} \to [0, +\infty)$. Given a reference value $\boldsymbol{r} \in \mathbb{V}$, a distance-based order is defined as follows for any $\boldsymbol{u} = (u_1, \ldots, u_n) \in \mathbb{V}$ and $\boldsymbol{v} = (v_1, \ldots, v_n) \in \mathbb{V}$:

$$\boldsymbol{u} \preccurlyeq_r \boldsymbol{v} \iff \begin{cases} d(\boldsymbol{v}, \boldsymbol{r}) <_{\mathbb{R}} d(\boldsymbol{u}, \boldsymbol{r}), \\ d(\boldsymbol{u}, \boldsymbol{r}) = d(\boldsymbol{v}, \boldsymbol{r}) \quad \text{and} \quad \boldsymbol{u} \leq_{\text{lex}} \boldsymbol{v}, \end{cases} \tag{3}$$

where the symbol "$\leq_{\text{lex}}$" denotes the usual lexicographical ordering, that is, $\boldsymbol{u} \leq_{\text{lex}} \boldsymbol{v}$ if and only if there exists $\ell \in \{1, \ldots, n\}$ such that $u_j = v_j$ for all $j < \ell$ and $u_\ell \leq_{\mathbb{R}} v_\ell$. Note that, using the distance-based order "$\preccurlyeq_r$", an element $\boldsymbol{v}$ is larger than or equal to another element $\boldsymbol{u}$ if $\boldsymbol{v}$ is closer to the reference $\boldsymbol{r}$ than $\boldsymbol{u}$. Furthermore, for any $\boldsymbol{u}, \boldsymbol{v} \in \mathbb{V}$ either $\boldsymbol{u} \preccurlyeq_r \boldsymbol{v}$ or $\boldsymbol{v} \preccurlyeq_r \boldsymbol{u}$ hold true. Hence, the relation "$\preccurlyeq_r$" given by (3) is a total order. Although a total order circumvent the false color problem [21], they are usually not regular with respect to a metric [6].

Suppose that $\mathbb{V} \subseteq \mathbb{R}^n$ is a complete lattice with the distance-based order given by (3). Accordingly, we define the distance-based erosion and dilation of a multivalued image $\boldsymbol{f} : \mathcal{D} \to \mathbb{V}$ by a structuring element B, denoted respectively by $\varepsilon_B^r(\boldsymbol{f})$ and $\delta_B^r(\boldsymbol{f})$, using (1) and (3). The dilation $\delta_B^r(\boldsymbol{f})(x)$ yields the element of $\boldsymbol{f}(B_x)$ closest to the reference $\boldsymbol{r}$. On the downside, the erosion $\varepsilon_B^r(f)$ does not have a simple interpretation [28]. Nevertheless, the distance-based morphological operators and their compositions have been effectively applied for image filtering, enhancement, and analysis [2]. In these applications, the reference $\boldsymbol{r} \in \mathbb{V}$ must be defined a priori by the user.

Apart from the two elementary operators, we define the distance-based gradient as the distance between dilation and erosion. The morphological gradient is given by

$$\varrho_B^r(\boldsymbol{f})(x) = d\big(\delta_B^r(\boldsymbol{f})(x), \varepsilon_B^r(\boldsymbol{f})(x)\big), \quad \forall x \in \mathcal{D}, \tag{4}$$

where d denotes a metric on $\mathbb{V}$. The morphological gradient can be used for image segmentation or edge detection. Since an effective application of the distance-based morphological operators depends on an appropriate choice of the reference $\boldsymbol{r} \in \mathbb{V}$, let us characterize them as a function of $\boldsymbol{r}$.

3   Characterization of Distance-Based Morphological Operators

The elementary distance-based morphological operators depend on the reference r. The effect of the reference on elementary distance-based morphological operators is related to the Voronoi diagram and its variation. Precisely, the distance-based dilation is directly related to the Voronoi diagram while the distance-based erosion is directly related to the farthest neighbor diagram. Using Voronoi's diagram and its variation, we can characterize the distance-based morphological operators as follows:

Theorem 1. *Let $\mathbb{V} \subseteq \mathbb{R}^n$, equipped with a metric $d : \mathbb{V} \times \mathbb{V} \rightarrow [0, +\infty)$, be a complete lattice with the distance-based order "$\preccurlyeq_r$" given by (3). Consider a multivalued image $\boldsymbol{f} : \mathcal{D} \rightarrow \mathbb{V}$ and a flat and finite structuring element B. Since B is finite, we can write $\boldsymbol{f}(B_x) = \{\boldsymbol{f}_1(x), \dots, \boldsymbol{f}_{k(x)}(x)\}$, where $k(x) \leq Card(B)$, for any $x \in \mathcal{D}$. Furthermore, let*

$$RV_\ell(x) = \{\boldsymbol{v} \in \mathbb{V} : d(\boldsymbol{f}_\ell(x), \boldsymbol{v}) <_\mathbb{R} d(\boldsymbol{f}_i(x), \boldsymbol{v}), \forall i \neq \ell\}, \tag{5}$$

and

$$RF_\ell(x) = \{\boldsymbol{v} \in \mathbb{V} : d(\boldsymbol{f}_j(x), \boldsymbol{v}) <_\mathbb{R} d(\boldsymbol{f}_\ell(x), \boldsymbol{v}), \forall j \neq \ell\}, \tag{6}$$

denote respectively the interior of the Voronoi region and the interior of the Voronoi region of the farthest neighbor which contains the value $\boldsymbol{f}_\ell(x) \in \boldsymbol{f}(B_x)$. The distance-based erosion, dilation, and gradient, given by (1) and (4), as a function of the reference $r \in \mathbb{V}$, satisfy the following equations for $x \in \mathcal{D}$:

$$\varepsilon_B^r(\boldsymbol{f})(x) = \boldsymbol{f}_i(x), \quad \text{if} \quad r \in RF_i(x), \tag{7}$$

$$\delta_B^r(\boldsymbol{f})(x) = \boldsymbol{f}_j(x), \quad \text{if} \quad r \in RV_j(x), \tag{8}$$

$$\varrho_B^r(\boldsymbol{f})(x) = d(\boldsymbol{f}_i(x), \boldsymbol{f}_j(x)), \quad \text{if} \quad r \in RF_i(x) \cap RV_j(x). \tag{9}$$

Proof. Due to page constrain, let us only show (7). The identity (8) can be derived in a similar fashion while (9) follows from (7) and (8). Let $\boldsymbol{f} : \mathcal{D} \rightarrow \mathbb{V}$ be a multivalued image and B a flat and finite structuring element. Given $x \in \mathcal{D}$, let us assume $r \in RF_i(x)$. From (6) we have $d(\boldsymbol{f}_j(x), r) < d(\boldsymbol{f}_i(x), r)$ for all $j \neq i$. Furthermore, from (3), we have $\boldsymbol{f}_i(x) \preccurlyeq_r \boldsymbol{f}_j(x)$ for all j. Thus, from (1), the following identities hold true where "$\curlywedge_r$" denotes the infimum obtained using the distance-based ordering "$\preccurlyeq_r$":

$$\varepsilon_B^r(\boldsymbol{f})(x) = \bigwedge_r \boldsymbol{f}(B_x) = \boldsymbol{f}_1(x) \curlywedge_r \boldsymbol{f}_2(x) \curlywedge_r \dots \curlywedge_r \boldsymbol{f}_{k(x)}(x) = \boldsymbol{f}_i(x).$$

We would like to point out that, for any $x \in \mathcal{D}$, $\varepsilon_B^r(\boldsymbol{f})(x)$ and $\delta_B^r(\boldsymbol{f})(x)$ assumes $k(x) \leq Card(\boldsymbol{f}(B_x))$ distinct values. In contrast, the gradient $\varrho_B^r(\boldsymbol{f})(x)$ admits at most $k(x)(k(x) - 1)/2$ values. The following example illustrates the distance-based morphological operators ε_B^r, δ_B^r, and ϱ_B^r as a function of r.

a) Voronoi diagram   b) Fartherst neighbor diagram

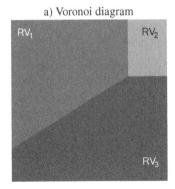 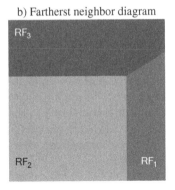

Fig. 1. Voronoi diagram and the diagram of the farthest neighbor of $\boldsymbol{f}(B_x)$.

Example 1. Consider $\mathbb{V} = [0,1]^2$ and let $\boldsymbol{f}(B_x) = \{\boldsymbol{f}_1(x), \boldsymbol{f}_2(x), \boldsymbol{f}_3(x)\}$, where $\boldsymbol{f}_1(x) = (0.6, 0.9)$, $\boldsymbol{f}_2(x) = (0.9, 0.9)$, and $\boldsymbol{f}_3(x) = (0.9, 0.4)$, be the set of values of a multivalued image $\boldsymbol{f} : \mathcal{D} \to \mathbb{V}$ selected by a finite structuring element B centered at $x \in \mathcal{D}$. The Voronoi regions $RV_1(x)$, $RV_2(x)$, and $RV_3(x)$ as well as the farthest neighbor regions $RF_1(x)$, $RF_2(x)$, and $RF_3(x)$ are shown in Fig. 1. Accordingly, the distance-based dilation and erosion satisfy respectively

$$\delta_B^r(\boldsymbol{f})(x) = \begin{cases} (0.6, 0.9), & r \in RV_1, \\ (0.9, 0.4), & r \in RV_2, \\ (0.9, 0.9), & r \in RV_3, \end{cases} \quad \text{and} \quad \varepsilon_B^r(\boldsymbol{f})(x) = \begin{cases} (0.6, 0.9), & r \in RF_1, \\ (0.9, 0.4), & r \in RF_2. \\ (0.9, 0.9), & r \in RF_3, \end{cases}$$

Furthermore, using the Euclidean distance, we obtain

$$\varrho_B^r(\boldsymbol{f})(x) = \begin{cases} 0.3, & r \in (RV_1 \cap RF_3) \cup (RV_3 \cap RF_1), \\ 0.5, & r \in (RV_2 \cap RF_3) \cup (RV_3 \cap RF_2), \\ 0.583, & r \in (RV_1 \cap RF_2) \cup (RV_2 \cap RF_1). \end{cases}$$

We would like to conclude this section remarking that Theorem 1 do not characterize the distance-based operators in the boundary ∂RV_ℓ of a Voronoi region and the boundary ∂RF_ℓ of a farthest neighbor region. According to (3), the values of $\varepsilon_B^r(\boldsymbol{f})(x)$, $\delta_B^r(\boldsymbol{f})(x)$, and $\varrho_B^r(\boldsymbol{f})(x)$ are determined by the lexicographical ordering if $r \in \partial RV_\ell$ or $r \in \partial RF_\ell$. Nevertheless, since the unions $\cup_{\ell=1}^{k(x)} \partial RV_\ell$ and $\cup_{\ell=1}^{k(x)} \partial RF_\ell$ of all the boundaries have both Lebesgue measure zero, the values of the reference r not covered by Theorem 1 are irrelevant in practical situations.

4   Statistics of Distance-Based Morphological Operators

In the last decades, different approaches to distance-based color MM have been proposed by prominent researchers including [1,2,8,14,20,26]. Despite successful applications, these approaches often face the difficult task of choosing a suitable reference for sorting vector values. In this section, we assume we are uncertain about the reference value $r \in \mathbb{V}$. Precisely, using probability theory, the uncertainty about the reference is expressed using a vector-valued random variable $\boldsymbol{R}$. Most importantly, we can use a descriptive statistic $\boldsymbol{S}$, such as a measure of central tendency or dispersion, to obtain relevant information on the distance-based morphological operators $\varepsilon_B^r(\boldsymbol{f})(x)$, $\delta_B^r(\boldsymbol{f})(x)$, and $\varrho_B^r(\boldsymbol{f})(x)$. Let us formalize this reasoning.

Let $\boldsymbol{f} : \mathcal{D} \to \mathbb{V}$ be a multivalued image and consider a multivariate random variable $\boldsymbol{R}$ on $\mathbb{V} \subset \mathbb{R}^n$. From Theorem 1, for each $x \in \mathcal{D}$, the distance-based operators ε_B^r and δ_B^r map a sample $r \in \mathbb{V}$ from $\boldsymbol{R}$ to an element of $\boldsymbol{f}(B_x) = \{\boldsymbol{f}_1(x), \ldots, \boldsymbol{f}_{k(x)}(x)\}$. Hence, given an image $\boldsymbol{f} : \mathcal{D} \to \mathbb{V}$, a structuring element B, and $x \in \mathcal{D}$, we can interpret $\varepsilon_B^r(\boldsymbol{f})(x)$ and $\delta_B^r(\boldsymbol{f})(x)$ as mappings from $\mathbb{V}$ to $\boldsymbol{f}(B_x)$. Precisely, we can define $\boldsymbol{H} : \mathbb{V} \to \{\boldsymbol{f}_1(x), \ldots, \boldsymbol{f}_{k(x)}\}$ by setting either $\boldsymbol{H}(\boldsymbol{r}) = \varepsilon_B^r(\boldsymbol{f})(x)$ or $\boldsymbol{H}(\boldsymbol{r}) = \delta_B^r(\boldsymbol{f})(x)$ for all $\boldsymbol{r} \in \mathbb{V}$. Furthermore, let $\boldsymbol{H}(\boldsymbol{R})$ denote the random variable obtained by applying $\boldsymbol{H}$ on $\boldsymbol{R}$ and let $\Pr[\boldsymbol{H}(\boldsymbol{R}) = \boldsymbol{f}_\ell(x)]$ denote the probability of $\boldsymbol{H}(\boldsymbol{R}) = \boldsymbol{f}_\ell(x)$, for $\ell \in \{1, \ldots, k(x)\}$. Using a descriptive statistic $\boldsymbol{S}$, we define new operators as follows in a point-wise manner where $r \in \mathbb{V}$ is interpreted as a sample from the random variable $\boldsymbol{R}$:

$$\boldsymbol{S}\varepsilon_B(\boldsymbol{f})(x) = \boldsymbol{S}\left(\varepsilon_B^r(\boldsymbol{f})(x)\right) \quad \text{and} \quad \boldsymbol{S}\delta_B(\boldsymbol{f})(x) = \boldsymbol{S}\left(\delta_B^r(\boldsymbol{f})(x)\right), \quad \forall x \in \mathcal{D}. \tag{10}$$

In the sequel, we review some vector-valued statistics used as measures of central tendency and dispersion of a (function of a) random variable [13,17].

The *expectation* of $\boldsymbol{H}(\boldsymbol{R})$, which is a measure of central tendency, is given by

$$E(\boldsymbol{H}(\boldsymbol{R})) = \sum_{j=1}^{k(x)} \boldsymbol{f}_j(x)\Pr\left[\boldsymbol{H}(\boldsymbol{R}) = \boldsymbol{f}_j(x)\right]. \tag{11}$$

As another measure of central tendency, the *mode* of $\boldsymbol{H}(\boldsymbol{R})$ is a value $\boldsymbol{f}_\ell(x)$ such that $\Pr[\boldsymbol{H}(\boldsymbol{R}) = \boldsymbol{f}_\ell(x)]$ is maximum.

The *scatter matrix* of $\boldsymbol{H}(\boldsymbol{R})$, denoted by $\boldsymbol{\Sigma}$, is determined by means of the equation

$$\boldsymbol{\Sigma} = E\left((\boldsymbol{H}(\boldsymbol{R}) - E(\boldsymbol{H}(\boldsymbol{R})))(\boldsymbol{H}(\boldsymbol{R}) - E(\boldsymbol{H}(\boldsymbol{R}))^T\right), \tag{12}$$

where $\boldsymbol{X}^T$ denotes the transpose of the column-vector $\boldsymbol{X}$. The scatter matrix $\boldsymbol{\Sigma}$ is used to define two measures of dispersion which generalize the notion of standard deviation. First, since $\boldsymbol{\Sigma}$ is a symmetric and semidefinite positive matrix, its determinant as well as its trace are non-negative real numbers. The square-root of the determinant of $\boldsymbol{\Sigma}$, denoted by $\text{std}_g(\boldsymbol{H}(\boldsymbol{R}))$, is called *generalized standard deviation* of $\boldsymbol{H}(\boldsymbol{R})$. The *total standard deviation* of $\boldsymbol{H}(\boldsymbol{R})$, denoted by $\text{std}_t(\boldsymbol{H}(\boldsymbol{R}))$, corresponds to the square-root of the trace of the scatter matrix $\boldsymbol{\Sigma}$. Both generalized and total standard deviation generalize the notion of standard deviation to multivalued random variables. Another measure of dispersion of the random variable $H(R)$ is the *Gini's mean difference*, denoted by $\text{gini}(H(R))$ and defined by

$$\text{gini}(H(R)) = \sum_{i,j=1}^{k} d(f_i(x), f_j(x)) \Pr[H(R) = f_i(x)] \Pr[H(R) = f_j(x)],$$
(13)

where d denotes a metric on $\mathbb{V} \subset \mathbb{R}^n$.

We would like to point out that a measure of central tendency of a distance-based morphological operator yields a vector-valued image. In contrast, a measure of dispersion of a distance-based morphological operator results a gray-scale image. In particular, note that the expectation of either a distance-based dilation or a distance-based erosion corresponds to a weighted average of the elements of $f(B_x)$. For example, the expectation of ε_B^r is given by the following spatially varying linear filter where $w(x,y) = \Pr[\varepsilon_B^r(f)(x) = f(y)]$ if $y \in B_x$ and $w(x,y) = 0$ otherwise:

$$E\varepsilon_B(f)(x) = \sum_{y \in \mathcal{D}} w(x,y) f(y), \quad \forall x \in \mathcal{D}.$$

In a similar way, we can compute statistics of the distance-based gradient $\varrho_B^r(f)(x)$ by interpreting the reference $r \in \mathbb{V}$ as a sample from the random variable R. In mathematical terms, given a descriptive statistic S, we define

$$S\varrho_B(f)(x) = S(\varrho_B^r(f)(x)), \quad \forall x \in \mathcal{D}.$$
(14)

Example 2. The aim of this example is to illustrate some statistics of distance-based morphological operators. Due to page constraints, let us focus on the distance-based dilation characterized in Example 1. Furthermore, let us assume that the random variable R is uniformly distributed on $\mathbb{V} = [0,1]^2$, that is, $R \sim U(\mathbb{V})$. From the maximum entropy principle, $R \sim U(\mathbb{V})$ means that we are totally ignorant about the choice of the reference used to define the distance-based morphological operators. Furthermore, when $R \sim U(\mathbb{V})$, we have $\Pr[\delta_B^r(f)(x) = f_\ell(x)] = \mu_L(RV_\ell)$, where $\mu_L(RV_\ell)$ denotes the Lebesgue measure (area, volume or hyper-volume) of the Voronoi region RV_ℓ. In this example, $\Pr[\delta_B^r(f)(x) = f_1(x)] = 0.43$, $\Pr[\delta_B^r(f)(x) = f_2(x)] = 0.09$, and $\Pr[\delta_B^r(f)(x) = f_3(x)] = 0.48$. Thus, we have

$$E\delta_B(f)(x) = 0.43f_1(x) + 0.09f_2(x) + 0.48f_3(x) = (0.77, 0.66),$$

and

$$\text{mode}\delta_B(f)(x) = f_3(x) = (0.9, 0.4).$$

Note that $E\delta_B(f)(x)$ is a weighted average of $f_1(x)$, $f_2(x)$, and $f_3(x)$. Moreover, we have $\text{mode}\delta_B(f)(x) = f_3(x)$, which means that $\text{mode}\delta_B(f)$ has no false colors. Similarly, we have

$$\text{std}_g\delta_B(f)(x) = 0, \quad \text{std}_t\delta_B(f)(x) = 0.29, \quad \text{and} \quad \text{gini}\delta_B(f)(x) = 0.15.$$

Accordingly, note that $E\delta_B(f)(x)$ and $\text{mode}\delta_B(f)(x)$ are vectors while $\text{std}_g\delta_B(f)(x)$, $\text{std}_t\delta_B(f)(x)$, and $\text{gini}\delta_B(f)(x)$ are scalars.

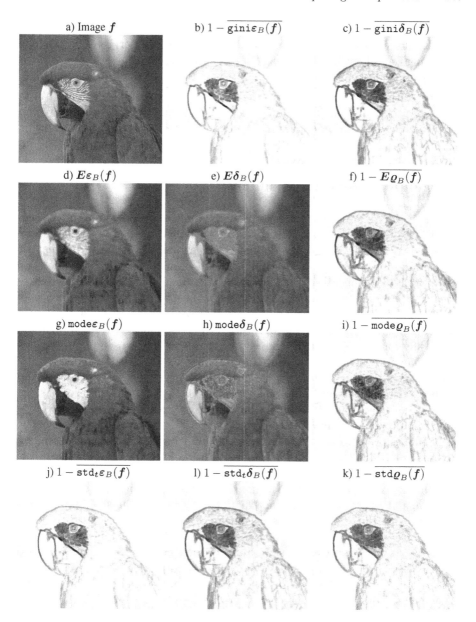

Fig. 2. Color image f and the corresponding statistics of distance-based morphological operators. (Color figure online)

Example 3. In this example, we provide a visual interpretation of some statistics of the distance-based morphological operators ε_B^r, δ_B^r, and $\overline{\varrho}_B^r$. To this end, consider the color image f depicted in Fig. 2(a). By considering the Euclidean distance and assuming $R \sim U(\mathbb{V}_{\mathrm{RGB}})$, which means we have no *a priori* information about the appropriate reference, we obtain the images shown in Fig. 2 using a 3×3 square structuring element. For a better visual interpretation, instead of showing a gradient-like image $g : \mathcal{D} \rightarrow [0, +\infty)$, we depicted the complement of its re-scaled version given by $(1 - \bar{g})(x) = 1 - \frac{g(x)}{\vee\{g(y):y\in\mathcal{D}\}}$, $\forall x \in \mathcal{D}$. Note that the statistics of the gradient as well as the measures of dispersion of $\varepsilon_B^r(\boldsymbol{f})$ and $\delta_B^r(\boldsymbol{f})$ resemble a gradient image. Like the morphological gradient, we can use these operators for edge detection and image segmentation. The following section confirms this claim.

5   Computational Experiment

Edge detection is an important task in many applications including image segmentation and object recognition. Broadly speaking, an edge is the boundary between two regions whose predominated values are significantly different. A gradient, combined with other techniques, can be used for edge detection [5]. In Example 3, we observed that the measures of dispersion of the distance-based morphological operators as well as the measures of central tendency of the distance-based gradient emphasize variations in intensity of a pixel in a neighborhood determined by the structuring element. Thus, we shall apply them empirically for edge detection.

In order to evaluate the performance of edge detectors based on a statistic of a distance-based operator, we used the Berkeley segmentation dataset and benchmark, University of California, Berkeley [16]. Briefly, this dataset has 300 natural color images and 1633 binary images that were manually segmented by a group of individuals. Like Canny's edge detector [5], given a natural color image $\boldsymbol{f} : \mathcal{D} \rightarrow \mathbb{V}_{\text{RGB}}$, we obtained a binary image $\beta \in \mathcal{D}$ by applying the non-maximum suppression (NMS) method and a hysteresis threshold on the gradient-like image g obtained from an appropriate statistics of a distance-based morphological operator. In our experiments, both NMS and hysteresis have been computed using the command NonMaxSupHyst from GNU Octave using the threshold parameters $T_1 = 0.01$ and $T_2 = 0.2$. Also, the Euclidean distance have been used to define the distance-based operators as well as to compute the Gini's mean difference. Furthermore, we used a 3×3 square structuring element S and we assumed $R \sim U(\mathbb{V}_{\text{RGB}})$, which means we have no *a priori* information on the reference.

Apart from the statistics of distance-based morphological operators, we also considered some other edge detectors from the literature. Namely, the gray-scale edge detectors of Sobel and Canny obtained using the GNU Octave commands edge and rgb2gray. We also considered the edge detectors obtained by applying non-maximum suppression and hysteresis on the following gradients: The usual Beucher gradient applied on the gray-scale image obtained using the command rgb2gray, the gradients ϱ_B^{Marg} and ϱ_B^{Lex} obtained using respectively the marginal and lexicographical RGB orderings [3], and the distance-based gradients ϱ_B^{black} and ϱ_B^{white} given by (4) with the reference fixed on black and white, respectively [2]. At this point, we would like to recall that some computational experiments using distance-based operators revealed that an achromatic color, such as black or white, has been recommended for an edge detection task if no information on the reference is available [26].

We quantitatively evaluated the performance of an edge detection technique by computing a normalized version of Pratt's figure of merit (FoM) as follows [18]: Let $\beta_i \subseteq \mathcal{D}$ be the binary image produced by the ith edge detector and $\tau \subseteq \mathcal{D}$ be an human segmented binary image, referred to as the ground truth, both with respect to a given color image f. The FoM value and normalized FoM value are given respectively by

$$\text{FoM}(\beta_i, \tau) = \frac{1}{M} \sum_{x \in \tau} \frac{1}{1 + \alpha d^2(x, \beta)} \quad \text{and} \quad \widehat{\text{FoM}}(\beta_i, \tau) = \frac{\text{FoM}(\beta_i, \tau) - \mu}{\sigma}, \quad (15)$$

where $M = \max\{\text{Card}(\beta), \text{Card}(\tau)\}$, α is a parameter, $d(x, \beta)$ denotes the distance from $x \in \tau$ to β, and μ and σ are the average and standard deviation of the FoM values produced by all edge detectors applied on the same image and its corresponding ground truth. The greater the normalized FoM value, the better the performance of the edge detector. In our experiments, we used $\alpha = 1/9$ and the Euclidean distance. Figure 3 shows the boxplot of the normalized FoM values produced by all the edge detectors on the entire dataset, resulting 1633 values for each method. Note that the edge detectors based on $\text{std}_t \varepsilon_B$, $\text{gini}\varepsilon_B$, $\text{std}\varrho_B$, ϱ_B^{Marg}, ϱ_B^{Lex}, ϱ_B^{black}, and ϱ_B^{white} yielded positive average normalized FoM values while the edge detectors of Sobel, Canny, and Beucher resulted negative

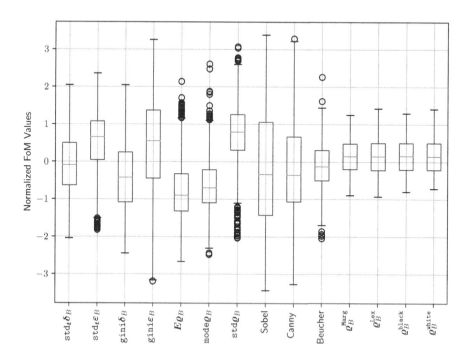

Fig. 3. Boxplot of the FoM values produced by edge detectors.

average normalized FoM values. The largest average normalized FoM values has been produced by the edge detectors based on the standard deviation of the gradient ($\mathtt{std}\varrho_B$), the total standard deviation and Gini's mean difference of the erosion ($\mathtt{std}_t\varepsilon_B$ and $\mathtt{gini}\varepsilon_B$). In fact, paired Student's hypothesis test with confidence level at 99% confirmed that the edge detector based on the standard deviation of the gradient outperformed all the other edge detectors considered in this computational experiment.

6   Concluding Remarks

In the last decades, several researchers have contributed with new approaches to extend mathematical morphology to multivalued images such as color and hyper-spectral images. In this paper, we focused on the distance-based approach whose ordering scheme depends on a single parameter – the reference $r \in \mathbb{V}$. Precisely, we characterized the operators of erosion, dilation, and gradient as a function of the reference using the diagrams of Voronoi and farthest neighbor.

In order to address uncertainties in the choice of the reference in the distance-based approach, we assumed that the reference is described by a random variable and used descriptive statistics such as measures of central tendency and dispersion to extract useful information on the distance-based morphological operators. For example, we can compute the expectation of distance-based erosions, the mode of distance-based dilations, or the standard deviation of the distance-based gradient, among others. Furthermore, we evaluated the performance of edge detectors based on some statistics of distance-based morphological operators using the image segmentation dataset of the University of California, Berkeley. Using Pratt's figure of merit (FoM) as a quantitative measure of the performance, we concluded that the edge detector based on the standard deviation of the gradient outperformed other approaches from the literature including the edge detector of Sobel and Canny.

Despite the promising result obtained in edge detection experiment, the statistics of distance-based morphological operators require further interpretation and theoretical background. Moreover, apart from descriptive statistics, we can use other probabilistic measures such as Shannon entropy to extract features or enhance our understanding of a multivalued image. Finally, we believe that the same reasoning can be extended to more general parametrized reduced ordering. Namely, we can assume that the parameters are samples from a random variable and use descriptive statistics or other probabilistic measures to obtain relevant information on the resulting morphological operators.

References

1. Al-Otum, H.M.: A novel set of image morphological operators using a modified vector distance measure with color pixel classification. J. Vis. Commun. Image Represent. **30**, 46–63 (2015)
2. Angulo, J.: Morphological colour operators in totally ordered lattices based on distances: application to image filtering, enhancement and analysis. Comput. Vis. Image Underst. **107**(1–2), 56–73 (2007). Special issue on Color Image Processing
3. Aptoula, E., Lefèvre, S.: A comparative study on multivariate mathematical morphology. Pattern Recogn. **40**(11), 2914–2929 (2007)
4. Birkhoff, G.: Lattice Theory, 3rd edn. American Mathematical Society, Providence (1993)
5. Canny, J.: A computational approach to edge-detection. IEEE Trans. Pattern Anal. Mach. Intell. **8**, 679–700 (1986)
6. Chevallier, E., Angulo, J.: The irregularity issue of total orders on metric spaces and its consequences for mathematical morphology. J. Math. Imaging Vis. **54**(3), 344–357 (2016)
7. Comer, M.L., Delp, E.J.: Morphological operations for color image processing. J. Electron. Imaging **8**(3), 279–289 (1999)
8. Deborah, H., Richard, N., Hardeberg, J.Y.: Spectral ordering assessment using spectral median filters. In: Benediktsson, J.A., Chanussot, J., Najman, L., Talbot, H. (eds.) ISMM 2015. LNCS, vol. 9082, pp. 387–397. Springer, Cham (2015). https://doi.org/10.1007/978-3-319-18720-4_33
9. Goutsias, J., Heijmans, H.J.A.M., Sivakumar, K.: Morphological operators for image sequences. Comput. Vis. Image Underst. **62**, 326–346 (1995)
10. van de Gronde, J.J., Roerdink, J.B.T.M.: Generalized morphology using sponges. Math. Morphol. Theory Appl. **1**(1), 18–41 (2016)
11. Heijmans, H.J.A.M.: Mathematical morphology: a modern approach in image processing based on algebra and geometry. SIAM Rev. **37**(1), 1–36 (1995)
12. Heijmans, H.J.A.M., Keshet, R.: Inf-semilattice approach to self-dual morphology. J. Math. Imaging Vis. **17**(1), 55–80 (2002)
13. Johnson, R.A., Wichern, D.W.: Applied Multivariate Statistical Analysis, 6th edn. Pearson Prentice Hall, Upper Saddle River (2007)
14. Ledoux, A., Richard, N., Capelle-Laizé, A.S., Fernandez-Maloigne, C.: Perceptual color hit-or-miss transform: application to dermatological image processing. Sig. Image Video Process. **9**(5), 1081–1091 (2015)
15. Lézoray, O.: Complete lattice learning for multivariate mathematical morphology. J. Vis. Commun. Image Represent. **35**, 220–235 (2016)
16. Martin, D., Fowlkes, C., Tal, D., Malik, J.: A database of human segmented natural images and its application to evaluating segmentation algorithms and measuring ecological statistics. In: Proceedings of the 8th International Conference on Computer Vision, vol. 2, pp. 416–423, July 2001
17. Oja, H.: Descriptive statistics for multivariate distributions. Stat. Probab. Lett. **1**(6), 327–332 (1983)
18. Pratt, W.: Digital Image Processing, 4th edn. Wiley, Hoboken (2007)
19. Ronse, C.: Why mathematical morphology needs complete lattices. Sig. Process. **21**(2), 129–154 (1990)
20. Sangalli, M., Valle, M.E.: Color mathematical morphology using a fuzzy color-based supervised ordering. In: Barreto, G.A., Coelho, R. (eds.) NAFIPS 2018. CCIS, vol. 831, pp. 278–289. Springer, Cham (2018). https://doi.org/10.1007/978-3-319-95312-0_24

21. Serra, J.: The "false colour" problem. In: Wilkinson, M.H.F., Roerdink, J.B.T.M. (eds.) ISMM 2009. LNCS, vol. 5720, pp. 13–23. Springer, Heidelberg (2009). https://doi.org/10.1007/978-3-642-03613-2_2
22. Soille, P.: Morphological Image Analysis. Springer, Berlin (1999)
23. Soille, P., Beucher, S., Rivest, J.F.: Morphological gradients. J. Electron. Imaging **2**(4)(12), 326–336 (1993)
24. Sussner, P., Valle, M.E.: Classification of fuzzy mathematical morphologies based on concepts of inclusion measure and duality. J. Math. Imaging Vis. **32**(2), 139–159 (2008)
25. Talbot, H., Evans, C., Jones, R.: Complete ordering and multivariate mathematical morphology. In: Proceedings of the Fourth International Symposium on Mathematical Morphology and Its Applications to Image and Signal Processing, ISMM 1998, pp. 27–34. Kluwer Academic Publishers, Norwell (1998)
26. Valle, M.E., Valente, R.A.: Mathematical morphology on the spherical CIELab quantale with an application in color image boundary detection. J. Math. Imaging Vis. **57**(2), 183–201 (2017)
27. Velasco-Forero, S., Angulo, J.: Supervised ordering in $\mathbb{R}^p$: application to morphological processing of hyperspectral images. IEEE Trans. Image Process. **20**(11), 3301–3308 (2011)
28. Velasco-Forero, S., Angulo, J.: Vector ordering and multispectral morphological image processing. In: Celebi, M.E., Smolka, B. (eds.) Advances in Low-Level Color Image Processing. LNCVB, vol. 11, pp. 223–239. Springer, Dordrecht (2014). https://doi.org/10.1007/978-94-007-7584-8_7

Mathematical Morphology on Irregularly Sampled Data Applied to Segmentation of 3D Point Clouds of Urban Scenes

Teo Asplund[1($\boxtimes$)], Andrés Serna[2], Beatriz Marcotegui[3], Robin Strand[1], and Cris L. Luengo Hendriks[4]

[1] Centre for Image Analysis, Division of Visual Information and Interaction, Uppsala University, Uppsala, Sweden
teo.asplund@it.uu.se
[2] Terra3D, 36 Boulevard de la Bastille, 75012 Paris, France
[3] MINES ParisTech, PSL Research University, CMM - Centre for Mathematical Morphology, 35 rue Saint Honoré, Fontainebleau, France
[4] Flagship Biosciences Inc., Westminster, Colorado, USA

Abstract. This paper proposes an extension of mathematical morphology on irregularly sampled signals to 3D point clouds. The proposed method is applied to the segmentation of urban scenes to show its applicability to the analysis of point cloud data. Applying the proposed operators has the desirable side-effect of homogenizing signals that are sampled heterogeneously. In experiments we show that the proposed segmentation algorithm yields good results on the Paris-rue-Madame database and is robust in terms of sampling density, i.e. yielding similar labelings for more sparse samplings of the same scene.

1 Introduction

Asplund *et al.* introduced morphological operators able to deal with irregularly sampled data. These operators, initially defined for 1D-signals, better approximate the continuous nature of the world [1,2] than regular, discrete morphology.

Large amounts of 3D data, acquired by Mobile Mapping System, are available nowadays (Oakland [6], KITTI [4], Rue Madame [10] datasets). Such data is of interest for a wide range of applications, for example: urban planning, documentation of cultural heritage, city modeling, among others [14]. 3D point clouds represent discrete versions of real world scenes. The sampling is by construction irregular, as it depends on the distance from the surface to the device, the angle of the surface, occlusions, vehicle speeds, etc. Thus, such data are particularly suitable to illustrate the potential of *irregular morphological operators* (i.e. operators on irregularly sampled data).

Previous morphological approaches to the problem of segmenting urban scenes have made use of regular, discrete morphology, by projecting the point cloud onto a regular grid [9,11]. A disadvantage of this approach is that the

© Springer Nature Switzerland AG 2019
B. Burgeth et al. (Eds.): ISMM 2019, LNCS 11564, pp. 375–387, 2019.
https://doi.org/10.1007/978-3-030-20867-7_29

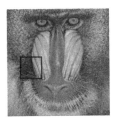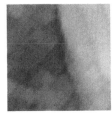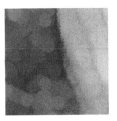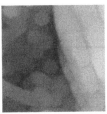

Fig. 1. From left to right: input image, regular dilation on sampled input, irregular dilation on sampled input, and regular dilation on non-sampled input.

segmentation depends on the coarseness of this grid (since nearby points can no longer be differentiated). Moreover, the projection loses information where the scene contains one part overshadowing another, for example where ground is occluded by a roof. A 3D morphological approach is proposed by Calderon and Boubekeur [3], that requires the estimation of the underlying surface.

In this paper we introduce a framework for applying 2D, real-valued morphology to irregularly sampled signals, extending previous work by Asplund *et al.* [1]. We show that our proposed approach can be applied to 3D point clouds with promising results for segmenting point clouds of urban scenes.

2   Morphology on Irregularly Sampled Data

Morphology on irregularly sampled data is related to morphology on graphs. Najman and Cousty provide a nice survey of a large number of papers regarding this issue [7]. Here we propose a new approach to morphology on point clouds.

Irregular morphology in the specific case of rectangular structuring elements (SE) for 2D real-valued signals was dealt with in a previous paper [1] focusing mostly on regularly sampled input. Moreover, the previously presented method has some limitations regarding the sampling of the transformed signal. The main issue is that edges end up being "fuzzy" because the number of samples that are generated near edges is too low. A proposed solution is to iterate operators. In this section we present a similar idea for irregular morphology, which however does not require such iterations.

Morphology on irregularly sampled signals has a number of advantages. For one, in the regular case, the SE depends on the sampling grid, i.e. a continuous shape must first be discretized, which creates artifacts. This is illustrated on a regularly sampled image in Fig. 1, where an input image is slightly blurred (Gaussian kernel with $\sigma = 1.0$ pixels) in order to remove high frequencies and then subsampled by taking every third pixel in every third row. The subsampled image is then dilated with a small disk ($r = 2$ pixels in the subsampled grid) using both regular, discrete morphology, as well as the proposed irregular approach detailed below. The figure also shows the result of applying a regular, discrete dilation to the original, blurred image (before subsampling), with a SE of corresponding radius. As can be seen, the small disk, after discretization, ends up looking more like a diamond. It is also worth noting that irregular morphology has a number of advantages over regular, discrete morphology when it comes to approximating continuous morphology on images in the discrete domain [1]. However, this is outside the scope of this paper.

We investigate the applicability of the proposed irregular morphological operators to point clouds. This seems a natural application, since they are naturally irregularly sampled.

2.1 Dilations of Irregularly Sampled Signals

The basic idea behind the proposed algorithm is to sample two disks at each input point. One with radius r, and one slightly larger with radius $r + \varepsilon$. Additionally, for each of these samples, examine whether a disk of radius $r + \varepsilon$ placed at any other input sample at a larger z-value overlaps this sample, and if so, discard it. The remaining samples from the larger disk are then projected downward, until they hit a disk (of radius r) placed at a lower sample, or falls through to $-\infty$. The remaining samples of the smaller disk are kept, defining a border between transitions in the value of the dilation. Here ε defines the width of this border.

We will use $B_{\mathbf{x}}$ to denote the set B translated by $\mathbf{x}$, i.e. $B_{\mathbf{x}} = \{\mathbf{b} + \mathbf{x} \mid \mathbf{b} \in B\}$. To perform a dilation by a disk of radius r of an irregularly sampled signal, we propose the following algorithm:

Choose some small $\varepsilon > 0$. For a disk shaped structuring element of radius r

$$D(r) = \{(x, y) \mid x^2 + y^2 \leq r\} \tag{1}$$

let $\check{D}$ denote a sampled version of D. In this paper, we will only consider the case where the disks are sampled uniformly along the border with 8 samples, and a ninth sample at the origin. This means that smaller circles are sampled more densely than larger ones. This keeps the number of samples down and larger SEs lead to larger plateaus, which probably do not need as many samples. However, less naive ways of sampling the disk are obviously of interest, but outside the scope of this paper.

To dilate some input $I = \{(x_i, y_i, z_i)\}_{i \in \{1, 2, \ldots, N\}}$ containing N samples at positions (x_i, y_i, z_i) do the following:

1. Let $I_\delta^0 = \emptyset$. Here I_δ^0 is empty, but will be filled with samples from the dilation as we process the samples in I.
2. In this step we iteratively select an unprocessed sample $(x_c, y_c, z_c) \in I$ and mark it as processed. This is analogous to sliding the SE over the image. In Fig. 2 the blue X is the sample being processed.
3. For each sample (x, y) of the translated disk $\check{D}_{(x_c, y_c)}(r)$, and each sample $(x_i, y_i, z_i) \in I \setminus \{x_c, y_c, z_c\}$, such that $z_i \geq z_c$, check if the distance from (x, y) to (x_i, y_i) is at most $r + \varepsilon$. If so, discard (x, y) (see the blue star in Fig. 2). Let $\check{D}_{(x_c, y_c)}(r)$ denote the remaining set of samples (blue circle). I.e. discard samples from the smaller disk that are overlapped.

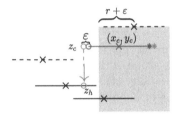

Fig. 2. Illustration of the algorithm from a side view. The disk shaped SE is located on top of each sample. The red circle is a sample (x, y) from $D(r + \epsilon)$. The black dashed disks do not belong to H since they are too far from (x, y) (and one of them is above z_c). Other two disks, placed below z_c, belong to H. z_h is found by projecting down until hitting the highest disk in H. In case no disk is hit, such a sample is discarded. The shaded area illustrates the part that is covered by a disk at a higher level. The samples that are in this shadow are discarded (blue and red stars as well as the blue X). The remaining samples are added to the dilation. This approach is similar to the paper by Asplund *et al.* [2]. (Color figure online)

4. Similarly to the previous step, for each sample (x, y) of the translated disk with radius $r + \varepsilon$, and each sample in $I \setminus \{x_c, y_c, z_c\}$ no lower than z_c, check if the distance from (x, y) to (x_i, y_i) is at most $r + \varepsilon$. If so, discard (x, y) (see the red star in Fig. 2). For each remaining sample (x, y) and each $(x_i, y_i, z_i) \in I$, such that $z_i < z_c$, let $H(x, y)$ be the set of values of samples for which the distance between (x, y) and (x_i, y_i) is at most r (see the red circle and black solid disks in Fig. 2), i.e.

$$H(x, y) = \{z_i \mid (x_i, y_i, z_i) \in I \wedge z_i < z_c \wedge \sqrt{(x - x_i)^2 + (y - y_i)^2} \leq r\} \quad (2)$$

Let $z_h(x, y) = \bigvee H(x, y)$. Conceptually, z_h is the value of a given sample (x, y) of the bigger disk, after projecting downward until hitting a disk at a lower value (or dropping out to $-\infty$). See red circle in Fig. 2.

5. Finally, let

$$\hat{D}^c_{(x_c, y_c)}(r) = \{(x, y, z_c) \mid (x, y) \in \hat{D}_{(x_c, y_c)}(r)\}, \quad (3)$$

$$\hat{D}^h_{(x_c, y_c)}(r + \varepsilon) = \{(x, y, z_h(x, y)) \mid (x, y) \in \hat{D}_{(x_c, y_c)}(r + \varepsilon)\}, \text{ and} \quad (4)$$

$$I^{i+1}_\delta = I^i_\delta \cup \hat{D}^c_{(x_c, y_c)}(r) \cup \hat{D}^h_{(x_c, y_c)}(r + \varepsilon) \quad (5)$$

I.e. add the samples spawned by the processed sample to I^{i+1}_δ, where i is the number of previously processed samples. These steps are then iterated until there are no unprocessed samples left.

Then I^N_δ is the dilated signal. That is, I^N_δ is the irregular dilation of I by a disk shaped structuring element of radius r. The parameter ε controls the distance between samples where translated disks overlap. This parameter should be small (as a rule, at least smaller than the distance between samples in the input).

Figure 3 illustrates an example of applying this algorithm to an input containing five irregularly sampled points when seen from above. The samples that fall

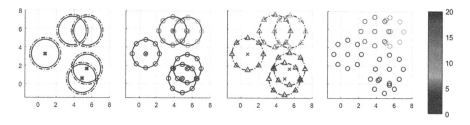

Fig. 3. A dilation of an irregularly sampled signal containing five samples (illustrated by crosses), by a disk shaped structuring element (solid black line). The first figure shows a SE placed on top of each input sample. The dashed circles illustrate the larger disk in step 3 (with radius $r + \varepsilon$). The second figure shows the sampled SEs (i.e. $\check{D}(r)$). The third figure shows the sampled larger disk (i.e. $\check{D}(r + \varepsilon)$). The last figure shows the final result (i.e. I_δ^N). Here the SE samples that fell under a large disk at a higher location (i.e. one of the disks in the third figure) have been suppressed, and the samples of the larger disks have been projected down, helping define the border between the sampled disks (see for example the bottom right of the last figure).

in the shadow of any SE placed at a higher sample are discarded. The remaining samples from the larger SE are projected down until they hit a SE at a lower input sample (or until they fall through to $-\infty$).

2.2 Erosions, Openings, and Closings

To get the erosion $\varepsilon(I)$ of an input signal I, one can simply compute the dilation $\delta(I^{C})$, where $I^{C} = \{(x_i, y_i, -z_i) \mid (x_i, y_i, z_i) \in I\}$, the erosion is then $\delta(I^{C})^{C}$. Openings and closings can be formed by composition [12].

3 Irregular Morphology for Point Clouds

Figure 4 shows some examples of different morphological operators applied to a simple point cloud depicting a signpost. The signpost is first eroded by a disk of radius 0.016 m, followed by a dilation with the same SE (yielding an opening). The result is interpolated back onto the original input positions, and then subtracted from the input, resulting in the tophat, which has enhanced the thin pole holding the sign. The value of ε is set to 10^{-6} m in all these cases.

We use nearest neighbor interpolation for a few reasons: Firstly, for the sake of simplicity. Secondly, empirically this seems to give good results. Finally the morphological operators will, in general, create cusps when applied to a signal, even if the input is smooth [1]. Therefore, a smooth interpolation of the opening is problematic. Some other possible candidates would be linear interpolation, or natural interpolation [13], however in some preliminary testing, these seem to give worse results for our data.

An interesting property of the irregular operators is their tendency to homogenize the sampling density. The sampling of the ground in the input is less dense

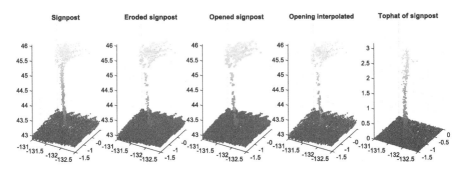

Fig. 4. Examples of irregular morphology applied to a simple point cloud of a signpost. Here, the SE is a disk with radius 0.016 m.

further from the camera (since these samples are further from the sensor). The eroded point cloud, on the other hand, has a similar sampling density close to, and further away from the camera. In essence, this behavior stems from not sampling inside the disks (except for one sample in the center) thereby reducing the sampling density where it is high (since disks are likely to overlap many nearby samples) accelerating further processing, while increasing the density where it is low (since one input sample gives rise to many output samples and there are fewer neighboring samples in the input that are likely to be suppressed). This homogenization may better define surfaces, enabling more accurately detecting object boundaries. Moreover, homogenizing the density introduces an implicit normalization of features and a more robust feature description, avoiding the excessive contribution of high density areas.

4 Application to Urban Point-Cloud Segmentation

As previously mentioned, the problem of segmenting point clouds of urban scenes has already been addressed using morphology on elevation images. However, using elevation images has some drawbacks. The images being projected onto a regular grid means that the segmentation will depend on the coarseness of said grid. Furthermore there is a question about how to deal with holes in the projected image. Thus, this problem seems like a natural candidate for applying irregular morphology.

We will consider a problem where the goal is to segment point clouds of urban scenes, to illustrate the applicability of our proposed method. In order to segment the point cloud, we first focus on separating out the facades and objects, followed by a refinement which yields the final segmentation. Figure 5 shows a diagram outlining the workflow.

We make use of point clouds from the Paris-rue-Madame dataset[1] [10], containing two point clouds from a 3D Mobile Laser Scanning (MLS) of a section of

---

[1] ©2014 MINES ParisTech. MINES ParisTech created this special set of 3D MLS data for the purpose of detection-segmentation-classification research activities.

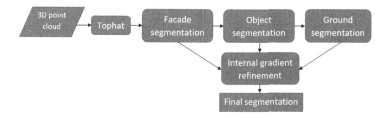

Fig. 5. Workflow diagram of the proposed algorithm. The output is a segmented point cloud with labels: ground, facade and object.

a street in Paris. Apart from the 3D-location, each point is also equipped with three additional pieces of data: reflectance, label, and class. However, we only use the position of the points for segmenting. We use the provided classes to test our segmentation. The reflectance and label information is ignored.

4.1 Tophat

An initial preprocessing step will be applied to the entire point cloud, the purpose being to flatten out the surface, removing large-scale variations in height (e.g. slopes and hills), while preserving local height differences (e.g. cars, pedestrians, facades). This preprocessing consists of a tophat applied to the point cloud. First we apply an opening to the set of points (by applying an erosion followed by a dilation), the result is then subtracted from the original signal (via interpolation as described in Sect. 3).

The size of the structuring element used by the opening controls the size of the support of height variations to be preserved. Based on the size of objects of interest one can make a decision on the radius of the SE. In this case, our objects of interest are on the scale of cars or smaller (we also capture facades, which are generally quite thin). Therefore, we set the radius of the disk to be 1.5 m, since a disk of this size cannot fit inside a normal vehicle. Figure 6 shows an example of the result of taking the tophat of a point cloud from the dataset [10].

4.2 Segmenting Facades

Facades are initially segmented by thresholding the tophat at a large value h_{facade}. We are assuming the highest objects in the scene are house facades. If this does not hold, one could use some more sophisticated feature, such as considering the elongation of objects [5,11], or learned features, but for our purposes, this assumption is fine. The value of h_{facade} is not very sensitive. The important part is that the threshold contains the top part of facades. We simply fix this value to be 5 m. This initial segmentation fails to include a large number of points that are actually parts of facades, however, it contains very few false positives (very tall objects might be included, e.g. lampposts, but they are easily eliminated). The purpose of this initial segmentation is to localize the facades.

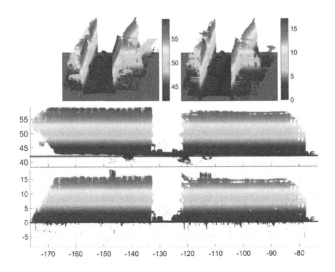

Fig. 6. The first row shows the tophat (right) of the input (left). Also shown is a horizontal plane in purple that is approximately at ground level for the input figure, and at 0 for the figure of the tophat, to illustrate slope. Rows two and three show the same point clouds from the side. Notice that the slope of the input has been removed in the tophat (bottom), and large scale variations in height have been suppressed. This makes thresholding for segmentation feasible. The tophat contains some negative values (points below the plain), which stems from interpolation artifacts. These negative points happen mostly inside buildings, where the sensor has sampled parts of the insides of buildings so that it has some samples from rooms on top of each other. (Color figure online)

Following this localization, we can find candidate facade points by thresholding the result of the tophat at a lower value h'_{facade} as described in Sect. 4.5. These candidates are then filtered based on their distance (in 2D) from the initially selected facade points.

Specifically, let

$$F_{h_{\text{facade}}} = \{(x, y, z) \in I \mid z > h_{\text{facade}}\}, \quad F_{h'_{\text{facade}}} = \{(x, y, z) \in I \mid z > h'_{\text{facade}}\} \quad (6)$$

where I is the set of input samples and $F_{h_{\text{facade}}}$ denotes the initial segmentation of the facades associated with h_{facade}. The refined segmentation F' is given by the following set of points:

$$F' = \{(x, y, z) \in F_{h'_{\text{facade}}} \mid d((x, y, z), F_{h_{\text{facade}}}) < \varepsilon_f\} \quad (7)$$

here $\varepsilon_f > 0$ is a parameter that controls the distance that the initial facade labels can be propagated. For our experiments we use $\varepsilon_f = 0.05\,\text{m}$. The function d denotes the minimum 2D-distance between all points in a set and some sample:

$$d((x, y, z), F) = \min_{(x', y', z') \in F} \sqrt{(x - x')^2 + (y - y')^2} \quad (8)$$

By restricting the potential facade points to those that are near the initially selected set of points, we can rule out a lot of false positives, meaning that we

Fig. 7. The left figure shows the initial facade segmentation in dark blue (other colors indicate height). On the right: points labeled facade after the process in Sect. 4.2. (Color figure online)

can push the threshold value h'_{facade} close to 0, i.e. close to the height of the ground, even though the number of ground points that are above the threshold increases as h'_{facade} approaches 0. Thresholding at $h'_{facades}$ (for reasonable values) will clearly find points that are parts of objects too (e.g. pedestrians), since the lower parts of facades are at the same height as objects. However, such points will generally not be immediately next to facades (although it is possible) and are therefore discarded, since they are too far ($\geq \varepsilon_f$) from the initial facade points. The remaining points are labeled "facade". Figure 7 illustrates this process.

4.3   Segmenting Objects

We now focus on the objects, i.e. everything that is neither facade nor ground. At this point, we only need consider the non-facade points. Again, we follow the procedure outlined in the previous section. Objects are initially segmented by selecting a threshold h_{object}. Again, this initial threshold is set at a height such that there are few non-object points above it, but many object points fall below it. To remedy this, a threshold of h'_{object} is selected (see Sect. 4.5) and applied to the remaining points. Finally, points that are too far from the initial set of object points (in 2D) are discarded (as in Eq. 7 with appropriate substitutions). The remaining points are labeled "object".

4.4   Segmenting Ground and Refining Labeling

The points that are not labeled "object" or "facade" are labeled "ground". However, there may be some points that are close to objects and facades that are mislabeled as "ground," since the thresholds h'_{facade} and h'_{object} cannot be set too close to ground level without introducing noisy labels close to the borders between ground and not ground. An additional step to refine the segmentation is performed by finding the internal gradient G_i [8] of the set of points, G, labeled ground. I.e. $G_i = G - \varepsilon_r(G)$, where ε_r denotes the erosion with a disk shaped structuring element of radius $r > 0$. Here r is set to a small value which depends on the average minimum distance from any ground point p to the rest of the ground points $G \setminus \{p\}$, i.e.

$$r = \frac{C}{|G|} \sum_{p \in G} d(p, G \setminus \{p\}) \qquad (9)$$

where $C \geq 1$ is a constant increasing the radius to give more context (we fix $C = 10$). In other words, this radius depends on the point cloud density. We want it to be small, in order to get a good approximation of the internal gradient, but it cannot be too small, since this would essentially lead to each point being looked at in isolation, which is a problem, since the gradient obviously requires information about values at nearby points. The internal gradient G_i enhances the edges, which are the ground points that are likely mislabeled. A final threshold h_{edge} is chosen as described in Sect. 4.5. The set of points above this threshold E, are then used instead of $F_{h'_{facade}}$ and $F_{h'_{object}}$ (see Eqs. (6)–(7)). As a result, more points close to the ground near edges can be correctly labeled, though this may lead to a small number of ground points being mislabeled.

4.5   Choosing the Threshold Value

For the upper thresholds (i.e. h_{facade} and h_{object}) we simply choose values by hand, based on the height of objects that we are interested in. The method is not particularly sensitive to the choice for these values. The main concern is to set values that capture the top part of facades and objects respectively. In this paper the values used are $h_{facade} = 5\,\text{m}$, and $h_{object} = 0.5\,\text{m}$. I.e., we are assuming that facades are higher than 5 m, and objects are at least half a meter tall.

For the lower thresholds (i.e. h'_{facade} h'_{object} and h_{edge}), ideally the value would be 0, but even after leveling the scene using the tophat, the ground is still not perfectly flat. For simplicity, we set the values of h'_{facade}, h'_{object}, and h_{edge} to the same value of 0.4 m. One may select a threshold automatically by examining the histogram of heights of the tophat for values between 0 and h_{object}. The histogram contains a large peak around the height of the ground, which can be used to set the other thresholds. It is possible to push the threshold value down, thus improving the labeling of the lower parts of objects and walls, at the cost of a somewhat noisier ground segmentation.

4.6   Results

The top of Fig. 8 shows the segmentation result for GT_Madame1_2 [9] for different sampling densities of the input (achieved by discarding every n:th sample). The result seems reasonably robust to changes in sampling density. To quantify the performance for different densities we consider the percentage of unchanged labels as more and more input samples are discarded (see bottom left of Fig. 8).

Additionally, we make use of the class labels provided in the dataset [9] to create a ground truth for our labeling by combining the class labels for facade, wall light, balcony plant, and wall sign into a single class label for the walls. Similarly, the ground label is comprised of the background class label and the ground class label, the remaining classes are considered objects. We then look at the percentage of correctly labeled samples for each of these three classes as the sampling density of the input is decreased (bottom right of Fig. 8).

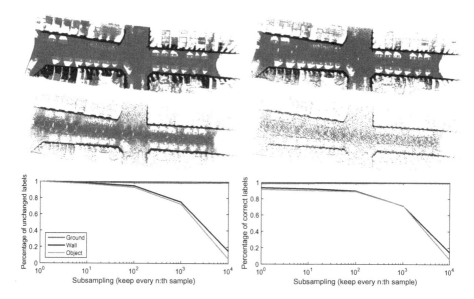

Fig. 8. Top: Labeled scene for different point cloud densities. From left to right and top to bottom, we have: 10^7, 10^6, 10^5, 10^4 samples. Red indicates object, blue indicates ground, and black indicates facade. Bottom left: The percentage of samples in the labeled point cloud, per class, that have changed when comparing less densely sampled inputs to the most densely sampled input. Bottom right: The percentage of correct labels for each class and different point cloud densities. The behavior of the curves for the ground label is explained by its simple nature. In general, the ground is relatively flat and does not need many samples to be well represented. (Color figure online)

5 Discussion and Conclusions

This is a first attempt at applying irregular morphology to 3D point cloud analysis. A previous morphological approach [9,11] to the problem studied in this paper first projected the point cloud onto a regularly sampled range image, which causes issues where multiple points in the point cloud correspond to the same pixel in the range image. Working directly on the irregular domain means that there is no need to resample onto a regular grid, thus avoiding this problem. It also allows for morphological operators of any scale, unrestricted by any grid.

We argue that another possible benefit of this approach is the homogenizing tendency of the proposed operators on the point cloud density. This can speed up further processing, since oversampled parts of the signal become sparser, while better defining the surface where the density is low, which may allow for more accurate detection of object boundaries.

In experiments we show that the proposed method is relatively robust, yielding similar labels for the remaining samples as we decrease the input sampling, thus keeping the percentage of correctly labeled samples high even as the density drops to a hundredth. The most robust class is ground, which is to be expected, since large and smooth surfaces do not require a large amount of samples.

Worth noting is that the proposed approach is slower than regular, discrete morphology. Computation time depends on a number of factors, including point density, size of the SE, and size of the input. Processing inputs on the order of 10^6 samples, using a k-d tree to find neighbors, takes seconds for smaller SEs (on an Intel Xeon 3.40 GHz), but may take several minutes if the SE is large (e.g. in the case of the large tophat used in our experiments), though further processing is normally faster, since the density is likely decreased, as previously described.

In the future, we would like to improve the sampling strategy. Instead of generating the same number of tentative output samples at each input sample, one could let this number depend on local properties of the signal (e.g. smoothness). Allowing for more general shapes of SEs is also of interest. This may also have an affect on the sampling strategy (e.g. pointy parts of the SE may need a denser sampling, while smoother parts can be sampled more sparsely). With regard to the segmentation approach, comparing our results should be the next step. It would be interesting to take into account some other features. Local thresholding could be useful. Finally, a postprocessing step could likely further improve the segmentation (e.g. via some regularization method).

Acknowledgement. Teo Asplund was funded through grant 2014-5983 from the Swedish Research Council.

References

1. Asplund, T., Luengo Hendriks, C.L., Thurley, M.J., Strand, R.: Mathematical Morphology on irregularly sampled signals. In: Chen, C.-S., Lu, J., Ma, K.-K. (eds.) ACCV 2016. LNCS, vol. 10117, pp. 506–520. Springer, Cham (2017). https://doi.org/10.1007/978-3-319-54427-4_37

2. Asplund, T., Luengo Hendriks, C.L., Thurley, M.J., Strand, R.: Mathematical morphology on irregularly sampled data in one dimension. In: Mathematical Morphology-Theory and Applications, vol. 2, no. 1, pp. 1–24 (2017)

3. Calderon, S., Boubekeur, T.: Point morphology. ACM Trans. Graph. (TOG) **33**(4), 45 (2014)

4. Geiger, A., Lenz, P., Urtasun, R.: Are we ready for autonomous driving? The KITTI vision benchmark suite. In: Conference on Computer Vision and Pattern Recognition (CVPR) (2012)

5. Morard, V., Decencière, E., Dokladal, P.: Geodesic attributes thinnings and thickenings. In: Soille, P., Pesaresi, M., Ouzounis, G.K. (eds.) ISMM 2011. LNCS, vol. 6671, pp. 200–211. Springer, Heidelberg (2011). https://doi.org/10.1007/978-3-642-21569-8_18

6. Munoz, D., Bagnell, J.A., Vandapel, N., Hebert, M.: Contextual classification with functional Max-Margin Markov networks. In: IEEE Conference on Computer Vision and Pattern Recognition, CVPR 2009, pp. 975–982. IEEE (2009)

7. Najman, L., Cousty, J.: A graph-based mathematical morphology reader. Pattern Recogn. Lett. **47**, 3–17 (2014)

8. Rivest, J.F., Soille, P., Beucher, S.: Morphological gradients. J. Electron. Imaging **2**(4), 326–337 (1993)

9. Serna, A., Marcotegui, B.: Detection, segmentation and classification of 3D urban objects using mathematical morphology and supervised learning. ISPRS J. Photogram. Remote Sens. **93**, 243–255 (2014)

10. Serna, A., Marcotegui, B., Goulette, F., Deschaud, J.E.: Paris-rue-Madame database: a 3D mobile laser scanner dataset for benchmarking urban detection, segmentation and classification methods. In: 4th International Conference on Pattern Recognition, Applications and Methods ICPRAM 2014 (2014)

11. Serna, A., Marcotegui, B., Hernández, J.: Segmentation of façades from urban 3D point clouds using geometrical and morphological attribute-based operators. ISPRS Int. J. Geo-Inf. **5**(1), 6 (2016)

12. Serra, J.: Image Analysis and Mathematical Morphology. Academic Press Inc., Orlando (1983)

13. Sibson, R.: A brief description of natural neighbour interpolation. In: Barnett, V. (ed.) Interpreting Multivariate Data (1981)

14. Vosselman, G., Maas, H.G.: Airborne and Terrestrial Laser Scanning. CRC, Boca Raton (2010)

Applications in Engineering

Attribute Filtering of Urban Point Clouds Using Max-Tree on Voxel Data

Florent Guiotte[1,2(✉)], Sébastien Lefèvre[2], and Thomas Corpetti[1]

[1] Univ. Rennes 2, LETG, 35000 Rennes, France
{florent.guiotte,thomas.corpetti}@univ-rennes2.fr
[2] IRISA, Univ. Bretagne Sud, 56000 Vannes, France
sebastien.lefevre@irisa.fr

Abstract. This paper deals with morphological characterization of unstructured 3D point clouds issued from LiDAR data. A large majority of studies first rasterize 3D point clouds onto regular 2D grids and then use standard 2D image processing tools for characterizing data. In this paper, we suggest instead to keep the 3D structure as long as possible in the process. To this end, as raw LiDAR point clouds are unstructured, we first propose some voxelization strategies and then extract some morphological features on voxel data. The results obtained with attribute filtering show the ability of this process to efficiently extract useful information.

Keywords: Point clouds · Max-tree · Rasterization · Voxel · Attribute filtering

1 Introduction

Thanks to the advances both in technologies such as laser scanning (LiDAR) and in methods from photogrammetry, digital point clouds form a popular object of study in many scientific fields such as geosciences (flow, erosion, rock deformations, ...), computer graphics (3D reconstruction), urban environments analysis from Earth Observation (detection of trees, roads, buildings, ...). In the context of urban scenes, they provide a rich 3D information w.r.t. digital 2D photographs.

Despite their growing interest, only limited studies have explored how to apply mathematical morphology on point clouds [2,13]. The usual approach remains to first rasterize the point cloud to obtain a digital image (also called a raster of pixels) on which standard morphological operators are applied [18]. This strategy was also recently followed for morphological hierarchies on LiDAR point clouds [8].

In this paper, we claim that discretizing a point cloud into a 2D raster leads to an oversimplification of the image that greatly reduces the potential of morphological operators. Thus, we consider here a 3D discretization in a voxel grid

This work was financially supported by Région Bretagne (CAMLOT doctoral project).

B. Burgeth et al. (Eds.): ISMM 2019, LNCS 11564, pp. 391–402, 2019.
https://doi.org/10.1007/978-3-030-20867-7_30

Fig. 1. LiDAR data projected into a voxel grid.

as illustrated in Fig. 1. Such an approach allows us to benefit from efficient algorithms that have been introduced for morphological hierarchies, while still maintaining the 3D information. We illustrate our solution with a very popular processing, namely attribute filtering, that is applied on an urban point cloud. The reported results show the relevance and the potential of this approach.

The paper is organized as follows. In Sect. 2 we review morphological approaches (especially those based on hierarchies) for point clouds and 3D processing. We then explain the different steps of our method in Sect. 3, before illustrating it in the urban remote sensing context in Sect. 4. Section 5 concludes this manuscript and provides directions for future research.

2   Related Work

2.1   Mathematical Morphology on Point Clouds

Conversely to digital images that are usually defined on a discrete 2D grid, a point cloud is characterized by a sparse set of points defined with continuous coordinates. So dealing with a 3D point cloud raises many issues including the lack of efficient processing algorithms. One of the most critical questions is the definition and fast computation of the neighborhood of a point, which is a fundamental concept in morphology.

As already stated, the usual approach is to project the 3D point cloud in a discrete grid (either 2D or 3D) where the neighborhood computation becomes straightforward. Thus, LiDAR point clouds were considered in [6] to characterize vegetation (trees). The authors use the point density to define voxel values, that are further processed with 3D adaptation of standard 2D morphological methods. In the astronomical context, the discretization proposed by [4] also relies on the density, this latter being estimated using adaptive kernel. A max-tree is then used to find local maxima that allow for identification of relevant subspaces for clustering the data.

More closely related to our study, a few attempts have been made to process urban point clouds. In [1], the segmentation and classification of an urban point cloud is achieved by means of super-voxels. They are created using a distance computed in the feature space between voxel properties such as distribution (by mean, variance) of spectral values (intensity or color information). Finally, a reprojection step is involved to obtain the resulting labeled point cloud. The same tasks are addressed by [18], with a different approach though. The point cloud is projected in a digital elevation model (DEM) before applying 2D morphology and classification. The results are then reprojected into a point cloud. More recently, the same authors have addressed segmentation of facades with attribute profiles [17]. Combination of DEM morphology and attribute filters (elongation) were considered on binary 3D images denoting an occupancy grid. Beyond these works on 2D and 3D rasters, a few works have been conducted directly on the continuous space of 3D points cloud, such as [2,13]. However the morphological methods introduced in these papers are dedicated to point clouds describing surfaces, and as such cannot be used with LiDAR data since a surface can not be reconstructed in each situation (points are also likely to belong to the inside of objects, for example in vegetation areas).

2.2   Morphological Hierarchies on 3D Images

The extension of morphological hierarchies (e.g. max-tree) from a 2D image to a 3D volume is rather straightforward. Unsurprisingly, it led to several works attempting to use it on 3D voxels, especially in the medical domain. Early work in [22] has for example used the max-tree to filter 3D images with volume and inertia attributes. Later on, the max-tree was used to filter and visualize the medical images [21] with three new 3D moment-based attributes (elongation, flatness and sparseness). Another 3D attribute was proposed in [10] to estimate the sphericity of objects. It was based on the computation of surface and volume of connected components and aims to be more efficient than the previous measures. Roundness of objects was estimated through another 3D attribute for max-tree filtering in [11]. Filtering medical images has also received a lot of attention until very recently, e.g. [3,7,12,20]. Finally, we can mention the work of [5] to compute the tree of shapes of nD images.

3   Method

In order to filter point clouds using hierarchical representations, we propose to rely on a prior discretization of the continuous domain into a regular 3D voxel grid (instead of a 2D raster). This intermediate representation allows us to use directly mathematical morphology with the 3D data. We then reproject the results of the morphological filtering into the continuous domain (i.e. as a new point cloud).

3.1   From Point Cloud to Voxel Grid

A raw dataset $\boldsymbol{X}$ issued from LiDAR acquisitions lives in $\mathbb{R}^3 \times \mathbb{R}$ where each data $\boldsymbol{x} = \{x, y, z, \mathcal{I}\} \in \boldsymbol{X}$ is such that the intensity taken in location (x, y, z) is $\mathcal{I}$. Though very interesting, the irregularity of available locations (x, y, z) prevents from the use of tools devoted to structured data with ordering relations as images or volumes. To cope this difficulty, we suggest to transform this dataset $\boldsymbol{X}$ into a structured volume. This "voxelization" step aims at defining the data on a regular 3D grid $\boldsymbol{E}_h \subset \mathbb{N}^3$ with a given spatial resolution h (for the sake of simplicity, we consider here isotropic resolutions but the method can be applied with anisotropic ones) such that the value taken in any point $(i, j, k) \in \boldsymbol{E}_h$ represents an information about initial data. This information can either be a boolean (related to the presence/absence of points in the voxel), the number of LiDAR points into the voxel, the average/standard deviation of associated intensities, the average/standard deviation of associated elevations, the majority label (for 3D labels), etc.

More formally, we apply a transformation $\mathcal{PV}_{h,f}$ (for "points to voxels", associated with a discretization step h and an information function f) defined as:

$$\mathcal{PV}_{h,f} : \mathbb{R}^3 \times \mathbb{R} \longrightarrow \boldsymbol{E}_h \times \mathbb{R}$$
$$\{x, y, z, \mathcal{I}\} \longmapsto \{i, j, k, \mathbf{I}\} \text{ with:}$$
$$\begin{cases} i & \text{s.t. } x_m + ih \leq x < x_m + (i+1)h \\ j & \text{s.t. } y_m + jh \leq y < y_m + (j+1)h \\ k & \text{s.t. } z_m + kh \leq z < z_m + (k+1)h \\ \mathbf{I} & = f(i, j, k, x, y, z, \mathcal{I}) \end{cases} \quad (1)$$

with x_m (resp. (y_m, z_m)) the minimum value of all points x (resp. y, z) in the dataset $\boldsymbol{X}$. The rule of function f is to associate to each voxel location (i, j, k) an information related to the original data points. Let us denote $\mathcal{I}_{i,j,k}$ the set of intensities $\mathcal{I}$ of points (x, y, z) inside a voxel (i, j, k) (i.e. fulfilling the 3 first conditions of (1)). Its cardinal is noted $|\mathcal{I}_{i,j,k}|$.

Many functions f can be defined, as for example:

– **Boolean** (noted f_b):

$$f_b(i, j, k, x, y, z, \mathcal{I}) = \begin{cases} 1 & \text{if } |\mathcal{I}_{i,j,k}| \geq 1 \\ 0 & \text{otherwise} \end{cases} \quad (2)$$

– **Density** (noted f_d, similar to [4]):

$$f_d(i, j, k, x, y, z, \mathcal{I}) = |\mathcal{I}_{i,j,k}| \tag{3}$$

– **Empirical average intensity** (noted f_a):

$$f_a(i, j, k, x, y, z, \mathcal{I}) = \frac{1}{|\mathcal{I}_{i,j,k}|} \sum \mathcal{I}_{i,j,k} \tag{4}$$

– **Empirical standard deviation of intensity** (noted f_s):

$$f_s(i, j, k, x, y, z, \mathcal{I}) = \sqrt{\frac{1}{|\mathcal{I}_{i,j,k}|} \sum (\mathcal{I}_{i,j,k} - f_a(i, j, k, x, y, z, \mathcal{I}))^2} \tag{5}$$

with f_a and f_s defined only if $f_b \neq 0$. Depending on the sought applications, many other functions can be used, for example related to the geometry (normal surface, main orientation, ...) or any other features of $\mathcal{I}_{i,j,k}$.

It should be outlined that empty cells can occur from two situations: (1) empty spaces into the scene or (2) missing data because of occlusions. Several approaches are possible to deal with such empty voxels (affecting a value 0, linear interpolation, ...). In this study, and without loss of genericity, we chose to assign them the null value.

Once the voxelization transformation $\mathcal{PV}$ is performed, our data $(i, j, k, \mathbf{I}) \in \mathbf{E}_h \times \mathbb{R}$ can be represented through a volume V such that:

$$\begin{aligned} V: \quad \mathbf{E}_h &\longrightarrow \mathbb{R} \\ (i, j, k) &\longmapsto \mathbf{I}. \end{aligned} \tag{6}$$

3.2   Attribute Filtering with the Max-Tree of Voxels

Max-Tree. Attribute filtering is a popular tool in mathematical morphology. It operates on connected components of an image (if binary) or of its level sets (if greyscale). As a connected filter, it does not shift object edges but proceeds by removing the components that do not fulfill a given criterion (related to the aforementioned attribute). It benefits from efficient implementation through the image representation as a max-tree.

As already stated, the usual definition of the max-tree for 2D images remains valid in case of 3D volumes. Only the connectivity needs to be updated, from 4- and 8-connectivity in 2D to 6-, 18- and 26-connectivity in 3D. The upper level sets of the volume V are obtained from successive thresholdings of the grey levels $l \in \mathbb{R}$ and noted

$$\mathcal{L}_l = \{(i, j, k) \in \mathbf{E}_h \mid V(i, j, k) \geq l\}. \tag{7}$$

We index by c the connected components within a level set, i.e. $\mathcal{L}_{l,c}$. These components are nested and form a hierarchy called the max-tree. The leaves of the tree correspond to the regional maxima while the root contains the whole volume.

Filtering. The max-tree structure provides an efficient way to filter its nodes (i.e. the connected components of the level sets). Such a filtering relies on some predefined criteria called attributes, whose values are usually computed for each node during the tree construction step.

We distinguish here between two kinds of attributes that are scale-dependent and scale-invariant, respectively. In the former category, we can mention the volume and surface (i.e. 3D counterparts of the 2D area and perimeter, respectively), as well as dimensions of the bounding box or the convex hull. Examples of scale-invariant attributes are distributions of grey levels (e.g. standard deviation, entropy), measures computed with moments of inertia (e.g. compactness, sphericity), or moment invariants (e.g. elongation, flatness). The interested reader is referred to [16] for more details.

We provide below a formal definition of the three attributes that have been used in the experiments reported in this paper:

– **Height** (noted A_h):

$$A_h(\mathcal{L}_{l,c}) = \max_{i,j,k}(k) - \min_{i,j,k}(k) \tag{8}$$

– **Volume** (noted A_v):

$$A_v(\mathcal{L}_{l,c}) = \left|(i,j,k)\right| \tag{9}$$

– **Extent** (also named "fill ratio" [9], noted A_e):

$$A_e(\mathcal{L}_{l,c}) = \frac{A_v(\mathcal{L}_{l,c})}{A_v(\mathcal{B}(\mathcal{L}_{l,c}))} \tag{10}$$

with $\mathcal{B}(\cdot)$ the bounding box of a set. Let us note that for the sake of conciseness, the condition $(i,j,k) \in \mathcal{L}_{l,c}$ was systematically omitted in the right part of the previous equations.

The aforementioned attributes are either increasing or non-increasing, depending if their value is increasing from leaves to root or not. The filtering simply consists in assessing each connected component by comparing its attribute value to a given threshold T, and retaining only the filtered set $\mathcal{L}' \subseteq \mathcal{L}$ defined as

$$\mathcal{L}'_l = \{\mathcal{L}_{l,c} \mid A(\mathcal{L}_{l,c}) \geq T\}. \tag{11}$$

While the filtering with an increasing attribute is achieved through pruning the tree (i.e. removing all descendant nodes of $\mathcal{L}_{l,c}$ if $A(\mathcal{L}_{l,c}) < T$), considering a non-increasing attribute leads to pruning and non-pruning strategies [15,19] that remove full branches or isolated nodes, respectively.

The final step of the filtering is to reconstruct the filtered volume F based on remaining nodes of the max-tree, i.e.:

$$\begin{aligned} F: \quad \boldsymbol{E}_h &\longrightarrow \mathbb{R} \\ (i,j,k) &\longmapsto \max_{l \in \mathbb{R}}\big((i,j,k) \in \mathcal{L}'_l\big). \end{aligned} \tag{12}$$

3.3   Reprojection to the 3D Point Cloud

After having performed attribute filtering (or any other morphological process-ing) on the 3D volume, the filtered volume $F(i, j, k)$ needs to be reprojected in the original continuous set of coordinates to produce a dataset $\{x, y, z, \mathcal{F}\} \in \mathbb{R}^3 \times \mathbb{R}$. To this end, we assign to all initial points embedded in each voxel (i, j, k) the value $F(i, j, k)$. Let us note that more complex functions could have been con-sidered here (e.g. interpolation taking into account the position of (x, y, z) in (i, j, k) and intensities of the neighboring voxels). Our choice mathematically reads as applying the inverse transformation function $\mathcal{VP}$ (for "voxels to points") defined as:

$$\mathcal{VP}: \qquad \boldsymbol{E}_h \times \mathbb{R} \longrightarrow \mathbb{R}^3 \times \mathbb{R}$$
$$\{i, j, k, F(i, j, k)\} \longmapsto \{x, y, z, \mathcal{F}\} \text{ with:}$$

$$
\begin{cases}
(x, y, z) = (x_e, y_e, z_e) \in \boldsymbol{X} \text{ s.t.} \\
\quad \begin{cases}
x_m + ih \le x_e < x_m + (i+1)h \\
y_m + jh \le y_e < y_m + (j+1)h \\
z_m + kh \le z_e < z_m + (k+1)h
\end{cases} \\
\mathcal{F} = F(i, j, k)
\end{cases}
\qquad (13)
$$

The proposed scheme allows us to consider the 3D information contained in the 3D point cloud by processing the associated volume. We will illustrate in the next section the relevance of such an approach.

4   Experiments

4.1   Dataset

Experiments have been carried out on the Paris Lille 3D dataset [14]. The LiDAR tiles considered here have been acquired by a mobile laser scanning (MLS) on a street of Paris. As the acquisition source is close to the ground, the point density varies greatly according to the distance of the scanned object (as a consequence, density based max-tree will tend to remove distant objects). However the point cloud density of this dataset is significantly high (between $1,000$ to $2,000$ per square meters), which gives flexibility in the choice of the spatial resolution h in the voxelization process. Additional data is available, with LiDAR intensity return and label associated with each point in the cloud.

4.2   Experimental Setup

As a first experiment, we have chosen to filter the labelled point cloud, which is illustrated in Fig. 2a. For this point cloud we have fixed the step of the voxel grid to $h = 10$ cm. The labels are given with the dataset and ordered as follows: void (value 0), unclassified (1), ground (2), road sign and traffic light (4), bol-lard (5), trash can (6), barrier (7), pedestrian (8), car (9) and vegetation (10).

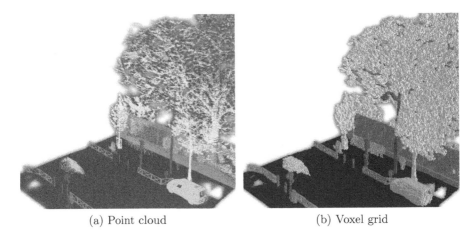

(a) Point cloud   (b) Voxel grid

Fig. 2. Visualization of the valued point cloud and the corresponding voxel grid. The 5 classes are represented as follows: road in purple, cars in green, fences in teal, trees in yellow and urban furniture in blue. (Color figure online)

This order has been used to construct max-tree on the [0,10] value range. We have represented for each cell of the grid the majority-class of the points (Fig. 2b).

We have built the max-tree considering 26-connectivity (i.e. two voxels (i, j, k) and (i', j', k') are neighbors if $\max(|i - i'|, |j - j'|, |k - k'|) = 1$). The tree is augmented with spectral features such as mean grey level and standard deviation and also with spatial features such as the volume, the bounding box and several geometric ratios. During the filtering process, we used the direct non-pruning strategy for the sake of simplicity and to retrieve all the objects corresponding to the required description on non-increasing criteria (extent in our case). Then the filtered max-tree is transformed back into a 3D volume and reprojected into a point cloud (therefore function F in (13) is the label value).

With this experiment we were able to interactively filter objects from the max-tree with geometric object attributes (e.g. volume, height, compactness) in order to choose the appropriate threshold values. The illustration given in Fig. 3a is the voxel grid result of a filtering with the volume criterion set as $1,000 < A_v \leq 5,000$. We then transfer the filtered result back into the point cloud (Fig. 3b).

An additional example of attribute filtering is given in Fig. 4. We can observe the relevance of the height attribute to extract tall objects (e.g. the lamp post is the only object with a height comprised between 10 and 13 m, Fig. 4a). Considering the extent attribute with value between 0.14 and 0.16 allows us to highlight road signs, cars and a few branches, as shown in Fig. 4b. Finally, it is possible to combine multiple attribute for a more precise filtering, e.g. objects with height between 1.5 and 3 m, a volume greater than 1,000 and an extent between 0.14 and 0.16 match cars in Fig. 4c.

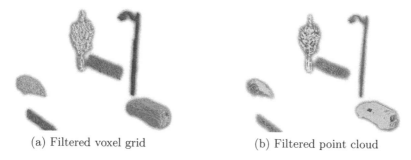

(a) Filtered voxel grid   (b) Filtered point cloud

Fig. 3. Visualization of the filtered voxel grid from (Fig. 2b) with an area criterion (a) and the reprojection into the point cloud (b).

Among the current limitations of the proposed approach, we have noticed that some small blocking artifacts were appearing at the border between two objects in the point cloud (see close-up view in Fig. 5). These artifacts are directly linked with the grid discretization method and depend on the voxel size. In our experiments, we observe that artifacts remain small with $h = 10$ cm.

The previous experiments showed that attribute filtering is useful to filter a labeled point cloud. The labels are either defined by visual expert analysis or by automatic classification of the raw LiDAR data. It is also possible to filter directly such raw data. We illustrate some results of preliminary experiments with LiDAR intensity in Fig. 6. We can see here the relevance of attribute filters to remove the noise in the point cloud. Indeed, the noise can be easily characterized by geometric attributes (e.g. small volume nodes correspond to points disconnected from the rest of the scene, see Fig. 6a). It is also relevant to remove extreme intensity values (i.e. outliers) with the max-tree.

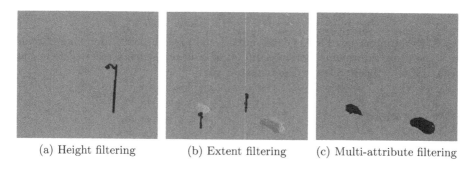

(a) Height filtering   (b) Extent filtering   (c) Multi-attribute filtering

Fig. 4. Different attribute filters on V: connected components characterized by (a) $A_h \in [100, 130]$, (b) $A_e \in [0.14, 0.16]$, (c) $A_h \in [15, 30]$, $A_v > 1,000$, and $A_e \in [0.14, 0.16]$.

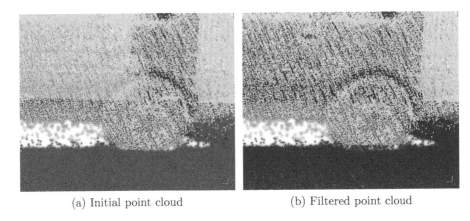

(a) Initial point cloud   (b) Filtered point cloud

Fig. 5. Close-up of the area where an object is in contact with the ground. In the filtered point cloud (b), blocking artifacts may appear at the boundary between objects (in this image part of the wheel is "blocked" in the road.

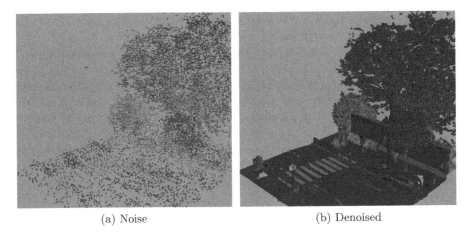

(a) Noise   (b) Denoised

Fig. 6. Denoising of raw LiDAR data: outliers and isolated points are removed from the scene using the volume attribute: (a) $A_v < 2$ identifies noise removed from the scene, (b) $A_v \geq 2$ performs the denoising.

5   Conclusion

While most approaches for applying mathematical morphology on point clouds are relying on a prior rasterization step into a 2D image, we explore here a different strategy. Indeed, we rather suggest a discretization of the space into voxels to build a 3D volume instead of a 2D image. This choice is motivated by the straightforward extension of morphological operators (including hierarchies) to nD data considering a connectivity of higher dimension. It allows us to benefit from a richer hierarchical representation where each node contains a set of voxels from which advanced features can be computed. We illustrate the relevance of such a framework with the popular attribute filtering, that is applied here on the voxel hierarchy before reconstructing a filtered point cloud. The results obtained on an urban point cloud show the performance of the proposed strategy, that goes far beyond the standard 2D processing usually considered in the literature.

Future work will include extending our proposal to other hierarchical models (e.g. min-tree, tree of shapes) as well as considering some other tree-based tools beyond the standard attribute filtering. Note than an alternative could be to a create nearest neighbour graph representation from the raw set of points and perform filtering using max-tree on this graph. Besides, multiscale description with attribute profiles and data segmentation with hierarchical cuts (to name a few) would be of great interest to deal with point clouds such as those considered in remote sensing. Finally, the blocking artifacts discussed in the paper call for some further studies.

References

1. Aijazi, A., Checchin, P., Trassoudaine, L.: Segmentation based classification of 3D urban point clouds: a super-voxel based approach with evaluation. Remote Sens. **5**(4), 1624–1650 (2013)
2. Calderon, S., Boubekeur, T.: Point morphology. ACM Trans. Graph. **33**(44) (2014)
3. Dufour, A., et al.: Filtering and segmentation of 3D angiographic data: advances based on mathematical morphology. Med. Image Anal. **17**(2), 147–164 (2013)
4. Ferdosi, B.J., Buddelmeijer, H., Trager, S., Wilkinson, M.H.F., Roerdink, J.B.T.M.: Finding and visualizing relevant subspaces for clustering high-dimensional astronomical data using connected morphological operators. In: IEEE Symposium on Visual Analytics Science and Technology, pp. 35–42 (2010)
5. Géraud, T., Carlinet, E., Crozet, S., Najman, L.: A quasi-linear algorithm to compute the tree of shapes of nD images. In: Hendriks, C.L.L., Borgefors, G., Strand, R. (eds.) ISMM 2013. LNCS, vol. 7883, pp. 98–110. Springer, Heidelberg (2013). https://doi.org/10.1007/978-3-642-38294-9_9
6. Gorte, B., Pfeifer, N.: Structuring laser-scanned trees using 3D mathematical morphology. Int. Arch. Photogrammetry Remote Sens. **35**(B5), 929–933 (2004)
7. Grossiord, E., Talbot, H., Passat, N., Meignan, M., Terve, P., Najman, L.: Hierarchies and shape-space for PET image segmentation. In: IEEE International Symposium on Biomedical Imaging, pp. 1118–1121 (2015)
8. Guiotte, F., Lefevre, S., Corpetti, T.: IEEE/ISPRS Joint Urban Remote Sensing Event (2019)
9. Hernández, J., Marcotegui, B.: Ultimate attribute opening segmentation with shape information. In: Wilkinson, M.H.F., Roerdink, J.B.T.M. (eds.) ISMM 2009. LNCS, vol. 5720, pp. 205–214. Springer, Heidelberg (2009). https://doi.org/10.1007/978-3-642-03613-2_19
10. Kiwanuka, F.N., Ouzounis, G.K., Wilkinson, M.H.F.: Surface-area-based attribute filtering in 3D. In: Wilkinson, M.H.F., Roerdink, J.B.T.M. (eds.) ISMM 2009. LNCS, vol. 5720, pp. 70–81. Springer, Heidelberg (2009). https://doi.org/10.1007/978-3-642-03613-2_7

11. Kiwanuka, F.N., Wilkinson, M.H.F.: Radial moment invariants for attribute filtering in 3D. In: Köthe, U., Montanvert, A., Soille, P. (eds.) WADGMM 2010. LNCS, vol. 7346, pp. 68–81. Springer, Heidelberg (2012). https://doi.org/10.1007/978-3-642-32313-3_5

12. Padilla, F.J.A., et al.: Hierarchical forest attributes for multimodal tumor segmentation on FDG-PET/contrast-enhanced CT. In: IEEE International Symposium on Biomedical Imaging, pp. 163–167 (2018)

13. Peternell, M., Steiner, T.: Minkowski sum boundary surfaces of 3D-objects. Graph. Models **69**(3–4), 180–190 (2007)

14. Roynard, X., Deschaud, J.E., Goulette, F.: Paris-Lille-3D: a large and high-quality ground truth urban point cloud dataset for automatic segmentation and classification. ArXiv e-prints (2017)

15. Salembier, P., Oliveras, A., Garrido, L.: Antiextensive connected operators for image and sequence processing. IEEE Trans. Image Process. **7**(4), 555–570 (1998)

16. Salembier, P., Wilkinson, M.: Connected operators. IEEE Signal Process. Mag. **26**(6), 136–157 (2009)

17. Serna, A., Marcotegui, B., Hernández, J.: Segmentation of facades from urban 3D point clouds using geometrical and morphological attribute-based operators. ISPRS Int. J. Geo-Inf. **5**(1), 6 (2016)

18. Serna, A., Marcotegui, B.: Detection, segmentation and classification of 3D urban objects using mathematical morphology and supervised learning. ISPRS J. Photogrammetry Remote Sens. **93**, 243–255 (2014)

19. Urbach, E., Wilkinson, M.: Shape-only granulometries and grey-scale shape filters. In: International Symposium on Mathematical Morphology, pp. 305–314 (2002)

20. Urien, H., Buvat, I., Rougon, N., Soussan, M., Bloch, I.: Brain lesion detection in 3D PET images using max-trees and a new spatial context criterion. In: Angulo, J., Velasco-Forero, S., Meyer, F. (eds.) ISMM 2017. LNCS, vol. 10225, pp. 455–466. Springer, Cham (2017). https://doi.org/10.1007/978-3-319-57240-6_37

21. Westenberg, M.A., Roerdink, J.B.T.M., Wilkinson, M.H.F.: Volumetric attribute filtering and interactive visualization using the max-tree representation. IEEE Trans. Image Process. **16**(12), 2943–2952 (2007)

22. Wilkinson, M.H.F., Westenberg, M.A.: Shape preserving filament enhancement filtering. In: Niessen, W.J., Viergever, M.A. (eds.) MICCAI 2001. LNCS, vol. 2208, pp. 770–777. Springer, Heidelberg (2001). https://doi.org/10.1007/3-540-45468-3_92

Complementarity-Preserving Fracture Morphology for Archaeological Fragments

Hanan ElNaghy$^{(\boxtimes)}$ and Leo Dorst

Computer Vision Lab, University of Amsterdam, Amsterdam, The Netherlands
{hanan.elnaghy,l.dorst}@uva.nl

Abstract. We propose to employ scale spaces of mathematical morphology to hierarchically simplify fracture surfaces of complementarily fitting archaeological fragments. This representation preserves complementarity and is insensitive to different kinds of abrasion affecting the exact fitting of the original fragments. We present a pipeline for morphologically simplifying fracture surfaces, based on their Lipschitz nature; its core is a new embedding of fracture surfaces to simultaneously compute both closing and opening morphological operations, using distance transforms.

Keywords: Boundary Morphology · Scale space ·
Fracture representation · Abrasion · Lipschitz · Extrusion

1 Introduction

The GRAVITATE H2020 project aims at providing archaeologists with the virtual tools to analyse digital artefacts, distributed across several collections (https://gravitate-project.eu/). The artefacts are digitally scanned by standard scanning techniques in the form of 3D meshes, capturing the geometrical properties of each object and some of its photometric proprieties. Reassembly of broken artefacts is one of the core objectives of the project.

The test case is on terracotta, which does not deform; but typical fragments are lacking material through abrasion and chipping. Each fragment undergoes a preliminary preprocessing step [6] which partitions its surface into significant sub-parts called '*facets*'. Each facet is characterized by its own geometrical properties of roughness of its surface and sharpness of its boundary, which in turn guides its categorization as either belonging to the fracture region or outside skin region of the fragment.

Our contribution to the project focuses on structuring the pair-wise alignment of promising fragments nominated for fitting by other selection modules. The computational approach is based purely on the geometrical properties of the fracture facets, ignoring other clues (such as possible continuity of decorative patterns).

This research was funded by the GRAVITATE project under EU2020-REFLECTIVE-7-2014 Research and Innovation Action, grant no. 665155.

© Springer Nature Switzerland AG 2019
B. Burgeth et al. (Eds.): ISMM 2019, LNCS 11564, pp. 403–414, 2019.
https://doi.org/10.1007/978-3-030-20867-7_31

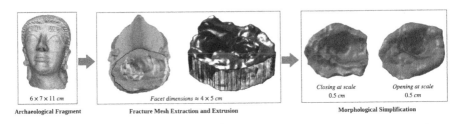

Fig. 1. Morphological simplification pipeline for archaeological fracture facets.

Optimal pair-wise alignment of potentially counter-fitting fragments is computationally expensive. We therefore take a hierarchical approach, considering the fragments in increasing resolution from simplified to detailed.

The common way of simplifying shapes represented as meshes is by various forms of linear filtering of their vertices; letting them move with a differentiable flow (such as based on the Poisson equation) to produce a smoother model [12]. Structure-aware mesh decimation to preserve the structure of the mesh by representing it as a set of pre-computed canonical proxies has been also studied by [16]. Such simplifications are strongly related to characterizing a surface by differential geometric features, such as those computed by some means of discrete differential geometry [5]. Indeed most methods to reassemble broken objects [8,10,14,15,19] are all based on such features. However, a small chip, resulting in a missing part of an object, will affect such differential measures in a *linear* fashion: the deeper the missing bit, the stronger the simplified shape is affected. This is undesirable: the actual match between part and counterpart is simply lacking some local evidence of a more binary nature. The effect of this gap should not depend on the depth of the hole (a gap with the same surface outline but twice as deep is not twice as bad); so our representation method should not be too sensitive to this either. For the same reason, applying ICP (Iterative Closest Point) algorithm to find a matching counterpart based on squared errors is not appropriate; some extensions of ICP that include distance transforms [7] are more in line of what we might employ.

In our hierarchical approach, we need a representation where a coarser, simplified version of the fracture surface maintains its potential fit with a coarser, simplified version of the counterpart, under abrasion-type noise. Mathematical morphology with its opening and closing scale spaces, of increasing size, is the natural choice for just that kind of representation. Such scale spaces have been studied in [1,2,11], and it has been demonstrated that they simplify the shape, when measured in terms of certain descriptive features such as local maxima [11], or zero-crossings of the curvature [4]. But our intended use for alignment and complementarity checking appears to be new.

In this paper, we therefore hone standard Mathematical Morphology (MM) to this novel application. First, we define scale-based complementarity of objects (rather than one object and its background). Then, we define morphological operations working only on the fracture region of the fragments. We implement this 'Boundary Morphology' by taking advantage of the Lipschitz nature of the individual fracture facets. Our extrusion method uses a single distance transform to simultaneously produce both scaled dilations and scaled erosions of a fracture facet (as in Fig. 1), leading to a simplified representation at every scale. We briefly discuss how the fracture Lipschitz property also naturally gives quantitative bounds on the detectable complementarity and collision-free alignment at each scale of our representation, even in the presence of abrasion.

2   Complementarity Preserving Scale Space

Consider a fracture between two fragments X and Y in perfect alignment (or, if you prefer, consider a newly developed crack in an object). Within a well-chosen mask M (such as the part of space the object occupies), the two fragments X and Y are almost complementary in the usual MM sense, with the fracture area being ambiguous (does it belong to X or Y?). In an implementation such an infinitely thin layer is not going to make a difference, so we arbitrarily choose X to be a closed set and Y to be open. Then the M-restricted parts of X and Y, $X_M(= X \cap M)$ and $Y_M(= Y \cap M)$ are two relatively complementary sets ($X_M^c = Y_M$). The properties of *exact complementarity* (no overlap, no gaps) can then be algebraically formulated as:

$$\text{No Overlap} \iff \text{Non-Intersection: } X_M \cap Y_M = \emptyset \tag{1}$$
$$\text{No Gaps} \iff \text{Completeness: } X_M \cup Y_M = M \tag{2}$$

In archaeology, we do not acquire the fragments to be in alignment, or in perfect condition. We choose MM scale spaces to process the fracture region of each of the fragments separately, producing a simplified representation for each fracture facet at a range of scales ρ. Morphological scale spaces are well suited to this, since they can simplify objects while closely maintaining local complementarity, even when objects are slightly damaged (as we will discuss in Sect. 4).

Performing erosions or dilations would change the objects considerably, so we prefer to use *openings* and *closings* to process the fragments (even though much essential structure is already contained in eroded and dilated versions). Because of the arbitrary orientation of presented objects, and an assumed isotropy of fracturing, our morphological scale space is built using structuring elements that are balls of radius ρ.

At a scale ρ, the opened version $\gamma_\rho(X)$ will be complementary (in the above sense) to the closed version $\phi_\rho(Y)$, and vice versa. Since a closing $\phi_\rho(X)$ simplifies the local geometry of the fracture at valleys, the complementarity with $\gamma_\rho(Y)$ becomes less specific in those areas (one can easily construct non-complementary counterparts Y' with the same opening $\gamma_\rho(Y') = \gamma_\rho(Y)$, that do not fit the original X). To maintain specificity of peaks and valleys of the common fracture surface, we must therefore compute both $\gamma_\rho(X)$ and $\phi_\rho(X)$ for each fragment X. Since both opening and closing are increasing, coarser levels of the scale space are guaranteed to contain less detail, thus enabling the hierarchical approach (as well as being a justification for calling the process a 'simplification').

In a schema, if $R > \rho$, we have the following complementarity and containment relationships:

$$\gamma_R(X_M) \subseteq \gamma_\rho(X_M) \subseteq X_M \subseteq \phi_\rho(X_M) \subseteq \phi_R(X_M)$$
$$\updownarrow c \qquad \updownarrow c \qquad \updownarrow c \qquad \updownarrow c \qquad \updownarrow c \qquad \qquad (3)$$
$$\phi_R(Y_M) \supseteq \phi_\rho(Y_M) \supseteq Y_M \supseteq \gamma_\rho(Y_M) \supseteq \gamma_R(Y_M)$$

Here c refers to complementarity within a well-chosen mask M containing the common fracture of X and Y. This 'well-chosen' actually hides some essential details, since masking does not commute with morphology (e.g., $\phi_\rho(X_M) \neq (\phi_\rho(X))_M$). It is not hard to show that when originally there was exact complementarity between X and Y within a mask M, then in the scale space exact complementarity still holds between $\phi_\rho(X)$ and $\gamma_\rho(Y)$, or between $\gamma_\rho(X)$ and $\phi_\rho(Y)$, but is only guaranteed within a doubly eroded mask $\varepsilon_{2\rho}(M)$.

3   Fracture Morphology

Consider a broken vase producing thin sherds as fragments. Volumetric morphological simplification of the fragments would be limited in scale by their thickness: an opening by a ball with a radius ρ of more than half this thickness would results in empty sets, which would only be complementary to their counterpart in a trivial way. Since the fractures themselves have informative morphological structure transcending this scale, we design in this section a way to focus the morphological processing purely on the fracture surface (rather than on fragment volumes). This should suffice for our puzzle, since whether two broken fragments can be refit locally depends on the complementary shape of their fractures only.

3.1   Boundary Morphology

In Sect. 2, we had masked objects X_M and Y_M, with complementarity in M (and a slight issue of how we treat their common boundary fracture F). Now consider only one of them, no longer in contact, and investigate how to compute its openings and closings. We are only interested in the effect on F, not on the remainder of the object volume within M.

F is a boundary, not a volume, and the classical MM does not apply immediately. We first define what we mean by applying mathematical morphology to a boundary. Let us call it '*Boundary Morphology*' and denote its operations by an over-bar. Consider a general object A of which we take the boundary ∂A (and ignore the masking effects for the moment). As we have seen, for true complementarity some objects may be open sets, others closed sets. We need to take the boundary of either, so we will always close the set and then apply the usual boundary operator ∂ to define our boundary operator (which we denote by $\overline{\partial}$):

$$\overline{\partial} A = \partial \overline{A} \qquad \qquad (4)$$

Thus $\overline{\partial} A$ returns the set of points of $\overline{A}$ that have neighbours in $\overline{A}^c$.

In order to distinguish erosion and dilation, we need to make the boundary oriented, for instance by denoting the outward pointing normal at each location. Then this orientation changes to its opposite if we consider the boundary of the complement. To more easily denote this, write the complement as the prefix operator c. We then have:

$$\bar{c}\,\overline{\partial}\,A \equiv \overline{\partial}\,c\,A\;(=\overline{\partial}\,A^c),\tag{5}$$

which defines the complementation $\bar{c}$ of the boundary.

We define what it means to dilate and erode a boundary, defining operators $\overline{\delta}$ and $\overline{\varepsilon}$ in terms of classical volumetric operators by:

$$\overline{\delta}\,\overline{\partial}\,A \equiv \overline{\partial}\,\delta\,A \quad\text{and}\quad \overline{\varepsilon}\,\overline{\partial}\,A \equiv \overline{\partial}\,\varepsilon\,A.\tag{6}$$

It now follows that we can perform an erosion as a dilation on the complemented (oppositely oriented) boundary:

$$\overline{\varepsilon}\,\overline{\partial}\,A = \overline{\partial}\,\varepsilon\,A = \overline{\partial}\,c\,\delta\,c\,A = \bar{c}\,\overline{\partial}\,\delta\,c\,A = \bar{c}\,\overline{\delta}\,\overline{\partial}\,c\,A = \bar{c}\,\overline{\delta}\,\bar{c}\,\overline{\partial}\,A.\tag{7}$$

and accordingly,

$$\overline{\varepsilon} = \bar{c}\,\overline{\delta}\,\bar{c} \quad\text{and}\quad \overline{\delta} = \bar{c}\,\overline{\varepsilon}\,\bar{c}.\tag{8}$$

In fact, this is merely the duality between dilation and erosion, extended to their boundary versions (with $\bar{c}$ playing the role of complementation). So once we can dilate a boundary, we can use that both to produce the dilation $\overline{\delta}$ and the erosion $\overline{\varepsilon}$ (by doing dilation on $\bar{c}\,\overline{\partial}\,A$, the oppositely oriented boundary):

This can be easily extended to boundary closing $\overline{\phi}$ and opening $\overline{\gamma}$ defined as:

$$\overline{\phi}\,\overline{\partial}\,A \equiv \overline{\varepsilon}\,\overline{\delta}\,\overline{\partial}\,A \quad\text{and}\quad \overline{\gamma}\,\overline{\partial}\,A \equiv \overline{\delta}\,\overline{\varepsilon}\,\overline{\partial}\,A\tag{9}$$

for it follows from Eqs. (7) and (8) that we can also perform an opening as a closing on the complemented boundary:

$$\overline{\gamma}\,\overline{\partial}\,A = \overline{\delta}\,\overline{\varepsilon}\,\overline{\partial}\,A = \bar{c}\,\overline{\varepsilon}\,\bar{c}\,\overline{\varepsilon}\,\overline{\partial}\,A = \bar{c}\,\overline{\varepsilon}\,\overline{\delta}\,\bar{c}\,\overline{\partial}\,A = \bar{c}\,\overline{\phi}\,\bar{c}\,\overline{\partial}\,A.\tag{10}$$

and accordingly,

$$\overline{\gamma} = \bar{c}\,\overline{\phi}\,\bar{c} \quad\text{and}\quad \overline{\phi} = \bar{c}\,\overline{\gamma}\,\bar{c}.\tag{11}$$

Now, the schema of Eq. (3) can be rewritten to describe the complementarity of two exactly fitting fracture boundaries (at scales R and ρ with $R > \rho$), as follows:

$$\overline{\gamma}_R\overline{\partial}\,(X_M) \subseteq \overline{\gamma}_\rho\overline{\partial}\,(X_M) \subseteq \overline{\partial}\,(X_M) \subseteq \overline{\phi}_\rho\overline{\partial}\,(X_M) \subseteq \overline{\phi}_R\overline{\partial}\,(X_M)$$
$$\updownarrow\bar{c} \qquad \updownarrow\bar{c} \qquad \updownarrow\bar{c} \qquad \updownarrow\bar{c} \qquad \updownarrow\bar{c} \tag{12}$$
$$\overline{\phi}_R\overline{\partial}\,(Y_M) \supseteq \overline{\phi}_\rho\overline{\partial}\,(Y_M) \supseteq \overline{\partial}\,(Y_M) \supseteq \overline{\gamma}_\rho\overline{\partial}\,(Y_M) \supseteq \overline{\gamma}_R\overline{\partial}\,(Y_M)$$

where $\bar{c}$ (complementarity) is taken within M.

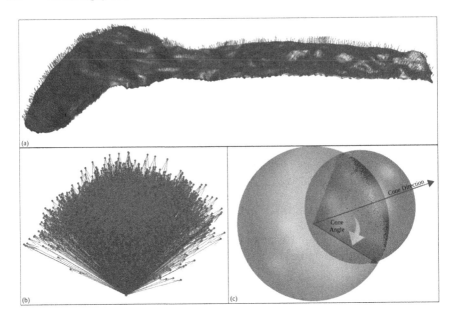

Fig. 2. (a) Fracture surface with blue local normal directions. (b) Mapping of the normals to a unit sphere of directions. (c) The minimum bounding sphere (in green) determines the minimum bounding cone on the unit sphere of directions (in cyan). (Color figure online)

3.2 Lipschitz Condition

Terracotta is uniform and brittle. It fractures rather simply such that there is at least one 3D direction from which the whole fracture facet is completely visible, i.e. a ray in that direction emanating from each point of the fracture will not encounter any other point of that local fracture facet. Accordingly, a given fracture facet can be represented as a function (a Monge patch); noting that another fracture facet of the same fragment may require a different visibility direction to be seen as a function. The visibility assumption can be efficiently characterized in terms of the 'Lipschitz condition' which frequently occurs in quantitative mathematical morphology [17].

For any pair of points on the graph of a slope-limited Lipschitz function with slope s, the absolute value of the slope of the line connecting them is always less than s (i.e., the Lipschitz constant). As a consequence, there exists a double cone whose vertex can be moved along any such Lipschitz-continuous function, so that it always remains entirely visible from the principal direction of the cone while the rest of the function is completely outside the cone. For a local fracture to meet the visibility condition, it should meet the Lipschitz condition over its entire domain.

To compute the Lipschitz slope s, we first collect the local normal vectors of the fracture facet on a unit sphere of directions, and then fit the largest possible cone inside this set (see Fig. 2). The axis of this cone we call the *cone direction*, and the opening angle we call the *cone angle*. Let us use the cone direction as the principal direction to align the fracture facet, so that it can be described as a Lipschitz function. This fracture function is then bounded in Lipschitz slope s, which is the cotangent of the cone angle.

In practice, the Lipschitz condition appears to intrinsically hold for each of our terracotta fractures. Even if it does not hold, we can always artificially separate the fracture regions by the *Faceting* preprocessing procedure [6] that delivers the piecewise Lipschitz facets of the fracture region.

3.3   Extrusion

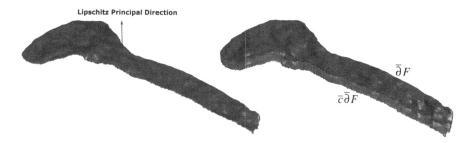

Fig. 3. Extrusion of a fracture facet $\overline{\partial} F$ along the Lipschitz principal direction, generating a volume (cylinder) with $\overline{\partial} F$ as its top red surface, and the oppositely oriented $\overline{c}\overline{\partial} F$ as its bottom aquamarine surface. (Color figure online)

The fracture surfaces, extracted using the Faceting preprocessing operation [6], are locally Lipschitz. The corresponding Lipschitz principal direction acts as the 'average breakage direction' for the whole fracture facet. Therefore, we can generate a thickened fracture volume by extruding the facet along such visibility direction without self-intersection. Such an extrusion effect is equivalent to making two copies of the fracture surface bounded by a generalized cylinder: one with original (non-inverted) normals $\overline{\partial} F$ and one with inverted normals $\overline{c}\overline{\partial} F$ (see Fig. 3). The outward propagation of the copy with non-inverted normals is the fracture surface dilation ($\overline{\delta}_\rho$), while the propagation of the one with inverted normals is the erosion ($\overline{\varepsilon}_\rho$), by Eq. (8). By a subsequent inward propagation of the resulting expanded surfaces, by the same amount ρ, one then acquires the closing ($\overline{\phi}_\rho$) and the opening ($\overline{\gamma}_\rho$) of the fracture.

Focusing our morphological simplification on the extruded fracture volume has the following advantages:

(a) Extending the functional morphological scale (ball radius ρ) beyond the minimum thickness of the archaeological object by only needing to do closing.
(b) Permitting simultaneous propagation of the fracture surface to both sides, thus allowing the production of MM opened and closed surfaces in one go.

(c) Avoiding the dilation of elements outside the mask (which could affect the outcome within the mask) by having the extruded surface surrounded by sufficient empty space completely isolated from outside influences.

(d) Avoiding needless processing of the non-fracture regions of the object.

3.4 Distance-Transform Based MM Implementation

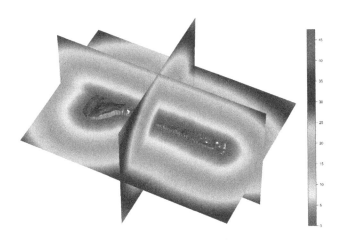

Fig. 4. Distance field computation, shown two cross sections.

We obtain our data as 3D meshes, so it would be natural to apply the operations of boundary morphology to fracture facets represented directly as extruded 3D meshes. However, applying morphological operations directly on a mesh representation is notoriously hard (the dilation and erosion operations lead to many additional vertices as the mesh faces start intersecting [9], and the ball structuring elements add many more). At this point of our narrative, it would seem that Point Morphology [3] is also a good candidate; we will see in Sect. 4 that we need more than just the simplified surfaces by themselves to determine our morphological features.

An alternative implementation of morphology, especially when done in a scale space context, is by means of the isosurfaces of distance functions. Since we do want to perform our operations at different scales of a ball-shaped structuring element, this is especially attractive. Two basic approaches exist: distance transform considered as a numerical level set in a space of 1 more dimension [18] or classic distance transform on a 3D grid. We adopted the latter approach which: (a) suffices as a straightforward demonstration of the validity of our concepts and (b) enables us to keep track of the original fracture points propagation in the distance field and their contribution to the generated MM surfaces through what we call the provenance map.

Our implementation relies on distance transforms in a volumetric representation [20], oriented with the Lipschitz principal direction along its z-direction. We convert the extruded mesh to binary representation by embedding the fracture surface in a 3D voxelized grid of a well chosen resolution, of step size g. We derived that the maximum outward and inward displacement that could take place due to discretization effects is no more than $\sqrt{3}g/2$. This implies that distances (and hence MM surfaces) are affected by no more than this amount. The generated grid is padded with an empty region sufficient to contain the maximally dilated versions of the fracture surfaces. For generating our distance field, the method proposed by [13] is employed, which calculates the euclidean distance transform in linear time on a binary voxelized representation of the object as shown in Fig. 4. The closing of the fracture volume with a ball of radius ρ (closing is all we need by virtue of Eq. (11)) is performed by first extracting the outward level set at distance ρ, then re-computing the distance field of the background and extracting the inward level set at distance ρ.

3.5   MM Surfaces

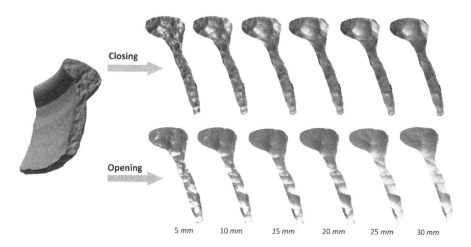

Fig. 5. Closing and opening scale spaces simplifying a fracture with increasing scale ρ.

Intuitively, morphologically closing the entire extruded volume is equivalent to rolling a ball on the fracture surface from both sides. The upper rolling (constrained by the fracture peaks) is equivalent to the closing effect, while the lower rolling (constrained by the fracture valleys) is equivalent to the desired opening effect.

Figure 5 shows the closed and opened simplified MM surfaces at 6 different scales of a given fracture surface with $5.8K$ vertices and $11K$ faces. The grid resolution is $g = 0.2$ mm of size $119 \times 323 \times 29$ and is padded with 151 voxels from each side to permit the outward propagation of the fracture surface up to 30 mm without losing information, thus producing a grid of size $421 \times 625 \times 331$.

The fracture Lipschitz computations together with the extrusion and embedding take less than 2 s for a mesh of size $10K$. The distance field computation time is more affected by the resolution of the grid: computing the distance transform for $500 \times 500 \times 500$ grid takes at most 1 min. Extracting the closed and opened MM surfaces takes less than 50 s for the same grid size. In the GRAVITATE system, such times are considered acceptable. We performed our experiments on an Intel(R) Xeon(R) CPU 2.9 GHz $\times$ 8 computer with 256 GB RAM.

4   Inexact Complementarity

In practice, there are many aspects to our data that prevents complementarity from being exact. However, using the Lipschitz condition for fracture surface characterization together with MM for hierarchical fracture simplification enables us to establish the bounds on the amount of inexactness. In this section, we briefly discuss the sources causing inexactness and how to treat their associated deviations.

- *Abrasion.* When a fractal surface as brittle as terracotta is rubbed, this will tend to take off the sharp local peaks without affecting the valleys. We propose that a reasonable model of such effect is that the object undergoes a morphological opening γ_α of a small size α.

  Assuming this, the opening scale space will be unaffected for $\rho > \alpha$, since $\gamma_\rho(\gamma_\alpha(X)) = \gamma_\rho(X)$ for $\rho \geq \alpha$. However, the closing *is* affected by the abrasion. Therefore, exact complementarity of $\gamma_\rho(X')$ and $\phi_\rho(Y')$ of Eq. (3) schema no longer holds for the abraded version $X' = \gamma_\alpha(X)$ and $Y' = \gamma_\alpha(Y)$, even if it did for the original X and Y. Nevertheless, the difference is bounded for the Lipschitz functions we are considering as the fracture facets. A simple sketch of a local sphere of radius α capped by a Lipschitz cone of slope s shows that the maximal deviation of a possible original surface and its α-opening is $(\sqrt{1+s^2} - 1) \simeq \frac{1}{2}\alpha s^2$. Since X' does not differ by more than $\frac{1}{2}\alpha s^2$ from X, also $\phi_\rho(X')$ will not differ from $\phi_\rho(X)$ by more than this amount. With that in mind, we can still test the possibility of complementarity, to within this bound, of the opened and closed abraded fractures.

- *Fracture Adjacency Effects.* Each facet is delineated by an outer border (contour). That border defines the complementarity zone of its processed MM surfaces. In practice, there are two different kinds of facet borders: *fracture-skin* border and *fracture-fracture* border. The former is the contour's segment that is adjacent to a facet which belongs to the exterior skin of the original object. By contrast, a fracture-fracture border occurs when the original fragment is further broken into sub-fragments, causing the adjacency of more than one fracture facet.

  If the facet has only fracture-skin borders, then complementarity with its counterpart still holds over the entire surfaces of its simplified MM versions across all scales. For facets with fracture-fracture borders, one should expect complementarity to hold within regions away from the border. But there is a scale-dependent zone adjacent to the fracture-fracture borders where the MM surfaces are no longer guaranteed to be complementary to those of its counterpart. In that zone, the representation of either part may have been affected by the secondary breaking. The bounds we can obtain on this zone under the general assumptions we made (such as Lipschitz) are weak. We have found that we can much more specifically delineate it through analysing the provenance map of the distance transform. Our current work is on efficient representation of the simplified scale space surfaces, taking this effect into account.

- *Misalignment.* Complementarity only exists in perfect alignment of the two counterparts; any offset in translation or orientation will destroy it. But this effect is not as boolean as Eqs. (1) and (2) makes it appear: the complementarity is quantifiable by means of the MM scale spaces. The distance transform method of producing the openings and closings at different scales can be employed to guide the alignment, from coarser to finer scales, with the separation between one surface's opening and closing at each scale giving an indication of reasonable bounds on the fine alignment at that scale.

5   Conclusion and Future Work

We have presented morphological simplification of the fracture facets of archaeological fragments based on their scanned 3D mesh representation, as a preparatory phase for optimal pair-wise alignment. This is a problem that lends itself very well to treatment by mathematical morphology, since that framework provides complementarity-preserving simplification of shapes in a manner that is insensitive to the kind of missing information and abrasion that we expect in the archaeological fragments. We showed how to perform morphology on boundaries in Sect. 3.1. The assumption that each fracture facet is a Lipschitz function allowed us to set up the extrusion method in Sect. 3.3. The duplicate internal and external copies of the local fracture facet surface enable computing the opening and closing scale spaces simultaneously, in a volumetric representation in Sect. 3.5.

The resulting surfaces are simpler, but still contain the essentials of complementarity. We are currently investigating the compact characterization of the MM scale space surfaces by means of characteristic scale space medial axis points, computed directly from the distance transform and its provenance map (specifying which points are influential at each location). This should allow the use of standard registration algorithms on those morphologically relevant feature points only. Moreover, we plan to use the provenance map of the distance transform to automatically control the reliability of those representations when parts of the original fracture facet are missing.

References

1. van den Boomgaard, R., Dorst, L.: The morphological equivalent of Gaussian scale space. In: Proceedings of Gaussian Scale-Space Theory, pp. 203–220. Kluwer (1997)
2. Bosworth, J.H., Acton, S.T.: Morphological scale-space in image processing. Digit. Signal Process. **13**(2), 338–367 (2003)
3. Calderon, S., Boubekeur, T.: Point morphology. ACM Trans. Graph. **33**(4), 1–13 (2014)
4. Chen, M.H., Yan, P.-F.: A multiscaling approach based on morphological filtering. IEEE Trans. PAMI **11**(7), 694–700 (1989)
5. Crane, K., de Goes, F., Desbrun, M., Schröder, P.: Digital geometry processing with discrete exterior calculus. In: ACM SIGGRAPH 2013 Courses, SIGGRAPH 2013. ACM, New York (2013)
6. ElNaghy, H., Dorst, L.: Geometry based faceting of 3D digitized archaeological fragments. In: 2017 IEEE ICCV Workshops, pp. 2934–2942 (2017)
7. Fitzgibbon, A.: Robust registration of 2D and 3D point sets. Image Vis. Comput. **21**, 1145–1153 (2003)
8. Funkhouser, T., et al.: Learning how to match fresco fragments. J. Comput. Cult. Herit. **4**(2), 1–13 (2011)
9. Hachenberger, P.: 3D Minkowski sum of polyhedra. In: CGAL User and Reference Manual. CGAL Editorial Board, 4.11.1 edn. (2018)
10. Huang, Q.X., Flöry, S., Gelfand, N., Hofer, M., Pottmann, H.: Reassembling fractured objects by geometric matching. In: ACM SIGGRAPH Papers, SIGGRAPH 2006, pp. 569–578. ACM, New York (2006)
11. Jackway, P.T., Deriche, M.: Scale-space properties of the multiscale morphological dilation-erosion. IEEE Trans. PAMI **18**(1), 38–51 (1996)
12. Kazhdan, M., Bolitho, M., Hoppe, H.: Poisson surface reconstruction. In: Proceedings of the Fourth Eurographics Symposium on Geometry Processing, SGP 2006, pp. 61–70. Eurographics Association, Aire-la-Ville (2006)
13. Maurer, C.R., Qi, R., Raghavan, V.: A linear time algorithm for computing exact Euclidean distance transforms of binary images in arbitrary dimensions. IEEE Trans. PAMI **25**(2), 265–270 (2003)
14. McBride, J.C., Kimia, B.B.: Archaeological fragment reconstruction using curve-matching. In: IEEE CVPR Workshops, vol. 1, p. 3 (2003)
15. Palmas, G., Pietroni, N., Cignoni, P., Scopigno, R.: A computer-assisted constraint-based system for assembling fragmented objects. In: 2013 Digital Heritage International Congress (DigitalHeritage), vol. 1, pp. 529–536 (2013)
16. Salinas, D., Lafarge, F., Alliez, P.: Structure-aware mesh decimation. Comput. Graph. Forum **34**, 211–227 (2015)
17. Serra, J.C.: Lipschitz lattices and numerical morphology. In: SPIE Conference on Image Algebra and Morphological Image Processing II, vol. 1568, pp. 54–56 (1991)
18. Sethian, J.: Fast marching methods and level set methods for propagating interfaces. In: Computational Fluid Dynamics. 29th Annual Lecture Series (1998)
19. Toler-Franklin, C., Brown, B., Weyrich, T., Funkhouser, T., Rusinkiewicz, S.: Multi-feature matching of fresco fragments. ACM Trans. Graph. **29**(6), 1–12 (2010)
20. Wu, Z., et al.: 3D shapenets: a deep representation for volumetric shapes. In: IEEE CVPR, pp. 1912–1920 (2015)

Pattern Spectra from Different Component Trees for Estimating Soil Size Distribution

Petra Bosilj[1,2]([✉]) [ID], Iain Gould[2] [ID], Tom Duckett[1] [ID], and Grzegorz Cielniak[1] [ID]

[1] Lincoln Centre for Autonomous Systems, School of Computer Science, University of Lincoln, Brayford Campus, Lincoln LN6 7TS, UK
[2] Lincoln Institute for Agri-food Technology, University of Lincoln, Riseholme Park, Lincoln LN2 2LG, UK
{pbosilj,tduckett,gcielniak,igould}@lincoln.ac.uk

Abstract. We study the pattern spectra in context of soil structure analysis. Good soil structure is vital for sustainable crop growth. Accurate and fast measuring methods can contribute greatly to soil management decisions. However, the current in-field approaches contain a degree of subjectivity, while obtaining quantifiable results through laboratory techniques typically involves sieving the soil which is labour- and time-intensive. We aim to replace this physical sieving process through image analysis, and investigate the effectiveness of pattern spectra to capture the size distribution of the soil aggregates. We calculate the pattern spectra from partitioning hierarchies in addition to the traditional max-tree. The study is posed as an image retrieval problem, and confirms the ability of pattern spectra and suitability of different partitioning trees to re-identify soil samples in different arrangements and scales.

Keywords: Pattern spectra · Inclusion trees · Partitioning trees · Soil structure

1 Introduction and Motivation

Soil structure concerns the arrangement of soil aggregates to provide an environment to facilitate good access to water, air and nutrients, and a suitable medium for root development [6]. Good soil structure is fundamental to sustainable crop growth, and can contribute to reducing the environmental impact of agriculture. Hence, robust and accurate methods to measure soil structure are important tools for informing soil management decisions.

A range of methods is available to assess soil structure, from in-field visual investigations [13,19], to lab-based assessments of soil aggregates and structure [1], both with their individual merits and drawbacks. Field-based methods can provide a rapid 'by eye' assessment by field practitioners, but as such rely on

© Springer Nature Switzerland AG 2019
B. Burgeth et al. (Eds.): ISMM 2019, LNCS 11564, pp. 415–427, 2019.
https://doi.org/10.1007/978-3-030-20867-7_32

some degree of subjectivity, limiting the reliability of comparison between different users. Conversely, laboratory techniques produce a quantifiable result for comparison of soil structure, but these methods tend to be time-consuming and labour-intensive. Developments in image analysis of soil aggregate distribution could provide a solution to some of these limitations, either by providing quantifiable, and comparable, units to in-field visual assessments, or by speeding up the laboratory process through replacement of manual sieving.

The laboratory techniques quantify the soil structure in terms of ratios of soil aggregates of certain sizes in the soil. In order to separate aggregates of different sizes, a stack of sieves of decreasing weight sizes is used. The residue in each sieve is then measured in terms of aggregate volume or mass, and used to categorise the soil into 5 categories ranging from friable (Sq1) to very compact (Sq5) according to the VESS (Visual Evaluation of Soil Structure) methodology. This physical process closely matches the algorithmic process of pattern spectra [15] where the image is filtered with a succession of openings of increasing sizes, often described as sieving. They can be calculated from a granulometry [5,16], where efficiency is achieved using a hierarchy such as a max-tree [22]. The amounts of image content removed by each attribute filter are used as global [25] or region descriptors [3]. We investigate the pattern spectra calculated on both inclusion trees, which are extrema-oriented, and partitioning trees, which capture intermediate level regions, and confirm their suitability for characterising soil structure through evaluation on an image retrieval problem.

In the next section, we outline the related literature on applying image processing and morphological techniques to calculate particle size distribution in soil and similar materials. The methodology for calculating the pattern spectra, as well as the different component trees used, are explained in Sect. 3. The dataset, collected by the authors for the study, is presented in Sect. 4. We elaborate our experiments and results in Sect. 5, and conclude in Sect. 6, outlining future directions of interest.

2   Related Work

Some of the earliest image processing techniques applied to estimation of aggregate size distribution focused on segmenting images of non-overlapping coarse aggregates (3 mm to 63 mm) [18]. Several segmentation-based techniques for working with overlapping particles of coarse-grained sediments, including those based on top-hat and watershed segmentation, are described in [12]. Size distribution of overlapping particles of coarse sands and gravel 0.7 mm to 20 mm) has also been analysed through statistical image properties [7], and correlated to sample distribution through regression over the images in a "look-up catalogue". In summary, these approaches are limited to cases of no to little particle overlap and samples comprising large aggregates, with further drawbacks including the reliance on segmenting the image into individual particles or a catalogue.

Granulometry [16], followed by pattern spectra [15], is one of the earliest morphological operations, and was developed as a tool for scale (size) and shape analysis of image content, with some of the earliest applications in petrography (studying the grain structure of rocks). Pattern spectra through opening and closing with reconstruction were used for the granulometric analysis of estuarine and marine sediments [11], as well as soil section images [10] (also including spectra based on area openings), however both applications focus on samples with mostly non-overlapping aggregates.

Pattern spectra based on area openings calculated on a max-tree were used to produce accurate grain size distributions of sands (smallest reported particle size 0.06 mm) [20]. Image granulometry is also discussed in the context of estimating the size distribution of stone fragments [21]. A recent study compares pattern spectra based on different structuring elements as well as area openings, closings and their combination for assessment of grain size for fine and coarse aggregates of sands and pebbles (0.125 mm to 16 mm) [2], obtaining the best results through attribute morphology. The mean grain size was estimated through regression on the training samples, by assuming a quadratic relation between measured grain size and image granulometry. The image granulometry was also related to the measured weight distribution of the samples. However, this work was validated on prepared samples with predetermined uniform grain size distribution, while processing partial images of very large samples (170 kg of soil).

We study the effectiveness of size pattern spectra in distinguishing soils in terms of their structure, i.e. soil aggregate distribution, posed as an image retrieval problem. The sample sizes used in our experiments are much smaller than those used in previous work (400 g samples, cf. Sect. 4 compared to 170 kg [2]), aiming at the eventual application of the technique to freshly dug samples on site. The image acquisition setup allows us to examine the effectiveness of this approach for cases of completely to partially overlapping or touching aggregates, which have not been examined before in the same study. We compare using 5 different component trees for pattern spectra calculation, including both inclusion [17,22] and partitioning hierarchies [23,24]. We propose to calculate the pattern spectra on partitioning trees by relying on the filtering rules outlined in [4]. Additionally, we investigate the influence of three different measures for defining a granulometry outlined in [8].

3   Methodology

In this section, we recall the concepts of granulometry and pattern spectra as well as attribute filtering, followed by a short overview of the component trees used in the study.

3.1   Granulometries and Pattern Spectra

Both granulometries [16] and pattern spectra [15] rely on openings and closings, and are used to capture the information on the distribution of image component sizes (later extended to shape [25,26]). While granulometries can be seen as size distributions of images, pattern spectra are the corresponding size histograms.

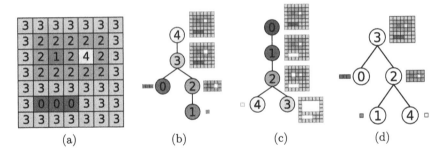

Fig. 1. The three different inclusion trees of a toy image (a). The min-tree is displayed in (b), while its dual max-tree is shown in (c). The self-dual tree of shapes is shown in (d). The (gray) levels of the nodes are displayed in the nodes, and the corresponding regions are shown beside the nodes.

When implemented through attribute filters, they interact directly with connected components of the image instead of pixels.

More formally, we define a grayscale image as $f : E \to \mathbb{Z}, E \subseteq \mathbb{Z}^2$, and the sets of images obtained by thresholding the image f at all possible values of its pixels, called its upper-level sets, as $\mathcal{L}^k = \{f \geq k\}$ with $k \in \mathbb{Z}$ (resp. lower-level sets $\mathcal{L}_k = \{f \leq k\}$). These sets are composed of their connected components, typically based on 4- or 8-connectivity, referred to also as peak components $\mathcal{P}$.

Attribute filtering is an operation applied to these peak components by evaluating a logical predicate on all the peak components by comparing an attribute α (e.g. area) with a threshold t (e.g. 300 px). An attribute, such as area, is increasing if, for two nested peak components $\mathcal{P}_A \subseteq \mathcal{P}_B$, its value is always greater for the larger region. The peak components not satisfying the logical predicate are removed from the input image. Filtering with an increasing attribute results in an attribute opening Γ_t (as it satisfies all the properties of an opening, i.e. the anti-extensivity, increasingness and idempotence).

A series of such openings with increasing size $\{\Gamma_{t_i}\}, t_{i+1} > t_i$ is called a *size granulometry* [5,16]. More components are removed from the image in every successive opening and the process can be seen as consecutively sieving the image with increasing mesh sizes. To quantify a granulometry $\{\Gamma_{t_i}\}$, the amount of detail remaining in the image is noted. A *pattern spectrum* is calculated from the granulometry by instead measuring the amount of detail removed between each successive pair of openings.

3.2   Component Trees

In attribute morphology, the typical way to interact with connected components of the image is to define them through a component tree, a hierarchical image representation. We distinguish inclusion hierarchies (examples in Fig. 1) comprising partial image partitions as cross-sections, which are typically extrema-oriented, and partitioning hierarchies (cf. Fig. 2) with nested image partitions as cross-sections, which are better at representing regions at intermediate values.

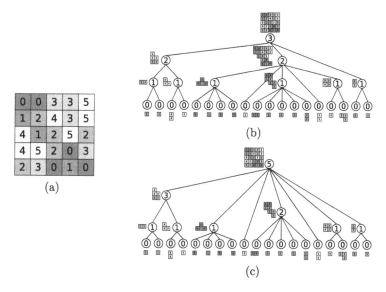

(a)

(b)

(c)

Fig. 2. For the toy image in (a), the α-tree is displayed in (b), while the constrained hierarchy (ω)-tree is shown in (c). The α (resp. (ω)) levels are displayed in the nodes and indicated by their height, with regions displayed besides the nodes.

The *min* and *max-trees* are dual hierarchies belonging to the class of inclusion trees and modelling the inclusion between the peak components of the lower and upper level sets of the image, and are well-suited for representing dark and bright components of the image, respectively. As such, they are the ones typically used for calculating the attribute openings and closings defining a pattern spectrum. Examples are shown in Fig. 1(b) and (c).

The *tree of shapes* (ToS) [17] unifies the representation of bright and dark image structures, producing a single self-dual image representation, treating bright and dark components equally based on their absolute contrast with their background. It comprises all the peak components of both upper and lower level sets with their holes filled, which also form an inclusion hierarchy. An example of a tree of shapes is shown in Fig. 1(d).

The α-*tree* is a partitioning tree based on the local range of its components [23,24] (also sometimes referred to as quasi-flat zone hierarchy [9]). The finest segmentation contained in the leaves of the tree comprises connected components of maximum extent of pixels at the same grey level, which are then merged according to the local neighbour similarity. As such, this hierarchy is capable of representing both bright, dark and intermediate level regions. However, due to the locality of the criterion used, the grey level variations within regions tend to be much higher than α when the grey levels in the image increase and decrease gradually, called the chaining effect [24]. An example of the hierarchy is shown in Fig. 2(b), with the chaining effect observable for $\alpha = 2$.

The most notable constrained connectivity hierarchy designed to deal with the chaining effect is the (ω)-*tree* [24], which rearranges the regions of the

Table 1. The different scale settings used.

Scale	Close	Middle	Far
Area [cm^2]	15×15	20×20	25×25
Camera height [cm]	60	78	94
Image resolution [px mm^{-1}]	16.7	12.5	10

α-tree according to their global intensity range, removing some of the regions but providing better grouping per level than just a local measure (cf. Fig. 2(c)).

A fast implementation of granulometry and pattern spectra has been made possible by their implementation on max-tree and min-tree image hierarchies [5,22], as attribute filtering can be achieved through removing nodes or branches of the tree. The attribute of interest is calculated for all the regions during tree construction, followed by determining the first opening from the sequence $\{\Gamma_{t_i}\}$ interacting with each region which corresponds to the bin i to which the region will contribute (cf. [25] for a more detailed description of this process). Different measures $\mathcal{M}$ can be used to describe the amount of detail contained in the nodes removed between consecutive attribute openings (i.e. remaining detail for the granulometry calculation), the most conventional being the *area of the removed regions* (in terms of number of pixels). We evaluate the pattern spectra calculated additionally on the other two measures proposed in the context of attribute profiles [8], the *number of changed regions*, as well as the *sum of gray level values*. To define attribute filters on the partitioning trees for granulometry and pattern spectra calculation, we follow the procedures outlined in [4].

4   Dataset Description

In this section, we describe the dataset collected by the authors to study the application of image processing techniques to soil structure assessment. Prior to taking images, we broke apart the soil structure into constituent aggregates in accordance with the VESS methodology [13]. To analyse a range of soils, we selected soils of different texture and structure. Each sample weighted approximately 400 g, with precise sample weight and aggregate size distribution noted down following the sieving process. Soil A was a calcareous sandy clay loam with a sub-angular to medium granular structure and occasional small stones (VESS category - Sq2). Soil B was a stone-free silt loam, sieved to 5 mm in order to create a uniform soil structure for the analysis (VESS category - Sq1). Soil C was a clay loam with a sub-rounded to medium granular structure (VESS category - Sq3). Soil D was a fine granular to single grained sandy silt loam with occasional stones (structureless). Examples of the soils are given in Table 2.

To create a uniform moisture content across samples, all samples were dried in an oven at 105 °C for 24 h, a standard procedure in soil science. Additionally, we have also collected images of the soils C and D as-dug, i.e. before drying them in the oven. Square surfaces of three different sizes (cf. Table 1) were drawn on

Table 2. Examples of all the soil samples A–D at the far scale (area $25 \times 25 \, \text{cm}^2$).

Soil	A	B	C	D
As-dug	—	—	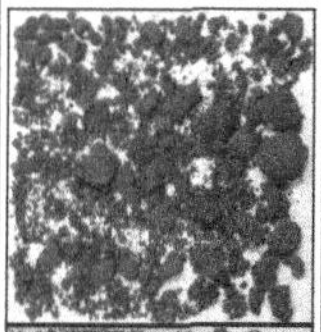	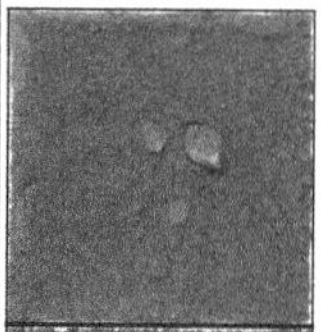
Dry	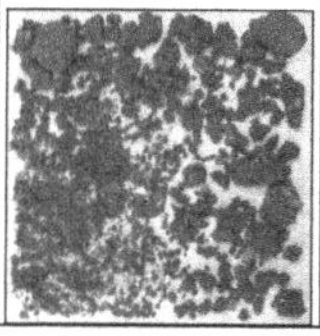		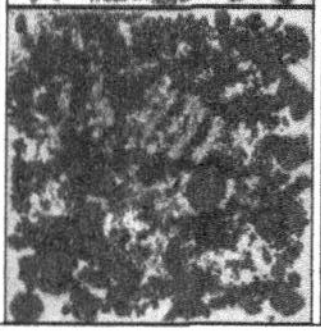	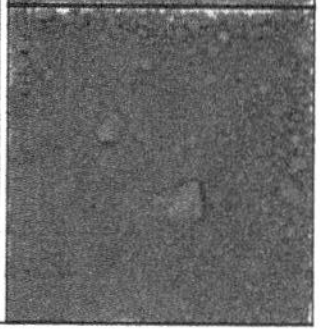

(a)   (b)   (c)

Fig. 3. The image acquisition setup: (a) the middle ($20 \times 20 \, \text{cm}^2$) square marked on the tray, (b) one of the original images of soil A, (c) the final image obtained by applying the rectifying homography.

a white tray (shown in Fig. 3(a)), the soils placed in the tray and then manipulated with brushes to fit the marked surface. This setup allowed us to produce images at different pixel resolutions as well as examine the influence of the visible background surface in the sample images. The images were taken with a Canon EOS 40D camera, which was placed at a fixed height for a top-down view of the samples and produced images of size $3888 \, \text{px} \times 2592 \, \text{px}$, cf. Fig. 3(b). The fixed height was empirically determined to allow for maximal pixel resolution for each of the three marked surface sizes (details in Table 1). Finally, after taking the images, the corners of the marked square were taken as markers for applying a homography to the images to produce a perfect top-down image as well as discard the parts of the image not picturing the sample. The resulting images (example in Fig. 3(c)) are of size $2500 \, \text{px} \times 2500 \, \text{px}$. The soils were fitted into each of the marked surfaces twice, resulting in two different arrangements of each soil at each scale for a total of 36 images in the dataset.

5   Experiments and Results

We design our experiments as an image retrieval problem, and assign two sets of ground truth matches (cf. Fig. 4) to examine the pattern spectra performance

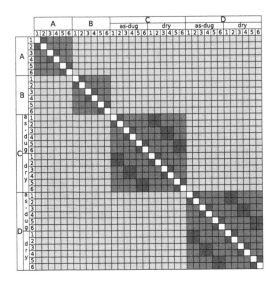

Fig. 4. The ground truth table for both image retrieval setups. Blue - same scale, green - all scales, grey - false matches, white - samples themselves. (Color figure online)

Table 3. The bin limits used (with estimated area in pixels).

Mesh size [mm]		0.212	2	3.15	5.0	9.5	25	50
Close	Side [px]	4	34	53	84	159	417	834
	Area [px]	13	1112	3757	6945	25 070	173 612	694 445
Middle	Side [px]	3	25	40	63	119	313	625
	Area [px]	8	625	1551	3907	14 102	97 657	390 626
Far	Side [px]	3	20	32	50	95	250	500
	Area [px]	5	400	993	2500	9025	62 500	250 000

in two distinct cases. In the first case (image pairs marked in blue on Fig. 4 considered correct matches), we examine the ability of pattern spectra to describe the soil samples and allow for their re-identification after rearranging the samples at the *same scale* only. In the second case (blue and green matches in Fig. 4), we consider re-identifying the samples at *all scales* present in the dataset. Posing this as an image retrieval problem has two advantages over the alternative setups. Firstly, we only need to change the evaluation metric (i.e. the ground truth), but not the retrieval setup, to examine these two distinct cases, while we would have had to train two different classifiers (one with 4 and one with 24 classes) had we posed it as a classification problem. Secondly, since our final goal is directly relating the pattern spectra to the physically measured soil size distributions, we want to avoid the aggregation step often present in classification and work with the descriptors directly.

The bin thresholds are chosen in two different ways and the performance of the resulting descriptors compared. The upper limit for the largest bin is set to the largest expected particle size of 50 mm (all the aggregates of all the samples A–D passed through a 50×50 mm^2 sieve). We firstly test the logarithmic binning, which is commonly used with pattern spectra descriptors [3]. Secondly, we also use the physical dimensions of each of the sieve meshes to calculate the area of grid openings in pixels for each of the scales (shown in Table 3), and use those as bin limits. The number of bins in the logarithmic binning is set to the same number of bins determined by the physical sieves used, $b = 7$.

We use 3 different inclusion trees (min and max-tree, tree of shapes) and 2 different partitioning trees (α and (ω)-tree) for pattern spectra calculation (outlined in Sect. 3.2), as well as the sum of the histograms obtained from the min- and max-trees (corresponding bins are summed to obtain a new histogram of the same length; denoted as $min+max$). Each image is described by its associated pattern spectrum, normalised so that the sum of all histogram values equals 1.

After calculating a pattern spectrum for every image, query results for an image are obtained by ranking the other database images according to descriptor similarity. We quantify our retrieval performance in terms of *mean Average Precision* (mAP), a measure designed specifically to evaluate ranked retrieval results [14] and provide a single measure of quality across all recall levels of a system for a set of multiple queries. For a single query image q, we can calculate *precision* and *recall* considering only the first m returned images in an unordered fashion. *Precision at m* is the ratio between the number of relevant images in the set of results and the total number of images retrieved at that point, m, while the *recall at m* is the ratio between the number of relevant images in the results and the total number of relevant images for that query. AP is the Average Precision of a query, and is equivalent to averaging the precision values obtained for the ordered retrieval results, after retrieving each new relevant result:

$$AP = \sum_{m=1}^{K} precision(m) \times \Delta \, recall(m) \tag{1}$$

$$= \frac{\sum_{m=1}^{K} precision(m) \times relevant(m)}{relevant Total},$$

where relevant(m) is an indicator variable indicating if the m-th retrieved image is relevant. The mAP is calculated as the mean value of the AP for all the queries.

While we evaluated the retrieval similarity for 5 different histogram similarity metric, we have found very similar trends between the results obtained by L_1 and L_2, as well as between the results obtained by χ^2 and Bhattacharyya, while the cosine distance underperformed in comparison to the other 4. Thus, the final results are presented using L_1 and χ^2 distances only.

The results confirm the suitability of pattern spectra for the analysis of soil structure, with achieving the mAP of 66.9% considering the samples at the same scales only (cf. Table 4), and 78.2% when considering all the scales (shown in Table 5). Out of the three measures of image content which were used, $\mathcal{M}_{area}$

Table 4. Performance of the retrieval system for the same scale ground truth. All results are expressed in terms of mAP as percentages [%].

Content measure	Sieve bins						Logarithmic bins					
	$\mathcal{M}_{volume}$		$\mathcal{M}_{area}$		$\mathcal{M}_{count}$		$\mathcal{M}_{volume}$		$\mathcal{M}_{area}$		$\mathcal{M}_{count}$	
Distance	L_1	χ^2	L_1	χ^2	L_1	χ^2	L_1	χ^2	L_1	χ^2	L_1	χ^2
α-tree	**58.1**	47.6	63.3	56.2	62.4	64.8	40.8	37.5	52.0	51.5	31.5	40.8
(ω)-tree	56.7	45.4	**65.3**	55.1	59.4	**66.9**	27.3	27.5	**52.2**	50.2	41.9	**64.7**
min-tree	45.8	45.6	46.5	44.3	42.7	54.4	**42.6**	42.1	43.9	42.4	35.4	39.8
max-tree	26.7	26.9	32.1	33.1	62.8	66.5	32.2	32.1	40.0	42.2	45.3	45.3
min + max	29.7	30.2	32.4	33.3	49.7	62.7	32.0	31.4	39.7	39.2	38.8	42.4
ToS	45.5	42.6	46.8	45.4	46.7	59.2	39.9	39.4	40.7	40.9	38.0	41.2

Table 5. Performance of the retrieval system for when all scales are considered in the ground truth. All results are expressed in terms of mAP as percentages [%].

Content measure	Sieve bins						Logarithmic bins					
	$\mathcal{M}_{volume}$		$\mathcal{M}_{area}$		$\mathcal{M}_{count}$		$\mathcal{M}_{volume}$		$\mathcal{M}_{area}$		$\mathcal{M}_{count}$	
Distance	L_1	χ^2	L_1	χ^2	L_1	χ^2	L_1	χ^2	L_1	χ^2	L_1	χ^2
α-tree	70.6	68.7	60.9	66.3	35.2	35.7	69.5	69.4	72.4	71.3	41.9	43.9
(ω)-tree	70.1	69.2	61.2	66.1	34.5	36.4	57.6	56.9	72.1	70.9	41.1	46.7
min-tree	70.8	70.8	69.9	70.07	36.9	42.3	69.5	68.5	68.3	67.5	54.0	**57.7**
max-tree	71.2	71.1	75.6	75.3	38.5	**45.1**	67.3	68.9	71.9	72.8	53.3	56.4
min + max	**75.4**	75.1	**78.2**	77.2	37.2	44.3	72.9	73.3	**75.9**	75.6	53.2	56.6
ToS	73.4	73.6	72.0	72.6	35.0	43.5	**73.5**	73.2	72.7	72.5	53.7	56.1

has the most consistency, performing well in both cases, with $\mathcal{M}_{volume}$ slightly underperforming in the single scale experiments and $\mathcal{M}_{count}$ resulting in a large performance drop of 30% in the experiments all the scales. We also notice a better performance of the L_1 over the χ^2 distance for $\mathcal{M}_{volume}$ and $\mathcal{M}_{area}$ which are measures of the region size, while the χ^2 is preferred for $\mathcal{M}_{count}$ where the regions are simply counted. Additionally, we can observe that using the sieve mesh sizes to define the bin limits improves the retrieval performance compared to the traditional logarithmic binning (except in the case of $\mathcal{M}_{count}$ across all scales, where performance is already low).

Finally, we can also observe that the partitioning trees greatly improve the retrieval results of soil images at the same scale for all the metrics, while the inclusion trees perform well only using the $\mathcal{M}_{count}$ metric. This was expected as we have observed the openings based on inclusion trees do not interact with aggregates positioned at the border with the background (cf. Fig. 5), which are correctly represented through partitioning trees and their ability to hold regions of intermediate level. This does not hold when considering all the scales, where the best performance was achieved using either the combination of min and max-trees, or the tree of shapes. The overall results reaffirm the claim that including

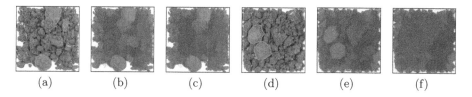

Fig. 5. Interaction of the tree of shapes with the aggregates positioned at the border with the background: (a) and (b) two different arrangements of soil A at close scale ($15 \times 15\,\mathrm{cm}^2$) with two largest aggregates outlined in green, (b) and (e) filtering with area 200 000 px, (b) and (e) filtering with area 600 000 px. The aggregates are correctly removed in (d) where they are positioned in the middle of the image, but not in (a) when bordering the background. (Color figure online)

dual information improves the descriptive power of pattern spectra [2], either by using the min and the max-tree together or one of the self-dual representations. The drop in performance of the partitioning trees compared to inclusion trees at all scales, which we plan to further investigate, is however not too severe, so in conclusion we recommend using partitioning tree based pattern spectra, and more specifically the ones calculated from the (ω)-tree for soil structure analysis.

6   Conclusions and Future Work

We have compared the suitability of pattern spectra calculated from a collection of component trees for soil structure analysis, framed as an image retrieval problem, as well as the influence of different measures of removed image detail and histogram distances. We specifically examine the pattern spectra based on partitioning trees, self-dual hierarchies able to represent image regions of intermediate grey levels. Our experiments are designed to work with small sample sizes, and examine the influence of scale and visible background. We confirm the stable performance of the partitioning trees, and find that pattern spectra calculated from the (ω)-tree using $\mathcal{M}_{\mathrm{area}}$ and compared with the L_1 distance result in the best and most stable performance on the examined retrieval problem.

As part of future work, we plan to further investigate the behaviour of partitioning trees when sample sizes at multiple scales are considered. Some techniques towards possible performance improvement include fully removing the background before tree construction and working on images with partial support, as well as trying different size attributes which would more accurately reflect particles passing through differently sized sieve meshes. The final goal is correlating the image-based granulometry to the granulometry obtained through physical measuring in a regression setup, as well as improving the robustness of the process for in-field deployment as opposed to working in laboratory setting.

References

1. Beare, M.H., Bruce, R.R.: A comparison of methods for measuring water-stable aggregates: implications for determining environmental effects on soil structure. Geoderma **56**(1–4), 87–104 (1993)
2. Bianconi, F., Di Maria, F., Micale, C., Fernández, A., Harvey, R.: Grain-size assessment of fine and coarse aggregates through bipolar area morphology. Mach. Vis. Appl. **26**(6), 775–789 (2015)
3. Bosilj, P., Aptoula, E., Lefèvre, S., Kijak, E.: Retrieval of remote sensing images with pattern spectra descriptors. ISPRS Int. J. Geo-Inf. **5**(12), 228 (2016)
4. Bosilj, P., Damodaran, B.B., Aptoula, E., Mura, M.D., Lefèvre, S.: Attribute profiles from partitioning trees. In: Angulo, J., Velasco-Forero, S., Meyer, F. (eds.) ISMM 2017. LNCS, vol. 10225, pp. 381–392. Springer, Cham (2017). https://doi.org/10.1007/978-3-319-57240-6_31
5. Breen, E.J., Jones, R.: Attribute openings, thinnings, and granulometries. Comput. Vis. Image Underst. **64**(3), 377–389 (1996)
6. Bronick, C.J., Lal, R.: Soil structure and management: a review. Geoderma **124**(1–2), 3–22 (2005)
7. Buscombe, D., Masselink, G.: Grain-size information from the statistical properties of digital images of sediment. Sedimentology **56**(2), 421–438 (2009)
8. Cavallaro, G., Falco, N., Dalla Mura, M., Benediktsson, J.: Automatic attribute profiles. IEEE Trans. Image Process. **26**(4), 1859–1872 (2017)
9. Cousty, J., Najman, L., Kenmochi, Y., Guimarães, S.: Hierarchical segmentations with graphs: quasi-flat zones, minimum spanning trees, and saliency maps. J. Math. Imaging Vis. **60**(4), 479–502 (2018)
10. Doulamis, A., Doulamis, N., Maragos, P.: Generalized multiscale connected operators with applications to granulometric image analysis. In: ICIP, vol. 3, pp. 684–687. IEEE (2001)
11. Frančišković-Bilinski, S., Bilinski, H., Vdović, N., Balagurunathan, Y., Dougherty, E.R.: Application of image-based granulometry to siliceous and calcareous estuarine and marine sediments. Estuar. Coast. Shelf Sci. **58**(2), 227–239 (2003)
12. Graham, D.J., Reid, I., Rice, S.P.: Automated sizing of coarse-grained sediments: image-processing procedures. Math. Geol. **37**(1), 1–28 (2005)
13. Guimarães, R.M.L., Ball, B.C., Tormena, C.A.: Improvements in the visual evaluation of soil structure. Soil Use Manag. **27**(3), 395–403 (2011)
14. Manning, C.D., Raghavan, P., Schütze, H.: Introduction to Information Retrieval. Cambridge University Press, Cambridge (2008)
15. Maragos, P.: Pattern spectrum and multiscale shape representation. IEEE Trans. Pattern Anal. **11**(7), 701–716 (1989)
16. Matheron, G.: Random Sets and Integral Geometry. Wiley, New York (1975)
17. Monasse, P., Guichard, F.: Scale-space from a level lines tree. J. Vis. Commun. Image Represent. **11**(2), 224–236 (2000)
18. Mora, C.F., Kwan, A.K.H., Chan, H.C.: Particle size distribution analysis of coarse aggregate using digital image processing. Cem. Concr. Res. **28**(6), 921–932 (1998)
19. Mueller, L., et al.: Visual assessment of soil structure: evaluation of methodologies on sites in canada, china and germany: Part I: comparing visual methods and linking them with soil physical data and grain yield of cereals. Soil Tillage Res. **103**(1), 178–187 (2009)
20. Pina, P., Lira, C., Lousada, M.: In-situ computation of granulometries of sedimentary grains-some preliminary results. J. Coast. Res. **64**, 1727–1730 (2011)

21. Salehizadeh, M., Sadeghi, M.T.: Size distribution estimation of stone fragments via digital image processing. In: Bebis, G., et al. (eds.) ISVC 2010. LNCS, vol. 6455, pp. 329–338. Springer, Heidelberg (2010). https://doi.org/10.1007/978-3-642-17277-9_34

22. Salembier, P., Oliveras, A., Garrido, L.: Anti-extensive connected operators for image and sequence processing. IEEE Trans. Image Process. **7**(4), 555–570 (1998)

23. Soille, P.: On genuine connectivity relations based on logical predicates. In: ICIAP, pp. 487–492. IEEE (2007)

24. Soille, P.: Constrained connectivity for hierarchical image partitioning and simplification. IEEE Trans. Pattern Anal. **30**(7), 1132–1145 (2008)

25. Urbach, E.R., Roerdink, J.B.T.M., Wilkinson, M.H.F.: Connected shape-size pattern spectra for rotation and scale-invariant classification of gray-scale images. IEEE Trans. Pattern Anal. **29**(2), 272–285 (2007)

26. Urbach, E.R., Wilkinson, M.H.F.: Shape-only granulometries and grey-scale shape filters. In: ISMM, pp. 305–314 (2002)

Contact Based Hierarchical Segmentation for Granular Materials

Olumide Okubadejo[1,2(✉)], Edward Andò[1], Laurent Bonnaud[2],
Gioacchino Viggiani[1], and Mauro Dalla Mura[2,3]

[1] Univ. Grenoble Alpes, CNRS, Grenoble INP,
3SR, 38000 Grenoble, France
olumide.okubadejo@3sr-grenoble.fr
[2] GIPSA-lab, Univ. Grenoble Alpes, CNRS, Grenoble INP, 38000 Grenoble, France
[3] Tokyo Tech WRHI, School of Computing,
Tokyo Institute of Technology, Tokyo, Japan

Abstract. Segmentation of tomography images of granular materials has mostly been done without any semantic notion of the materials these images represent. Poor segmentation; undersegmentation or oversegmentation can be mitigated through the introduction of a prior that defines what a valid boundary between two labels should be.

In this paper, we present an approach to multiscale tomography segmentation of granular materials, based on hierarchical segmentation using minima. A 3D tomography image of granular materials is segmented with the introduction of a contact prior combined with other extinction for the purpose of formulating the hierarchy. The contact prior based hierarchy is biased towards a segmentation with a realistic mechanically valid contact network. The proposed method is observed to significantly increase the segmentation accuracy of these grains.

Keywords: Mathematical morphology · Watershed hierarchies · Prior-based segmentation · Granular material segmentation

1 Introduction

X-ray tomography is used in the study of granular materials, and has enabled significant findings. Such studies include the study of particle morphology and its evolution as a result of mechanical loading [1,11]. Image segmentation methods have been used extensively for grain labelling on the acquired grayscale volume images (in which each voxel is associated to a scalar value) [1,11]. The most notable of segmentation procedures used in tomography imaging for granular materials is morphological watershed, since these materials can often be scanned with sufficient contrast so that image thresholding can meaningfully be applied

Grenoble INP—Institute of Engineering Univ. Grenoble Alpes.
WRHI—World Research Hub Initiative.

to identify the grain phase. In some extreme cases of mechanical loading, the grains themselves can break in different ways. This is also of major interest.

Hierarchical segmentation has become a major trend [4,5,8,16] due to its multiscale solution. Multiscale in this context, is a subset of the study of scale-spaces seen extensively in literature [20]. A hierarchical segmentation is composed of a sequence of segmentation; from fine to a coarse scale (i.e., small to large regions) [9,10,12]. Consequently, the resulting hierarchical segmentation is not a single partition of the image pixels into sets but rather, a single multiscale structure comprised of segmentation (*partition sets*) at increasing scales. Many studies have formulated watershed based on graphs as a hierarchical scheme [7,17].

The base scale of the hierarchy, can be an initial segmentation which based on a roughly defined set of minima, or the image pixel set. In the minima based approach, the hierarchical scale space is built by the successive removal of minima that rank lowest, according to an attribute ordering function (*extinction function*) [9,12]. Once removed, the image object (*region*) it represents, is merged to the closest (*in terms of a dissimilarity measure*) region [10]. The successive removal of minima, creates a scale-space (*hierarchy*), with the fine scale at the bottom and a single partition at the top. A partition can be extracted from the hierarchy, at a defined scale. In this paper, the starting point is always an initial segmentation based on a roughly defined minima set.

The properties of granular materials have been studied extensively in literature e.g. contact between grains [6]. These studies relate the contact to the predominant size and shape of the material assembly. Thus we formulate a hierarchical segmentation process that takes into account a contact model based on the known interaction of granular materials, to aid grain segmentation in images of granular materials.

The main contributions of this paper is the investigation of the effect of combining known attribute ordering functions with attribute ordering based on the physical properties of the granular assembly, such as contact interaction. We focus on the effect of a contact based model. This models the implicit contact properties of grains and should lead to better segmentation.

We show also, how artificially generated datasets using spheres generated with kalisphera [18], can offer some insight into algorithms' performance on real datasets when noise and blur are added in the style of [19].

The paper is organised as follows. First, we review the fundamentals on hierarchies, attribute filtering and saliency maps. Next, the experimental study is presented and the results are evaluated.

2   Background

Graphs and Notations: A 3-dimensional tomography image can be represented as a weighted digraph $G = (V, E, w)$, whose vertices V are image pixels in the image $I \subset Z^3$, and edges E are adjacent pixel pairs as defined by an adjacency relation $\mathcal{A}$ [12,17]. A pixel x is adjacent to a pixel y, if x is in the neighbourhood of y. An edge between x and y is denoted by $e_{x,y}$.

Partitions and Hierarchy: A *partition* P of a finite set V is a set of nonempty disjoint subsets of V whose union is V. Any element of the partition P of V is called a *region* of P and represents a region of connected pixels/superpixels in the image. A pixel x an element of the set V, uniquely belongs to an element (*region*) of P [12,17]. This unique relationship is denoted as $[P]_x$. Given two different partitions P and P' of a set V, we say that P' is a refinement of P if any region of P' is included in a region of P [12,17].

A *hierarchy* (on V) is a sequence $\mathcal{H} = (P_0, \ldots, P_l)$ of indexed partitions of V such that P_{i-1} is a refinement of P_i, for any $i \in 1, \ldots, l$. The integer l is called the depth of $\mathcal{H}$.

Saliency Maps and Ultrametric Maps: The cut of P (for the graph G) denoted by $\phi(P)$, is the set of edges of the graph G whose two vertices belong to different regions of P. The saliency map of $\mathcal{H}$ is a mapping $\Phi(\mathcal{H})$ from the edge set E to $0, \ldots, l$ such that each edge is represented by the maximum partition depth λ in which it belongs to the cut set [2,3]. An ultrametric contour map is an image representation of a saliency map [2,3].

Hierarchical Segmentation: For each 3D image, a fine partition set can be the 3D pixel points or can be produced by an initial segmentation (as in a set of superpixels). Figure 1 shows the minimum spanning forest (MST)/fine partition. This fine partitioning, contains all image contours. A dissimilarity measure is defined between adjacent regions of this fine partition set. The superpixels are the nodes/vertexes of the graph structure. Adjacent pixels are connected together by edges with weight w. w is computed according to a dissimilarity measure which we will define.

Starting with a base minima set, the lowest ranked minimum is progressively removed from the minima set according to an attribute ranking function (*extinction function*). This results in an indexed hierarchy of partitions (H, λ), with $\mathcal{H}$ a hierarchy of partitions and $\lambda : \mathcal{H} \rightarrow R^+$. When a minimum is removed, the image region associated with it is merged to the closest adjacent region (*defined by the dissimilarity function*). The creation of partition sets, as a function of decreasing minima set, results in a hierarchy.

This hierarchy, can be made to emphasise implicit image characteristic by choosing an attribute ranking function to reflect this implicit statistic. At each level set of the hierarchy (*partition set*), each region is defined by a minima of the level minima set.

On the resulting hierarchy, the saliency map and consequently the ultrametric contour map (UCM) is computed [2]. The saliency and ultrametric contour maps show how resistant to merging an edge is i.e. the edge strength. Stronger edges are edges that persist in the scale-space and thus are more likely to be chosen in the final contour set that defines the labels or grains.

Minima Ranking: A function $f : r_i \mapsto \mathbb{R}$ maps each region in the partition to a scalar value. The functional mapping is based on a defined criteria (volume, area, dynamics) on the region corresponding each minima. The scalar attribute values are sorted and the minima corresponding to lowest ranked attribute is removed. Its corresponding region is merged with the closest region.

Here we add a contact function to the attribute ranking function. This enables valid granular contacts to persist in the hierarchy space. Segmentation is extracted from the contour maps using the $\alpha - \omega$ method described in [17]. Here, we impose combinations by biasing the extinction function to conserve multiple statistics.

Algorithm 1. Hierarchical Segmentation

Data: gradientImage
Result: Ultrametric contour map
minima = generateMinima(*gradientImage*)
segments = generateInitialMSF(*minimas, gradientImage*)
while countMinima() $\geq$ 2 **do**
   removeMinima(*minima, attributeFunction*)
   updateMST(*minima, segments*)
   updateHierarchy(*segments*)
end

An algorithm showing the process for hierarchical segmentation [9] is shown in Algorithm 1.

3   Proposed Contact Based Attribute Function

The contact based attribute function requires a separation between foreground and background. This is so because contact interaction between two grains is different from contact interaction between grain and void. Thus, we first separate background. This can be achieved using image thresholding methods such as Otsu [14].

Recall that a cut of a partition set P, is a subset of the edge set E such that both vertex of an edge, $e_{x,y} \in E$, v_x and v_y belong to adjacent regions.

Consider two adjacent regions in a partition P, r_i and r_j. The contact set C is thus a partitioning of the cut set, such that $\forall c_{i,j} \in C$, $c_{i,j}$ is a set containing edges between two adjacent regions r_i, r_j in contact.

We propose a contact based attribute function to highlight two contact properties necessary for a granular material assembly spherical contact model.

– **Length** With the background node excluded, the total contact surface area per region surface area should be minimised. Let C_i be a set containing all the contacts, $c_{i,j}$ of a given region r_i

$$K = \frac{1}{|C_i|} \sum |c_{i,j}| \quad \forall c_{i,j} \in C_i$$

For two spheres in contact, the contact surface relative to surface area should be minimal. Minimising grain to grain contact length thus constrains the segmentation to small contact lengths. Large contact length most times leads to oversegmentation.

– **Flatness** Contact surfaces can be approximated by a plane. As such, we penalise contacts that deviate from this description. We compute how closely a contact surface can be approximated by a plane, by looking at the minimum of the contact bounding box dimension. This approximation reflects the prevailing shape of the contact.

We find the enclosing cuboid (w, h, d), aligned with the image axis, for the set of points.

The minimal enclosing cuboid is computed as the bounding box on the contact points which is the bounding box on the convex hull of the set of points. We denote flatness of a contact $c_{i,j} \in C_i$ as;

$$u(c_{i,j}) = min(w, h, d)$$

Thus for a contact set of a region (minima), the flatness score is denoted as, the sum of values $u(c_{i,j})$ over the contact set, normalised by the cardinality of the set $|\{u(c_{i,j}), u(c_{i,k}), ...\}|$. This flatness measure is denoted by U.

In reality, this approximation might not be sufficient for contacts of grain with more complex shapes.

We combine these properties into an affine objective function, allowing larger penalisation. The overall function to be minimised is,

$$O(r) = \frac{1}{U\sqrt{K}} \tag{1}$$

where $O(\cdot)$ is the objective function to be minimised evaluating a region r. To obtain this formulation, we examine each parameter defined. The K parameter calculates the average contact length. Thus its parametric maximum is the surface area of a grain. The parametric limit of U, is the radius of a grain. Assuming, a spherical model, we normalise K to refer to its radius, thus ensuring the same scale for both independent measures. This results in Eq. 1.

Contact attribute descriptor is a *soft-feature*. This implies, that although it can improve the accuracy of descriptor statistics such volume or dynamics, it is not sufficiently reliable as a stand alone statistic for an attribute function. In other words, contact attribute function should be combined with other descriptor based attribute functions to improve the accuracy of segmentation. The cost function space of contact-alone based attribute function, is more difficult to navigate due to its non-linearity and non-convexity i.e., it does not satisfy the increasingness criterion as defined in [9]. As such it is not monotonic across scales. However, it can aid other attribute functions towards having a well defined minimum in the cost function space. To combine the contact function with other attribute functions, we design a minima ranking function as seen in Algorithm 2. A feature such as volume is computed on the minima set. Each minimum is given

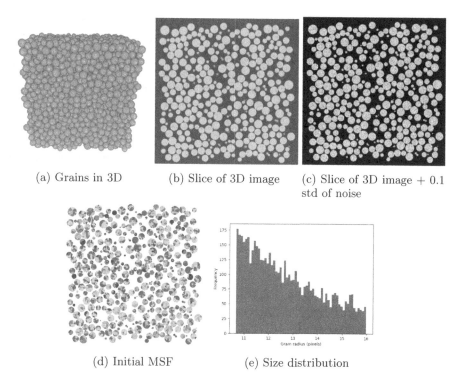

(a) Grains in 3D   (b) Slice of 3D image   (c) Slice of 3D image + 0.1 std of noise

(d) Initial MSF   (e) Size distribution

Fig. 1. Given structural properties; radius and 3D position, kalisphera generates a 3D image containing 5521 grains as shown in Fig. 1a. A slice of this 3D image is shown in Fig. 1b. Due to near image uniformity, the image is augmented with 0.1 standard deviation of additive gaussian noise as shown in Fig. 1c. After adding noise, the maximum pixel value was thus 255 and the minimum 0. An initial segmentation with 200, 000 segments is obtained using simple linear iterative clustering.

a rank. The contact index is also computed with each minimum earning a rank. Both ranks are combined using

$$N_r = N_m + \psi N_c$$

where N_c is the contact ranking, N_m is the ranking from an attribute like volume, area or dynamics and ψ is a weight.

This combined value N_c, defines a score for each minimum. The least scoring minimum is removed during the process of extinction.

Algorithm 2. Minima ranking

Data: minima, image

minima = **generateMinimas**(*gradientImage*)

segments = **generateInitialMSF**(*minima, gradientImage*)

ranksA = **rankMinima**(*minima, segments, attributeFunctionA*)

ranksB = **rankMinima**(*minima, segments, attributeFunctionB*)

ranks = ranksA · ψ ranksB

---

4   Experimental Study

4.1   Datasets

Synthetic Dataset: Our method is assessed using synthetic images of spheres (with analytically-calculated partially-filled voxels on the edges) generated by the kalisphera software [18]. A data set was obtained from [19], consisting in a time-series of 7 mechanically stable configurations of 5522 spheres undergoing compression (using a discrete element modelling (DEM) simulation). The data set is entirely described by 5522 positions and radii, and is then rendered into a 3D image using kalisphera, having only to choose the pixel size relative to the sphere size. The pixel size is $15e^{-6}$ m/pixel. The size distribution is shown in Fig. 1e.

The result is 7 volumes of $500 \times 500 \times 500$ pixels with 5522 grains. The dynamic range of each image is 8-bit, with the background (pores) at 64 and the foreground (grains) at 192. 0.1 standard deviation of noise is added to each image. An example of kalisphera generated grains is shown in Fig. 1.

Real Dataset: Our method is also evaluated on an idealised granular material imaged using x-ray tomography imaging. Sapphire spheres, as shown in 4 is a spherical material. The imaged sample consists of regular spheres ranging from 300 mm to 800 mm. In image sample used, the image is a $256 \times 256 \times 256$ volume containing 109 grains.

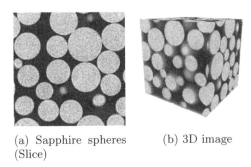

(a) Sapphire spheres (Slice)   (b) 3D image

Fig. 2. Slice of sapphire spheres and the full 3D image sample

4.2   Assessment Method

To account for performance, we use a supervised assessment strategy outlined in [15]. The likelihood of extracting optimal segmentation is evaluated with respect to a ground truth. The ground truth for the real dataset is drawn by hand. The quality of the extracted segmentation is measured using the Bidirectional Consistency Error (BCE) [13]. We denote the resulting segmentation as S_g and the reference segmentation as S_r. The bidirectional consistency error as defined in [13] is used to measure the fidelity of the obtained segmentation. This criterion is defined between 0.0 and 1.0 with 1 indicating a perfect match.

Hierarchy Evaluation: The accuracy potential of a hierarchy is computed as the curve, BCE vs the fragmentation level. Fragmentation level is measured as

$$fragmentation = \frac{expectedSegments}{actualSegments}$$

This measures the tendency of an hierarchical process to make the right merging decisions leading towards the best possible segmentation; close enough to the ground truth segmentation.

4.3   Experimental Set-Up

Our hierarchical watershed is compared against a morphological watershed which is consistently used in processing images of granular materials. For hierarchical segmentation, a single scale segmentation is extracted by varying the α, ω parameter as iterated in [17]. The values of α, ω is progressively increased (α and ω always have the same value), till the image is a single label or region. For each extracted segmentation, we compute the BCE score and the fragmentation level.

Two variants of morphological watershed is considered; gradient-based and distance map-based. In the gradient approach, we compute the gradient magnitude of the 3D image. The minima on the gradient magnitude surface is computed and both are used by the watershed transform to generate a label image. The distance map approach however, starts with a thresholding on the image using Otsu thresholding [14]. A distance map is computed on the threshold image using the euclidean distance map function. The minima of the inverted distance map are used by the watershed function to generate also a corresponding label image. The connectivity used in both gradient and distance map computations is the 26-neighbourhood connection (fully connected). To have the same basis for comparison, we use a masking layer to separate foreground from background in morphological approaches.

We also verify the effect of using a contact based function. This is done by comparing a contact based function combined with a volume and dynamics function against a functions solely based on single criterion; volume and dynamics. In effect, we compare baseline functions against functions combined with a contact prior model. We use the combination strategy already outlined.

5 Result and Evaluation

5.1 Evaluation of Hierarchies on Kalisphera

To evaluate the improvement offered by the addition of a contact prior, we determine optimal ψ values for the attribute combination function. The ψ parameter is the weighting on the combination function which determines the contribution of the contact ranking function to the overall ranking. We observe that optimal values for the combination of dynamics and contact function range between 0.3 and 0.4 as shown in Fig. 3a. As seen in Fig. 3a, ψ values and BCE do not exhibit a linear relationship.

Methods	BCE						
	$\psi = 0.1$	$\psi = 0.2$	$\psi = 0.3$	$\psi = 0.4$	$\psi = 0.5$	$\psi = 0.6$	$\psi = 0.7$
Dynamics + contact	0.680	0.726	0.831	**0.860**	0.713	0.622	0.601
Volume + contact	0.581	0.623	0.719	0.751	**0.752**	0.617	0.563

(a) BCE values at different ψ combination values evaluated for both dynamics and volume

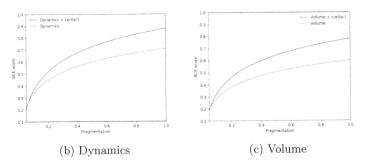

(b) Dynamics   (c) Volume

Fig. 3. Fragmentation plots for volume and dynamics, before and after the use of contact attribute function

In Fig. 3, With dynamics and volume, it can be observed that the introduction of a contact prior, allows merging to reach a more optimal segmentation. This proves a significant improvement over dynamics or volume alone as a metric for minima extinction.

5.2 Comparison Against Morphological Watershed

We compare our hierarchical segmentation method with morphological watershed methods (gradient and distance map based methods). We iterate through the hierarchy (varying α, ω), calculating the best BCE value from the hierarchy. We compare this BCE against BCE values obtained using morphological watershed variants. Results are shown in Fig. 4b.

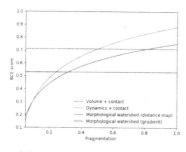

(a) BCE-Fragmentation curves

	BCE
Morphological watershed (gradient)	0.53
Morphological watershed (distance)	0.71
Dynamics + contact ($\psi = 0.36$)	0.88
Volume + contact ($\psi = 0.48$)	0.75

(b) Best BCE score

Fig. 4. Bidirectional consistency error of morphological variants compared against hierarchical based methods using the defined contact model

Gradients amplify noise. This effect, coupled with the effect due to partial volume, mostly lead to poor segmentation in noisy images. This has noticeably informed the trend of the use of distance maps in the processing of images of granular materials. The results obtained using morphological watershed with gradients, is consistent with distance maps, producing better performance in the segmentation of granular materials.

The distance map based watershed, although more resistant to noise as compared to the gradient variant, is still affected by it. A dynamics based contact extinction procedure, has significant descriptive power as it uses both the topological landscape and a contact prior. This leads to significantly better performance.

5.3   Evaluation of Unsupervised Hierarchies on Natural Grains

We apply our method for the segmentation of an assembly of sapphire sphere grains. In the UCM, weak gradients at grain contact points are more resistant to being merged prematurely, thus resulting in higher saliency values. This reduces the possibility of undersegmentation. This is shown in Fig. 5.

This shows that the segmentation evolves to find a contact configuration that is realistic in terms of the defined model. It is therefore less attracted to oversegmentation or grain configurations with contacts that do not fit the defined model.

To verify, we observe the segmentation of two grains in contact. As shown in Fig. 6, when a dynamics contact extinction is applied, the best obtainable segmentation undersegments both grains, merging them as one due to the weak contact between them. However a combination with the attribute contact model yields a clear separation between both grains in the extracted segmentation. The combination clearly improves the ability to segment grains in contact with weak delineating gradients.

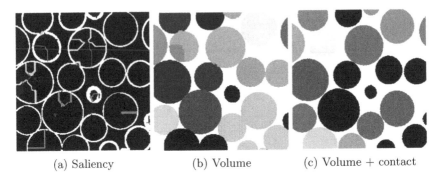

| (a) Saliency | (b) Volume | (c) Volume + contact |

Fig. 5. Volume extinction vs contact extinction (saliency): shows the saliency values at the contact of granular materials.

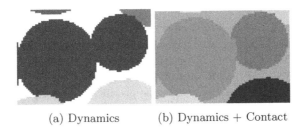

| (a) Dynamics | (b) Dynamics + Contact |

Fig. 6. Two sapphire grain segmentation: In sapphire spheres, this shows how a contact aware segmentation can reduce or minimise the oversegmentation of granular materials

6   Conclusion

The combination of a contact model with dynamics and volume extinction functions is shown to increase the accuracy of grain segmentation in 3D tomography images. saliency values of these contact regions are augmented, thus making them more resistant to merging. Although, the contact prior is simple and can be said to apply to spherical grains, the improved result, validates the combination with a contact model describing its mechanical contact properties.

Despite the morphological simplicity of the spheres studied in this example, there is clear interest in applying these techniques to data sets with more complex natural grains shapes such as angularity. The parameters used for the proposed segmentation procedure are not strictly related to spheres and therefore can in principle be applied to different-shaped grains, although significantly different contact topology may affect performance. The application to more angular shapes is ongoing work, whose performance is challenging to measure due to the difficulty in establishing the ground truth.

References

1. Ando, E.: Experimental investigation of microstructural changes in deforming granular media using x-ray tomography. Ph.D. thesis, Université de Grenoble (2013)
2. Arbelaez, P.: Boundary extraction in natural images using ultrametric contour maps. In: Conference on Computer Vision and Pattern Recognition Workshop 2006, CVPRW 2006, pp. 182–182. IEEE (2006)
3. Arbelaez, P., Maire, M., Fowlkes, C., Malik, J.: From contours to regions: an empirical evaluation (2009)
4. Arbelaez, P., Maire, M., Fowlkes, C., Malik, J.: Contour detection and hierarchical image segmentation. IEEE Transact. Pattern Anal. Mach. Intell. **33**(5), 898–916 (2011)
5. Arbeláez, P., Pont-Tuset, J., Barron, J.T., Marques, F., Malik, J.: Multiscale combinatorial grouping. In: Proceedings of the IEEE Conference on Computer Vision and Pattern Recognition, pp. 328–335 (2014)
6. Azéma, E., Radjai, F.: Force chains and contact network topology in sheared packings of elongated particles. Phys. Rev. E **85**(3), 031303 (2012)
7. Cousty, J., Bertrand, G., Najman, L., Couprie, M.: Watershed cuts: minimum spanning forests and the drop of water principle. IEEE Transact. Pattern Anal. Mach. Intell. **31**(8), 1362–1374 (2009)
8. Deng, J., Dong, W., Socher, R., Li, L.J., Li, K., Fei-Fei, L.: ImageNet: a large-scale hierarchical image database. In: IEEE Conference on Computer Vision and Pattern Recognition 2009, CVPR 2009, pp. 248–255. IEEE (2009)
9. Guigues, L., Cocquerez, J.P., Le Men, H.: Scale-sets image analysis. Int. J. Comput. Vis. **68**(3), 289–317 (2006)
10. Guimarães, S.J.F., Cousty, J., Kenmochi, Y., Najman, L.: A hierarchical image segmentation algorithm based on an observation scale. In: Gimel'farb, G., et al. (eds.) Joint IAPR International Workshops on Statistical Techniques in Pattern Recognition (SPR) and Structural and Syntactic Pattern Recognition (SSPR), pp. 116–125. Springer, Heidelberg (2012). https://doi.org/10.1007/978-3-642-34166-3_13
11. Hall, S., et al.: Discrete and continuum experimental study of localised deformation in hostun sand under triaxial compression using x-ray μct and 3D digital image correlation. Géotechnique **60**(5), 315–322 (2010)
12. Maia, D.S., de Albuquerque Araujo, A., Cousty, J., Najman, L., Perret, B., Talbot, H.: Evaluation of combinations of watershed hierarchies. In: Angulo, J., Velasco-Forero, S., Meyer, F. (eds.) ISMM 2017. LNCS, vol. 10225, pp. 133–145. Springer, Cham (2017). https://doi.org/10.1007/978-3-319-57240-6_11
13. Martin, D.R., Malik, J., Patterson, D.: An Empirical Approach to Grouping and Segmentation. Computer Science Division, University of California, Berkeley (2003)
14. Otsu, N.: A threshold selection method from gray-level histograms. IEEE Transact. Syst. Man Cybern. **9**(1), 62–66 (1979)
15. Perret, B., Cousty, J., Guimarães, S.J.F., Maia, D.S.: Evaluation of hierarchical watersheds. IEEE Transact. Image Process. **27**(4), 1676–1688 (2018). https://doi.org/10.1109/TIP.2017.2779604
16. Ren, Z., Shakhnarovich, G.: Image segmentation by cascaded region agglomeration. In: Proceedings of the IEEE Conference on Computer Vision and Pattern Recognition, pp. 2011–2018 (2013)

17. Soille, P.: Constrained connectivity for hierarchical image partitioning and simpli-fication. IEEE Transact. Pattern Anal. Mach. Intell. **30**(7), 1132–1145 (2008)
18. Tengattini, A., Andò, E.: Kalisphera: an analytical tool to reproduce the partial volume effect of spheres imaged in 3D. Measur. Sci. Technol. **26**(9), 095606 (2015)
19. Wiebicke, M., Andò, E., Herle, I., Viggiani, G.: On the metrology of interparticle contacts in sand from X-ray tomography images. Measur. Sci. Technol. **28**(12), 124007 (2017)
20. Witkin, A.P.: Scale-space filtering. In: Readings in Computer Vision, pp. 329–332. Elsevier (1987)

Detecting Branching Nodes of Multiply Connected 3D Structures

Xiaoyin Cheng[1]([✉]), Sonja Föhst[1,2], Claudia Redenbach[2], and Katja Schladitz[1]

[1] Fraunhofer-Institut für Techno- und Wirtschaftsmathematik, Kaiserslautern, Germany
xiaoyin.cheng@itwm.fraunhofer.de
[2] Technische Universität Kaiserslautern, Kaiserslautern, Germany

Abstract. In order to detect and analyze branching nodes in 3D images of micro-structures, the topology preserving thinning algorithm of Couprie and Zrour is jointly used with a constraint set derived from the spherical granulometry image. The branching points in the thus derived skeleton are subsequently merged guided by the local structure thickness as provided by the spherical granulometry. This enables correct merging of nodes and thus a correct calculation of the nodes' valences. The algorithm is validated using two synthetic foam structures with known vertices and valences. Subsequently, the algorithm is applied to micro-computed tomography data of a rigid aluminium foam where the valence is known, to the pore space of polar firn samples, and to corrosion casts of mice livers.

Keywords: Topology preserving skeletonization ·
Spherical granulometry · Maximal inscribed spheres ·
Branch counting · Connectivity analysis

1 Introduction

Detecting branching points of structures and determining their valence is a frequently arising task in 3D image analysis. Based on topologically correct skeletons, branching pixels can be detected by classifying local pixel configurations using Betti numbers (and combinations of them) for foreground or background neighboring regions w.r.t. varying adjacency systems, see [8] and references therein. Due to discretization, this approach yields an overwhelming majority of branching pixels with exactly three neighbors. In order to remedy this, the concept of a junction is introduced in [5]. In [8], thick junctions are determined by an iterative post-processing step.

Considering junctions is a special way of merging skeleton nodes erroneously separated due to discretization effects. A wide variety of strategies is suggested in the context of reconstructing pore spaces for flow simulation by pore-throat networks: In [11], medial axis points are joined based on the Euclidean distance between the points and of the points to the solid structure. More precisely, the

B. Burgeth et al. (Eds.): ISMM 2019, LNCS 11564, pp. 441–455, 2019.
https://doi.org/10.1007/978-3-030-20867-7_34

points are merged if one of them is included in the maximal inscribed sphere centered at the other one. This approach has been altered in order to solve ambiguities w.r.t. where a node/pore ends and a throat starts, in particular so-called snowballing [18] – merging of a whole chain of nodes because of overlapping maximal inscribed spheres. E. g. Dong and Blunt [2] merge pores iteratively starting from the largest, while Silin and Patzek [19] build a hierarchy of maximal inscribed pores for the same purpose. Building on [19] and [2], Hormann et al. [3] suggest a partitioning of an amorphous pore space starting from the centers of local maxima in the granulometry image.

We aim at a generally applicable and fully automatic method that is not tailor-suited to a particular structure type like e. g. [8]. Our only crucial assumption on the structure to be analyzed is that its thickness increases in the vicinity of branching nodes. In order to detect this, we require a minimal local structure thickness of three pixels to ensure correct branching analysis. Our method is based on Euclidean distance measurements. Thus a proper binarization is crucial for a correct result. In particular, noise has to be eliminated before applying any of the algorithms used here. Finally, our implementations of both skeletonization and granulometry assume a cubic lattice although in principle a generalization to cuboidal lattices is possible.

We detect branching nodes in micro-structures based on the topologically correct yet rather smooth skeleton of [1]. Skeleton branching points form the candidates set that is thinned using local structure thickness as provided by the spherical granulometry. This enables correct merging of nodes and thus also correct counting of the branches associated to each node. The algorithm is validated using two synthetic foam structures for which true vertices and their valences are known. Moreover, for a real foam, the algorithm's results are compared with those derived in [10]. Finally, we apply our algorithm to the pore space of polar firn samples and to corrosion casts of mice livers.

2   Algorithm

The algorithm exploits two types of spheres – the regional maximal spheres (RMSs) in the spherical granulometry image and the maximal inscribed spheres (MISs) derived from the squared Euclidean distance image. That means, a RMS is a regional maximum in the morphological sense [20] in the spherical granulometry image. A MIS is induced by a regional maximum in the squared Euclidean distance image. More precisely, let m be the locally maximal squared Euclidean distance, observed in a set P of connected pixels. The corresponding MIS is then the set of all pixels y with $d(P, y)^2 \leq m$, where d denotes Euclidean distance. Note that in both cases, non-spherical shapes are possible. E. g. for a perfectly grid aligned capsule (spherocylinder) of a diameter of an odd pixel number, both RMS and MIS are exactly that capsule again. Denote by R the set of all RMSs and by M the set of all MISs, by X the structure in the foreground, by $MA(X)$ the medial axis of X consisting of the centers of maximal inscribed spheres, and by $Y(X, C)$ the minimal skeleton of X under the constraint set $C \subseteq X$ as defined below.

First, the 3D image is binarized. That is, the structure to be analyzed is segmented to be foreground. Second, the skeleton is computed using the 3D topology preserving thinning algorithm from [1]. Informally, a simple point p of X is a pixel which can be removed without changing the topology of X. The thinning procedure iteratively removes simple points from X, starting from those closest to the background progressing into the structure. A constraint set $C \subseteq X$ can be used to ensure that the skeleton retains essential structural information like extreme points. The minimal skeleton $Y(X, C)$ under constraint C, i.e. $C \subset Y(X, C)$, is called the ultimate homotopic skeleton. That is, there are no simple points left in the set $Y \setminus C$. Note that for a single connected component X without any holes or tunnels, $Y(X, \emptyset)$ consists of just one point. Thus, prior to skeletonization, C has to be set with care to preserve the branching information. An intuitive option for C is the medial axis $MA(X)$. In 3D, $MA(X)$ may contain surfaces (e. g., the medial axis of a cuboid) that are consequently preserved in the skeleton. However, 2D parts within the skeleton affect the location of branching points. As a remedy for this, Couprie and Zrour [1] suggest to prune $MA(X)$ guided by the so-called bisector function assigning each pixel x of $MA(X)$ the maximal angle between two disjoint background points with minimal distance to x. This distance and the angle control the branching: Pixels with small distance or small bisector angle are removed from $MA(X)$. However, for a good result, thresholds for pruning have to be chosen for every sample individually.

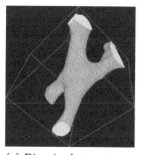  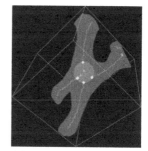

(a) Binarized structure   (b) Skeleton with branching points by method from [1]   (c) The RMS and the 2 branching points inside

Fig. 1. Example demonstrating the suggested treatment of branching points inside the RMS. 2 branching points of the skeleton fall into the same RMS. Instead of counting branches emerging from these nodes, the RMS is checked for branches by counting the skeleton entry points on its surface. As a result, the proposed method finds 1 node and 4 branches instead of 2 nodes with 3 branches each.

Here, we follow [4] to construct an alternative constraint set C_{rm} composed of the centers of RMSs [20] in the granulometry image. The set R consists of spheres inscribed in the locally thickest parts of the structure. Using the centers of these spheres as the constraint set ensures $Y(X, C_{rm})$ to reach all the locally thickest parts of X, where branching may take place. Since C_{rm} contains just isolated points, $Y(X, C_{rm})$ is composed of 1D curves without surface patches.

Considering the 26 adjacency system, a point is identified as a skeleton branching point if it has more than two neighbours. The naive solution to just use these points as branching nodes leads to a strong overestimation of node density due to surface roughness and discretization effects. Thus, skeleton branching points have to be merged. In order to do this in a generally applicable way, we exploit the regional maximal spheres inscribed locally into the structure. Finally, the branch number is determined by counting the number of intersection points between the skeleton and the surface of these spheres, see Fig. 1. The branching points and the intersection points are dilated for visualization purposes.

RMSs located at the locally thickest parts of X are naturally the appropriate branching node candidates. A branching node shall include at least one skeleton branching point. Despite the centers of the RMSs forming the constraint set for the skeletonization, the set of skeleton branching points does not necessarily coincide with C_{rm}. In order to identify RMSs including branching points during the merging, we apply a morphological reconstruction by dilation [20] using the branching points as marker set and the RMSs image as mask. That way, branching points within the same RMS are merged. RMSs not containing any branching points are omitted.

In the next step, we determine the number of branches with respect to a RMS as the number of skeleton lines entering: First, a RMS is dilated such that its radius is increased by n pixels, forming an n pixel wide dilation shell. To avoid connectivity paradoxes [6], we choose $n = 2$. Second, the connected components of the intersection of the skeleton and the shell are labeled. The number of connected components, i. e. the number of the skeleton entry points, is set to be the branch number of that RMS. The RMSs containing more than two branches are considered to be true branching nodes, all others are omitted.

Due to the particular local structure shape and discretization effects, skeleton branching points can be outside the union of all RMS R, see for instance the blood vessel systems investigated in Sect. 4.3 below. In this application, the cross-sectional shape of the vessels may strongly deviate from spherical shape causing skeleton branching points located outside R.

To merge these remaining branching points outside R too, we exploit MISs centered at the branching points as given by the Euclidean distance transform. MISs are obtained by adaptive dilation of the branching points, using the corresponding values from the Euclidean distance transform as radii. The number of branches entering each MIS is counted in the same way as in the RMSs case.

Introducing MIS may result in snow-balling as shown in Fig. 2. In order to diminish this effect, we apply sphere segmentation prior to counting branches. The image of all RMSs including at least one branching point is added to the MISs image. Then we compute the spherical granulometry on the union image and differentiate individual spheres by their radii in the granulometry image. Thus the largest sphere representing the local structure best, preserves its shape and includes as many branching points as possible. The remaining incomplete

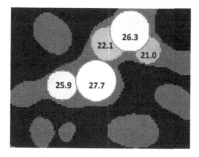

Fig. 2. 2D demonstration of a severe snow-balling case. A low gray-value is assigned to the foreground and added to the granulometry image of the overlapping spheres. Spheres have different pixel intensities indicating different diameters, where the true numbers are drawn on the image. There are two RMSs marked by the blue circles, all others are MISs. (Color figure online)

spheres with a radius of at least three pixels are treated as branching node candidates: The number of branches based on the two-pixel wide dilation shell having the same incomplete spherical shape is checked. Those with more than two branches are considered the true branching nodes, all others are omitted. Spheres with radius less than three pixels are removed since they are too small to separate different skeleton entry points on the dilation shell.

Both skeleton and granulometry are based on the medial axis. An efficient implementation of our algorithm would exploit this. Consequently, the computation time is dominated by the calculation of the granulometry image. We use the efficient algorithm from [15, 16] ensuring linear complexity in the number of pixels as long as the pre-computed look up table covers the maximal structure thickness. Nevertheless, each pixel's value in the granulometry image has to be updated several times, depending on the structure thickness. Thus the constant is high and the algorithm might take hours for an image of $1\,000^3$ pixels or larger.

Detecting branching nodes allows us to characterize and compare micro-structures in various aspects, including but not limited to: Segmentation of individual struts in open foams by subtracting the slightly dilated branching nodes from the foam structure; analyzing the distribution of the distance between two branching nodes based on the measured radii of nodes and the structure's skeleton; evaluation of the position and radius distribution of detected branching nodes, revealing homo- or heterogeneity in connectivity; and application of point process statistics to the set of branching points for complex, multiply-connected micro-structures. For instance, Häbel et al. [4] perform an anisotropy analysis using point process methods. Further examples are given in Sect. 4.

3   Validation

The proposed method has been validated using the synthetic open foam structures shown in Figs. 3(a) and 6(a). The first is the dilated strut system of a

Weaire-Phelan foam [22] with equal volume cells and planar faces as described in [7] and [17]. By design, there are 315 vertices and 4 struts per vertex. All struts are dilated to a constant thickness of 15 pixels. The essential steps of the proposed branching point detection method are illustrated in Figs. 3 and 4. Nodes and corresponding branching points at the image borders are not taken into account since their connectivity can not be observed completely.

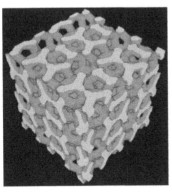 

(a) Dilated edge system of the Weaire-Phelan structure

(b) Granulometry image and derived RMSs (highlighted)

Fig. 3. The Weaire-Phelan foam structure [12, 22] with straight struts of diameter 15 pixels in an image of 200 × 200 × 200 isotropic pixels and the corresponding granulometry image.

We analyze the branch numbers by our new method and by simply using the skeleton branching points and counting their neighbourhood pixels. By the latter, naive method, we find 1 087 nodes – about 2 times too many. The average number of branches per point is 3.5 and thus significantly lower than expected. The purely skeleton based method finds 621 three-branch nodes, 466 four-branch nodes, no nodes with more than four branches.

Our method detects 314 nodes and the average number of branches per node is 4. We find no three-branch node, 313 four-branch nodes, no five-branch nodes, 1 six-branch node, no nodes with more than six branches.

The slight deviation from the expected numbers is solely due to 2 four-branch nodes close to each other being merged into the six-branch node, see Fig. 5(b).

The second synthetic structure used for validation is the realization of a stochastic geometry model for an open copper foam. The cell structure is modeled by a random Laguerre tessellation generated by a force-biased packing of spheres as described in [9, 14]. The edge thickness is subsequently fit to the thickness distribution observed in the real foam by locally adaptive morphological dilation as described in [10]. The model used here is derived by the refined method from [8]. There are 1 518 vertices excluding those at image borders. The branch numbers are analyzed again by both the proposed method and by the

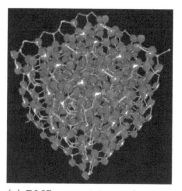

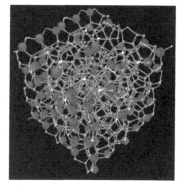

(a) RMSs containing at least one branching point

(b) MISs containing at least one branching point not incontaining RMSs

Fig. 4. Skeleton and RMSs and MISs, respectively, for the Weaire-Phelan foam structure from Fig. 3.

naive method. By the naive, skeleton based method, we find 4 580 nodes with an average of 3.4 branches. More detailed, the naive method yields 2 999 three-branch nodes, 1 488 four-branch nodes, 88 five-branch nodes, 5 six-branch nodes, and no nodes with more than six branches. Thus only 32% of all found nodes feature the expected 4 branches.

Our method detects 1 193 nodes with the mean branch number 4.14. We find 186 three-branch nodes, 792 four-branch nodes, 127 five-branch nodes, 55 six-branch nodes, 17 seven-branch nodes, 14 eight-branch nodes, 1 nine-branch node, 1 ten-branch node, and no nodes with more than ten branches.

Thus 66% of all found nodes feature the expected 4 branches. Moreover, 1 144 (96%) of the detected nodes are true ones while only 49 (4%) false positives are detected. (Here a node is considered to be true if a true vertex can be found inside the RMS or MIS corresponding to the node.) The set of false positives consists of 37 three-branch nodes, 5 four-branch nodes, 3 five-branch nodes, 2 six-branch nodes, 1 seven-branch node, and 1 eight-branch node. Without the false nodes, the average branch number increases to 4.17. Hence, the false nodes influence the branch number evaluation only slightly. An example of erroneously detected nodes is shown in Fig. 7(a). There, the skeletonization generated a small additional facet, leading to the unexpected branching points and thus derived nodes. Furthermore, 123 vertices are not detected. Figure 7(b) shows one reason for this error: the synthetic foam structure can have tiny or acute-angled facets that are completely filled by the applied locally adaptive dilation of the edges [21]. As a consequence, the skeleton does not branch there. Another cause for overlooked nodes is explained in Fig. 7(c): vertices are close enough to be included in a single node.

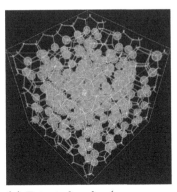
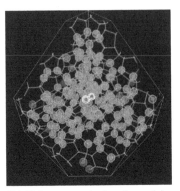

(a) Detected nodes (gray spheres) and the true vertices (red, dilated)

(b) Volume rendering from (a) clipped to show the six-branch node (highlighted)

Fig. 5. Results of the node detection for the Weaire-Phelan foam from Fig. 3. (Color figure online)

These validation results for both synthetic data sets confirm that the proposed method identifies true branching nodes more reliable than the method solely based on the skeleton branching points.

4   Applications

The proposed method has been applied to CT image data from three different fields of application: an open cell aluminum foam, the air pore system within a Greenland ice core, and the blood vessel system in mice livers.

4.1   Open Cell Aluminum Foam

The tomographic image of the open cell aluminum foam has $630 \times 1\,370 \times 1\,370$ isotropic pixels with an edge length of $14.16\,\mu m$. We use an arbitrarily selected sub-volume of $314 \times 500 \times 890$ pixels, corresponding to a sample size of $4.4\,mm \times 7.1\,mm \times 12.6\,mm$. The high gray value contrast of the image allows us to separate the foam from background using Otsu's global gray value thresholding method [13] without pre-processing. Small holes in the strut system were filled morphologically to prepare the binarized data set for the analysis.

The binarized sample and its skeleton are shown in Fig. 8 together with the image of the detected branching nodes (green spheres). Given that we deal with data of a real sample, there is no ground truth. It is however an open rigid foam. Thus, solely 4-branch nodes are expected. We analyze the branch numbers using three methods: 1st the skeleton branching points based method; 2nd our method, and 3rd the struts segmentation method from [10]. For this 3rd method, to be consistent with [10], the number of branches per node is considered to be the

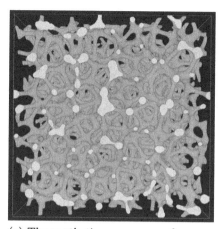
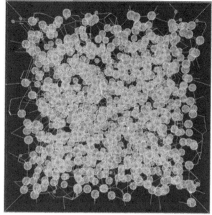

(a) The synthetic open copper foam structure

(b) True vertices (red), detected nodes (gray spheres), and skeleton (lines)

Fig. 6. Result of the branching node detection for the synthetic open copper foam structure [8]. The foam was discretized to 656^3 isotropic pixels. The struts are straight with thickness varying from 29 pixels in the vertices to 13 pixels in the strut centers. (Color figure online)

number of struts connected to it. For all three methods, we exclude the nodes at image borders from the analysis.

The resulting histograms are compared in Fig. 9. Similarly to what we have seen in Sect. 3, the mean branch numbers per node are 3.4 by method 1, 3.9 by our method, and 4.0 by the method from [10]. The latter two agree very well both with respect to mean number as well as proportion of four-branch nodes (88% and 86%). They disagree however considerably in the number of branching nodes found. As explained in [10], method 3 can miss some struts/vertices due to structural peculiarities in the aluminum foam. Moreover, this method was designed for obtaining a representative sample of struts, not for finding all vertices. Thus it is not surprising that we find more branching nodes.

4.2 Greenland Firn

Due to accumulation of snow on the top of polar ice sheets, the ice in deeper layers is compressed and undergoes a sintering process. In a certain stage of this process, the ice is called firn and contains a connected network of irregularly shaped air pores. With increasing depth, both the porosity and the connectivity of the pore system decrease. After the firn-ice transition, the pore system will consist of isolated air bubbles.

In this work, we consider firn samples from an ice core drilled at the Renland ice cap in Greenland in the framework of the REnland ice CAP project (RECAP). We compare firn samples from different depths. The depth is measured in bags where a bag corresponds to 0.55 m. The pore system of the ice is

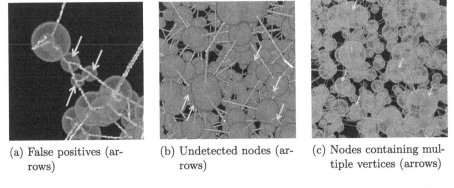

(a) False positives (arrows)

(b) Undetected nodes (arrows)

(c) Nodes containing multiple vertices (arrows)

Fig. 7. Enlarged sub-regions with detection errors from the result for the synthetic open copper foam from Fig. 6. (Color figure online)

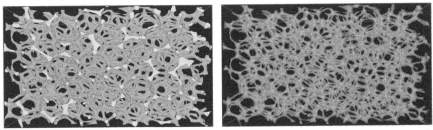

(a) Binarized aluminum foam sample

(b) Skeleton and detected nodes (green)

Fig. 8. Result of the branching node detection for an aluminum open foam. Physical size: $4.4\,mm \times 7.1\,mm \times 12.6\,mm$ corresponding to $314 \times 500 \times 890$ isotropic pixels at pixel edge length $14.16\,\mu m$. (Color figure online)

analyzed using μCT images provided by the Alfred-Wegener-Institute for Polar and Marine Research in Bremerhaven, Germany.

We analyze the air pore system at bags 50, 55, 60, and 65, corresponding to firn samples at depths of 27.5 m, 30.25 m, 33.0 m, and 35.75 m. The gray value μCT images were smoothed by a $5 \times 5 \times 5$ pixel median filter. Owing to the high resolution, we can coarsen the lattice by a factor of two in each direction without changing the general topology of the structures. The resulting images have a pixel edge length of $44(\pm0.08)\,\mu m$. A volume of size 600^3 pixels is used, corresponding to a cubic sample of edge length 26.4 mm. The air pores were extracted for analysis by slightly adjusting Otsu's threshold for each data set.

In Fig. 10, we inspect the distribution of nodes in x, y, and z directions. Each point in the plot corresponds to the number of nodes in slices of thickness 50 pixels. We exclude the first and the last 50 pixels in all directions to avoid edge effects. In general, the number of nodes decreases along the z direction but not significantly in x or y direction. This result confirms, the number of nodes decreases as depth increases. We also observe that the dominant branch number

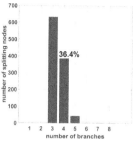

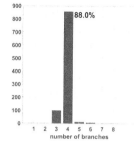

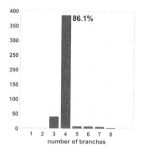

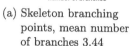

(a) Skeleton branching points, mean number of branches 3.44

(b) Suggested method, mean number of branches 3.93

(c) Strut segmentation method, mean number of branches 4.02

Fig. 9. Histograms of the numbers of branches in the aluminum foam computed by skeleton branching point, our method, and struts segmentation [10].

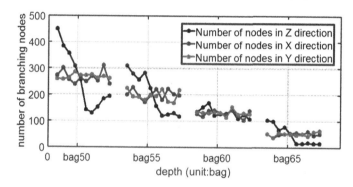

Fig. 10. Distributions of branching nodes in x (blue), y (red), and z (black) directions in firn samples bag 50, 55, 60, and 65. (Color figure online)

of the air pore system is 3 per node, independent of the changes in depth. More specifically, the average branch numbers per node are $3.13(\pm0.39)$, $3.12(\pm0.36)$, $3.05(\pm0.25)$ and $3.07(\pm0.26)$ for bags 50, 55, 60, and 65, respectively.

4.3 Mice Livers

Liver fibrosis is a disease characterized by scar formation in the liver through an increased deposition of collagenous connective tissue. Scars induce a reorganization of the capillary blood vessels which have to grow around the scars. In this section we investigate the morphology of the blood vessel system in mice where liver fibrosis was induced by either chronic injections of thioacetamide (TAA) or gene manipulation (MDR2). A third scan was taken from an animal from the control group. Capillary blood vessels were prepared as vascular corrosion casts, i.e. the blood vessels were filled with resin. After removing the surrounding tissue, the casts were scanned at the μCT station of the Materials Science

Beamline TOMCAT at the Swiss Light Source of the Paul-Scherrer-Institut (Villigen, Switzerland) at a pixel size of 325 nm. See visualizations in Fig. 11.

We apply our method on the binarized images of the three samples. Again, branching nodes at the image borders are excluded to avoid edge effects. The average branch number and the total numbers of detected nodes are compared. The three-branch nodes are found to be dominant, leading to an average of 3.1 branches per node in all samples. The samples with fibrosis induced by TAA injection, MDR2, and the control group have foreground volumes of $1.8e - 04\,mm^3$, $3.9e - 04\,mm^3$, and $3.7e - 04\,mm^3$, respectively. The corresponding numbers of nodes per mm^3 are $3.0e + 05$, $6.2e + 05$ and $7.3e + 05$.

The TAA sample with severe fibrosis has the least branching nodes per volume unit (less than 50% compared to the control group), indicating that the connectivity of blood vessels may be deteriorated by fibrosis. The intensity of branching nodes could be a characteristic to quantify the severity of liver fibrosis.

Fig. 11. Volume renderings of three mice liver samples. From left to right: fibrosis induced by TAA, fibrosis induced by MDR2, control group. Physical size: $590\,\mu m \times 590\,\mu m \times 700\,\mu m$ corresponding to $1\,810 \times 1\,810 \times 2\,160$ pixels at pixel size 325 nm.

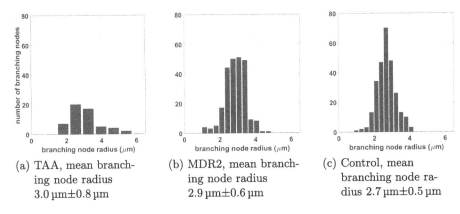

(a) TAA, mean branching node radius $3.0\,\mu m \pm 0.8\,\mu m$

(b) MDR2, mean branching node radius $2.9\,\mu m \pm 0.6\,\mu m$

(c) Control, mean branching node radius $2.7\,\mu m \pm 0.5\,\mu m$

Fig. 12. Histograms showing the distributions of the radii of detected branching nodes in the three mice liver samples.

In addition, the radii of detected branching nodes are characterized in Fig. 12. Interesting findings are: the TAA liver showing the worst fibrosis has on average the largest branching nodes, where the mean radius reaches $3.0(\pm0.8)$ μm. The remaining livers with less or without fibrosis have smaller branching nodes and the mean radii are $2.9(\pm0.6)$ μm and $2.7(\pm0.5)$ μm.

5 Discussion

We presented a fully automatic method to detect branching nodes based on the skeleton of 3D structures. The proposed method has the advantage to be applicable to structures with a natural network structure like the synthetic and real open foams as well as to more amorphous structures like the air pore system in firn. Two types of spheres (RMS and MIS) are considered as candidates of branching nodes. An alternative is to compute only MISs at all branching points. This is considered computationally cheaper since the computation of MISs does not involve the expensive granulometry operation. However, we introduced RMSs in the consideration of: 1st the granulometry and RMS are readily computed for the constraint set C_{rm}; 2nd due to discretization effects, branching points with respect to the same structure node are often found scattered inside the locally thickest area. MISs may form a large group of overlapping spheres depending on the number and the distribution of the branching points. The RMSs, on the other hand, have a fixed size no matter how many or how scattered the branching points are. This advantage of RMSs effectively decreases the risk of snow-balling. We never observed severe cases of it in our synthetic and real structures, not even in the Weaire-Phelan foam structure that has by design struts of constant thickness. Thus, no additional precautions to prevent it had to be applied.

It does however happen that a RMS consists of several intersecting spheres of constant radius. In this case, the union set is considered a branching node. We observe that most prominently in the open copper foam where a RMS consisting of 4 spheres occurs. Nevertheless, in all examples considered here, no additional separation criterion had to be applied. Strategies for separating connected spheres could exploit the distances between centers of overlapping spheres or geometric features like the volume or the maximal width of the union set. In any case, these would have to be chosen and parameterized application-driven.

Acknowledgements. We thank André Liebscher for providing the foam data, Johannes Freitag and Tetsuro Taranczewski for the firn data, and Maximilian Ackermann and Willi Wagner for the mice liver data. This work was supported by the Fraunhofer High Performance Center for Simulation- and Software-Based Innovation, by the Deutsche Forschungsgemeinschaft (DFG) in the framework of the priority programme "Antarctic Research with comparative investigations in Arctic ice areas", grant RE 3002/3-1, and by the German Federal Ministry of Education and Research (BMBF), grant 05M16UKA (AMSCHA).

References

1. Couprie, M., Zrour, R.: Discrete bisector function and euclidean skeleton. In: Andres, E., Damiand, G., Lienhardt, P. (eds.) DGCI 2005. LNCS, vol. 3429, pp. 216–227. Springer, Heidelberg (2005). https://doi.org/10.1007/978-3-540-31965-8_21
2. Dong, H., Blunt, M.J.: Pore-network extraction from micro-computerized-tomography images. Phys. Rev. E **80**, 036307 (2009)
3. Hormann, K., Baranau, V., Hlushkou, D., Höltzel, A., Tallarek, U.: Topological analysis of non-granular, disordered porous media: determination of pore connectivity, pore coordination, and geometric tortuosity in physically reconstructed silica monoliths. New J. Chem. **40**, 4187–4199 (2016)
4. Häbel, H., et al.: A three-dimensional anisotropic point process characterization for pharmaceutical coatings. Spat. Stat. **22**, 306–320 (2017)
5. Klette, G.: Branch voxels and junctions in 3D skeletons. In: Reulke, R., Eckardt, U., Flach, B., Knauer, U., Polthier, K. (eds.) IWCIA 2006. LNCS, vol. 4040, pp. 34–44. Springer, Heidelberg (2006). https://doi.org/10.1007/11774938_4
6. Kong, T.Y., Rosenfeld, A.: Digital topology: introduction and survey. Comput. Vis. Graph. Image Process. **48**, 357–393 (1989)
7. Lautensack, C., Sych, T.: A random Weaire-Phelan foam. In: 8th International Conference on Stereology and Image Analysis in Materials Science STERMAT 2008, Zakopane, Poland (2008)
8. Liebscher, A.: Stochastic modelling of foams. Ph.D. thesis, Technische Universität Kaiserslautern (2014)
9. Liebscher, A., Redenbach, C.: 3D image analysis and stochastic modelling of open foams. Int. J. Mater. Res. **103**(2), 155–161 (2012)
10. Liebscher, A., Redenbach, C.: Statistical analysis of the local strut thickness of open cell foams. Image Anal. Stereology **32**(1), 1–12 (2013)
11. Lindquist, W.B., Venkatarangan, A., Dunsmuir, J., Wong, T.-F.: Pore and throat size distributions measured from synchrotron X-ray tomographic images of Fontainebleau sandstones. J. Geophys. Research **105B**, 21508–21528 (2000)
12. Ohser, J., Schladitz, K.: 3D Images of Materials Structures - Processing and Analysis. Wiley VCH, Weinheim (2009)
13. Otsu, N.: A threshold selection method from gray level histograms. IEEE Trans. Syst. Man Cybern. **9**, 62–66 (1979)
14. Redenbach, C.: Microstructure models for cellular materials. Comput. Mater. Sci. **44**, 1397–1407 (2009)
15. Remy, E., Thiel, E.: Look-up tables for medial axis on squared euclidean distance transform. In: Nyström, I., Sanniti di Baja, G., Svensson, S. (eds.) DGCI 2003. LNCS, vol. 2886, pp. 224–235. Springer, Heidelberg (2003). https://doi.org/10.1007/978-3-540-39966-7_21
16. Remy, E., Thiel, E.: Exact medial axis with euclidean distance. Image Vis. Comput. **23**(2), 167–175 (2005)
17. Schladitz, K., Redenbach, C., Sych, T., Godehardt, M.: Model based estimation of geometric characteristics of open foams. Methodol. Comput. Appl. Probab. **14**, 1011–1032 (2012)
18. Sheppard, A., Sok, R., Averdunk, H.: Improved pore network extraction methods. In: International Symposium of the Society of Core Analysts (2005)
19. Silin, D., Patzek, T.: Pore space morphology analysis using maximal inscribed spheres. Physica A Stat. Theoret. Phys. **371**, 336–360 (2006)

20. Soille, P.: Morphological Image Analysis. Springer, Heidelberg (1999). https://doi. org/10.1007/978-3-662-05088-0
21. Vecchio, I., Redenbach, C., Schladitz, K.: Angles in Laguerre tessellation models for solid foams. Comput. Mater. Sci. **83**, 171–184 (2014)
22. Weaire, D., Phelan, R.: A counter-example to Kelvin's conjecture on minimal surfaces. Philos. Mag. Lett. **69**(2), 107–110 (1994)

Segmenting Junction Regions Without Skeletonization Using Geodesic Operators and the Max-Tree

Andrés Serna[1($\boxtimes$)], Beatriz Marcotegui[1,2], and Etienne Decencière[2]

[1] Terra3D, 36 Boulevard de la Bastille, 75012 Paris, France
andres.serna@terra3d.fr
[2] MINES ParisTech, PSL Research University,
35 rue Saint Honoré, Fontainebleau, France
{beatriz.marcotegui,etienne.decenciere}@mines-paristech.fr
http://www.terra3d.xyz/
http://cmm.mines-paristech.fr

Abstract. In a 2D skeleton, a junction region indicates a connected component (CC) where a path splits into two or more different branches. In several real applications, structures of interest such as vessels, cables, fibers, wrinkles, etc. may be wider than one pixel. Since the user may be interested in the junction regions but not in the skeleton itself (e.g. in order to segment an object into single branches), it is reasonable to think about finding junction regions directly on objects avoiding skeletonization. In this paper we propose a solution to find junction regions directly on objects, which are usually elongated structures, but not necessary thin skeletons, with possible protrusions and noise. Our method is based on geodesic operators and is intended to work on binary objects without holes. In particular, our method is based on the analysis of the wavefront evolution during geodesic propagation. Our method is generic and does not require skeletonization. It provides an intuitive and straightforward way to filter short branches and protuberances. Moreover, an elegant and efficient implementation using a max-tree representation is proposed.

Keywords: Triple points · Bifurcation pixel · Junction region · Skeleton · Branches · Geodesic distance · Geodesic extremities · H-maxima · Max-tree

1 Introduction

In a 2D skeleton, a triple point or bifurcation pixel is a point where the skeleton splits into three different branches. If it splits into four or more branches, the more generic term of junction point is used. Finding junction points in skeletons is a classic problem, usually solved by applying iterative hit-or-miss operators with some special structuring elements [17, 18].

© Springer Nature Switzerland AG 2019
B. Burgeth et al. (Eds.): ISMM 2019, LNCS 11564, pp. 456–467, 2019.
https://doi.org/10.1007/978-3-030-20867-7_35

Extracting the skeleton from a given object is a well studied problem in the literature [2,3,7–9,16,21]. However, as far as we know, there is no universally established best skeleton method.

In several applications, the user may be interested in the junction regions but not in the skeleton itself (e.g. in order to segment an object into single branches). In such a case, the skeleton computation is not needed.

In this work, we propose a method to find junction regions of an object without holes avoiding skeletonization. Our method is based on geodesic operators and provides an intuitive and straightforward way to filter out spurious short branches and protuberances. The paper is organized as follows: Sect. 2 reminds some basic notions. Section 3 introduces our proposed method to find junction regions on binary objects without holes. An elegant and efficient implementation using a max-tree is also proposed. Section 4 presents some results and a comparison with respect to a classic method based on skeletonization. Finally, Sect. 5 presents the conclusions and perspectives of the work.

2   Basic Notions

The goal of this Section is reminding some basic notions related to the Euler characteristic, geodesic operators and the max-tree structure. These concepts will be used in the proposed method explained later in Sect. 3.

2.1   Euler Characteristic

Let I_b be a digital binary image $I_b\colon D \to V_b$, with $D \subset Z^2$ the image domain with a given connectivity and $V_b = \{0,1\}$ the set of binary values. The Euler characteristic refers to the number of connected components (CC) of image I_b minus the number of holes of these CCs [6]. In this paper, we call binary object each single CC in image I_b. Considering the Euler characteristic of a single CC allows us to classify binary objects into two categories: porous and non-porous. On the one hand, non-porous objects (without holes) have an Euler characteristic equal to one. On the other hand, porous objects (with holes) have an Euler characteristic lower than or equal to zero. This work focuses on non-porous objects. In this context, non-porous objects are also known as simply connected sets or objects without holes.

2.2   Geodesic Operators: Distance, Extremities and Diameter

The geodesic propagation consists in an iterative displacement of a wavefront within an object from some starting seeds until the whole object is reached. The shape of the wavefront is defined by a given connectivity. Let X be a single binary object and let p and q be two points belonging to X. The minimum number of geodesic iterations required to reach q starting from p (or vice versa) is called the geodesic distance $d_X(p,q)$. In other words, the geodesic distance $d_X(p,q)$ is the minimum length of a path joining p and q and included in X [10].

Figure 1 presents an example of geodesic distance from given pixels inside a synthetic binary object using 8-connectivity. The propagation starts from the two pixels marked with a green square in the center of Fig. 1(a). In the first iteration, the geodesic distance for the starting points is set to zero. In the second iteration, the distance for the first neighboring points within the object is set to one. In the third iteration, the distance to the second neighbors is set to two, and so on. The process ends when the whole object has been processed. In this example, the process finishes at the 75th iteration at the end of the spiral (See Fig. 1(b)). The geodesic distance is only computed for the points belonging to the object, thus pixels outside the object are set to *infinite* (or an invalid distance value, -1 in this example). Note that the geodesic propagation can be implemented as iterative geodesic dilations. However, a more efficient implementation is possible using priority queues [22].

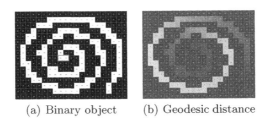

(a) Binary object   (b) Geodesic distance

Fig. 1. Geodesic distance from the highlighted pixels in the spiral.

We can compute $d_X(p,q)$ for all $p,q \in X$. Any couple of points where $d_X(p,q)$ reaches its global maximum are the geodesic extremities of X. To find the geodesic extremities of X, we use an efficient approximation based on the barycentric diameter proposed in [11, 12], which can be summarized as follows:

1. Let binary object X be a single CC of image I_b and its barycenter B.
2. Let x be the farthest point of X from B. Since B does not necessary belong to X, the Euclidean distance is used to look for the farthest point x.
3. Compute the geodesic distance from x inside X. The maximum value of this geodesic distance is $l^1(X)$ and the point where it is reached is y_1.
4. Compute the geodesic distance from y_1 inside X. The maximum of this new geodesic distance is $l^2(X)$ and the point where it is reached is y_2.

Based on this processing, y_1 and y_2 are estimations of the geodesic extremities of X while $l^2(X)$ corresponds to its barycentric diameter. The authors [12] have shown that the barycentric diameter is a very good approximation of the geodesic diameter with an average error lower than 0.25% in a database of 51400 binary shapes. Figure 2 illustrates the computation of the barycentric diameter in order to estimate two geodesic extremities and the geodesic diameter of an object.

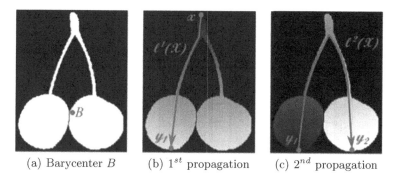

(a) Barycenter B (b) 1st propagation   (c) 2nd propagation

Fig. 2. Computation of the barycenter diameter in order to find two geodesic extremities of object X.

2.3   Max-Tree

Let I be a digital gray-scale image $I \colon D \to V$, with $D \subset Z^2$ the image domain and $V = \{0, ..., R\}$ the set of gray-scale levels. A decomposition of I can be obtained considering successive thresholds $T_t(I) = \{p \in D \mid I(p) \geq t\}$ for $t = \{0, ..., R\}$. Since this decomposition satisfies the inclusion property $T_t(I) \subseteq T_{t-1}(I), \forall t \in [1, ..., R]$, it is possible to build a tree with the CCs of the sets $T_t(I)$. The max-tree (resp. min-tree) is a data structure in which each leaf corresponds to a maximum (resp. minimum) of the image and each node corresponds to a CC of a threshold of the image. This structure is particularly useful in the domain of connected operators [15]. The construction of the tree is based on the inclusion rule [5,14]. Several max-tree implementations are proposed in the literature [5,13,14,23]. In this paper, we use an improved version of the fast max-tree implementation proposed in [4].

Let us explain the max-tree construction with the toy example of Fig. 3. For this example 4-connectivity is used. Consider the Fig. 3(a) containing 8 flat-zones enumerated from a to h (in lower-case letters). Gray regions correspond to zero valued pixels. Figures 3(d) to (i) present the successive thresholds from $T_1(I)$ to $T_6(I)$, respectively. Figure 3(c) presents the corresponding max-tree built from those successive thresholds. Note that a given threshold may be composed by several CCs, enumerated from A to H (in upper-case letters). For example, threshold $T_4(I)$ has two CCs D and G. Moreover, a CC may be related to several flat-zones. For example, node D in the max-tree is associated to flat-zones d, e and f in the input image. The max-tree construction starts at the lowest level $t = 0$, corresponding to the root of the tree. The first level is composed by threshold $T_1(I)$ containing CC A in Fig. 3(d). Threshold $T_2(I)$ is composed by single CC B while $T_3(I)$ is composed by C. Following the inclusion rule, $T_3(I) \subseteq T_2(I) \subseteq T_1(I)$, we add B and C to the main trunk of the tree. Node C contains two children since threshold $T_4(I)$ in Fig. 3(g) contains two CCs D and G. The construction of these two branches is straightforward since $F \subseteq E \subseteq D$ and $H \subseteq G$.

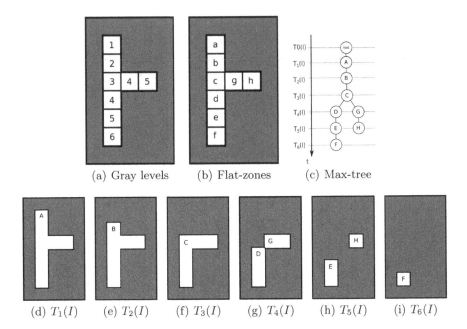

Fig. 3. Max-tree construction. The image is composed by 8 flat-zones enumerated from a to h. Each node corresponds to a CC, enumerated from A to H, of threshold $T_t(I)$. The max-tree contains two leafs corresponding to the two maxima at f and h.

In addition, it is possible to associate a flat-zone in I to a given node in the tree. Note that a node might be associated to several flat-zones. In fact, a given flat-zone at level t is contained in the subset of $T_t(I)$ where gray level is equal to t in the input image. For example, flat-zone c in Fig. 3(b) corresponds to the pixels of C in Fig. 3(f) where the gray level is equal to 3 in the input image : $c = \{p \in C \mid I(p) = 3\}$.

In general, several filters can be applied to the tree (i.e. pruning leaves) and the output image is the restitution from the filtered tree. There exist several filtering rules to recompose the output image from the filtered max-tree [1,14, 19,20]. The choice of the filtering rule depends on the application. In this paper, we will focus on the direct rule: all CCs $T_t(I)$ holding a given criterion are stacked to recompose the output image.

Let us illustrate the filtering process based on max-tree with the example of Fig. 4. In this case, we want to filter branches with contrast lower than or equal to 2 (this is equivalent to a h-Maxima filter with $h = 2$). The contrast of a branch is defined as the gray-level difference between the first and the last node of the branch. This leads to remove the nodes G and H in Fig. 4(a). Figure 4(b) shows the output image recomposed from the filtered max-tree using the direct rule. Figure 4(c) shows the corresponding flat-zones filtered in the process.

Figure 5 presents another example of image filtering using the max-tree structure. In this case, we want to remove nodes with less than 2 children. This leads to

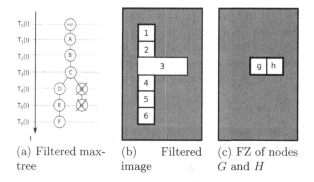

(a) Filtered max-
tree

(b)   Filtered
image

(c) FZ of nodes
G and H

Fig. 4. Max-tree based filtering on image of Fig. 3(a). Branches with contrast lower than or equal to 2 are removed. The output image is recomposed from the filtered max-tree using the direct rule. This is equivalent to a h-Maxima filter with $h = 2$.

remove all the nodes except C and the root of the tree (see Fig. 5(a)). Figure 5(b) shows the output image recomposed from the filtered max-tree using the direct rule. Figure 5(c) shows the flat-zone c corresponding to remaining node C.

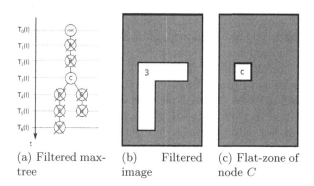

(a) Filtered max-
tree

(b)   Filtered
image

(c) Flat-zone of
node C

Fig. 5. Max-tree based filtering on image of Fig. 3(a). Nodes with less than 2 children are removed. Output image is recomposed from the filtered max-tree. Figure 5(c) shows the flat-zone c corresponding to node C in the input image.

3   Finding Junction Regions

A classic method to find junction regions can be summarized in Fig. 6 as follows:

1. Compute the skeleton of the input image (Fig. 6(a)). A great number of methods are proposed in the literature for this step [2, 3, 7–9, 16, 21].
2. Ideally, each junction region in the skeleton representation corresponds to an interesting intersection in the input object. In practice, there are often spurious junction regions associated to unwanted skeletal branches. If needed, these branches can be removed using a parametric pruning in order to get a clean skeleton (Fig. 6(b)).

3. Individual skeletal branches are obtained by suppressing junction regions from the skeleton (magenta points in Fig. 6(b)).
4. Junction regions are computed using the skeleton by influence zones (SKIZ) of the remaining skeleton (Fig. 6(c)) within the input object.
5. Finally, the SKIZ allows the segmentation in individual branches (Fig. 6(d)).

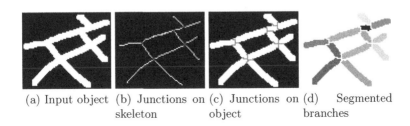

(a) Input object   (b) Junctions on   (c) Junctions on   (d)   Segmented
                       skeleton   object   branches

Fig. 6. Extraction of junction regions using a method based on skeletonization.

In general, this classic method works well for thin and clean structures, but may fail on thick objects with bent branches and protrusions. A critical step is the choice of both the appropriate skeletonization algorithm and the pruning strategy in order to keep interesting branches. In several applications, the user is not particularly interested in the skeleton but in the junction regions. Thus, we propose to find junction regions directly on binary objects avoiding skeletonization. Our method is based on geodesic operators since the wavefronts of the geodesic propagation provide interesting information about junction regions. It is intended to work not only on thin objects but also on elongated structures with possible thick branches, protrusions and noise. Moreover, it provides an intuitive and straightforward way to filter out short branches and protuberances.

Consider a binary object without holes X, a geodesic propagation starting from a geodesic extremity y and a max-tree T built from the geodesic distance function $d_X(y)$. Let analyze the object of Fig. 3. In such a case, the propagation starts from the upper pixel of the object. When the propagation reaches a junction region, the wavefront splits from a single flat-zone c at level $t = 3$ to two separated flat-zones d and g at level $t = 4$. This is represented by a node with two children in the max-tree. It is noteworthy that the flat-zone c, corresponding to node C, is a junction region. Informally, the number of branches in a junction region is equivalent to the number of children at its corresponding node in the max-tree representation. Since nodes with more than one child correspond to junction regions, we propose the following methodology to find junction regions on a binary object without holes X:

1. Find a geodesic extremity y of object X,
2. Compute the geodesic distance from y inside X, called $d_X(y)$,
3. Construct a max-tree T from $d_X(y)$,
4. (Optional) Compute a filtered tree T' by removing branches shorter than h,

5. Compute a filtered tree T'' by removing nodes with less than two children,
6. Junction regions J are obtained as the flat-zones of I corresponding to remaining nodes in filtered tree T'',
7. Individual branches B are obtained by removing junction regions J from the object X: $B = X \backslash J$.

Note that step 4 is optional and equivalent to applying a h-Maxima filter to the geodesic distance function. For the sake of efficiency, this filter can be computed directly on the max-tree by computing the contrast of each branch.

Figure 7 illustrates our proposed method on a synthetic binary object. In this example 8-connectivity is used. The propagation starts from the lower left extremity of the object (green square in Fig. 7(a)). At the beginning, between iteration 1 and 12 in Fig. 7(b), each wavefront is composed of a single flat-zone. When the propagation reaches the first bifurcation at iteration 13, the wavefront splits into two separate flat-zones. This bifurcation indicates the presence of a junction region. If we continue to the right of the image, we can see another junction region at iteration 16. In this case, the wavefront splits producing three different flat-zones at value $t = 17$.

It is noteworthy that each maximum in the geodesic distance function corresponds to an extremity of the object. As explained in Subsect. 2.3, the contrast of a branch corresponds to its length. The h-Maxima operator can be used to filter maxima less contrasted than a given threshold h. This is a straightforward way to remove short and spurious extremities. Figures 7(c) to (f) present four examples of finding junction regions after h-Maxima filtering. Green points indicate the initial point of the geodesic distance computation, blue points indicate the regional maxima, and magenta points indicate the detected junction regions. In Fig. 7(c), the geodesic distance function contains four maxima, corresponding to the four extremities of the object (the starting point is of course an extremity, but it is a minimum because it is the starting point of the propagation). The object contains two junction regions at levels $t = 13$ and $t = 16$, respectively. In Fig. 7(d), branches shorter than 3 pixels (h-Maxima filter with $h = 3$) are filtered out. Note that the second extremity is not a maximum anymore and has been merged with the junction region. Please note here the impact of the connectivity on the branch length estimation. Although the upper left and the middle spikes of the object seem to have the same length, they are not filtered for the same value of h. If we use 4-connectivity instead of 8-connectivity, both extremities would be filtered at the same value of h. In Fig. 7(e), branches shorter than 4 pixels (h-Maxima filter with $h = 4$) are filtered out. Note that only two extremities are still regional maxima of the image and only one junction region is detected. Finally, Fig. 7(f) shows a h-Maxima filter with $h = 8$. Note that only the longest branch is still a maximum of the geodesic distance function. In this case, no junction regions are extracted.

4   Results

To evaluate our method, we present three tests on synthetic and real images. Our method is qualitatively compared to a classic method based on skeletonization

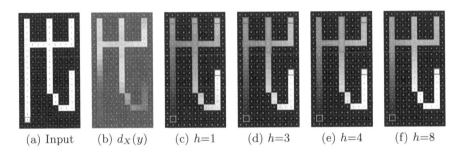

| (a) Input | (b) $d_X(y)$ | (c) $h=1$ | (d) $h=3$ | (e) $h=4$ | (f) $h=8$ |

Fig. 7. Junction regions on an object without holes. Green point indicates the initial seed of the geodesic distance, blue points indicate the regional maxima, and magenta points indicate the detected junction regions for several h parameters. (Color figure online)

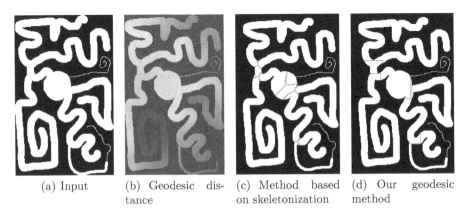

| (a) Input | (b) Geodesic distance | (c) Method based on skeletonization | (d) Our geodesic method |

Fig. 8. Junction regions using a classical method based on skeletonization and our proposed method. In both methods, branches shorter than 30 pixels are filtered out.

(explained in Sect. 3). In order to make a fair comparison, the classical method is based on a "perfect" hand-made skeleton.

Figure 8 illustrates an example of detection of junction regions on a binary object containing thin and thick branches. Figure 8(c) presents the result using a classic method based on skeletonization. Figure 8(d) presents the result of our proposed method. In both methods, branches shorter than 30 pixels have been filtered out. Note that our method is able to manage thick objects and the filtering strategy to remove non interesting branches is intrinsically taken into account in the max-tree representation. Note also that both methods are able to extract correctly the five junction regions, generating 10 individual branches. However, our method does not require any skeleton computation.

Figure 9 illustrates a real top-view image of electric cables in a railway environment. Cables are interconnected at several locations. The goal is to segment the image into individual cable segments. Figure 9(b) presents the segmentation result with our geodesic method considering all geodesic extremities ($h = 1$).

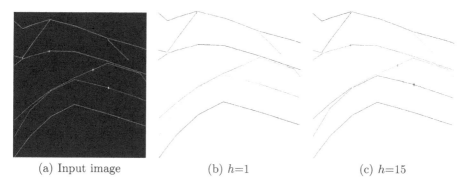

(a) Input image   (b) h=1   (c) h=15

Fig. 9. Segmentation of individual electric cables on a railway environment. Junction regions are extracted with our proposed geodesic method using two different h parameters. Each color represents an individual cable.

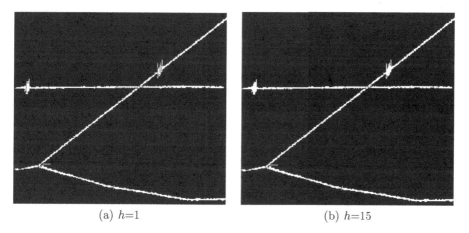

(a) h=1   (b) h=15

Fig. 10. Segmentation of individual electric cables on a railway environment. Two different h parameters have been used.

Note that several junction regions are extracted due to short protrusions in the image. In fact, those protrusions correspond to cable insulators. In order to filter out these insulators, the segmentation is carried out considering only branches longer than 15 pixels, as shown in Fig. 9(c).

Figure 10 presents another example of electric cables in a railway environment. Figure 10(a) presents the junction regions considering all branches ($h = 1$). Figure 10(b) presents the junction regions considering only branches longer than 15 pixels ($h = 15$). In this case, cable insulators are not detected as junction regions, resulting in a correct segmentation of individual cables.

5 Conclusions and Perspectives

We have proposed a method to find junction regions in objects without holes based on geodesic operators. In particular, our method is based on the analysis of the wavefront evolution in a geodesic propagation. Our method is generic and does not require skeletonization. It provides an intuitive and straightforward way to filter short extremities and protuberances. Moreover, an elegant and efficient implementation using a max-tree representation is proposed.

The problem with objects with holes is that the basic hypothesis: "the maxima on the geodesic distance correspond to the extremities of the object" does not hold anymore. In fact, a loop introduces the union of wavefronts coming from different branches, which introduces additional maxima on the geodesic distance function. We are currently working in a more generic method to extract junction regions on any binary object. The key step of such a method is replacing the max-tree structure by a more generic geodesic graph.

References

1. Breen, E.J., Jones, R.: Attribute openings, thinnings, and granulometries. Comput. Vis. Image Underst. **64**(3), 377–389 (1996)
2. Chaussard, J.: Topological tools for discrete shape analysis. Ph.D. thesis, Université Paris-Est, December 2010
3. Chaussard, J., Couprie, M., Talbot, H.: Robust skeletonization using the discrete lambda-medial axis. Pattern Recogn. Lett. **32**(9), 1384–1394 (2011)
4. Fabrizio, J., Marcotegui, B.: Fast implementation of the ultimate opening. In: ISMM 2009, 9th International Symposium on Mathematical Morphology, Groningen, The Netherlands, pp. 272–281 (2009)
5. Garrido, L., Oliveras, A., Salembier, P.: Motion analysis of image sequences using connected operators. In: Proceedings of SPIE - The International Society for Optical Engineering, vol. 3024, pp. 546–557, January 1997
6. Gray, S.B.: Local properties of binary images in two dimensions. IEEE Trans. Comput. **C-20**(5), 551–561 (1971)
7. Hesselink, W.H., Visser, M., Roerdink, J.B.T.M.: Euclidean skeletons of 3D data sets in linear time by the integer medial axis transform. In: Ronse, C., Najman, L., Decencière, E. (eds.) Mathematical Morphology: 40 Years On, vol. 30, pp. 259–268. Springer, Dordrecht (2005). https://doi.org/10.1007/1-4020-3443-1_23
8. Kimmel, R., Shaked, D., Kiryati, N., Bruckstein, A.M.: Skeletonization via distance maps and level sets. Comput. Vis. Image Underst. **62**(3), 382–391 (1995)
9. Lantuéjoul, C.: Skeletonization in quantitative metallography, vol. 34, pp. 107–135. Sijthoff and Noordhoff (1980)
10. Lantuéjoul, C., Beucher, S.: On the use of the geodesic metric in image analysis. J. Microsc. **121**(1), 39–49 (1981)
11. Morard, V., Decencière, E., Dokládal, P.: Geodesic attributes thinnings and thickenings background: attribute thinnings. In: ISMM 2011, 10th International Conference on Mathematical Morphology, pp. 200–211 (2011)
12. Morard, V., Decencière, E., Dokládal, P.: Efficient geodesic attribute thinnings based on the barycentric diameter. J. Math. Imaging Vis. **46**(1), 128–142 (2013)

13. Najman, L., Couprie, M.: Building the component tree in quasi-linear time. IEEE Trans. Image Process. **15**(11), 3531–3539 (2006)
14. Salembier, P., Oliveras, A., Garrido, L.: Anti-extensive connected operators for image and sequence processing. IEEE Trans. Image Process. **7**(4), 555–570 (1998)
15. Salembier, P., Serra, J.: Flat zones filtering, connected operators, and filters by reconstruction. IEEE Trans. Image Process. **4**, 1153–1160 (1995)
16. Serra, J.: Image Analysis and Mathematical Morphology, vol. 1. Academic Press, Orlando (1982)
17. Serra, J.: Image Analysis and Mathematical Morphology: Theoretical Advance, vol. 2. Academic Press, Orlando (1988)
18. Soille, P.: Morphological Image Analysis: Principles and Applications. Springer, Secaucus (2003). https://doi.org/10.1007/978-3-662-05088-0
19. Urbach, E.R., Roerdink, J.B.T.M., Wilkinson, M.H.F.: Connected shape-size pattern spectra for rotation and scale-invariant classification of gray-scale images. IEEE Trans. Pattern Anal. Mach. Intell. **29**(2), 272–285 (2007)
20. Urbach, E.R., Wilkinson, M.H.F.: Shape-only granulometries and gray-scale shape filters. In: ISMM 2002, 8th International Symposium on Mathematical Morphology, pp. 305–314 (2002)
21. Vincent, L.: Efficient computation of various types of skeletons. In: Medical Imaging V: Image Processing, vol. 1445, pp. 297–312. International Society for Optics and Photonics (1991)
22. Vincent, L.: Morphological grayscale reconstruction in image analysis: efficient algorithms and applications. IEEE Trans. Image Process. **2**, 176–201 (1993)
23. Huang, X., Fisher, M., Smith, D.J.: An efficient implementation of max tree with linked list and hash table. In: DICTA (2003)

Applications in (Bio)medical Imaging

Multilabel, Multiscale Topological Transformation for Cerebral MRI Segmentation Post-processing

Carlos Tor-Díez[1]([✉]), Sylvain Faisan[2], Loïc Mazo[2], Nathalie Bednarek[3,4], Hélène Meunier[4], Isabelle Bloch[5], Nicolas Passat[3], and François Rousseau[1]

[1] IMT Atlantique, LaTIM U1101 INSERM, UBL, Brest, France
`carlos.tordiez@imt-atlantique.fr`
[2] ICube UMR 7357, Université de Strasbourg, CNRS, FMTS, Illkirch, France
[3] Université de Reims Champagne-Ardenne, CReSTIC, Reims, France
[4] Service de médecine néonatale et réanimation pédiatrique,
CHU de Reims, Reims, France
[5] LTCI, Télécom ParisTech, Université Paris-Saclay, Paris, France

Abstract. Accurate segmentation of cerebral structures remains, after two decades of research, a complex task. In particular, obtaining satisfactory results in terms of topology, in addition to quantitative and geometrically correct properties is still an ongoing issue. In this paper, we investigate how recent advances in multilabel topology and homotopy-type preserving transformations can be involved in the development of multiscale topological modelling of brain structures, and topology-based post-processing of segmentation maps of brain MR images. In this context, a preliminary study and a proof-of-concept are presented.

Keywords: Homotopic deformation · Multilabel topology · Multiscale modelling · Segmentation · MRI · Brain

1 Introduction

Topological methods generate a growing interest in the field of medical image processing and analysis. Indeed, ensuring the coherence of structural properties of the organs and tissues in 3D medical data is a cornerstone e.g. for registration, modelling or visualization tasks. In particular, topological concepts developed in the field of discrete imagery, and in particular digital topology, can allow for the development of efficient approaches that take into account not only quantitative and morphometric information carried by anatomical objects of interest, but also more intrinsic properties related to their structure [21].

In particular, the brain has received a specific attention. Indeed, by contrast with other organs, it exhibits an important interindividual variability from shape and size points of view. In the meantime, it is organized into many distinct subparts and tissues with a strong topological invariance. Taking into account

© Springer Nature Switzerland AG 2019
B. Burgeth et al. (Eds.): ISMM 2019, LNCS 11564, pp. 471–482, 2019.
https://doi.org/10.1007/978-3-030-20867-7_36

topological priors is then a relevant hypothesis for guiding and/or regularizing image processing procedures [19].

This article presents a contribution in the field of topology-based brain structure segmentation. More especially, we focus on topological post-processing of multi-label segmentation maps obtained beforehand with efficient, but non-topologically guided, methods. Our purpose is then to build, from such segmentation maps, a corrected output consolidated by topological priors.

This research area is not new; related works are recalled in Sect. 2. However, the embedding of topological priors in segmentation paradigms is a complex task, a fortiori when we consider n-ary segmentation, with $n > 2$, i.e. more than one object vs. its background. In this context, our contributions are the following. First, we rely on a topological framework introduced a few years ago by some of the authors, that allows to correctly model digital images by considering the topology of n labels but also that of the combinations within their power lattice [16,18]. Second, we consider a multiscale approach for topological modelling of the cerebral structures. Indeed, we assume that the topological assumptions that should guide a segmentation process actually depend on the level of details of the observed structures. Based on these two, multilabel and multiscale, paradigms, we develop a generic, homotopic deformable model methodology that progressively refines a segmentation map with respect to the data and the associated topological priors.

In Sect. 3, we recall the involved topological framework and how it can be used for multiscale, multilabel topological modelling of anatomical structures. In Sect. 4, we describe the algorithmic scheme that allows us to progressively refine an initial segmentation with respect to this multiscale topological modelling. Experimental results are proposed in Sect. 5; at this stage, they have mainly an illustrative value. Section 6 concludes this article by presenting the main perspectives offered by the proposed approach and the remaining challenges to be tackled.

2   Related Works

Topology-based segmentation methods dedicated to brain structures can be mainly divided into two families [19]: on the one hand, the methods that consider a topological prior for guidance from the very beginning of the process; on the other hand, the methods that aim at recovering a posteriori some correct topological properties. The first lie in the family of topological deformable models; the second in the family of topological correction methods.

In most of these methods, the considered structures of interest are the main three classes of tissues, namely grey matter (GM) (mainly, the cortex), white matter (WM), and cerebrospinal fluid (CSF). A majority of the proposed methods focus on the cortex, that presents a complex geometry, with strong folding and a low thickness, leading to a high curvature 2D-like thin ribbon. Cortex segmentation is generally presented as a binary segmentation problem (cortex vs. other structures) or a ternary segmentation problem (GM, WM, CSF).

The associated topological hypotheses are often a simplified version of the anatomical reality, and the different classes of tissues are assumed nested: the central class is simply connected (i.e. a full sphere) with successively nested hollow spheres. Based on these simplified assumptions, it is possible to develop segmentation strategies from binary digital topology [13], and in particular homotopic transformations based on simple points [8] (less frequently, alternative topological models were proposed, for instance cellular complexes [7,9]).

Since the pioneering works proposed in [15], different variants of such topological deformable models have been proposed, in fully discrete paradigms [5] or by coupling continuous and digital models [11]. However, the hard topological constraints imposed by the model can lead to deadlocks that are difficult to handle (see [10] for a discussion on that topic). Based on the same topological hypotheses, many topology correction methods have been proposed for the cortex. They mainly consist of identifying the tunnels/handles generated on the cortical surface, and removing them based on ad hoc strategies [3,14]. These methods aim at reformulating the topological problem to be solved by considering a binary, simply connected topological model, which has the virtue to be easy to handle, but the drawback of poorly modelling the anatomical structures.

More recently, new ways were explored for tackling the issue of real multilabel segmentation and/or potentially complex topologies. Handling complex topology can be done by relying on less constraining—but also less robust—topological invariants, such as the genus [22]. Homotopic transformations based on simple points however remain the gold standard for carrying out model transformation. In order to avoid topological deadlocks, non-monotonic transformation processes of complex, multilabel topological models were investigated [4,20]. Although promising, these approaches suffer from various theoretical weaknesses [2,20,23] (see [18] for a discussion). They also work at a unique scale, with induced difficulties to ensure simultaneously spatial/geometrical and topological reliability.

3   Multilabel, Multiscale Topological Modelling

3.1   Theory

In [16], Mazo proposed a framework for modelling the topology of multilabel digital images. The main ideas of this framework are as follows.

1. The multilabel image is splitted in a collection of binary images such that each binary image represents a region of interest, that is a region that has been previously labelled or represents a meaningful union of some labeled regions. The unions are labeled thanks to a lattice structure: the label of a union of regions is the supremum of the labels of the regions.
2. Each binary digital image is embedded in a partially ordered set (poset) by adding intervoxels elements (pointels, linels, surfels). Such elements are assigned to the foreground or the background thanks to the minimum (6-adjacency) or maximum (26-adjacency) membership rule. Doing so, the connected components are preserved and the digital fundamental groups, as

defined by Kong [12], are mapped, through isomorphisms, to the fundamental groups of the Alexandrov topology [17].

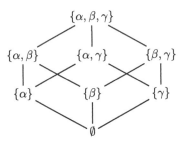

Fig. 1. The power lattice $\Lambda = 2^L$ for the set of labels $L = \{\alpha, \beta, \gamma\}$ of the three considered classes of cerebral tissues.

In this framework, a homotopic relabelling boils down to a collection of simple point moves in binary images with the guarantee to preserve the underlying topological structures of the involved posets, and in particular their homotopy-types.

3.2   Application

In the applicative context of this work, our purpose is to post-process ternary classification maps of MR images defined as

$$\left| \begin{array}{l} F : \Omega \to [0,1]^3 \\ x \mapsto (p_{\text{CSF}}, p_{\text{GM}}, p_{\text{WM}}) \end{array} \right. \tag{1}$$

where Ω is a part of $\mathbb{Z}^3$ corresponding to the support of an MR image, and for each point x, p_{CSF}, p_{GM} and p_{WM} are the probabilities that x belong to the cerebrospinal fluid, the grey matter (cortex) and the white matter, respectively. In particular, we have $\sum_\ell p_\ell = 1$.

Such maps F are obtained by a fuzzy segmentation process described in [24] (but they may be obtained via any other similar segmentation procedure, e.g. [6]). A specificity of this segmentation strategy—and many others—despite its good accuracy from quantitative and geometric points of view, lies in the fact that it is not guided by topological constraints. As a consequence, its output needs to be post-processed for any further application requiring topological guarantees (for instance mesh generation or differential cortical surface analysis).

Our purpose is then to build, from F, a new crisp segmentation map

$$\left| \begin{array}{l} T : \Omega \to \{0,1\}^3 \\ x \mapsto (c_{CSF}, c_{GM}, c_{WM}) \end{array} \right. \tag{2}$$

such that for each point x, c_{CSF}, c_{GM} and c_{WM} are equal to either 0 or 1 depending on the (unique) class of x, i.e. with $\sum_\ell c_\ell = 1$.

In addition, T should satisfy—unlike F—some topological priors related to the structure of the different tissues.

3.3   Multiscale Topological Modelling

The framework proposed in [16] allows us to model the topological structure of the three classes of cerebral tissues of interest, namely CSF, GM and WM, further noted γ, β and α, respectively, for the sake of concision. It also allows us to model any combinations of these labels, leading to the whole power lattice $\Lambda = 2^L$; see Fig. 1.

In particular, in [16], the notion of topology preservation relies on the definition of simple points that preserve the homotopy-type of *all* the labels of Λ. However, it is possible to consider the notion of simpleness for only a given (strict) subset of labels of Λ. In such a case, it is sufficient to guarantee that the relabelling of a (simple) point fulfills the required topological conditions for the chosen subset of labels, while the topology of the objects/complexes induced by the other labels are allowed to evolve.

This strategy, that only focuses on specific (combinations of) labels of interest, permits to define a multiscale topological modelling. Indeed, the notion of topology in discrete imaging is strongly related to the scale of observation of the objects of interest. For instance, fine topological details that are relevant at a high resolution, become useless (and sometimes incorrect) at a coarser resolution, due to the loss of precision induced by partial volume effects.

Based on this assumption, we consider two distinct topological models of the brain tissues, according to their scale (see Fig. 2):

- **Model 1 – Coarse/intermediate scales (S_1 and S_2):** At these scales, we assume that the WM, GM and CSF are successively nested, which is the hypothesis currently considered in the literature. In other words, we aim at preserving the homotopy-type of three labels: $\{\alpha\}$ (simply connected); $\{\beta\}$ (hollow sphere); and γ (hollow sphere); but also the homotopy-type of their four combinations.
- **Model 2 – intermediate/fine scales (S_2 and S_3):** At these scales, we assume that the cortex (GM) is no longer a hollow sphere. Indeed, on the lower part of the encephalus, the hypothesis of GM surrounding WM is not satisfied (due to the connection of the encephalus to the brainstem). Based on this hypothesis, we aim at preserving the homotopy-type of five labels: $\{\alpha\}$ (simply connected); $\{\alpha, \beta\}$ (simply connected); $\{\beta, \gamma\}$ (hollow sphere); $\{\gamma\}$ (hollow sphere); and $\{\alpha, \beta, \gamma\}$ (simply connected). In particular, this allows to make the topology of the cortex evolve whereas remaining coherent with respect to its neighbouring structures (CSF and WM).

4   Multilabel, Multiscale Topology-Controlled Deformation

We now describe how the topological assumptions modeled above can be considered for designing, at each scale, well-fitted homotopy-type preserving deformation processes.

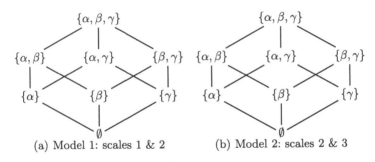

(a) Model 1: scales 1 & 2   (b) Model 2: scales 2 & 3

Fig. 2. Topological modelling of the classes of brain tissues, at the coarse/intermediate (a) and intermediate/fine scales (b). The labels coloured in red and purple are those for which the topology has to be preserved. The red labels need to be explicitly handled; the topology of the purple labels is preserved as a corollary of the topological preservation of the red ones. The topology of the black labels is authorized to evolve during the transformation process. (Color figure online)

4.1   Grid Refinement

We work on a space Ω defined as a subset of $\mathbb{Z}^3$. In order to carry out the deformation process at each scale, the space Ω has to be adapted to the current scale. It is important to note that the notion of simple points and homotopy-type are compliant with respect to digital grid refinement. In other words, a digital object X defined in $\mathbb{Z}^3$ (and more generally in $\mathbb{Z}^n$) has the same homotopy-type as its up-sampled analogue X_2 in $\mathbb{Z}^3$, defined by $x \in X_2 \subset \mathbb{Z}^3 \Leftrightarrow \lfloor x/2 \rfloor \in X$. In addition, x is a simple point for X iff there exists a sequence of successively simple points for X_2 composed from the 8 points of $2x + \{0,1\}^3$. This topology-preserving octree refinement [1] remains trivially valid for up-samplings at any other (discrete) resolution.

In particular, we consider a scale factor $k \in \mathbb{N}$, $k > 1$, between the grids of each scale. The set $\Gamma_1 \subset \mathbb{Z}^3$ is the grid considered at the coarse scale S_1. It is refined, at scale S_2 into a second grid Γ_2 such that $\Gamma_1 = k\Gamma_2$, which means that one point of Γ_1 corresponds to k^3 points of Γ_2. Finally, Γ_2 is refined into a third grid Γ_3 such that $\Gamma_2 = k\Gamma_3$; this means that one point of Γ_1 corresponds to k^3 points of Γ_2 and k^6 points of Γ_3. Note that we have $\Omega = \Gamma_3$. It is then convenient to assume (without loss of generality) that Ω is defined as a Cartesian product $\prod_{i=1}^{3} [\![0, N_i - 1]\!] \subset \mathbb{Z}^3$ such that N_i is a multiple of k^2 for any $i = 1, 2, 3$.

4.2   Metrics

In order to guide the topology-controlled transformation process, we need to define metrics for assessing the error between the current (evolving) segmentation map and the target segmentation map. In particular, by minimizing a cost function associated to this error, our purpose is to make our segmentation map progressively converge onto the target, whereas correctly handling the topology.

In this context, we consider two metrics. The first is a classification metric, noted M_d, defined by the classification error at each point of the image. The second, noted M_Δ, is a distance-based metric, defined by the L_1 distance between the misclassified points and the corresponding, correctly classified region in the target image.

In order to build these two metrics, we first need to define fuzzy classification maps F_i from F (see Eq. (1)) at each scale i ($i = 1, 2, 3$). This is done with a standard mean policy:

$$F_i(x) = \frac{1}{k^{3 \cdot (3-i)}} \sum_{a=0}^{k^{(3-i)}-1} \sum_{b=0}^{k^{(3-i)}-1} \sum_{c=0}^{k^{(3-i)}-1} F(k^{(3-i)}x + (a, b, c)). \tag{3}$$

In particular, we have $F_3 = F$.

For a given (crisp) segmentation map C, the first metric M_d assessing the error between C and F_i is then defined as:

$$M_d(C, F_i) = \sum_x \|C(x) - F_i(x)\|_2. \tag{4}$$

For defining the second metric M_Δ, we build crisp classification maps G_i from the fuzzy maps F_i by a standard majority voting process, i.e. for any x, we set $G_i(x) = \arg_{\{\alpha, \beta, \gamma\}} \max F_i(x)$.

For a given (crisp) map C, the second metric M_Δ assessing the error between C and G_i is then defined as:

$$M_\Delta(C, G_i) = \sum_x \Delta(x, G_i^{-1}(C(x))) \tag{5}$$

where $\Delta(x, X) = \min_{y \in X} \|x - y\|_1$ is the L_1 distance between the point x and the set X, whereas $G_i^{-1}(v) = \{x \mid G_i(x) = v\}$. In particular, we have $\Delta(x, G_i^{-1}(C(x))) = 0$ iff $C(x) = G_i(x)$.

4.3 Initialization and Optimization Process

The process is iterative, and proceeds from the coarse scale S_1 up to the fine scale S_3. For each step, the current input segmentation map is the output of the previous step (possibly up-sampled, if we switch between S_i and S_{i+1}). The only explicit initialization is then required for the very first step of the process. At this stage, we simply consider a three-layer nested sphere model, with a central simply connected full sphere of label α surrounded by a first hollow sphere of label β and finally a second hollow sphere of label γ.

At each step of the optimization process, we consider either the metric M_d or M_Δ. In order to make M_d decrease, we build a map that defines, for each point x and each possible relabelling $\ell_1 \to \ell_2$ (with ℓ_1 the current label at x), the associated benefit with respect to M_d, namely $\|\ell_2 - F_i(x)\|_2 - \|\ell_1 - F_i(x)\|_2$. Then, we iteratively carry out the relabelling of (simple) points with the maximal (non-negative) benefit, until stability.

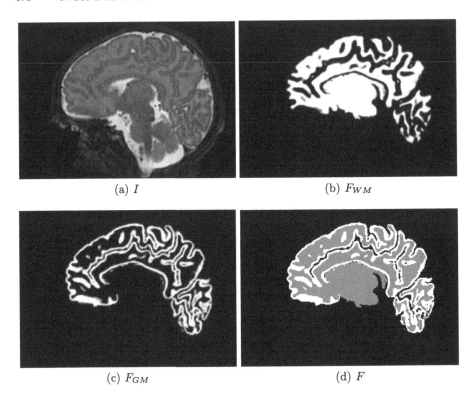

(a) I

(b) F_{WM}

(c) F_{GM}

(d) F

Fig. 3. (a) MR image. (b) Fuzzy classification of the WM from (a). (c) Fuzzy classification of the GM from (a). (d) Crisp classification map obtained from the fuzzy classification maps of (a) (WM in grey; GM in white; CSF + background in black). (a–d) Sagittal slices.

In order to make M_Δ decrease, we build a map that defines, for each point x and each possible relabelling $\ell_1 \to \ell_2$ (with ℓ_1 the current label at x), the associated benefit with respect to M_Δ, as $\Delta(x, G_i^{-1}(\ell_2))$ if $\Delta(x, G_i^{-1}(\ell_2)) < \Delta(x, G_i^{-1}(\ell_1))$ and 0 otherwise. Then, we iteratively carry out the relabelling of (simple) points with the maximal benefit, until stability.

5   Experiments and Results

We present some results computed with the proposed method, for topological correction of fuzzy segmentation maps obtained from the method proposed in [24] (or equivalent methods). An example of such map F obtained from an MR image is illustrated in Fig. 3.

Here, we carried out a 5 step iterative procedure with the meta-parameters summarized in Table 1.

The initial images I are defined on $\Omega = [\![0, 291]\!] \times [\![0, 291]\!] \times [\![0, 203]\!]$. We use as scale factor $k = 2$. The successive segmentation maps (and the associated

Table 1. Meta-parameters for the successive steps of the process.

Step	Scale	Topology	Metric
1	S_1	Model 1	M_Δ
2	S_2	Model 1	M_Δ
3	S_2	Model 1	M_d
4	S_2	Model 2	M_Δ
5	S_3	Model 2	M_Δ

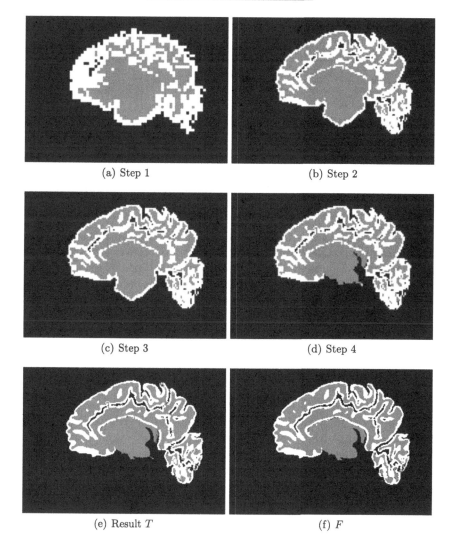

(a) Step 1   (b) Step 2

(c) Step 3   (d) Step 4

(e) Result T (f) F

Fig. 4. (a–d) Intermediate results obtained by the successive steps of the process (see Table 1). (e) Final segmentation map. (f) Crisp classification map (see also Fig. 3(d)) used as reference for guiding the process.

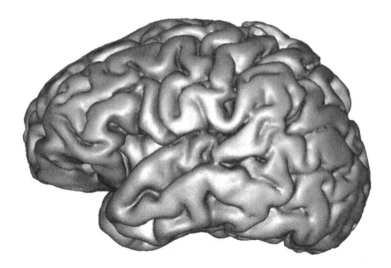

Fig. 5. Surface mesh generated from the final topological segmentation map T of Fig. 4(e).

classification maps F_3, F_2 and F_1) are then defined on $\Gamma_3 = [\![0, 291]\!] \times [\![0, 291]\!] \times [\![0, 203]\!]$, $\Gamma_2 = [\![0, 145]\!] \times [\![0, 145]\!] \times [\![0, 101]\!]$, and $\Gamma_1 = [\![0, 72]\!] \times [\![0, 72]\!] \times [\![0, 50]\!]$, respectively, i.e. with $|\Gamma_3| \simeq 1, 7.10^7$, $|\Gamma_2| \simeq 2, 2.10^6$ and $|\Gamma_1| \simeq 2, 7.10^5$.

An example of the successive steps of the topological transformation process is illustrated in Fig. 4. One can observe the progressive convergence of the model toward the classification map, whereas controlling the topology of the segmentation, in particular on the cortex (preserved on the upper part of the encephalus), whereas the GM is relevantly removed on the lower part, in accordance with the initial classification map and with the topological model.

A 3D mesh visualization of the resulting map T is illustrated in Fig. 5, qualitatively emphasizing the topological correctness of the result.

6   Conclusion

This preliminary study provides a proof of concept for the relevance of using the multilabel topological framework proposed in [16] for developing a multiscale, topological modelling of the cerebral structures, allowing one to either preserve or relax topological constraints over the power lattice of a set of elementary semantic labels.

In particular, this framework can be efficiently used for carrying out multiscale, topology-controlled deformation of label maps based on the concept of simple points, here in the context of topological correction of fuzzy segmentation maps computed beforehand.

Among numerous perspective works, we will further investigate (1) more sophisticated metrics for guiding the deformation process; (2) the possibility to carry out deformation processes at a superpixel resolution and/or to use cubical complex models [18] for topological modelling; and (3) a richer modelling of the brain with more anatomical structures, e.g. for atlas-based segmentation.

Acknowledgements. The research leading to these results has been supported by the ANR MAIA project (http://recherche.imt-atlantique.fr/maia), grant ANR-15-CE23-0009 of the French National Research Agency; INSERM and Institut Mines Télécom Atlantique (*Chaire "Imagerie médicale en thérapie interventionnelle"*); the *Fondation pour la Recherche Médicale* (grant DIC20161236453); and the American Memorial Hospital Foundation.

References

1. Bai, Y., Han, X., Prince, J.L.: Digital topology on adaptive octree grids. J. Math. Imaging Vis. **34**(2), 165–184 (2009)
2. Bazin, P.-L., Ellingsen, L.M., Pham, D.L.: Digital homeomorphisms in deformable registration. In: Karssemeijer, N., Lelieveldt, B. (eds.) IPMI 2007. LNCS, vol. 4584, pp. 211–222. Springer, Heidelberg (2007). https://doi.org/10.1007/978-3-540-73273-0_18
3. Bazin, P.L., Pham, D.L.: Topology correction of segmented medical images using a fast marching algorithm. Comput. Methods Programs Biomed. **88**(2), 182–190 (2007)
4. Bazin, P.L., Pham, D.L.: Topology-preserving tissue classification of magnetic resonance brain images. IEEE Trans. Med. Imaging **26**(4), 487–496 (2007)
5. Caldairou, B., et al.: Segmentation of the cortex in fetal MRI using a topological model. In: International Symposium on Biomedical Imaging (ISBI), Proceedings, pp. 2045–2048 (2011)
6. Caldairou, B., Passat, N., Habas, P.A., Studholme, C., Rousseau, F.: A non-local fuzzy segmentation method: application to brain MRI. Pattern Recognit. **44**(9), 1916–1927 (2011)
7. Cointepas, Y., Bloch, I., Garnero, L.: A cellular model for multi-objects multi-dimensional homotopic deformations. Pattern Recognit. **34**(9), 1785–1798 (2001)
8. Couprie, M., Bertrand, G.: New characterizations of simple points in 2D, 3D, and 4D discrete spaces. IEEE Trans. Pattern Anal. Mach. Intell. **31**(4), 637–648 (2009)
9. Damiand, G., Dupas, A., Lachaud, J.O.: Fully deformable 3D digital partition model with topological control. Pattern Recognit. Lett. **32**(9), 1374–1383 (2011)
10. Faisan, S., Passat, N., Noblet, V., Chabrier, R., Meyer, C.: Topology preserving warping of 3-D binary images according to continuous one-to-one mappings. IEEE Trans. Image Process. **20**(8), 2135–2145 (2011)
11. Han, X., Xu, C., Prince, J.L.: A topology preserving level set method for geometric deformable models. IEEE Trans. Pattern Anal. Mach. Intell. **25**(6), 755–768 (2003)
12. Kong, T.Y.: A digital fundamental group. Comput. Graph. **13**(2), 159–166 (1989)
13. Kong, T.Y., Rosenfeld, A.: Digital topology: introduction and survey. Comput. Vis. Graph. Image Process. **48**(3), 357–393 (1989)
14. Kriegeskorte, N., Goebel, N.: An efficient algorithm for topologically correct segmentation of the cortical sheet in anatomical MR volumes. NeuroImage **14**(2), 329–346 (2001)
15. Mangin, J.F., Frouin, V., Bloch, I., Régis, J., López-Krahe, J.: From 3D magnetic resonance images to structural representations of the cortex topography using topology preserving deformations. J. Math. Imaging Vis. **5**(4), 297–318 (1995)

16. Mazo, L.: A framework for label images. In: Ferri, M., Frosini, P., Landi, C., Cerri, A., Di Fabio, B. (eds.) CTIC 2012. LNCS, vol. 7309, pp. 1–10. Springer, Heidelberg (2012). https://doi.org/10.1007/978-3-642-30238-1_1

17. Mazo, L., Passat, N., Couprie, M., Ronse, C.: Digital imaging: a unified topological framework. J. Math. Imaging Vis. 44(1), 19–37 (2012)

18. Mazo, L., Passat, N., Couprie, M., Ronse, C.: Topology on digital label images. J. Math. Imaging Vis. 44(3), 254–281 (2012)

19. Pham, D.L., Bazin, P.L., Prince, J.L.: Digital topology in brain imaging. IEEE Signal Process. Mag. 27(4), 51–59 (2010)

20. Poupon, F., Mangin, J.-F., Hasboun, D., Poupon, C., Magnin, I., Frouin, V.: Multi-object deformable templates dedicated to the segmentation of brain deep structures. In: Wells, W.M., Colchester, A., Delp, S. (eds.) MICCAI 1998. LNCS, vol. 1496, pp. 1134–1143. Springer, Heidelberg (1998). https://doi.org/10.1007/BFb0056303

21. Saha, P.K., Strand, R., Borgefors, G.: Digital topology and geometry in medical imaging: a survey. IEEE Trans. Med. Imaging 34(9), 1940–1964 (2015)

22. Ségonne, F.: Active contours under topology control - genus preserving level sets. Int. J. Comput. Vis. 79(2), 107–117 (2008)

23. Siqueira, M., Latecki, L.J., Tustison, N.J., Gallier, J.H., Gee, J.C.: Topological repairing of 3D digital images. J. Math. Imaging Vis. 30(3), 249–274 (2008)

24. Tor-Díez, C., Passat, N., Bloch, I., Faisan, S., Bednarek, N., Rousseau, F.: An iterative multi-atlas patch-based approach for cortex segmentation from neonatal MRI. Comput. Med. Imaging Graph. 70, 73–82 (2018)

Hierarchical Approach for Neonate Cerebellum Segmentation from MRI: An Experimental Study

Pierre Cettour-Janet[1](✉), Gilles Valette[1], Laurent Lucas[1], Hélène Meunier[2], Gauthier Loron[1,2], Nathalie Bednarek[1,2], François Rousseau[3], and Nicolas Passat[1]

[1] Université de Reims Champagne-Ardenne, CReSTIC, Reims, France
`pierre.cettour-janet@univ-reims.fr`
[2] Service de médecine néonatale et réanimation pédiatrique, CHU de Reims, Reims, France
[3] IMT Atlantique, LaTIM U1101 INSERM, UBL, Brest, France

Abstract. Morphometric analysis of brain structures is of high interest for premature neonates, in particular for defining predictive neurodevelopment biomarkers. This requires beforehand, the correct segmentation of structures of interest from MR images. Such segmentation is however complex, due to the resolution and properties of data. In this context, we investigate the potential of hierarchical image models, and more precisely the binary partition tree, as a way of developing efficient, interactive and user-friendly 3D segmentation methods. In particular, we experiment the relevance of texture features for defining the hierarchy of partitions constituting the final segmentation space. This is one of the first uses of binary partition trees for 3D segmentation of medical images. Experiments are carried out on 19 MR images for cerebellum segmentation purpose.

Keywords: 3D segmentation · Binary partition tree · Texture features · MRI · Cerebellum · Premature neonates

1 Introduction

Each year, about 15 million babies are born premature, i.e. before 37 weeks of gestation. A part of them will develop cerebral palsy [9] while others will experiment troubles in cognitive performance or behavioral issues [19]. This motivates an active clinical research devoted to predict, at an early stage, the most probable pathologies. In particular, specific information on newborn brain structures can

The research leading to these results has been supported by the ANR MAIA project (http://recherche.imt-atlantique.fr/maia), grant ANR-15-CE23-0009 of the French National Research Agency; and the American Memorial Hospital Foundation.

© Springer Nature Switzerland AG 2019
B. Burgeth et al. (Eds.): ISMM 2019, LNCS 11564, pp. 483–495, 2019.
https://doi.org/10.1007/978-3-030-20867-7_37

be used as biomarkers. For instance, a correlation between early life cerebellar anatomy and neurodevelopmental disorders was shown recently [32].

In this context, Magnetic Resonance Imaging (MRI) provides an efficient way of observing the newborn brain in a safe fashion. However, MR images of newborns are of lower quality than with adults. Indeed, the used coils are often not adapted to babies. Moreover, newborns are likely to move during MR acquisitions. As a consequence, specific adaptations are made on standard MR sequences; but they generally result in images with a lower resolution, a decreased signal-to-noise ratio, and possible artifacts. In addition, due to a non-complete maturity of tissues, contrast variations of MR signal and partial volume effects can occur, for instance caused by uncomplete myelination.

An accurate segmentation of brain structures is a prerequisite for carrying out efficiently morphometric measures, further used as biomarkers. While MRI brain structure segmentation in adults has been an active research area for decades, the case of the newborn is more recent. Dedicated (in general, automatic) segmentation methods can be classified into four types [17]: unsupervised [8,21]; parametric [15]; classification [3,18,20,22,36]/atlas fusion [5,12,31]; and deformable models [34,35]. These methods aim at segmenting the whole brain, cerebral tissues, or specific structures (e.g., cortex, ventricles, hippocampus), but none of them focused on the cerebellum. Indeed, to our best knowledge, cerebellum segmentation was only considered in adults, in a few papers. In particular, the related approaches were based on texture analysis [28], deformable models [2,11,16], or atlases [13,26,37].

Cerebellum segmentation from neonate MR images is a difficult task, both in an automated and in a manual way. On the one hand, automated segmentation is hardly tractable due to the image properties (low resolution, poor signal-to-noise ratio), and the methods initially designed for adult MRI reach their limits in neonate images. On the other hand, manual segmentation remains a complex and error-prone task: it is time-consuming, has to be carried out by an expert, and generally leads to inter- and intra-operator variability. Based on these considerations, we investigate mixed approaches that consist of selecting, in a first, automated step, a reduced subspace of potential segmentation results. In a second, interactive step, the expert-user is then allowed to navigate in this sub-space in order to interactively select and tune a segmentation result.

To reach that goal, we consider hierarchical image models, mainly developed in the field of mathematical morphology, and used for developing connected operators [30]. More precisely, we focus on a specific data-structure, the Binary Partition Tree (BPT, for brief) [29]. The BPT is relevant for two main reasons. First, it enables one to model the image in a multiscale way, from fine anatomical structures to larger classes of tissues. Second, it is built by taking into account not only the intrinsic image information, but also prior knowledge [23] that can be specifically geared towards the selection of structures of interest. As many morphological hierarchies, the BPT is a partially ordered subset of partitions of the image support, organized as a tree structure. It is then possible to easily navigate inside a BPT for interactively defining a cut of the tree, that directly leads to a segmentation result.

In this article, we focus on the first step of our two-step approach, namely the way of building a BPT dedicated to cerebellum segmentation from neonate MR images. Beyond the applicative interest of this work, namely the ability to segment neonates cerebellum, the methodological novelties are manifold. First, we consider BPTs for 3D imaging, while they were mainly dedicated to 2D imaging, until now. Second, we propose space/time cost reduction strategies, by coupling BPT construction with a superpixel preprocessing step (Sect. 2.2). Third, we investigate the relevance of texture features for BPT construction (Sect. 2.4). Four, we propose a way of assessing the relevance of a BPT with respect to ground-truth via a Pareto front paradigm (Sect. 3.2). Section 2 describes our methological pipeline. Section 3 provides an experimental discussion. Concluding remarks on perspective works are finally provided in Sect. 4.

2 Building a BPT from Neonate MR Images

In this section, we describe the pipeline that allows us to build a BPT, i.e. a subspace of potential partitions, from the MR image of a neonate, for cerebellum segmentation purpose. We describe the data and the preprocessing steps; we recall the BPT construction algorithm; and we present the chosen features that enable to guide the construction process for finally obtaining a cerebellum-oriented BPT.

2.1 MR Images and Preprocessing

We worked on two MRI datasets: the Neonatal Brain Atlas (ALBERTs[1]) [6,7], and the French Epirmex[2] dataset; see Fig. 1. ALBERTs data are acquired on a 3T Philips Intera scanner. Images have a resolution of $0.82 \times 0.82 \times 1.60$ mm, and dimensions $155 \times 155 \times 189$ voxels, with TE = 17 ms and TR = 4.6 ms. Epirmex data are T1 MR images acquired in Reims Hospital (Neonatology Service), on a 3T Philips MR Imaging DD005 scanner. Images have a resolution of $3 \times 3 \times 3$ mm, and dimensions $560 \times 560 \times 90$ voxels, with TE = 17 ms and TR = 1019 ms.

In the sequel, an MR image is noted I. It is handled as a function $I : \Omega \to V$ that associates to each point x of the space Ω a value $I(x)$ of the set V. In our specific case, Ω is a matrix of voxels, namely a subset of $\mathbb{Z}^3$ of the Cartesian grid. The set V is an interval of grey-level values within $\mathbb{N}$.

The MR images are preprocessed. In particular, a denoising by non-local means is performed, based on the BTK library [27]. Then, a bias field correction is applied, based on the N4 algorithm [33]. Finally, in order to reduce the spatial complexity of the BPT construction, an intracranial mask registration

---

is performed, for subdividing the support Ω of the image I into intra- and extracranial volumes (by abuse of notation, from now on, we will call Ω the only intracranial area).

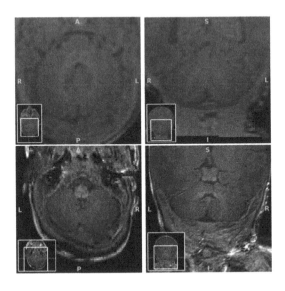

Fig. 1. Examples of MR images (left: axial view; right: coronal view). First row: Epirmex. Second row: ALBERTs.

2.2   Initial Partitioning: Waterpixels

The BPT models a family of partitions of an initial set (in general, the image support Ω of I). The size of the BPT is $\Theta(n)$ where n is the size of this initial set, and the time cost of its construction is $\mathcal{O}(n \log n)$. In practice, these (quasi-)linear space/time complexities can lead to high costs, when considering 3D images composed of several millions of voxels.

In order to reduce the space cost of Ω, we investigated superpixel paradigms for simplifying the initial partition induced by Ω into a reduced partition Λ. In particular, we considered the waterpixels [14], namely a superpixel approach that applies a seeded watershed on a mixed gradient/distance, saliency map. More precisely, we developed a 3D version of this 2D superpixel model, namely the watervoxels [4], in order to allow for a relevant partitioning of an MR image. The main two parameters of waterpixels are the density of seeds (that determines the number of elements of the partition Λ) and the trade-off between the gradient map and the distance map. The experimental setting of these parameters is discussed in Sect. 3.1.

2.3   BPT Construction: Algorithmics

A BPT is a tree; each of its nodes is a connected region of the image support Ω. A node is either a leaf representing an element of Λ, or the union of two other neighbouring regions; the root node corresponds to the whole image support Ω.

The BPT construction is a bottom-up process. It proceeds from the leaves, up to the root. This is done by choosing and merging iteratively two adjacent regions that minimize a criterion reflecting their similarity. This choice relies on two main notions:

- a region model $M(N)$ which specifies how a region N is characterized (e.g. intensity, texture, geometry); and
- a merging criterion $O(N_1, N_2)$ which defines the similarity of neighbouring regions N_1, N_2 and thus the merging order.

A strategy commonly adopted for representing each region is, for instance, to consider their average intensity, and to merge adjacent regions being the most similar.

To model the fact that two points x and y of Ω are neighbours, we define an adjacency relation on Ω. In practice, in $\mathbb{Z}^3$, the standard 6 or 26-adjacencies defined in digital topology are considered. For the partition Λ of Ω, we define an adjacency relation A_Λ inherited from that of Ω. We say that two distinct sets $N_1, N_2 \in \Lambda$ are adjacent if there exist $x_1 \in N_1$ and $x_2 \in N_2$ such that x_1 and x_2 are adjacent in Ω. Then, $G_\Lambda = (\Lambda, A_\Lambda)$ is a non-directed graph.

The construction of a BPT is indeed a graph collapsing process. More precisely, the BPT is the data-structure that describes the progressive collapsing of G_Λ onto the trivial graph $(\{\Omega\}, \emptyset)$. This process consists of defining a sequence $(G_i = (\Gamma_i, A_{\Gamma_i}))_{i=0}^n$ (with $n = |\Lambda| - 1$) as follows. First, we set $G_0 = G_\Lambda$. Then, for each i from 1 to n, we choose the two nodes N_{i-1} and N'_{i-1} of G_{i-1} linked by the edge $(N_{i-1}, N'_{i-1}) \in A_{\Gamma_{i-1}}$ that minimizes the chosen merging criterion. We define G_i such that $\Gamma_i = (\Gamma_{i-1} \setminus \{N_{i-1}, N'_{i-1}\}) \cup \{N_{i-1} \cup N'_{i-1}\}$; in other words, we replace these two nodes by their union. The adjacency A_{Γ_i} is defined accordingly from $A_{\Gamma_{i-1}}$. We remove the edge (N_{i-1}, N'_{i-1}), and we replace each edge (N_{i-1}, N''_{i-1}) and/or (N'_{i-1}, N''_{i-1}) by an edge $(N_{i-1} \cup N'_{i-1}, N''_{i-1})$. In particular, two former edges may be fused into a single.

The BPT T is the Hasse diagram of the partially ordered set $(\bigcup_{i=0}^n \Gamma_i, \subseteq)$. It is built in parallel to the progressive collapsing from G_0 to G_n. In particular, T stores the node fusion history. More precisely, we define a sequence $(T_i)_{i=0}^n$ as follows. We set $T_0 = (\Gamma_0, \emptyset) = (\Lambda, \emptyset)$. Then, for each i from 1 to n, we build T_i from T_{i-1} by adding the new node $N_{i-1} \cup N'_{i-1}$, and the two edges $(N_{i-1} \cup N'_{i-1}, N_{i-1})$ and $(N_{i-1} \cup N'_{i-1}, N'_{i-1})$. The BPT T is then defined as T_n.

2.4   Features for BPT Construction

The construction of a BPT requires to choose features involved in the definition of the region model (M), and a merging criterion (O) for assessing the relevance of merging two nodes/regions.

Region Model. For defining the region model that describes each region/node of the BPT, we first consider the mean grey-level value (μ). This feature is classically chosen, as it provides a necessary criterion for merging two nodes (i.e. two nodes with very different mean values will not be merged in priority).

We also consider texture features. Indeed, our hypothesis is that the cerebellum has a specific structure that may be characterized by texture analysis. In particular, we focus on texture indices computed from Haralick co-occurrence matrix [10], that carries second-degree statistics on the image. This co-occurrence matrix provides, for each couple of grey-levels of the MR images, the probability of co-occurence between voxels located at a given distance in a chosen direction. In our case, we consider the matrices $\mathcal{M}_{\mathbf{e}_i}$ ($i = 1, 2, 3$) corresponding to co-occurence between neighbouring voxels in the 3 principal orientations, i.e. the 6-adjacent voxels. More precisely, for any $v, w \in V$ with $V \subset \mathbb{N}$ the interval of MRI values, we have:

$$\mathcal{M}_{\mathbf{e}_i}(v, w) = \frac{1}{|\Omega|} \left| \left\{ x \in \Omega \mid I(x) = v \land I(x + \mathbf{e}_i) = w \right\} \right| \tag{1}$$

where $\{\mathbf{e}_1, \mathbf{e}_2, \mathbf{e}_3\}$ is the canonical orthonormal basis of $\mathbb{Z}^3$. From these matrices, we consider the homogeneity (H), contrast (C) and entropy (E) indices, as texture information:

$$H = \sum_{i=1}^{3} \sum_{v \in V} \sum_{w \in V} \frac{\mathcal{M}_{\mathbf{e}_i}(v, w)}{1 + |v - w|} \tag{2}$$

$$C = \sum_{i=1}^{3} \sum_{v \in V} \sum_{w \in V} |v - w|^2 \cdot \mathcal{M}_{\mathbf{e}_i(v,w)} \tag{3}$$

$$E = -\sum_{i=1}^{3} \sum_{v \in V} \sum_{w \in V} (v - w) \cdot \log(\mathcal{M}_{\mathbf{e}_i}(v, w)) \tag{4}$$

Merging Criterion. For each pair of adjacent regions N_1, N_2, endowed with one of the four features described above (μ, H, C, E), we compute the absolute difference of feature values between both regions, noted $O(N_1, N_2)$. The lower this value, the more similar the two regions, and the higher the priority for their merging during the BPT construction.

3   Experiments and Results

We carried out experiments on ALBERTs (10 MR images) and Epirmex (9 MR images). ALBERTs data are endowed with ground-truth; in particular, the cerebellum region is obtained by union of various labeled subregions. For Epirmex, the 9 MR images were manually segmented by a medical expert in neonatology, for providing cerebellum ground-truth.

3.1   Watervoxel Partition: Parameter Setting

As a preprocessing step to the BPT construction, a simplification of the initial image support Ω is first carried out. This is done with watervoxels, for subdividing Ω into a partition Λ. This partition Λ should ideally be composed of regions either fully inside or fully outside the cerebellum, in order to avoid initialization errors that the construction of the BPT will not be able to correct (indeed, it merges regions, but never split them).

For a given partition Λ of the image support Ω, we define the associated error δ based on a measure called discordance, introduced in [25]. For the ground-truth segmentation $C \subset \Omega$ of the cerebellum, and the partition $\Lambda \subset 2^{\Omega}$, we define δ as follows:

$$\delta = \frac{1}{|C|} \sum_{L \in \Lambda} \min\{|L \setminus C|, |L \cap C|\} \tag{5}$$

In other words, the discordance measure is defined as the relative quantitative error on the false positives and false negatives with respect to C induced by Λ, and more precisely by the nodes that partially intersect G.

Note that from the very definition of δ, the set $\widehat{S} \subseteq \Lambda$ that best matches C with respect to the discordance measure is such that:

$$\forall L \in \Lambda, L \subseteq C \Rightarrow L \in \widehat{S} \tag{6}$$
$$\forall L \in \Lambda, L \subseteq \overline{C} \Rightarrow L \notin \widehat{S} \tag{7}$$

In other words, the discordance error depends only on the acceptance or rejection of the k regions L of Λ partly meeting C. Due to the additive/separable formulation of Eq. (5), the computation of δ is carried out in linear time $\Theta(k)$, whereas other (similar, although non-equivalent) metrics, e.g. the Dice error, would lead to an exponential $\mathcal{O}(2^k)$ time.

Practically, the partition Λ is controlled by two parameters: (1) the density $d \in [0, 1]$ of seeds in the image support, that defines the number of regions in the partition as $d.|\Omega|$; and (2) the trade-off parameter $\alpha \in [0, 1]$ between the (normalized) distance map $\Delta(\Omega)$ and the (normalized) gradient ∇I of the MR image, that leads to a map $(1 - \alpha).\nabla I + \alpha.\Delta(\Omega)$. Then, the error defined in Eq. (5) is a 2-dimensional function $\delta(d, \alpha) : [0, 1]^2 \to \mathbb{R}$. (Note that for $d = 1$, we have the trivial partition $\Lambda = \{\{x\} \mid x \in \Omega\}$ and then, $\delta(1, .) = 0$.)

Based on the experimental results summarized in Fig. 2, we set $d = 2, 32.10^{-3}$ (resp. $d = 1, 56.10^{-2}$) and $\alpha = 0, 1$ (resp. $\alpha = 0, 6$) for Epirmex (resp. ALBERTs). With these parameters, the mean value of discordance is $\delta = 0, 311$ (resp. $\delta = 0, 156$) for Epirmex (resp. ALBERTs). These values are indeed sufficient for further computing BPTs based on the partition Λ, with respect to the provided ground-truth of the 9 (resp. 10) MR images. It is however worth mentioning that the discordance error with Epirmex data is much higher than that of ALBERTs. One can then expect worse BPT quality results with clinical images from Epirmex than with ALBERTs images.

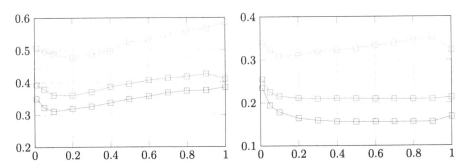

Fig. 2. Discordance δ (y-axis) depending on the trade-off parameter α between the gradient map and the distance map (x-axis), for various densities d. Left: Epirmex dataset. In blue: $d = 2,32.10^{-3}$ (initial Voronoi cells of $12 \times 12 \times 3$ voxels). In red: $d = 1,11.10^{-3}$ ($15 \times 15 \times 4$ voxels). In yellow: $d = 5,54.10^{-4}$ ($19 \times 19 \times 5$ voxels). In green: $d = 3,15.10^{-5}$ ($23 \times 23 \times 6$ voxels). Right: ALBERTs dataset. In blue: $d = 1,56.10^{-2}$ (initial Voronoi cells of $4 \times 4 \times 4$ voxels). In red: $d = 8,00.10^{-3}$ ($5 \times 5 \times 5$ voxels). In yellow: $d = 2,92.10^{-3}$ ($7 \times 7 \times 7$ voxels). In green: $d = 1,95.10^{-3}$ ($8 \times 8 \times 8$ voxels). (Color figure online)

3.2 Segmentation/BPT Evaluation

We then assess the ability of a BPT to provide a correct segmentation. As observed in Sect. 2.3, the nodes of a BPT are regions of the image support Ω. By construction, two nodes N_1 and N_2 of the BPT are either disjoint or included: $N_1 \cap N_2 \neq \emptyset \Rightarrow N_1 \subseteq N_2 \vee N_2 \subseteq N_1$. A cut of a BPT is a subset $\mathcal{C} = \{N_\star\}$ of pairwise disjoint nodes. Their union $\mathcal{S} = \cup N_\star \subseteq \Omega$ provides a segmentation of the image I.

The quality of a cut $\mathcal{C}$/segmentation $\mathcal{S}$ is characterized from two points of view. First, $\mathcal{S}$ has to fit at best the target (here the cerebellum C), i.e. it should maximize a quality measure [24]. Here, we consider the Dice index, defined as:

$$D(\mathcal{S}, C) = \frac{2|\mathcal{S} \cap C|}{|\mathcal{S}| + |C|} \qquad (8)$$

Second, $\mathcal{C}$ should contain a minimal set of nodes [25]. Indeed, the less numerous the nodes, the easier the task of defining the cut by automated and/or interactive investigation of the BPT. Then, we consider a second quality measure, namely the size $|\mathcal{C}|$ of the cut.

The best segmentation $\widehat{\mathcal{S}}$, given by the best cut $\widehat{\mathcal{C}}$ of a BPT, should optimize both measures, that is:

$$\widehat{\mathcal{S}} = \arg\max D(\mathcal{S}, C) \qquad (9)$$

$$\widehat{\mathcal{C}} = \arg\min |\mathcal{C}| \qquad (10)$$

But, in practice, these are antagonistic.

Thus, in order to assess the relevance of a given BPT, we compute the Pareto front for Eqs. (9–10), that is the set of optimal elements for the couple $(D(\mathcal{S}, C), |\mathcal{C}|)$ in the space $[0, 1] \times \mathbb{N}$, for the order relation $(\geq_{[0,1]}, \leq_{\mathbb{N}})$.

For each cut $\mathcal{C}$ of the BPT, associated to a segmentation $\mathcal{S}$ of Ω, three information are useful: $a = |\mathcal{S} \cap C|$, $b = |\mathcal{S}|$, and $c = |\mathcal{C}|$. From this triple (a, b, c), one can recover the Dice and cardinality information of the cut, as $\frac{2.a}{b + |C|}$ and c. A cut can be (1) empty; (2) equal to a singleton set corresponding to the root of the BPT; or (3) formed by the union of two cuts, i.e. one in each subtree of the BPT. Based on these facts, the Pareto front can then be built in a recursive, bottom-up fashion as follows.

Let N be a node of the BPT, and N_1, N_2 its children nodes (if N is not a leaf). If N is a leaf, only two cuts can be built from the associated subtree, corresponding to $\{N\}$ and $\emptyset$. They correspond to the triples $(|N \cap C|, |N|, 1)$ and $(0, 0, 0)$, respectively, that lead to the points of coordinates $(2|N \cap C|/(|N| + |C|), 1)$ and $(0, 0)$ in the Pareto space, respectively. These two triples are stored in a set $\mathcal{T}_N$. If N is not a leaf, for any triple (a_1, b_1, c_1) of $\mathcal{T}_{N_1}$ and any (a_2, b_2, c_2) of $\mathcal{T}_{N_2}$, one can build a new triple $(a, b, c) = (a_1 + a_2, b_1 + b_2, c_1 + c_2)$ that corresponds to a unique cut of the (sub-)BPT of root N. This includes in particular the empty cut. A supplementary cut, namely $\{N\}$, is not formed by union of two cuts. It corresponds to the triplet $(a, b, c) = (a_1 + a_2, b_1 + b_2, 1)$ where (a_1, b_1, c_1) (resp. (a_2, b_2, c_2)) is the triple of maximal value c_1 (resp. c_2) in $\mathcal{T}_{N_1}$ (resp. $\mathcal{T}_{N_2}$).

Following this divide-and-conquer algorithmic process, one may build all the points of the Pareto space. However, each computation at a node N creates new triples by Cartesian product between those of the two children nodes. This leads to an exponential space and time complexity, that makes an exhaustive computation practically untractable.

To tackle this issue, three remarks can be made. First, it is not required to process the whole BPT. Indeed, the nodes fully outside C and those with a parent node fully inside C are useless for computing the Pareto front. Then, one can omit them when scanning the BPT, and thus process an ad hoc subtree of the BPT, with a limited subset of leaves.

Second, the computation of the Pareto front for high values of cardinality $|\mathcal{C}|$ is not practically relevant. Indeed, the quality of a BPT lies in its ability to allow for the computation of cuts leading to correct segmentations. However, such cuts should have a reasonably low cardinality for making their selection tractable, both for a human user or an optimal cut procedure. As a consequence, at each stage of the computation, all the triples (a, b, c) with a value $c > t$ for a given threshold value t corresponding to the maximally allowed cardinality may be removed. Note that this optimization requires that each node N stores the two values $|N \cap C|$ and $|N|$, in addition to its set of triples, for allowing the computation of the singleton cut $\{N\}$. This information can be computed in an additive bottom-up fashion, as for the triples (a, b, c).

Third, even with such optimizations, a combinatorial explosion may occur. As a consequence, it is better to compute a (limited but relevant) subset of the points in the Pareto space by only preserving, for each node, a list of $t + 1$ triples that lead to the best Dice scores, i.e. one triple for each of the $k \in [\![0, t]\!]$ distinct size of cuts $|\mathcal{C}|$. Doing so, the size of the list of triples is bounded by $t + 1$ at each node, and the time cost of generation of such list is also bounded by a constant value $(t + 1)^2$. The obtained Pareto front is not guaranteed to be optimal, but the way to choose the preserved triples allows us to expect a near-to-optimal approximated result.

We investigated the BPTs built for each of the four textures μ, H, C, E (Sect. 2.4). The average results are depicted in Fig. 3. We restricted our study to cuts of maximal cardinality $t = 30$.

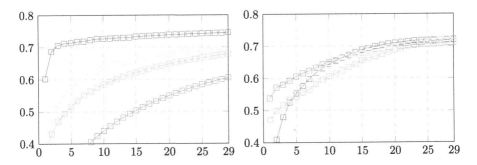

Fig. 3. Pareto front (Dice D in y-coordinate vs. size of the cut $|\mathcal{C}|$ in x-coordinate) for the BPTs constructed with the four texture features μ (in blue), H (in red), C (in yellow) and E (in green). The results are obtained by averaging the Dice values over the datasets Left: Epirmex dataset. Right: ALBERTs dataset. (Color figure online)

4   Conclusion and Perspective Works

The analysis of the Pareto fronts allows us to observe that the resolution of the MR images has a strong influence on the relevance of using texture features for BPT construction. In particular, for Epirmex data, due to a low resolution, texture features are not able to guide the BPT construction process better than the standard mean feature (μ). In the case of ALBERTs, i.e. with MR images of higher resolution, one can observe that Homogeneity (H) provides better results than the standard mean feature (μ) up to cuts of 10 nodes. For cuts of higher cardinalities, the behaviour of the four features are quantitatively comparable.

This tends to prove that (well-chosen) texture features are indeed relevant for carrying out the segmentation of the cerebellum based on BPT paradigms. This robustness remains, however, to be confirmed. Indeed, on the one hand, the current Pareto fronts are built in a near-to-optimal, but non-exhaustive, fashion. On the other hand, the obtained Dice values for a cut of the BPT remain lower than expected for a fully satisfactory segmented cerebellum.

his emphasizes two perspective works. First, we will investigate BPTs built from several features, instead of one. For that purpose, we may rely on recent works on multi-feature BPT [23]. This will allow us to couple texture features, but also to mix them with prior knowledge modeled by atlases. On the other hand, we will experiment various paradigms of gradient, in order to improve the ability of the watervoxel preprocessing to provide initial partitions with greater voxel sizes. As a side effect, we may be able to produce BPT cuts with better Dice scores for a same cardinality.

Further works will also consist of developing visualization/interaction paradigms to allow a medical expert to easily navigate within a BPT, and observe/define in real time the segmentation associated to a given cut in 3D MR images.

References

1. Ancel, P.Y., Goffinet, F.: EPIPAGE 2 writing group: EPIPAGE 2: a preterm birth cohort in France in 2011. BMC Pediatr. **14**, 97 (2014)
2. Bogovic, J.A., Bazin, P.L., Ying, S.H., Prince, J.L.: Automated segmentation of the cerebellar lobules using boundary specific classification and evolution. In: IPMI, Proceedings, pp. 62–73 (2013)
3. Cardoso, M.J., et al.: AdaPT: an adaptive preterm segmentation algorithm for neonatal brain MRI. NeuroImage **65**, 97–108 (2013)
4. Cettour-Janet, P., et al.: Watervoxels. HAL Research Report hal-02004228 (2019)
5. Coupé, P., Manjón, J.V., Fonov, V., Pruessner, J., Robles, M., Collins, D.L.: Patch-based segmentation using expert priors: application to hippocampus and ventricle segmentation. NeuroImage **54**, 940–954 (2011)
6. Gousias, I.S., et al.: Magnetic resonance imaging of the newborn brain: manual segmentation of labelled atlases in term-born and preterm infants. NeuroImage **62**, 1499–1509 (2012)
7. Gousias, I.S., et al.: Magnetic resonance imaging of the newborn brain: automatic segmentation of brain images into 50 anatomical regions. PLoS One **8**, e5999 (2013)
8. Gui, L., Lisowski, R., Faundez, T., Hüppi, P., Lazeyras, F., Kocher, M.: Morphology-based segmentation of newborn brain MR images. In: MICCAI Neo-BrainS12, Proceedings, pp. 1–8 (2012)
9. Hack, M., Fanaroff, A.A.: Outcomes of children of extremely low birthweight and gestational age in the 1990s. Semin. Neonatol. **5**, 89–106 (2000)
10. Haralick, R.M., Shanmugam, K., Dinstein, I.: Textural features for image classification. IEEE Trans. Syst. Man Cybern. **SMC–3**, 610–621 (1973)
11. Hwang, J., Kim, J., Han, Y., Park, H.: An automatic cerebellum extraction method in T1-weighted brain MR images using an active contour model with a shape prior. Magn. Reson. Imaging **29**, 1014–1022 (2011)
12. Kim, H., Lepage, C., Evans, A.C., Barkovich, J., Xu, D.: NEOCIVET: extraction of cortical surface and analysis of neonatal gyrification using a modified CIVET pipeline. In: MICCAI, Proceedings, pp. 571–579 (2015)
13. van der Lijn, F., et al.: Cerebellum segmentation in MRI using atlas registration and local multi-scale image descriptors. In: ISBI, Proceedings, pp. 221–224 (2009)
14. Machairas, V., Faessel, M., Cárdenas-Peña, D., Chabardes, T., Walter, T., Decencière, E.: Waterpixels. IEEE Trans. Image Process. **24**(11), 3707–3716 (2015)
15. Mahapatra, D.: Skull stripping of neonatal brain MRI: using prior shape information with graph cuts. J. Digit. Imaging **25**, 802–814 (2012)
16. Makris, N., et al.: MRI-based surface-assisted parcellation of human cerebellar cortex: an anatomically specified method with estimate of reliability. NeuroImage **25**, 1146–1160 (2005)

17. Makropoulos, A., Counsell, S.J., Rueckert, D.: A review on automatic fetal and neonatal brain MRI segmentation. NeuroImage **170**, 231–248 (2017)
18. Makropoulos, A., et al.: Automatic tissue and structural segmentation of neonatal brain MRI using expectation-maximization. In: MICCAI NeoBrainS12, Proceedings, pp. 9–15 (2012)
19. Marlow, N., Wolke, D., Bracewell, M.A., Samara, M., Group, E.S.: Neurologic and developmental disability at six years of age after extremely preterm birth. N. Engl. J. Med. **352**, 9–19 (2005)
20. Melbourne, A., Cardoso, M.J., Kendall, G.S., Robertson, N.J., Marlow, N., Ourselin, S.: NeoBrainS12 challenge: adaptive neonatal MRI brain segmentation with myelinated white matter class and automated extraction of ventricles I–IV. In: MICCAI NeoBrainS12, Proceedings, pp. 16–21 (2012)
21. Péporté, M., Ghita, D.E.I., Twomey, E., Whelan, P.F.: A hybrid approach to brain extraction from premature infant MRI. In: SCIA, Proceedings, pp. 719–730 (2011)
22. Prastawa, M., Gilmore, J.H., Lin, W., Gerig, G.: Automatic segmentation of MR images of the developing newborn brain. Med. Image Anal. **9**, 457–466 (2005)
23. Randrianasoa, J.F., Kurtz, C., Desjardin, E., Passat, N.: Binary partition tree construction from multiple features for image segmentation. Pattern Recognit. **84**, 237–250 (2018)
24. Randrianasoa, J.F., Kurtz, C., Gançarski, P., Desjardin, E., Passat, N.: Evaluating the quality of binary partition trees based on uncertain semantic ground-truth for image segmentation. In: ICIP, Proceedings, pp. 3874–3878 (2017)
25. Randrianasoa, J.F., Kurtz, C., Gançarski, P., Desjardin, E., Passat, N.: Intrinsic quality analysis of binary partition trees. In: ICPRAI, Proceedings, pp. 114–119 (2018)
26. Romero, J.E.: CERES: a new cerebellum lobule segmentation method. NeuroImage **147**, 916–924 (2017)
27. Rousseau, F., et al.: BTK: an open-source toolkit for fetal brain MR image processing. Comput. Methods Programs Biomed. **109**, 65–73 (2013)
28. Saeed, N., Puri, B.: Cerebellum segmentation employing texture properties and knowledge based image processing: applied to normal adult controls and patients. Magn. Reson. Imaging **20**, 425–429 (2002)
29. Salembier, P., Garrido, L.: Binary partition tree as an efficient representation for image processing, segmentation, and information retrieval. IEEE Trans. Image Process. **9**, 561–576 (2000)
30. Salembier, P., Wilkinson, M.H.F.: Connected operators: a review of region-based morphological image processing techniques. IEEE Signal Process. Mag. **26**, 136–157 (2009)
31. Serag, A., et al.: Accurate learning with few atlases (ALFA): an algorithm for MRI neonatal brain extraction and comparison with 11 publicly available methods. Sci. Rep. **6**, 23470 (2016)
32. Stoodley, C.J., Limperopoulos, C.: Structure-function relationships in the developing cerebellum: evidence from early-life cerebellar injury and neurodevelopmental disorders. Semin. Fetal Neonatal Med. **21**, 356–364 (2016)
33. Tustison, N.J., et al.: N4ITK: improved N3 bias correction. IEEE Trans. Med. Imaging **29**, 1310–1320 (2010)
34. Wang, L., Shi, F., Lin, W., Gilmore, J.H., Shen, D.: Automatic segmentation of neonatal images using convex optimization and coupled level sets. NeuroImage **58**, 805–817 (2011)
35. Wang, L., Shi, F., Yap, P.T., Gilmore, J.H., Lin, W., Shen, D.: 4D multi-modality tissue segmentation of serial infant images. PLoS One **7**, e44596 (2012)

36. Xue, H., et al.: Automatic segmentation and reconstruction of the cortex from neonatal MRI. NeuroImage **38**, 461–477 (2007)
37. Yang, Z., et al.: Automated cerebellar lobule segmentation with application to cerebellar structural analysis in cerebellar disease. NeuroImage **127**, 435–444 (2016)

Atlas-Based Automated Detection of Swim Bladder in Medaka Embryo

Diane Genest[1,2(⊠)], Marc Léonard[2], Jean Cousty[1], Noémie de Crozé[2], and Hugues Talbot[1,3]

[1] Université Paris-Est, Laboratoire d'Informatique Gaspard Monge, CNRS, ENPC, ESIEE Paris, UPEM, 93162 Noisy-le-Grand, France
diane.genest@esiee.fr
[2] L'OREAL Research and Innovation, 93600 Aulnay-sous-Bois, France
[3] CentraleSupelec, Université Paris-Saclay, équipe OPIS-Inria, 91190 Gif-sur-Yvette, France
diane.genest@rd.loreal.com

Abstract. Fish embryo models are increasingly being used both for the assessment of chemicals efficacy and potential toxicity. This article proposes a methodology to automatically detect the swim bladder on 2D images of medaka fish embryos seen either in dorsal view or in lateral view. After embryo segmentation and for each studied orientation, the method builds an atlas of a healthy embryo. This atlas is then used to define the region of interest and to guide the swim bladder segmentation with a discrete globally optimal active contour. Descriptors are subsequently designed from this segmentation. An automated random forest classifier is built from these descriptors in order to classify embryos with and without a swim bladder. The proposed method is assessed on a dataset of 261 images, containing 202 embryos with a swim bladder (where 196 are in dorsal view and 6 are in lateral view) and 59 without (where 43 are in dorsal view and 16 are in lateral view). We obtain an average precision rate of 95% in the total dataset following 5-fold cross-validation.

1 Introduction

Fish embryo models are widely used in both environmental and human toxicology and for the assessment of chemicals efficacy [6]. According to the international animal welfare regulations, fish embryos are ethically acceptable models for the development of toxicological assays, with the complexity of a complete organism [23,26]. As an aquatic organism, fish embryos can provide relevant information about the environmental impact of chemicals. The study of early developmental stage malformations provides relevant information on developmental toxicity of chemicals [15,27,28]. The absence of swim bladder, an internal gas-filled organ that allows the embryo to control its buoyancy is one of the most sensitive marker of developmental toxicity (Fig. 1) [1,9]. In this article, we focus on the

© Springer Nature Switzerland AG 2019
B. Burgeth et al. (Eds.): ISMM 2019, LNCS 11564, pp. 496–507, 2019.
https://doi.org/10.1007/978-3-030-20867-7_38

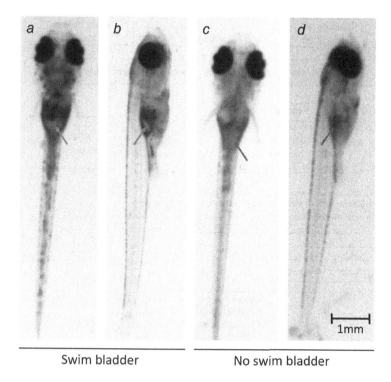

Swim bladder   No swim bladder

Fig. 1. Medaka embryos with or without a swim bladder. The blue arrows indicate the swim bladder location for embryos with a swim bladder (a and b) and the red arrows indicate the location where a swim bladder should be present for embryos without a swim bladder (c and d). a and c: embryos seen in dorsal view; b and d: embryos seen in lateral view. (Color figure online)

automated detection and extraction of the swim bladder on 2D images of 9 days post fertilization (dpf) medaka embryos (*Oryzias latipes*).

The proper detection of the swim bladder depends on the orientation of fish embryos. This implies to manually place the fish embryo before the image acquisition, which is tedious and time consuming. To overcome this difficulty, we developed a method capable of detecting the swim bladder regardless of the fish embryo position. Here, a methodology based on morphological operators is proposed to automatically detect the swim bladder on 2D images of fish embryos seen either in dorsal view or in lateral view [18,22]. After a pre-processing step that includes the automated determination of the embryo orientation [20], the methodology consists of the generation of an atlas representative of a healthy embryo [8,13,21], of swim bladder segmentation using this atlas, of descriptors calculation and of embryos classification according to the presence or absence of a swim bladder. Using this method on a dataset of 261 images, we obtain an accuracy of 95%. The successive steps of the method are represented in Fig. 2.

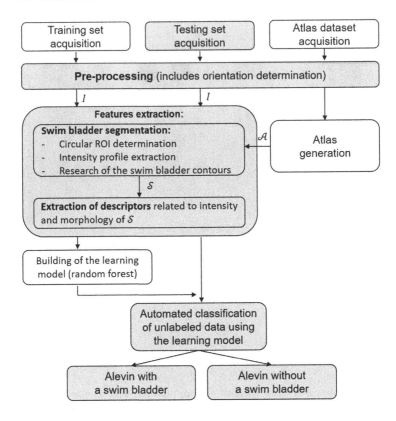

Fig. 2. Flowchart of the swim bladder detection assay.

Section 2 introduces the swim bladder localization step, including the atlas generation in Sect. 2.1, the identification, using this atlas, of a region of interest (ROI) for the search of the swim bladder, and the swim bladder segmentation by a globally optimal active contour algorithm applied on this ROI in Sect. 2.2 [2]. Then Sect. 3 describes how the segmented shape is characterized with intensity and morphological descriptors extraction. Finally, Sect. 4 presents the process of embryos classification by a random forest classifier according to the presence or absence of a swim bladder [3].

2   Swim Bladder Localization

As a swim bladder can be present or absent in embryos images, the aim of this part is to find the most probable contours of a swim bladder in such images. For practical reasons, the swim bladder segmentation method is described in this part with respect to the embryo orientation that appears the most frequently in our database, which is the dorsal view. However, this methodology is adaptable to embryos seen in lateral view with minimal parameter changes. As we also have developed a method to automatically determine the orientation of the embryo based on orientation related descriptors, it is feasible to fully automate the swim bladder detection process.

2.1   Swim Bladder Atlas Generation

The objective here is to build a median image I_{med} representative of a typical healthy embryo with respect to the diversity of embryos that exists in experimental conditions, and a probability function p_{sb} defined on the ensemble of I_{med} pixels coordinates and that represents the likelihood of each pixel of the represented image to belong to the swim bladder. In the following, such a pair (I_{med}, p_{sb}) is called an atlas of the swim bladder for medaka embryo images and will be denoted by $\mathcal{A}$.

In order to build the atlas, n images of healthy embryos are selected and their swim bladder are manually segmented. Among these images, one is randomly chosen as being the fixed reference image I_F. Then, the $n-1$ remaining images, called the moving images hereafter, are aligned on I_F by applying an affine image registration algorithm. It consists of finding, for each moving image, an affine transformation that minimizes a similarity measure between the fixed image and the transformed moving image [14,17]. The similarity measure used here is the mutual information. This maximizes the measure of the mutual dependence between the pixel intensity distributions of the fixed and the moving images. Mutual information is defined by Thevenaz and Unser as:

$$MI(I_F, I_M) = \sum_m \sum_f p(f,m) log_2(\frac{p(f,m)}{p_F(f)p_M(m)}) \qquad (1)$$

where f and m are the intensities of the fixed image and moving image respectively, p is the discrete joint probability, and p_F and p_M are the marginal discrete probabilities of the fixed image I_F and the moving image respectively I_M [16,25]. The multiresolution affine registration algorithm from the Elastix toolbox is used to perform this process [12]. The median I_{med} of the n registered images is calculated (Fig. 3). For each moving image I_M, the optimal transformation μ is also applied to the corresponding manual segmentation of the swim bladder. The average I_{av} of the n registered swim bladder segmentations is calculated. We define as the probability function p_{sb} the mapping that, to each pixel of coordinates (x,y) of a new image I with the same dimensions as I_{med}, associates the value I_{av} (x,y). The atlas $\mathcal{A} = (I_{med}, p_{sb})$ will then be used in order to identify, in a new image I, the ROI where to search the swim bladder.

2.2   Swim Bladder Segmentation

After registering the atlas on a new image I, we observe that the registered atlas does not segment the contour of the swim bladder with precision, as shown in Fig. 4. As our methodology relies on the characterization of the swim bladder most probable contours to distinguish embryos with and without a swim bladder, we need to obtain a more accurate delineation of the swim bladder, when it is present.

Fig. 3. Atlas $\mathcal{A} = (I_{med}, p_{sb})$ obtained for fish embryos seen in dorsal view. The three lines show the iso-contours that delimit the areas where the pixels have a probability equal to 1 (green), to 0,5 (pink), and to 0,05 (blue) to belong to the swim bladder. (Color figure online)

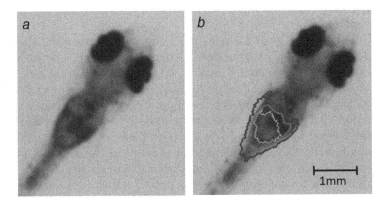

Fig. 4. Projection of the atlas $\mathcal{A}$ on an embryo image. The three lines delimit the areas where pixels have a probability equal to 1 (green), to 0.5 (pink) and to 0.05 (blue) to belong to the swim bladder according to the projected probability function p'_{sb} of the atlas. (Color figure online)

ROI Localization. The atlas $\mathcal{A} = (I_{med}, p_{sb})$ is used on an embryo image I in order to identify the region of interest in which the swim bladder will be searched. To this aim, we search for the transformation μ' to apply to I_{med} that optimally registers I_{med} to the analyzed image I. We apply the same affine registration process as described in Sect. 2.1. This transformation μ' is then applied to the definition ensemble of p_{sb}, leading to a new transformed probability function p'_{sb}. We compute the binary mask M that contains the pixels of I where the probability of finding the swim bladder, if a swim bladder is present, is equal to 1. The barycenter of this binary image M is extracted and considered as the center C of the ROI. The image ROI is then defined as the circle of center C and of diameter 40 pixels, experimentally determined (Fig. 5).

Extraction of the Swim Bladder Most Probable Contours. As shown in Fig. 1, the swim bladder is characterized by a high contrast between dark contours and light inner part that cannot be too small. By contrast, embryos

Primal frame ROI:

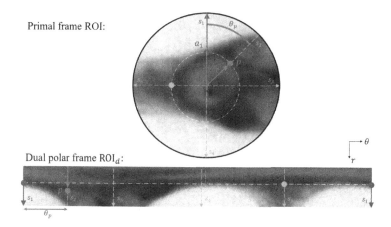

Dual polar frame ROI_d:

Fig. 5. Representation of the image ROI in the primal frame and of its associated image ROI_d in the dual polar frame.

without a swim bladder present a homogeneous body in the location where the swim bladder should be present. For this reason, the swim bladder detection method relies on the determination of a circular shortest path extracted from the image ROI represented in a dual polar frame defined as follow. Considering ROI of center $C = (x_C, y_C)$ and of radius r in the primal frame of the image I, we define its associated representation in a dual polar frame ROI_d, as the image that is the concatenation of all ROI radial sections, starting from a radial section s_1 that is perpendicular to the embryo skeleton obtained during pre-processing.

Each radial section s_θ of the image ROI, defined by an angle θ in degrees from the initial section, is then precisely the θth column of the dual image representation ROI_d (Fig. 5). Hence, to each pixel $p = (\theta_p, r_p)$ in ROI_d, we associate the intensity $c(p)$ of the pixel $p' = (r_p cos(\theta_p), r_p sin(\theta_p)$ in the primal image ROI.

We then consider circular shortest paths in the image ROI_d, i.e., paths corresponding to contours of minimum energy in the primal image [2,5]. For this purpose, the image ROI_d is equipped with a directed graph such that a pair (a, b) of two pixels of ROI_d is a directed arc if $a_1 = b_1 - 1$ and $|a_2 - b_2| \le 1$, where $a = (a_1, a_2)$ and $b = (b_1, b_2)$. In this graph, a circular path is a sequence $(p_0, \ldots, p_l)$ of pixels of ROI_d such that:

- for any i in $\{1, \ldots, l\}$, the pair (p_{i-1}, p_i) is an arc;
- the first coordinate of p_0 is equal to 0;
- the first coordinate of p_l is equal to 360 (i.e. the maximal possible value); and
- the second coordinate of p_0 and of p_l are the same.

The energy cost $EC(\pi)$ of a circular path $\pi = (p_0, \ldots, p_l)$ is defined as the sum of the intensities of the pixels in the path: $EC(\pi) = \sum_{i \in \{0, \ldots, l\}} c(p_i)$. A circular path $\pi = (p_0, \ldots, p_l)$ is called optimal whenever the energy cost of π is less than or equal to the energy cost of any circular path from p_0 to p_l. Circular

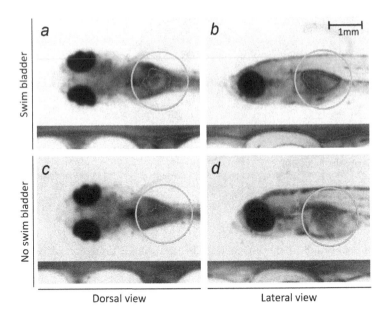

Fig. 6. Swim bladder segmentation results on the primal frame of the image and associated shortest path in the dual polar frame. The yellow circle delimits the ROI in which the red line shows the contour of the segmented shape $\mathcal{S}$ in case of embryos with a swim bladder (a and b) and embryos without a swim bladder (c and d). a and c: embryos seen in dorsal view. b and d: embryos in lateral view. (Color figure online)

optimal path can be computed with any graph shortest path algorithm such as the one of Dijkstra [5,24].

In order to obtain the most probable contour of the swim bladder, we start by selecting the most peripheral local minimum of the first radial section r_1, called a_1. We also define $rmin$ as the minimal radius of the ROI below which the shortest path must not be searched. It is experimentally set to 10 pixels. We then consider a circular shortest path π starting at a_1. This circular shortest path, found in the image ROI_d, corresponds to a closed contour $\mathcal{S}$ in the image ROI which surrounds the center C and which is of minimal energy. Such optimal contour $\mathcal{S}$ is hereinafter referred to as the most probable contour of the swim bladder (Fig. 6). This methodology expects to segment the swim bladder if present, or a random part of the embryo body otherwise. Swim bladder characterization will now allow to distinguish both cases.

3   Swim Bladder Characterization

We now need to identify if the segmented shape $\mathcal{S}$ corresponds to a swim bladder or not, basing on descriptors of the swim bladder.

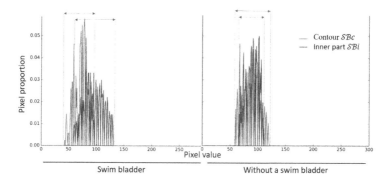

Fig. 7. Histograms of the inner part $\mathcal{S}_i$ and the contour $\mathcal{S}_c$ of the segmented shape $\mathcal{S}$. The shift between the two histograms is highlighted by the arrows.

3.1 Intensity Descriptors

A swim bladder is characterized by a high contrast between a dark contour and a lighter inner part, contrary to an embryo without any swim bladder that presents a more homogeneous intensity in the delimited shape. On the histograms of both the inner part $\mathcal{S}_i$ and the contour $\mathcal{S}_c$ of the segmented shape $\mathcal{S}$, this means that a shift is visible between both distributions of pixel intensities in case of an embryo with a swim bladder (Fig. 7). The following intensity-related parameters are extracted from both histograms: the minimal, maximal, average and median intensities, the intensity mode, and their piecewise differences are calculated in order to characterize the contrast between $\mathcal{S}$ inner part and contour. We also measure the two ranges, *i.e.*, the difference between the maximal and the minimal intensities, and the ratio between them. We finally compute the pixel intensity variance of $\mathcal{S}$, and the covering, defined as follow. We calculate the difference between the maximal value of $\mathcal{S}_c$ and the minimal value of $\mathcal{S}_i$ in one hand, and the difference between the maximal value of $\mathcal{S}_i$ and the minimal value of $\mathcal{S}_c$ on the other hand. The covering is defined as the ratio between both differences. These intensity-related descriptors will then be combined with other descriptors characteristic of the swim bladder morphology which are described in the following subsection.

3.2 Morphological Descriptors

In order to characterize the swim bladder shape, the following descriptors are extracted from $\mathcal{S}$. We refer to the convexity of a swim bladder, as the set difference between the convex hull of $\mathcal{S}$ and $\mathcal{S}$ itself [10]. We refer to the concavity of a swim bladder as the area of the deleted component after an opening of $\mathcal{S}$ by a disk-shape structuring element of size 5. We furthermore consider the elongation of $\mathcal{S}$ defined by $\frac{4\pi area(\mathcal{S})}{perimeter(\mathrm{S})^2}$. All descriptors related to intensity and morphological characterization of $\mathcal{S}$ are summarized in Table 1.

Table 1. List of descriptors used for swim bladder characterization.

Name	Description	Name	Description
Intensity descriptors:			
$variance_i$	Variance of pixel intensity	min_diff	$min_{S_i} - min_{S_c}$
min_i	Minimum intensity	max_diff	$max_{S_i} - max_{S_c}$
max_i	Maximum intensity	av_diff	$average_{S_i} - average_{S_c}$
$average_i$	Average intensity	$mode\_diff$	$mode_{S_i} - mode_{S_c}$
$mode_i$	Histogram mode	$median\_diff$	$median_{S_i} - median_{S_c}$
$median_i$	Median intensity	$rrange$	$\frac{range_{S_i}}{range_{S_c}}$
$range_i$	$max_i - min_i$	$covering$	$\frac{max_{S_c} - min_{S_i}}{max_{S_i} - min_{S_i}}$
for i in either $\mathcal{S}_i$ or $\mathcal{S}_c$			
Morphological descriptors:			
$convexity$	convex_hull$(\mathcal{S}) - \mathcal{S}$	$elongation$	$\frac{4\pi area(\mathcal{S})}{perimeter(S)^2}$
$concavity$	$area(\gamma_5(\mathcal{S}) - \mathcal{S})$		

4 Fish Embryo Classification

We detail in this section the classification process used to assess the methodology.

4.1 Experimental Setup

On day 0, fish eggs are placed on a 24-well plate, one egg per well, in 2 mL of incubation medium with or without chemicals [11]. The ninth day, embryos are anesthetized by adding tricaine to the incubation medium (final concentration: 0,18 g/L) and one image of each embryo is recorded. Each embryo is then observed in all possible orientations under a microscope by an expert. The expert annotates each fish embryo as having or not a swim bladder. This observation constitutes the ground truth.

With this protocol, we set up a total database of 287 images of embryos, where 259 are seen in dorsal view (91.5%) and 28 in lateral view (8.5%). A subset of this database is used in order to generate the atlas described in Sect. 2.1. For reasons linked to the unbalanced proportions of both orientations in our total available database, we select $n = 20$ images of healthy embryos seen in dorsal view for dorsal atlas generation, and $n = 6$ healthy embryos seen in lateral position for lateral atlas generation. The remaining 261 images constitute the validation dataset. Among them, 202 present a swim bladder according to the ground truth, and 59 do not present a swim bladder. In particular, the subset of 202 embryos with a swim bladder is composed of 196 images of fish embryos seen in dorsal view and 6 seen in lateral view. The subset of 59 embryos without a swim bladder is composed of 43 seen in dorsal view and 16 in lateral view.

4.2   Classifier Description

A random forest classifier is defined with the following parameters [3,29]: the number of estimators is set to 50, the maximal depth to 30, the minimal number of samples required to split an internal node to 5 and the minimal number of samples required to be at a leaf node to 2. The entropy criterion is chosen.

4.3   Classification Results and Discussion

The results of our classification process after a 5-fold cross validation are presented in Table 2 in the form a confusion matrix that shows the distribution between embryos with and without a swim bladder according to the ground truth and to the prediction results. We consider as a true positive a correctly detected anomaly, *i.e.* absence of a swim bladder. A correctly detected swim bladder is referred to as a true negative. The classifier reaches an accuracy score of 95% in the total dataset. The sensitivity is 90% and the specificity is 96%.

Table 2. Classification results after 5-fold cross validation.

Ground truth	Prediction	
	Swim bladder	No swim bladder
Swim bladder	195	7
No swim bladder	6	53

This article aimed to describe a method for the automated detection of the swim bladder in medaka fish embryos. As a screening test, we assess the accuracy of the method by calculating the overall accuracy, the sensitivity and the specificity. With an overall accuracy of 95% measured, this study reveals the feasibility of an automated detection of the swim bladder from 2D images of medaka embryos. We also observe a high specificity of 96% and an acceptable sensitivity of 90%, which suggests that the use of this method is accurate for the global screening test that is to be developed. Indeed, the swim bladder detection assay is intended to be a part of a series of morphological abnormalities detection assays whose each specificity needs to be maximized in order to maximize the accuracy of the whole detection assay and not eliminate a too important number of chemicals [7,20]. The sensitivity of each individual test is expected to improve the sensitivity of the whole test.

The experimental protocol of our method presents the advantage to not require the manual positioning of each embryo in the well. After anesthesia, embryos remained in the incubation medium in their well and can have any possible orientation before images are acquired and treated. However, this protocol does not permit to control the orientation. In our database, a disproportion between natural positioning of embryos exists. Without any control on embryo positioning, we obtain highly unbalanced datasets between dorsal and lateral views. In our total dataset of 261 images, we only have 8.5% of embryos seen in lateral view vs. 91.5% of embryos seen in dorsal view. Among the lateral embryos, 90,9% are correctly classified, vs. 95,4% for the dorsal embryos. These results reveal that the presence or the absence of swim bladder can be detected with a satisfying accuracy in embryos images, regardless of their orientation, with an adaptive swim bladder detection method. As we are also able to detect automatically the embryos orientation, that means this process can be fully automated. In a future work, we plan to adapt this method to the intermediary orientations, by performing a 3D reconstruction of a healthy embryo from dorsal and lateral atlases interpolation [4,19].

5   Conclusion

This article proposes a methodology to automatically classify images of medaka embryos according to the presence or the absence of swim bladder. It relies on the use of a healthy embryo atlas. This atlas is used to define the research area of the swim bladder in the analyzed embryo. The most probable contour of the swim bladder is segmented in this research area and descriptors are extracted to characterize the segmented shape. The final classification using a random forest classifier yields an accuracy of 95% in the total dataset, which is satisfying, in particular with respect to the major advantage of the method of being orientation-independent. In terms of experimental manipulations, being able to automatically analyze embryos regardless to their orientation would save significant time as the protocol would not require neither any manual positioning nor any visual inspection of embryos to detect abnormalities. To increase the precision and analyze intermediary orientations, we plan to generate a 3D atlas.

References

1. Ali, S., Aalders, J., Richardson, M.K.: Teratological effects of a panel of sixty water-soluble toxicants on zebrafish development. Zebrafish 11(2), 129–141 (2014)
2. Appleton, B., Talbot, H.: Globally optimal geodesic active contours. J. Math. Imaging Vis. 23(1), 67–86 (2005)
3. Breiman, L.: Random forests. Mach. Learn. 45(1), 5–32 (2001)
4. Bryson-Richardson, R.J., et al.: Fishnet: an online database of zebrafish anatomy. BMC Biol. 5(1), 34 (2007)
5. Dijkstra, E.W.: A note on two problems in connexion with graphs. Numer. Math. 1(1), 269–271 (1959)
6. Embry, M.R., et al.: The fish embryo toxicity test as an animal alternative method in hazard and risk assessment and scientific research. Aquat. Toxicol. 97(2), 79–87 (2010)
7. Genest, D., Puybareau, E., Léonard, M., Cousty, J., De Crozé, N., Talbot, H.: High throughput automated detection of axial malformations in medaka embryo. Comput. Biol. Med. 105, 157–168 (2019)
8. Iglesias, J.E., Sabuncu, M.R.: Multi-atlas segmentation of biomedical images: a survey. Med. Image Anal. 24(1), 205–219 (2015)

9. Iwamatsu, T.: Stages of normal development in the medaka oryzias latipes. Mech. Dev. **121**(7–8), 605–618 (2004)
10. Jarvis, R.A.: On the identification of the convex hull of a finite set of points in the plane. Inf. Process. Lett. **2**(1), 18–21 (1973)
11. Kinoshita, M., Murata, K., Naruse, K., Tanaka, M.: Medaka: Biology, Management, and Experimental Protocols. Wiley, Ames (2009)
12. Klein, S., Staring, M., Murphy, K., Viergever, M.A., Pluim, J.P.: Elastix: a toolbox for intensity-based medical image registration. IEEE Trans. Med. Imaging **29**(1), 196–205 (2010)
13. Klein, S., Van Der Heide, U.A., Lips, I.M., Van Vulpen, M., Staring, M., Pluim, J.P.: Automatic segmentation of the prostate in 3D MR images by atlas matching using localized mutual information. Med. Phys. **35**(4), 1407–1417 (2008)
14. Lester, H., Arridge, S.R.: A survey of hierarchical non-linear medical image registration. Pattern Recogn. **32**(1), 129–149 (1999)
15. Lovely, C.B., Fernandes, Y., Eberhart, J.K.: Fishing for fetal alcohol spectrum disorders: zebrafish as a model for ethanol teratogenesis. Zebrafish **13**(5), 391–398 (2016)
16. Maes, F., Vandermeulen, D., Suetens, P.: Medical image registration using mutual information. Proc. IEEE **91**(10), 1699–1722 (2003)
17. Mattes, D., Haynor, D.R., Vesselle, H., Lewellen, T.K., Eubank, W.: PET-CT image registration in the chest using free-form deformations. IEEE Trans. Med. Imaging **22**(1), 120–128 (2003)
18. Najman, L., Talbot, H.: Mathematical Morphology: From Theory to Applications. Wiley, London (2013)
19. Narayanan, R., et al.: Adaptation of a 3D prostate cancer atlas for transrectal ultrasound guided target-specific biopsy. Phys. Med. Biol. **53**(20), N397 (2008)
20. Puybareau, É., Genest, D., Barbeau, E., Léonard, M., Talbot, H.: An automated assay for the assessment of cardiac arrest in fish embryo. Comput. Biol. Med. **81**, 32–44 (2017)
21. Rohlfing, T., Brandt, R., Menzel, R., Maurer Jr., C.R.: Evaluation of atlas selection strategies for atlas-based image segmentation with application to confocal microscopy images of bee brains. NeuroImage **21**(4), 1428–1442 (2004)
22. Serra, J.: Image Analysis and Mathematical Morphology. Academic Press Inc., Orlando (1983)
23. Strähle, U., et al.: Zebrafish embryos as an alternative to animal experiments a commentary on the definition of the onset of protected life stages in animal welfare regulations. Reprod. Toxicol. **33**(2), 128–132 (2012)
24. Sun, C., Pallottino, S.: Circular shortest path in images. Pattern Recogn. **36**(3), 709–719 (2003)
25. Thévenaz, P., Unser, M.: Optimization of mutual information for multiresolution image registration. IEEE Trans. Image Process. **9**(12), 2083–2099 (2000)
26. Union, E.: Directive 2010/63/EU of the European parliament and of the council of 22 september 2010 on the protection of animals used for scientific purposes. Off. J. Eur. Union **276**, 33–79 (2010)
27. Yamashita, A., Inada, H., Chihara, K., Yamada, T., Deguchi, J., Funabashi, H.: Improvement of the evaluation method for teratogenicity using zebrafish embryos. J. Toxicol. Sci. **39**(3), 453–464 (2014)
28. Yang, X., et al.: Developmental toxicity of synthetic phenolic antioxidants to the early life stage of zebrafish. Sci. Total Environ. **643**, 559–568 (2018)
29. Zhang, C., Ma, Y.: Ensemble Machine Learning: Methods and Applications. Springer, New York (2012). https://doi.org/10.1007/978-1-4419-9326-7

Self-dual Pattern Spectra for Characterising the Dermal-Epidermal Junction in 3D Reflectance Confocal Microscopy Imaging

Julie Robic[1,2(✉)], Benjamin Perret[2], Alex Nkengne[1], Michel Couprie[2], and Hugues Talbot[3]

[1] Laboratoires Clarins, Pontoise, France
julie.robic@clarins.com
[2] Université Paris-Est, Laboratoire d'Informatique Gaspard-Monge UMR 8049, UPEMLV, ESIEE Paris, ENPC, CNRS, Noisy-le-Grand, France
[3] CentraleSupélec, Centre de Vision Numérique, Inria, Gif-sur-Yvette, France

Abstract. The Dermal-Epidermal Junction (DEJ) is a 2D surface separating the epidermis from the dermis which undergoes multiple changes under pathological or ageing conditions. Recent advances in reflectance confocal microscopy now enables the extraction of the DEJ from in-vivo imaging. This articles proposes a method to automatically analyse DEJ surfaces using self-dual morphological filters. We use self-dual pattern spectra with non-increasing attributes and we propose a novel measure in order to characterize the evolution of the surface under the filtering process. The proposed method is assessed on a specifically constituted dataset and we show that the proposed surface feature significantly correlates with both chronological ageing and photo-ageing.

Keywords: Pattern spectrum · Tree-of-Shapes · 2d surface · Skin

1 Introduction

We address the problem of automatically characterizing a specific structure of the skin, the Dermal-Epidermal Junction (DEJ), in in-vivo Reflectance Confocal Microscopy (RCM) 3D images. The DEJ is a complex, surface-like, 2D boundary, separating the epidermis from the dermis, which undergoes multiple changes under pathological or ageing conditions (see Fig. 1). In particular, alterations of the DEJ modify the regularity and the depth of its peaks and valleys, called dermal papillae [6,12]. Automated analysis of RCM images of the skin is a challenging task due to the very low signal to noise ratio and the important blur present in the acquisitions.

In previous work [13,15], we have presented a method to obtain reliable segmentation of the DEJ in RCM images using 3D conditional random fields

© Springer Nature Switzerland AG 2019
B. Burgeth et al. (Eds.): ISMM 2019, LNCS 11564, pp. 508–519, 2019.
https://doi.org/10.1007/978-3-030-20867-7_39

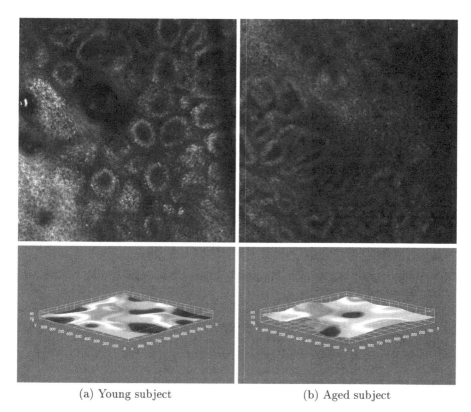

(a) Young subject   (b) Aged subject

Fig. 1. DEJ ageing. First row: en-face sections of 2 RCM stacks coming respectively from a young (a) and an older (b) subject. The DEJ corresponds to the brightest rings. Second row: automatically segmented DEJ surfaces from a young (a) and an older (b) subject using the method desribed in [15].

embedding biological prior (see Fig. 1). Moreover, the proposed method guarantees that the segmented region of the 3D RCM image is indeed a 2D topological surface defined on a regular grid, *i.e.*, an elevation/topographic map.

Besides this, granulometries and pattern spectra [10] are standard morphological tools for characterizing the content of an image. Their principle is to iteratively apply a sequence of increasing filters and to measure the evolution of the filtered image. Their extension to attribute connected filters [17], is widely used to obtain efficient and powerful multi-scale features using tree based representations of images that enable a region-based analysis, rather than a pixel-based one. In particular, attribute profiles [4] are now a standard approach to compute pixel-wise features for the semantic segmentation of high-resolution remote sensing images.

Moreover, their self-dual extension [3,9], using the Tree-of-Shapes image representation [11] (also called tree of level lines or inclusion tree), have been shown to increase the performances of classifiers. In such a representation, each level line (or iso-contour) of the surface is the boundary of a region of interest. Then, the inclusion relation between the various level lines of the surface forms a tree structure: this representation is called the Tree-of-Shapes. This representation benefits from several useful properties: it is invariant to scaling, rotations, translations, and monotone contrast modifications. This last invariance also implies that the Tree-of-Shapes is a self-dual representation, *i.e.*, that the inversion of the topographic surface (where peaks become valleys and conversely) does not modify the representation.

The objective of this article is to provide a method for automatically characterizing a DEJ surface with the objective to quantify the ageing process. As our DEJ surfaces are indeed elevation maps, it is thus possible to analyse them using morphological representations such as the Tree-of-Shapes. An example of a DEJ topographic surface, its level lines, and the corresponding Tree-of-Shapes are presented in Fig. 2. From the DEJ surface, we next compute self-dual pattern spectra, with the following desirable properties:

1. the ability to select the attribute guiding the filtering according to the dermatologists' analysis;
2. providing an analytic formulation of the algorithm generalizing the subtractive filtering rule [17] to non increasing attribute filters on the Tree-of-Shapes proposed in [2];
3. the definition a novel measure in order to characterize the filtered surfaces.

The proposed method is assessed on a specifically constituted dataset involving 15 persons divided in two age groups. Our results show that the proposed characterization of the DEJ surface is significantly correlated with both chronological ageing and photo-ageing.

This article is organized as follows. Section 2 recalls the definition of the Tree-of-Shapes. Section 3 defines subtractive self-dual attribute filters and related pattern spectra. The parameters of the method and their biological significance are discussed in Sect. 4. Finally, Sect. 5 presents the assessment of the proposed method and discusses the results.

2   Tree of Shapes

In this section, we review the definition of the Tree-of-Shapes. We mainly follow the formalism proposed in [1].

A DEJ surface is modelled by a *topographic map*, *i.e.*, a function f from a non empty finite rectangular subset E of $\mathbb{Z}^2$ to $\mathbb{R}$. An element of E is called *a point*. The value of f at a point x of E is called the *elevation or level of f at x.*

Let $X \subseteq E$, we denote by $CC_x(X)$ the connected component of X that contains the pixel x using the classical 4-adjacency on the discrete grid. We denote by $CC(X)$ the set of connected components of X: $CC(X) = \{CC_x(X),\ x \in X\}$. Moreover, we assume that E is connected: *i.e.*, $CC(E) = \{E\}$.

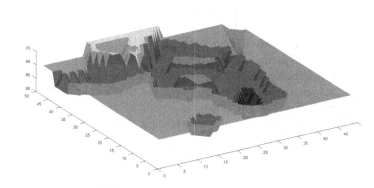

(a) Segmented DEJ as a 2D surface

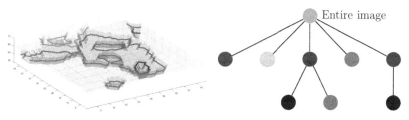

(b) Level lines of the 2D surface in (a)   (c) Tree-of-Shapes shown in (b)

Fig. 2. Example of DEJ segmentation as a topographic surface. In (a), the DEJ segmentation is presented in 3D. Each level line is the contour of a level set of the surface (b). In (c), the Tree-of-Shapes induced by the inclusion relationship between the level lines.

Given a topographic map f, the *upper-level set* (respectively *lower-level set*) of f at level k in $\mathbb{R}$, denoted by $[f > k]$ (respectively $[f < k]$), is the set of pixels where the elevation of f is larger than k (respectively lower than k):

$$[f > k] = \{x \in E \mid f(x) > k\}, \text{and} \tag{1}$$
$$[f < k] = \{x \in E \mid f(x) < k\}. \tag{2}$$

While the set of connected components of the upper-level (respectively lower-level) sets of f enables to define the classical Max-Tree [16] (respectively Min-Tree) which is oriented toward the maxima (respectively the minima) of f, considering the union of both enables to define a new structure, the Tree-of-Shapes, that represents equally the minima and the maxima of f.

Let X be a connected subset of E, the *saturation of* X denoted sat(X), is the set defined as the union of X and its *holes*. Formally, one can define sat(X) equals to $CC_{p_\infty}(X^c)^c$, where p_∞ is an arbitrary point designing the exterior of E (usually taken on the border of E) and where, given a subset Y of E, Y^c is the complement of Y in E, i.e., $Y^c = \{x \in E \mid x \notin Y\}$. The saturation of a connected component of a lower/upper level set of a topographic map of

f is called a *shape of* f and the set of shapes of f is denoted by $\mathcal{S}(f)$. The level lines of the topographic map f are the contours of the shapes $\mathcal{S}(f)$. Note that, to ensure that each level line of f is an isolated closed curve, we used the intermediate continuous multivalued function representation proposed in [5].

Let f be a topographic map. The inclusion relation on the set of shapes $\mathcal{S}(f)$ induces a tree structure called the *Tree-of-Shapes* (see Fig. 3). Given two distinct shapes s_1 and s_2 in $\mathcal{S}(f)$, we say that s_2 is *a child of* s_1, and that s_1 is *the parent of* s_2, if s_2 is included in s_1 and if it is maximal for this condition: *i.e.*, for any shape s_3 in $\mathcal{S}(f)$ such that $s_2 \subseteq s_3 \subset s_1$ then $s_2 = s_3$. For any shape s in $\mathcal{S}(f)$ different from E, the unique parent of s is denoted Par(s). By abuse of notation, we set Par(E) equals to E. Given a shape s in $\mathcal{S}(f)$, the set of children of s is denoted Ch(s).

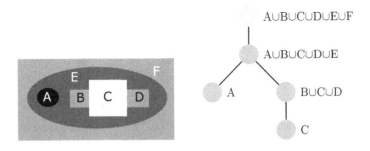

Fig. 3. Tree-of-Shapes of a topographic map. The root of the tree is at the top represented by a red circle. (Color figure online)

Let f be a topographic map. Given a shape s in $\mathcal{S}(f)$, the *proper points of s*, denoted by Pro(s) is the non empty subset of points of s that do not belong to any child of s: *i.e.*, Pro$(s) = s \backslash \bigcup$ Ch(s). Given a shape s in $\mathcal{S}(f)$, the elevation of f is constant on Pro(s): it is called *the level of s* and it is denoted by Lvl(s).

3   Attribute Profiles and Pattern Spectra

An attribute profile of a topographic map is obtained by the application of a sequence of *increasing* morphological filters [10,17]: the sequence of results then provides a multi-scale characterization of the map highlighting the features selected by the filters. While attribute profiles can be used directly as features for point-wise segmentation [4], global features can be defined by considering the evolution of some measure, such as the volume or the area, on the filtered maps: this is then called a pattern spectrum of the map [10,17]. In the context of this article, we define attribute profiles and pattern spectra in the particular case of self-dual attribute connected filters; more general definitions can be found in the previously cited articles.

Let f be a topographic map. An attribute on $\mathcal{S}(f)$ is a function A from $\mathcal{S}(f)$ to $\mathbb{R}^+$. In general A is not an increasing function and some care must be taken in order to define the result of a filtering of f by A: this issue has been studied in the context of the Max-Tree in [16] and, for the construction of attribute profiles, the subtractive rule is usually chosen [17]. The authors of [2] proposed an algorithm to extend the subtractive filtering rule to the case of the Tree of Shapes, for which we give an analytic formulation in the followings paragraph.

In the original definition of the subtractive rule in [17], the level of a point in the filtered map is obtained by counting the number of non deleted components under it in the original map. To generalize this idea to a Tree-of-Shapes based filter, one must take into account that node levels are not necessarily increasing in such trees: counting nodes is thus not a useful method to retrieve meaningful levels. Instead, for any shape s, we consider the *contrast of s*, denoted by $\mathrm{cont}(s)$, and defined as the difference between the level of the shape s and the level of its parent, which can be positive or negative: *i.e.*, $\mathrm{cont}(s) = \mathrm{Lvl}(s) - \mathrm{Lvl}(\mathrm{Par}(s))$. Then, the new level of a filtered shape s is equal to its original level minus the contrast of every deleted ancestors of this shape. Formally, the subtractive self-dual attribute filter $\sigma_A(f, k)$ of f by A at threshold $k \in \mathbb{R}^+$ is defined by:

$$\forall x \in E, \sigma_A(f, k)(x) = \mathrm{Lvl}(\min\{s \in \mathcal{S}(s) \mid x \in s, \text{and } A(s) \geq k\})$$

$$- \sum_{\substack{q \in \mathcal{S}(s) \\ s \subset q, A(q) < k}} \mathrm{cont}(q), \quad (3)$$

i.e., the level of the smallest non deleted shape containing x minus the contrast of every deleted shapes containing x. Applications of the subtractive self-dual attribute filter are illustrated in Fig. 4.

The *self-dual attribute profiles of the topographic map f* for the subtractive self-dual attribute filter σ_A by the attribute A is then defined as the partial function application $\sigma_A(f, \cdot)$: *i.e.*, the function that associates the result of the subtractive self-dual attribute filter $\sigma_A(f, k)$ to a threshold value $k \in \mathbb{R}^+$.

Finally, given a measure ξ that associates a single real value to any topographic map and a self-dual attribute profiles $\sigma_A(f, \cdot)$, the pattern spectrum $\rho_A^\xi(f)$ of the topographic map f for A is defined as the function:

$$\rho_A^\xi(f) = \frac{d(\xi \circ \sigma_A(f, \cdot))}{dk}, \quad (4)$$

i.e., the derivative of the function that associates the measure of $\sigma_A(f, \cdot)$ for ξ to a threshold value $k \in \mathbb{R}$.

4   Choice of the Attribute and of the Measure

Two parameters need to be determined to define a pattern spectrum of a DEJ: the attribute guiding the filtering process and the measure characterizing the result of the filters.

Concerning the choice of the filtering criterion, visual analysis made by dermatologist [7,8] have shown that the regularity of the dermal papillae, *i.e.*, the

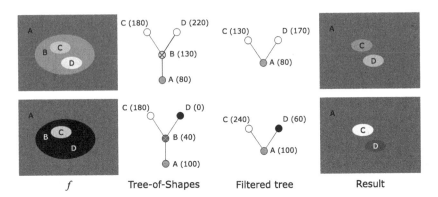

Fig. 4. Self-dual attribute filter with the subtractive rule. Each row shows the filtering of a map f composed of 4 regions (first column). The second column shows the Tree-of-Shapes of f with the node levels in parenthesis. The third column shows the filtered tree, and finally, the last column shows the filtered map. In each case, the shape corresponding to the region B is removed by the filter. In the first case, the Tree-of-Shapes is equal to the Max-Tree of f and the result is identical to the one obtained with the classical definition of the subtractive rule [17].

peaks and the valleys of the DEJ, decreases with age. Therefore, we propose to use the compactness attribute during the filtering process. Formally, the compactness comp(s) of a shape $s \subseteq E$ is defined as:

$$\text{comp}(s) = \frac{4\pi \, \text{area}(s)}{\text{perimeter}(s)^2}, \tag{5}$$

where area(s) is the number of points in s and perimeter(s) is the length of the perimeter of s. The compactness is a scale and rotation invariant measure that is maximal for a circle with a value of 1 and that tends toward 0 as the shape regularity decreases. An illustration of the DEJ surface evolution undergoing a filtering process with a compactness criterion is presented in Fig. 5.

Then, we need to determine the measure to be extracted at each filtering step. As the DEJ is a 2D surface, separating the epidermis from the dermis, we aim to compute an attribute in connection with the DEJ anatomy. Among all possible criteria, the surface area is a good candidate to obtain global information on the complexity of the surface as it is directly linked to the number, depth and shape of the peaks and the valleys in the DEJ.

We define the *surface area of a shape* s, denoted by sarea(s), recursively as:

$$\text{sarea}(s) = \text{area}(s) + \text{perimeter}(s) \times \text{cont}(s) + \sum_{c \in \text{Ch}(s)} (\text{sarea}(c) - \text{area}(c)), \tag{6}$$

where the second part of the equation adds up the surface area of the children Ch(s) of s minus their area, which is already counted in the area of s. An example of the surface area attribute is presented in Fig. 6. The surface area attribute

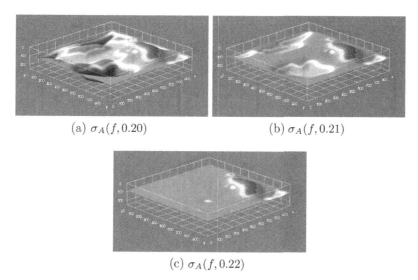

(a) $\sigma_A(f, 0.20)$ (b) $\sigma_A(f, 0.21)$

(c) $\sigma_A(f, 0.22)$

Fig. 5. Filtering of a DEJ surface f with a compactness criterion. In each case, the threshold value is indicated in the figure caption.

differs from the area attribute in the way that it takes into account the level of a node which, in our case, encodes the location in depth of the corresponding region in the 3D segmentation. The surface area attribute can be computed efficiently in linear time using Algorithm 1. The surface area attribute is calculated from the leaves up to the root of the tree. First, the product of the perimeter and contrast is calculated from the leaves to the root of the tree. Second, the area of each node (with its holes filled) is added to their corresponding surface area attribute. One can note that all the parameters of the algorithm (*e.g.*, the perimeter length of the shapes) can also be computed efficiently [18].

Algorithm 1. Surface area attribute.

   Input   : A Tree-of-Shapes *tree*, and the three functions area, perimeter and
                     cont on the nodes of the tree.
   Output: The surface area attribute.
1 **for** *all nodes i of tree* **do** sarea$(i) := 0$;
2 **for** *all nodes i of the tree from the leaves to the root (excluded)* **do**
3 | sarea$(i)+ = $ cont$(i) \times$ perimeter(i) ;
4 | sarea$(\mathrm{Par}(i))+ = $ sarea(i) ;
5 **end**
6 **for** *all nodes i of tree* **do** sarea$(i)+ = $ area(i) ;
7 **return** S;

One can note that the surface area of the root (*i.e.*, the surface area of the whole topographic map) is equal to the discrete variation of the map plus the area

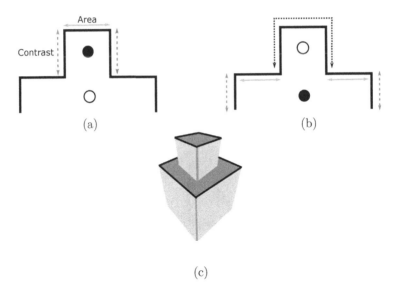

Fig. 6. Computation of the surface area. The surface area is the sum of the area (red arrows) and the contrast × perimeter (blue arrows). In (a) the attribute is calculated on a leaf of the tree. In (b) it is calculated on its parent. A 3D representation of the surface area attribute is shown in (c). The red areas correspond to the area of the corresponding nodes. The contrast (blue lines) × perimeter (black lines) is calculated to obtain the grey areas which are added to red areas to finally obtain the surface area attribute. (Color figure online)

of its domain. Moreover, as the pattern spectrum $\rho_{\text{comp}}^{\text{sarea}}$ measures the evolution of the surface area, the area terms cancel each other out, and the pattern spectrum then measures the evolution of the total variation of the surface.

The whole pattern spectrum of a topographic map f can be computed exactly and efficiently: for each shape by increasing value of compactness λ, subtract the product of the perimeter by the contrast of the shape to the surface area of the root node in order to obtain $\xi \circ \sigma_A(f, \lambda)$.

5   Experiments

In the following experiments, we assess the variability of the proposed DEJ characterisation between two groups of patients with different ages. Fifteen healthy volunteers with fair skin were enrolled in this study. Volunteers were assigned to two groups: a 7-person group aged from 18 to 25 and another 8-person group aged from 55 to 65. Our investigations were carried out on the cheek to assess chronological ageing, the dorsal forearm and the volar arm to assess photo-ageing. No cosmetic products or skin treatment were allowed on the day of the acquisitions. Appropriate consent was obtained from all subjects before imaging.

Distributions are compared using a two-sample Kolmogorov-Smirnov test and the statistics significance is defined as follow:

1. *: $0.01 <$ P-values ≤ 0.05;
2. **: $0.001 <$ P-values ≤ 0.01;
3. ***: P-values ≤ 0.001.

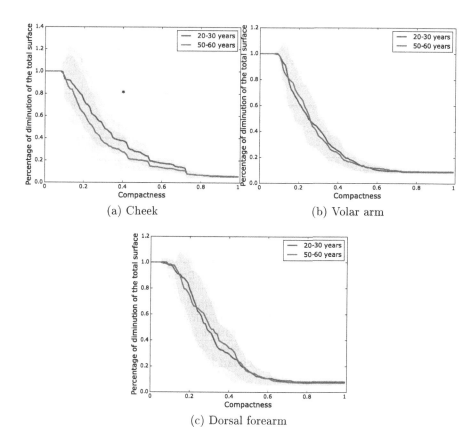

(a) Cheek

(b) Volar arm

(c) Dorsal forearm

Fig. 7. Mean and standard deviation of the pattern spectra $\rho_{\mathrm{comp}}^{\mathrm{sarea}}$ of the two age groups on the three acquisition locations.

In order to reduce the inter-subject variability, we study the pattern spectra expressed as a percentage of decrease of the surface area attribute. The average pattern spectra $\rho_{\mathrm{comp}}^{\mathrm{sarea}}$ computed for each group at each location are presented in Fig. 7. The lines represent the mean values and the shadowed areas the standard deviations for each population (20–30 years old and 55–65 years old). The pattern spectra for the epidermal surface on the cheek are statistically different. We do not find statistical differences on the other areas.

Finally, to obtain an aggregated descriptor of the DEJ shape, we study the area under the curve (AUC), of the pattern spectra. We present box-and-whisker plots of the AUC measure of the pattern spectra for the three acquisition locations (cheek, volar arm and dorsal forearm) in Fig. 8. Parametric data are analyzed using Student's unpaired t-test and non-parametric data are subjected to a Mann-Whitney test. One can observe that the AUC of the pattern spectra is higher in the younger group than in the older group on the cheek. When comparing the volar arm (photo-protected) and the dorsal forearm (photo-exposed), the photo-ageing effect is quantified among both populations with a significant decrease of the AUC. Therefore, the AUC of the pattern spectra of the DEJ with a compactness attribute and a surface area measure is able to assess the chronological ageing on the cheek and the photo-ageing for our two populations.

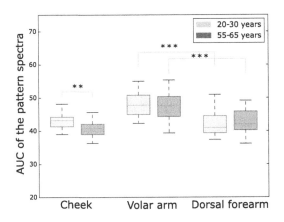

Fig. 8. Area under the curve of the pattern spectra. For each location and age group, we see: (1) the median AUC (central bar), (2) the first and third quartile (extremities of the box), and (3) the lowest datum still within 1.5 inter quartile range (difference between the third and first quartile) of the lower quartile, and the highest datum still within 1.5 inter quartile range of the upper quartile range (bottom and top extremities).

These results support our previous findings using local topological descriptors of the Tree-of-Shape regarding the chronological ageing on the DEJ on the cheek and photo-ageing among the young population [13]. However, the use of the pattern spectra brings more significant characteristics regarding the photo-aging among the older population compared to the local descriptors.

The proposed method also enriches the skin ageing descriptor focused on the automatic characterization of the epidermal cells in RCM images [14]. In future works, we plan to develop a novel method to automatically characterize the third significant ageing feature observable in skin RCM images: the dermal fibers.

References

1. Carlinet, E., Géraud, T.: MToS: a tree of shapes for multivariate images. IEEE Trans. Image Process. **24**(12), 5330–5342 (2015)
2. Cavallaro, G., Mura, M.D., Benediktsson, J.A., Plaza, A.: Remote sensing image classification using attribute filters defined over the tree of shapes. IEEE Trans. Geosci. Remote Sens. **54**(7), 3899–3911 (2016)

3. Dalla Mura, M., Benediktsson, J.A., Bruzzone, L.: Self-dual attribute profiles for the analysis of remote sensing images. In: Soille, P., Pesaresi, M., Ouzounis, G.K. (eds.) ISMM 2011. LNCS, vol. 6671, pp. 320–330. Springer, Heidelberg (2011). https://doi.org/10.1007/978-3-642-21569-8_28

4. Dalla Mura, M., Benediktsson, J., Waske, B., Bruzzone, L.: Morphological attribute profiles for the analysis of very high resolution images. IEEE Trans. Geosci. Remote Sens. **48**(10), 3747–3762 (2010)

5. Géraud, T., Carlinet, E., Crozet, S., Najman, L.: A quasi-linear algorithm to compute the tree of shapes of nD images. In: Hendriks, C.L.L., Borgefors, G., Strand, R. (eds.) ISMM 2013. LNCS, vol. 7883, pp. 98–110. Springer, Heidelberg (2013). https://doi.org/10.1007/978-3-642-38294-9_9

6. Lavker, R.M., Zheng, P., Dong, G.: Aged skin: a study by light, transmission electron, and scanning electron microscopy. J. Invest. Dermatol. **88**, 44–51 (1987)

7. Longo, C., Casari, A., Beretti, F., Cesinaro, A.M., Pellacani, G.: Skin aging: in vivo microscopic assessment of epidermal and dermal changes by means of confocal microscopy. J. Am. Acad. Dermatol. **68**(3), e73–e82 (2013)

8. Longo, C., Casari, A., Pace, B., Simonazzi, S., Mazzaglia, G., Pellacani, G.: Proposal for an in vivo histopathologic scoring system for skin aging by means of confocal microscopy. Skin Res. Technol. **19**(1), e167–e173 (2013)

9. Luo, B., Zhang, L.: Robust autodual morphological profiles for the classification of high-resolution satellite images. IEEE Trans. Geosci. Remote Sens. **52**(2), 1451–1462 (2014)

10. Maragos, P.: Pattern spectrum and multiscale shape representation. IEEE Trans. Pattern Anal. Mach. Intell. **11**(7), 701–716 (1989)

11. Monasse, P., Guichard, F.: Fast computation of a contrast-invariant image representation. IEEE Trans. Image Process. **9**(5), 860–872 (2000)

12. Rittié, L., Fisher, G.J.: Natural and sun-induced aging of human skin. Cold Spring Harb. Perspect. Med. **5**(1), a015370 (2015)

13. Robic, J.: Automated quantification of the skin aging process using in-vivo confocal microscopy. Theses, Université Paris Est, June 2018. https://hal.archives-ouvertes.fr/tel-01884978

14. Robic, J., Nkengne, A., Perret, B., Couprie, M., Talbot, H.: Automated quantification of the epidermal aging process using in-vivo confocal microscopy. In: IEEE ISBI, pp. 1221–1224 (2016)

15. Robic, J., Perret, B., Nkengne, A., Couprie, M., Talbot, H.: Classification of the dermal-epidermal junction using in-vivo confocal microscopy. In: IEEE ISBI, pp. 252–255 (2017)

16. Salembier, P., Oliveras, A., Garrido, L.: Antiextensive connected operators for image and sequence processing. IEEE Trans. Image Process. **7**(4), 555–570 (1998)

17. Urbach, E., Roerdink, J., Wilkinson, M.: Connected shape-size pattern spectra for rotation and scale-invariant classification of gray-scale images. IEEE Trans. Pattern Anal. Mach. Intell. **29**(2), 272–285 (2007)

18. Xu, Y., Carlinet, E., Géraud, T., Najman, L.: Efficient computation of attributes and saliency maps on tree-based image representations. In: Benediktsson, J.A., Chanussot, J., Najman, L., Talbot, H. (eds.) ISMM 2015. LNCS, vol. 9082, pp. 693–704. Springer, Cham (2015). https://doi.org/10.1007/978-3-319-18720-4_58

Spherical Fluorescent Particle Segmentation and Tracking in 3D Confocal Microscopy

Élodie Puybareau[1,2(✉)], Edwin Carlinet[1], Alessandro Benfenati[2], and Hugues Talbot[2,3]

[1] EPITA Research and Development,
14-16 rue Voltaire, 94276 Le Kremlin-Bicetre, France
elodie@lrde.epita.fr
[2] Université Paris-Est, ESIEE Paris, 2 boulevard Blaise-Pascal,
93162 Noisy-le-Grand, France
[3] CentraleSupelec, Centre de Vision Numérique, Inria équipe OPIS,
3 rue Joliot-Curie, 91190 Gif-sur-Yvette, France

Abstract. Spherical fluorescent particle are micrometer-scale spherical beads used in various areas of physics, chemistry or biology as markers associated with local physical media. They are useful for example in fluid dynamics to characterize flows, diffusion coefficients, viscosity or temperature; they are used in cells dynamics to estimate mechanical strain and stress at the micrometer scale. In order to estimate these physical measurements, tracking these particles is necessary. Numerous approaches and existing packages, both open-source and proprietary are available to achieve tracking with a high degree of precision in 2D. However, little such software is available to achieve tracking in 3D. One major difficulty is that 3D confocal microscopy acquisition is not typically fast enough to assume that the beads are stationary during the whole 3D scan. As a result, beads may move between planar scans. Classical approaches to 3D segmentation may yield objects are not spherical. In this article, we propose a 3D bead segmentation that deals with this situation.

Keywords: Segmentation · Shaping

1 Introduction

The use of fluorescent beads as physical markers is common in many research problems and applications related to fluid mechanics [9], rheology [4], and cell biology [12].

Spherical bead tracking in confocal microscopy is a classical and well-studied problem in image analysis. Recently, an ISBI challenge was organized to compare state-of-the-art bead segmentation and tracking methods [3]. However, this very challenge has exhibited a number of shortcomings.

B. Burgeth et al. (Eds.): ISMM 2019, LNCS 11564, pp. 520–531, 2019.
https://doi.org/10.1007/978-3-030-20867-7_40

First among them is the common assumption that beads are very small, similar in diameter to the optical limit of resolution of the microscope, i.e. 0.2–0.4 μm. In reality, beads can be much larger than this. 1–2 μm diameters are commonly used. This is due to the fact that larger beads are less expensive, more visible and better adapted to some problems, for example when the surface of the bead needs to be activated. In particular, tracking of very small beads is difficult when there are many beads. Due to their larger dimension, a more precise location of their center is also possible, whereas smaller beads are typically only made of a few pixels and so a precise location of their center is imprecise.

A second assumption is that since beads are expected to be very small, they are located in a single plane and are so acquired in a single pass. Unfortunately, larger beads span several planes and need to be acquired over several scans, which takes time.

Thirdly, a common assumption is that scanning time is negligible compared with the studied motion. This is often true in mechanical cell studies. However, in fluid analysis, beads are subjected to Brownian motion [2] and can move significantly even in the absence of flow. This is a problem even during a single 3D acquisition since a bead can visible move as the confocal scan is taking place.

In this article, we study and propose solutions for comparatively large bead segmentation and tracking. Due to Brownian motion, tracking must take place both at the intra and inter-volume level. We present an application where we measure the diffusion coefficient of the Brownian motion. This is useful for measuring viscosity or temperature in fluid media [10].

2   Particle Tracking

Our objective is to track relatively large beads (0.8–2 μm in diameter) in 3D confocal microscopy. We do not assume that the diameter is known precisely. Typically most bead production processes allow a standard deviation of ±10%. Traditional confocal microscopy proceeds by scanning a 3D volume slice-by-slice. Scanning a small area of a single slice, such as the few lines that may include a bead, is fast enough that the intersection of a scanning plane with any bead can be assumed to be a disk. However, scanning a whole slice can be slow, and so the particle may move slightly in between slice acquisition.

We do not assume that the particle forms a perfect sphere. We do assume that deviations from the sphere are relatively small. We also assume that the slice separation is small enough so that each bead is scanned multiple times. This is typically an adjustable parameter of the microscope.

Figure 1 summarize the pipeline we developed. We named "2D" the slices as shown in Fig. 2. These 5 slices composed the 3D volume for a fixed time.

2.1   2D Segmentation

The first step of our pipeline is a 2D segmentation of the beads and uses both standard and modern mathematical morphology techniques [8]. The objective is

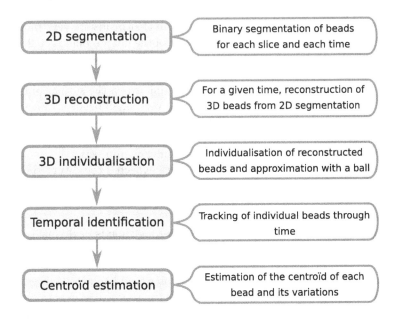

Fig. 1. Flowchart of our pipeline

to obtain a binary image of the beads suitable for masking and labelling. The steps are the following:

1. Supression of small local minimas: this is performed using a closing by reconstruction with an initial dilation by a disk of radius 4. This yields image g.
2. Calculation of the min-tree of g.
3. Filtering of nodes according to multiple criteria: size must be smaller than 500, the grey-level intensity must be lower than 100, and the contrast of the component greater than 80. We keep the maximal components meeting all these criteria.
4. Regularisation of the selected blobs: holes are filled, and the components are dilated with a 4-connected structuring element and opened with a disk of radius 5.

An example of results of this procedure is shown in Fig. 3.

2.2   3D Regularization

For each acquisition time, we stack the corresponding 2D segmented slices. As the segmentation is performed in 2D, and the beads may move significantly between slices, some irregularities may appear at the bead boundaries, especially near the bead poles. The first step is hence a regularization in the z axis to reconnect these incomplete beads by applying a 2D closing with a disk of radius 1 in all the xz planes.

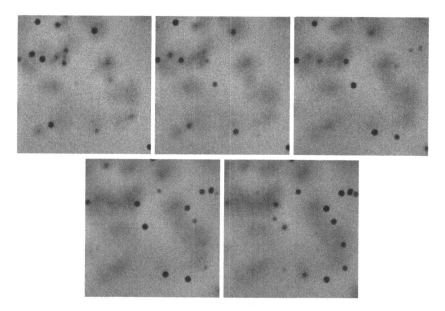

Fig. 2. Consecutive depth (z) slices corresponding to a single time.

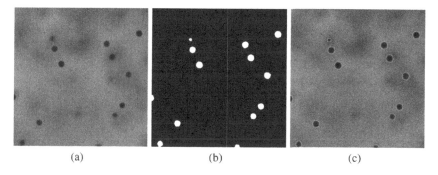

(a)   (b)   (c)

Fig. 3. Illustration of the segmentation procedure: (a) initial image, (b) result of the segmentation described in the text, (c) overlay of the segmentation on the initial image.

2.3   3D Individualization

As some beads can touch, we separated them using a classical watershed procedure. After noise removal with an opening on the binary image, we extract beads markers using the distance transform with a threshold and the background marker. A watershed is then applied. The result is cleaned by removing small objects using a volume criterion. All the steps are illustrated in Fig. 4. Morphological parameters were empirically optimized on a small subset of the data.

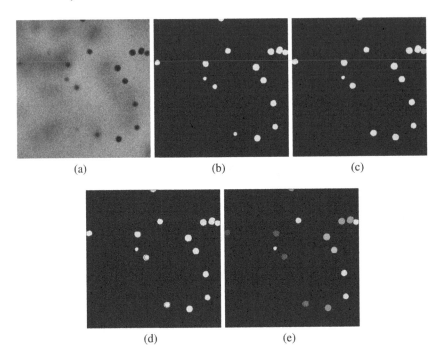

(a)   (b)   (c)

(d)   (e)

Fig. 4. Results of the successive steps: (a) initial image, (b) initial segmentation, (c) 3d regularization, (d) bead separation, (e) individualization and labeling.

2.4 Temporal Identification

To label the beads at time t, we propagate the labels of the beads at the previous time $t - 1$ by finding the intersection between the beads at t and $t - 1$. These intersections are the markers used for the reconstruction by dilation. We hence obtain for time t labels for the beads also present at time $t - 1$. The remaining unlabeled beads are given new labels. Overall, this procedure is simple but effective, as we will see in the result section.

3 Position Estimation

In the previous section, 2D bead segmentation is performed and label clustered in 3D. We consequently assume that a uniquely labeled object in 3D constitutes a single bead.

Estimating the 3D position of each bead from their 2D slice is a challenge. We propose to estimate 2D positions first, and derive the 3D position from an estimation of the radius of the bead.

3.1 Centroid Estimation in 2D

Various methods can be used to estimate the centroid of a disk. A least-square best-fit disk can for example be used [5,6]. However these methods are more appropriate when only a subset of the disk contour is known. Since the beads are small, we can assume that most of them do not lie near the border of the field of view, and so are totally visible. In this case, a moments-based method seems more appropriate [7]. Let M_{pq} be a moment of order $p + q$ defined as follows:

$$M_{pq} = \sum_x \sum_y x^p y^q I(x, y), \tag{1}$$

where $I(x, y)$ is the intensity of the image at position x, y. If we restrict $I(x, y)$ to be equal to 1 for the segmented bead with label l and 0 elsewhere, then

$$\bar{x}_l = \frac{M_{10}}{M_{00}} \; ; \; \bar{y}_l = \frac{M_{01}}{M_{00}}, \tag{2}$$

yield the coordinates of the centroid of bead l. Moments of order 0 and 1 are not affected by isotropic blur or partial volume effects and are relatively resistant to noise, and so this estimation is rather robust, as we will see later in simulations. Moreover, we can also estimate the radius via the second order centered moments

$$\mu_{20} = M_{20} - \bar{x} M_{10} \; ; \; \mu_{02} = M_{02} - \bar{y} M_{01}, \tag{3}$$

and then for a disk $\mu_{20} = \mu_{02} = \frac{\pi}{8} r^4$, or more simply from the disk area equation $r = \sqrt{M_{00}/\pi}$.

3.2 Centroid Estimation in 3D with Known Bead Radius

If we know the radius R of a bead to a good precision, we can estimate the depth p_R by noting that

$$p_R = \pm\sqrt{R^2 - r^2}, \tag{4}$$

with r the estimated radius of the disk, as seen on Fig. 5.

Unfortunately there is a sign uncertainty with this estimation, so a correct estimation must include information from the other slices associated with the same bead. Figure 6 illustrates what happens, in a 2D $x + z$ simplified setting.

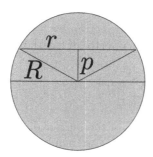

Fig. 5. Bead depth estimation: R is the real radius, r the observed radius, and p the estimated depth.

Each intersection between the bead and a scanning plane yields a disk (a line in the drawing), from which two depths can be estimated, only one of which is correct. The beads is moving between acquisitions, so there are two depths estimations per slice. We assume the correct ones are those that minimize the overall distance between them.

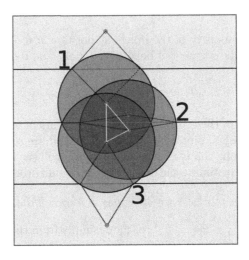

Fig. 6. When estimating the depth of the bead centroid over multiple slice acquisition (labeled 1, 2 and 3 in the drawing), as the bead is moving during the acquisition, the centroid depth varies and can be one of two possibilities. We select the depth estimations that minimizes the overall Euclidean distance between them (i.e. corresponding to the perimeter of the yellow triangle). (Color figure online)

More formally, for every slice s, for every bead with label l we estimate two centroid positions

$$c_s^{l,0} = (\bar{x}_l, \bar{y}_l, d(s) - p_R(s)) \tag{5}$$

$$c_s^{l,1} = (\bar{x}_l, \bar{y}_l, d(s) + p_R(s)), \tag{6}$$

with $d(s)$ the depth of slice s and p given by (4). For every slice s we have two choices. If $v \in \{0,1\}^{n_s}$ is a binary vector of length n_s, the total square distance is

$$d_{v,R}^2 = \sum_{s=0}^{n_s-1} \|c_s^{l,v[s]} - c_s^{l,v[(s+1) \bmod n_s]}\|^2, \tag{7}$$

where $a \bmod b$ is the remainder of the integer division of a by b, and assuming without loss of generality that slices are numbered 0 to $n_s - 1$.

We propose to select the vector v corresponding to the combination that minimizes d_v. Consequently, if there are n_s slices for a bead, we need to test 2^{n_s} combinations. In our experiments, we have 3–6 slices for each bead, meaning up to 64 tests per bead, which is manageable. If we had more slices, this could be a problem. In this case, a simplification would be to select the middle slice, and for both centroid estimations related to this slice, compute the distance to the two centroid position estimations to every other slices. In other words we now minimize $d_{v,R}''^2$ defined as follows:

$$d_{v,R}''^2 = \sum_{s=0}^{n_s-1} \|c_s^{l,v[s]} - c_s^{l,v[\lfloor n_s/2\rfloor]}\|^2, \tag{8}$$

where $\lfloor a \rfloor$ is the integer part of a. The difference here is that each term of this sum can be minimized independently, and so there are only $4n_s$ combinations to test, which is beneficial as soon as $n_s \geq 4$.

3.3 Centroid Estimation in 3D with Imprecise Bead Radius

If the radius is not known precisely and varies from bead to bead, which is often the case, the problem is more difficult. We assume that the bead radius for each bead is a deviate from a known Gaussian distribution of mean $\bar{R}$ and standard deviation σ_R. Typically, $\sigma_r \approx 0.1\bar{R}$.

$$R_l \sim G(\bar{R}, \sigma_R) \tag{9}$$

Given the known observations for each bead, we also know that r_l is greater than the largest measured radius $r_{\max} = \max\{r_s, s \in [\![0..n_s - 1]\!]\}$. We solve the following mixed problem:

$$R_l^*, v^* = \arg\min_{R,v} d_{v,R}^2 + \frac{1}{2\lambda}\|R - \bar{R}\|^2 \tag{10}$$

$$\text{s.t. } R \geq r_{\max}, \tag{11}$$

where λ is a Lagrangian parameter, and the second term of (10) is the log-likelihood of a Gaussian distribution for R, thus we expect λ to be proportional to σ_R^2. The tighter the distribution, the more important this term becomes.

For a given v binary vector, (4) is convex due to the constraint (11), the constraint itself is linear since $r_{\max}$ is an observation, but both (7) and (8) are smooth differences of convex functions, so this problem cannot be solved by standard constrained convex optimisation methods [1]. However, it can be tackled by DC methods [11]. In addition, the problem remains combinatorial in v, as seen above, so we need to run several such DC problems, up to 2^{n_s} or $4n_s$ depending on whether we choose to optimise (7) or (8).

Since DC method solvers are complex to handle, we propose to solve (10) using gradient descent with the d' measure of (8), but taking as reference the slice with the largest measured radius $r_{\max}$ instead of the middle slice. We set $\max\{r_{\max}, \bar{R}\}$ as the initialization for R and $\lambda = \sigma_R$. In practice, the true R^* is very close to the largest observed radius.

4   Results on Simulations

To test our method, we simulate N discrete sphere beads at high resolution with a random radius sampled from a Gaussian distribution as described in the previous section. The beads are subjected to isotropic Brownian motion and depth-ordered slice acquisition. We down-sample the result to ensure sub-pixel accurate positions. We estimate the radius of the intersection disk between each slice and the bead via the area method, and the $\bar{x}$, $\bar{y}$ position via moments. The z position and the bead radius is estimated as described above using the d' method.

Fig. 7. A single simulated bead acquired over several layers subjected to Brownian motion.

For this simulation, we simulate high-resolution beads 80 pixels in radius, in a 350^3 cube, acquired through 7 layers and subjected to Brownian motion with an isotropic diffusion coefficient of 0.1 relative to the size of the cube, or 35 pixels at high resolution. Low resolution images were down-sampled by a factor of 10 with nearest-neighbor interpolation. The diffusion coefficient of the Brownian motion can be interpreted as the standard deviation of the the components of the position of the center of the bead from one slice to the next. With Brownian motion, the average deviation is zero.

Figure 7 shows the intersection of a single bead through successive slices. The Brownian motion influence is significant and clearly seen. We simulated 100 beads, which generated 540 slices. Via our optimisation procedure, each slice provides a 3D center estimate from which various measurements can be derived. In particular, we compare the estimation of the center position to the simulated ground truth. The results are shown in Table 1.

Table 1. Positional RMS relative error

	x	y	z
Relative error	$1.6 \cdot 10^{-3}$	$1.6 \cdot 10^{-3}$	$8.2 \cdot 10^{-3}$

On this table, we see that the average Euclidean position error in x and y is very low. Indeed, a pixel error in 350 (the high-resolution size of the cube) would translate into $2.8 \cdot 10^{-3}$. However, the z estimation error is $5\times$ greater, which is still sub-pixel accurate at the low sub-sampled resolution, and so still quite acceptable. We note that these estimates are the same as our estimate of the average Brownian deviation, which should be zero.

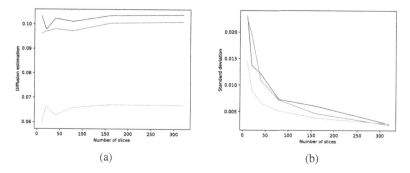

Fig. 8. Convergence of the diffusion estimator to a stable value (a) and its standard deviation as a function of the number of slices (b).

On Fig. 8, we show the estimation of the relative Brownian diffusion coefficient as a function of the sample size. We expect a relative coefficient of 10^{-1}, i.e. the one we simulated. The x and y are very close to this estimate, but the z diffusion estimation appears biased with a value about 70% as much as it should be. All three estimators converge quickly and can be deemed reliable from about 50 slices, i.e. 10 beads or so, with a standard deviation of the estimate of diffusion coefficient close to 10^{-2}, ie. 10% error. With 40 beads, or 200 slices, the standard deviation of the diffusion coefficient drops down to $5 \cdot 10^{-3}$ or 5% relative error.

Currently, this means that our optimisation method is indeed good enough for positional estimation in 3D, but that for more subtle measurements, like diffusion coefficient estimations, 2D measurements are sufficient. We note that we are able to measure this diffusion coefficient from single 3D acquisition only.

5   Results on Real Data

We applied our method on a video of $3\,\mu\text{m}$ beads, with $\Delta t = 0.5\,\text{s}$. Resolution is $0.125\,\mu\text{m}$ per pixel in x and y, and only $0.5\,\mu\text{m}$ per slide in z. Comparison with manual labelling results show that we manage to correctly detect and label all beads, as well as track each bead individually with zero label error between 3D volumes. However, we overestimate beads presence in the z-axis due to the point spread function (PSF) of the microscope. In practice this means that beads faintly appear in slices above and below their true position. This overestimation is depicted in Fig. 9(c). On slice 0, we tend to find beads that are not supposed to be segmented (false positives), i.e. the central bead of the illustration. This "oversegmentation" is nonetheless acceptable as it remains coherent in 3D and does not imply bias in 2D. In 3D, this can cause the bead radius to be overestimated, and will need to be corrected in future work.

On this sample, using our 2D method, we estimate an average diffusion coefficient of $4.3 \cdot 10^{-2}$ relative to the beads average diameter. Currently we do not have the ground truth for this measurement.

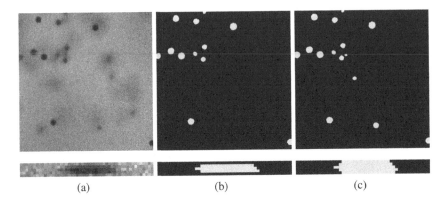

Fig. 9. Comparison of segmentation results: (a) original frame, (b) results of the first step of the 2D segmentation procedure before regularization, and (c) after 3D regularization. Top images are the entire xy slices, and the bottom is a xz view along the z-axis

6 Conclusion

In this article we tacked the problem of tracking fluorescent beads in confocal microscopy, when we do not assume that beads are small enough to be assimilated to point sources, but instead span several slices; when they move between slices due to Brownian motion; and when their radius is not precisely known. Starting from 2D-based segmentation and 3D clustering using shape-based mathematical morphology, we have shown that estimating bead radius and true 3D positions from a small number of 2D slices is a difficult, but tractable problem. We have also shown that tracking these beads between 3D acquisitions is possible with a high degree of accuracy. One benefit of our approach is that we can obtain good estimates of the diffusion coefficient of Brownian motion from single 3D acquisition, which is useful for various physical measurements.

In future work, we plan to improve the quality and speed of our optimisation method, to make better use of prior knowledge in 3D particularly for initial segmentation, and we will release our software as a free, open-source package.

References

1. Boyd, S., Vandenberghe, L.: Convex Optimization. Cambridge University Press, Cambridge (2004)
2. Brown, R.: A brief account of microscopical observations made in the months of June, July and August, 1827, on the particles contained in the pollen of plants; and on the general existence of active molecules in organic and inorganic bodies. Phil. Mag. **4**, 161–173 (1828)
3. Chenouard, N., et al.: Objective comparison of particle tracking methods. Nat. Methods **11**(3), 281 (2014)
4. de Gennes, P.G.: Scaling Concepts in Polymer Physics. Cornell University Press, Ithaca and London (1979)

5. Fitzgibbon, A., Pilu, M., Fisher, R.B.: Direct least square fitting of ellipses. IEEE Trans. Pattern Anal. Mach. Intell. **21**(5), 476–480 (1999)
6. Gander, W., Golub, G.H., Strebel, R.: Least-squares fitting of circles and ellipses. BIT Numer. Math. **34**(4), 558–578 (1994)
7. Hu, M.K.: Visual pattern recognition by moment invariants. IEEE Trans. Inf. Theory **8**(2), 179–187 (1962)
8. Najman, L., Talbot, H. (eds.): Mathematical Morphology: From Theory to Applications. ISTE-Wiley, London, UK, September 2010. ISBN 978-1848212152
9. Park, J., Choi, C., Kihm, K.: Temperature measurement for a nanoparticle suspension by detecting the Brownian motion using optical serial sectioning microscopy (OSSM). Meas. Sci. Technol. **16**(7), 1418 (2005)
10. Puybareau, É., Talbot, H., Gaber, N., Bourouina, T.: Morphological analysis of Brownian motion for physical measurements. In: Angulo, J., Velasco-Forero, S., Meyer, F. (eds.) ISMM 2017. LNCS, vol. 10225, pp. 486–497. Springer, Cham (2017). https://doi.org/10.1007/978-3-319-57240-6_40
11. Tao, P.D., et al.: The DC (difference of convex functions) programming and DCA revisited with DC models of real world nonconvex optimization problems. Ann. Oper. Res. **133**(1–4), 23–46 (2005)
12. Wang, N., Butler, J.P., Ingber, D.E.: Mechanotransduction across the cell surface and through the cytoskeleton. Science **260**(5111), 1124–1127 (1993)

Combining Mathematical Morphology and the Hilbert Transform for Fully Automatic Nuclei Detection in Fluorescence Microscopy

Carl J. Nelson[1] , Philip T. G. Jackson[2] , and Boguslaw Obara[2]([✉])

[1] University of Glasgow, Glasgow, UK
[2] Durham University, Durham, UK
boguslaw.obara@durham.ac.uk

Abstract. Accurate and reliable nuclei identification is an essential part of quantification in microscopy. A range of mathematical and machine learning approaches are used but all methods have limitations. Such limitations include sensitivity to user parameters or a need for pre-processing in classical approaches or the requirement for relatively large amounts of training data in deep learning approaches. Here we demonstrate a new approach for nuclei detection that combines mathematical morphology with the Hilbert transform to detect the centres, sizes and orientations of elliptical objects. We evaluate this approach on datasets from the Broad Bioimage Benchmark Collection and compare it to established algorithms and previously published results. We show this new approach to outperform established classical approaches and be comparable in performance to deep-learning approaches. We believe this approach to be a competitive algorithm for nuclei detection in microscopy.

Keywords: Nuclei detection · Hilbert transform ·
Mathematical morphology · Nuclei counting

1 Introduction

Quantitative biology relies heavily on the analysis of fluorescence microscopy images. Key to a large number of experiments is the ability to detect and count the number of cells or nuclei in an image or region of interest. Frequently, such as for high-throughput and large-scale imaging experiments, this is the first step in an automated analysis pipeline. As such, the scientific community requires novel, robust and automatic approaches to this essential step.

Manual cell counting and annotation, by visual inspection, is difficult, labour intensive, time consuming and subjective to the annotator involved. As such, a large number of groups have proposed a variety of approaches for automatic

© Springer Nature Switzerland AG 2019
B. Burgeth et al. (Eds.): ISMM 2019, LNCS 11564, pp. 532–543, 2019.
https://doi.org/10.1007/978-3-030-20867-7_41

nuclei detection in fluorescent microscopy images. In recent years proposed methods have included:

- variations upon thresholding-based segmentation, which are limited in performance due to intensity inhomogeneity and nuclei/cell clustering [5];
- H-minima transforms [9];
- voting-based techniques [19], which show good results but are sensitive to parameters;
- gradient vector flow tracking and thresholding [10,11], such PDE-based methods require strong stopping and reinitialisation criteria, set in advance, to achieve smooth curves for tracking;
- using Laplacian of Gaussian filters [18], which is low computational complexity but struggles with variation in size, shape and rotation of objects within an image;
- graph-cut optimisation approaches [13], which requires the finding of initial seed points for each nucleus;
- arc segment based methods such as [8], which are fast but rely on accurate edge point detection;
- convolutional neural networks, which generally require copious amounts of manually labelled training data, and struggle to separate overlapping objects [7].

In our experience, nuclei size (scale) is an important contributor to 1. whether or not an algorithm is successful, 2. how robust an algorithm is to image variation and 3. to the running time of an algorithm. We have also found that many excellent algorithms exist for detecting small blobs, on the scale of a few pixels diameter, that fail or are less reliable for medium or large blobs, *i.e.* over ten pixels diameter, when such objects tend to display significant eccentricity. At this and larger scales the number of algorithms available decreases, and stronger assumptions are made on the input data.

In this paper we apply a new ellipse detection algorithm based on mathematical morphology and the Hilbert transform to the task of counting medium blobs, *i.e.* nuclei, on the scale of 10 to 50 pixels in fluorescence microscopy images. This new algorithm requires no preprocessing and no training. Simplistically, the algorithm performs three steps: 1. Mathematical morphology is used to create an object sparse search space over different orientations and scales; 2. The Hilbert transform is used to search this space and range object edges and for determining the ellipse/nucleus geometry at all pixels of the image; 3. Prominent ellipses/nuclei are extracted and post-detection pruning is used to eradicate over detection. We call this method Hilbert Edge Detection and Ranging (HEDAR). All code is available on GitHub at https://github.com/ChasNelson1990/Ellipse-Detection-by-Hilbert-Edge-Detection-and-Ranging.

In the rest of this paper we introduce HEDAR (Sect. 2) and compare this method to several traditional and new algorithms for blob/nuclei counting in fluorescence microscopy (Sect. 3). We have used the Broad Bioimage Benchmark Collection [12] throughout this paper to ensure comparison with previously published results, which we have reported where appropriate.

2   Methods

Our new algorithm, Hilbert Edge Detection and Ranging, is a three step process. First, we erode the original 2D, greyscale image I (Fig. 1a) with a bank of line end-point structuring elements $SE_{d,\theta}$ of different integer pixel lengths $d \in [1, d_{\max}]$ and rotations $\theta \in [0, 180)$. We then stack the resulting eroded images into a four dimensional array M, such that $M(x, y, d, \theta) = (I \ominus SE_{d,\theta})(x, y)$. We note that in 2D pixel-space we can define the limit on the number of rotations as $N_\theta = 2d$, *i.e.* the perimeter of the bounding box of the half-circle covered by all (symmetric) structuring elements of length d. Thus, for a given x, y, θ where the point (x, y) lies inside a bright ellipse, $M(x, y, :, \theta)$ yields a morphological intensity signal that drops off at the distance s from (x, y) to the boundary of the ellipse along the direction θ.

Next, we calculate the 1D Hilbert transform over all morphological signals - that is, along the d dimension of M (Fig. 1b). The transform will be maximal at a step in the morphological signal from high to low (and minimal at a step from low to high). By identifying the maxima in each signal we can range the distance s from the current pixel to the nearest edge along that angle. Mathematically, we can define this as,

$$s(x, y, \theta) = \underset{d}{\operatorname{argmax}} \left(\text{Hilbert} \left(M(x, y, d, \theta) \right) \right). \tag{1}$$

The major axis a of any ellipse, *i.e.* nucleus, centred at (x, y) in the image will be,

$$a(x, y) = \max_{\theta}(s(x, y, \theta)). \tag{2}$$

We can concurrently identify the orientation ϕ of any nucleus centred at (x, y) in the image as,

$$\phi(x, y) = \underset{\theta}{\operatorname{argmax}}(s(x, y, \theta)). \tag{3}$$

We then extract the distance to any edge at $90°$ to the major axis a to determine the minor axis b of any nucleus centred at (x, y) in the image. Using the Hilbert transform makes this process robust to noise and blurring.

Like other mathematical morphological image processing approaches, such as neuriteness in vessel enhancement, so far the algorithm has assumed the presence of an ellipse at all points (*c.f.* assuming a vessel in neuriteness), the next step is to identify likely ellipses/nuclei. To accomplish this we element-wise multiply the minor and major axes at every pixel. This produces a heat-map P (Fig. 1c) with high prominence peaks at the centres of any possible nuclei.

The prominence map is masked with a minimum threshold size t for Hilbert steps in morphological signals. This acts similarly to local thresholding of the image: as each step must be at least as large as the threshold, we only consider pixels where the possible nucleus is brighter than its local background by at least t. This suppresses false positives caused by background noise. After smoothing the masked prominence map with a (3×3) pixels median filter we then identify the regional maxima. Each maximum represents a nucleus and

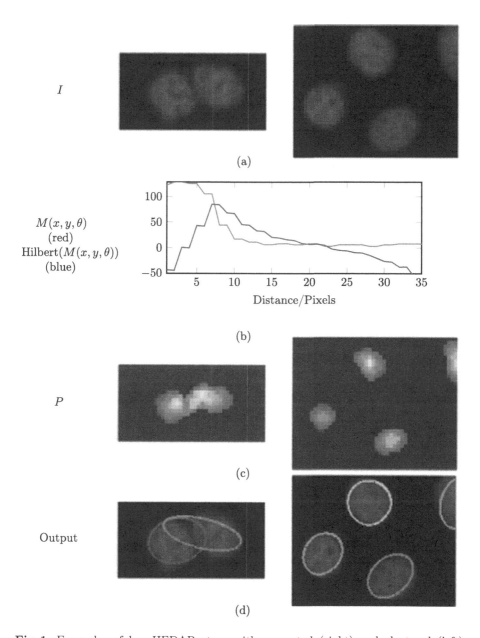

Fig. 1. Examples of key HEDAR steps with separated (right) and clustered (left) nuclei. (a) Regions of an example input image. (b) Example signal from M centred on a nucleus in orientation $\theta = 0$ (red) and the Hilbert transform of that signal (blue). (c) Equivalent regions of the prominence map used to detect ellipse centroids. Note how clustered nuclei show extended regions of maxima that effect the final output. (d) Example output ellipses for these regions. (Color figure online)

we determine the major axis, minor axis and orientation from previous steps (Fig. 1d).

In post-processing, we cycle through all detected ellipses, removing the more overlapped ellipse of any pair with intersection-over-union greater than 50%.

We report results on image sets BBBC001v1 [3], BBBC002v1 [3], BBBC004v1 [16] and BBC039v1 [2] of the Broad Bioimage Benchmark Collection [12].

3   Results

3.1   Combined Mathematical Morphology and Hilbert Transform for Accurate Nuclei Counting

Combined mathematical morphology and the Hilbert transform in our Hilbert Edge Detection and Ranging algorithm enables accurate nuclei counting in fluorescence microscopy, as show in Fig. 2. Most nuclei are accurately counted giving a low mean percentage error. HEDAR does fail to detect a small number of cells, specifically irregular cells and some clustered or overlapping cells.

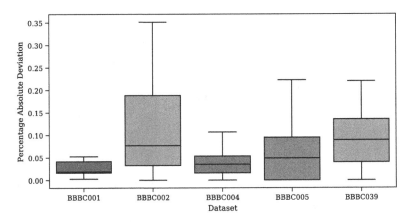

Fig. 2. Hilbert Edge Detection and Ranging enables accurate nuclei counting in fluorescent microscopy. Distributions of percentage absolute deviation from known number of nuclei for HEDAR across five fluorescence microscopy data sets with ground truth counts.

By comparing the counting results of our automated algorithm to the counts by two human observers (see Sect. 3.4 for details of datasets), we can show the accuracy of our HEDAR algorithm (Fig. 3d and e). The sample Pearson's correlation coefficient (r value) gives a measure of the correlation between the results of our automated HEDAR algorithm and human observers—an r value closer to 1 represents better correlation, *i.e.* agreement. For the BBBC001 datatset, the r value between the two human observers is $r = 0.92$. With r values of 0.98 (with human observer #1) and 0.98 (with human observer #2) we clearly show that our HEDAR algorithm has excellent performance for counting nuclei in fluorescent microscopy on dataset BBBC001. Likewise, for the BBBC002 dataset we achieve r values of 0.991 and 0.992 (with human observer #1 and #2, respectively); the r value between the two human observers is 0.992.

This can be further compared using the mean percentage absolute deviation from the human mean, as reported by other papers (Fig. 3b and c). We note that, as seen in Fig. 2, the distribution of percentage absolute deviation is often skewed; as such, when comparing to full results from comparison methods we report the median and not the mean, which can be affected by outliers.

For completeness, we compared the count accuracy of our HEDAR algorithm against all the Broad Bioimage Benchmark Collections datasets with 2D fluorescence microscopy images on fluorescently labelled nuclei (see Sect. 3.4 for details of datasets). These datasets also enable the exploration of how our HEDAR algorithm is able to cope with overlapping nuclei (Fig. 4a), increasing blur (Fig. 4b) and relative focus (Fig. 4b). Our Hilbert Edge Detection and Ranging algorithm consistently enables accurate counting of nuclei in fluorescence microscopy images under these challenging conditions. Figure 4 shows the median percentage absolute difference compared to ground truth (BBBC004 and BBBC005; see Sect. 3.4).

3.2   HEDAR is Able to Accurately Extract Ellipse Parameters from Detected Ellipses

To investigate the accuracy of ellipse parameters using our HEDAR algorithm we have emulated the synthetic experiments first shown in [4]. We ran our method and two ellipse detection algorithms on synthetic images of a single ellipse without noise. First, we compare to a well-cited, efficient, randomised elliptical Hough transform [17]. Second, we compare to a more recent but established non-iterative, geometric method for ellipse fitting: Ellifit [15]. Note that these algorithms are not optimised specifically for nuclei but have been shown to accurately extract nuclei in a variety of real world images; comparing to these methods in synthetic data sets a high standard for comparing our HEDAR algorithm.

Figure 5 shows how robust these methods are to size and eccentricity Fig. 5a as well as orientation (Fig. 5b). In these plots each 'pixel' intensity value corresponds to the Jaccard similarity index [6] between the input image and the result of a given algorithm. Hence, a yellow pixel (with a value closer to 1.0) indicates that the given method was able to correctly identify the ellipse, *i.e.* accurately extract ellipse parameters. A deep blue pixel (with a value of 0.0) indicates that no ellipse was detected or that more than one ellipse was detected, under which situations a Jaccard similarity of zero is assigned. Any pixel between these two extremes can be considered to have found an ellipse with more (yellow) or less (blue) accurate parameters. Note that the Jaccard similarity index was chosen as it satisfies the triangle inequality.

538 C. J. Nelson et al.

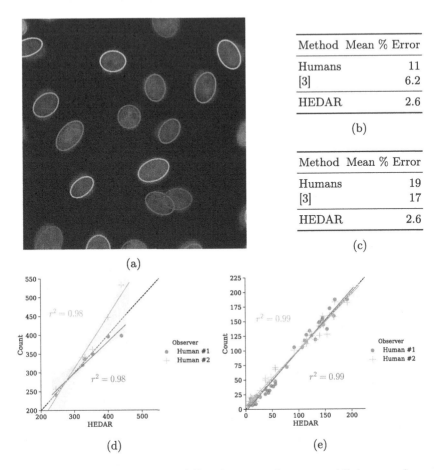

Method	Mean % Error
Humans	11
[3]	6.2
HEDAR	2.6

(b)

Method	Mean % Error
Humans	19
[3]	17
HEDAR	2.6

(c)

Fig. 3. Hilbert Edge Detection and Ranging strongly agrees with human observers in counting nuclei in fluorescence microscopy images. (a) Examples of successfully detected nuclei in the BBBC001 dataset of Hoechst 33342-labelled human HT28 colon cancer cells [3]. (b) The percentage absolute deviation is reported for our HEDAR algorithm and other published results on the BBBC001 dataset. The deviation between the two human observers is also reported. (c) The percentage absolute deviation is reported for our HEDAR algorithm and other published results on the BBBC002 dataset. The deviation between the two human observers is also reported. (d) Comparison of nuclei count from HEDAR to the counts by two human observers on the BBBC001 dataset. (e) Comparison of nuclei count from HEDAR to the counts by two human observers on the BBBC002 dataset [3].

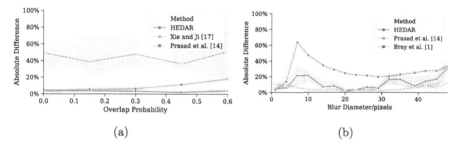

Fig. 4. Hilbert Edge Detection and Ranging is able to cope with overlapping nuclei, increasing blur and out-of-focus images. (a) Comparison of nuclei count from HEDAR to the known counts for dataset BBBC004 [16] across the different overlap probabilities used to create the synthetic images. (b) Comparison of nuclei count from HEDAR to the known counts for dataset BBBC005 [12] across the different blur diameters used to create the synthetic images.

In Fig. 5a each pixel represents the minimum Jaccard similarity for all ellipses of a given major axis length a and axis ratio (minor axis over major axis; b/a) over all rotations. Although HEDAR does get as consistently high a Jaccard similarity index as the Ellifit algorithm, HEDAR covers a much greater range of major axis and axis ratios, more like the range covered by the elliptical Hough transform. These results show that HEDAR is able to accurately extract ellipse parameters over a wide range of ellipse sizes and axes ratios (*c.f.* eccentricity).

In Fig. 5b each pixel represents the minimum Jaccard similarity for ellipses of a given orientation (θ) and axis ratio b/a for $a \leq 32$ pixels. HEDAR shows consistently higher Jaccard similarity indices over small axes ratios (*i.e.* high eccentricity) when compared to both the Ellifit and elliptical Hough algorithms.

3.3   HEDAR is Able to Accurately Count Ellipses

Not only is the overall accuracy of any method important for its application but so too is the method's ability to deal with many objects. Figure 6 shows how HEDAR is able to correctly detect the number of ellipses in an image. We ran HEDAR on a set of synthetic images with a known number of small ellipses. Each trial image was of size (256×256) and contained n small ellipses of varied size, position and orientation. The plot shows how, as the number of ellipses increases, both the elliptical Hough transform and Ellifit struggle to correctly count the number of objects, whilst HEDAR is able to maintain a correct count for most values of n.

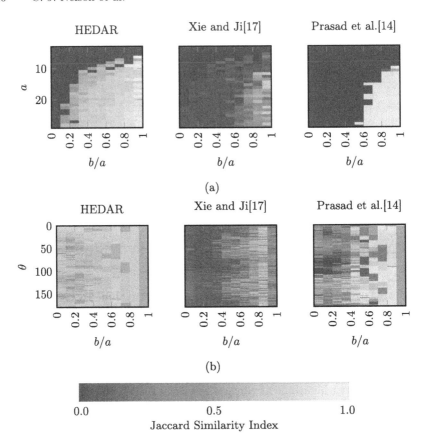

(a)

(b)

Jaccard Similarity Index

Fig. 5. Hilbert Edge Detection and Ranging detects ellipses with high accuracy of axis lengths and orientation. (a) Accuracy maps showing how HEDAR and other ellipse detection algorithms cope with a range of major axis lengths a and axes ratios a/b. (b) Accuracy maps showing how HEDAR and other ellipse detection algorithms cope with a range of orientations θ and axes ratios a/b. (Color figure online)

3.4   Details of Benchmarking Data

We have tested our new ellipse detection algorithm, Hilbert Edge Detection and Ranging, on five image datasets of fluorescent nuclei, all freely available from the Broad Bioimage Benchmark Collection [12]. Full details of the datasets can be retrieved online but are summarised in Table 1. All datasets are two dimensional.

Optimal parameters for HEDAR were manually determined using a single image from each subset of data, *e.g.* each z-focus in BBBC004, and those parameters were used to analyse each subset. Parameters are also shown in Table 1.

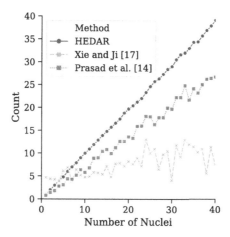

Fig. 6. Hilbert Edge Detection and Ranging can robustly detect the number of ellipses in an image. Mean detected number of ellipses against actual number of non-overlapping, fixed-size ellipses for HEDAR and comparator methods. Each point shows the mean of ten randomly generated synthetic scenarios.

Table 1. BBBC datasets used to evaluate HEDAR and compare to existing algorithms. Table indicates the number of images used from the dataset, whether the dataset was synthetic data (as opposed to real data) and whether the ground truth provided is a count of nuclei or a labelled, segmented image. Indicative HEDAR parameters (d_{max} and t) are also provided.

Accession	Citation	Number	Synthetic	Truth	Count	Labels	d_{max}	t
BBBC001v1	[3]	6		Human	✓		20	0.25
BBBC002v1	[3]	50		Human	✓		45	0.5
BBBC004v1	[16]	100	✓	Known	✓	✓	40	0.2
BBBC005v1	[1,12]	320[a]	✓	Known	✓	✓	30	0.6
BBBC039v1	[2]	50		Human	✓[b]	✓	60	0.5

[a] Although the whole dataset includes 9600 images of nuclei, only 20 randomly selected images per blur were used.
[b] Nuclei count has been extracted from labelled ground truth.

4   Conclusions

These results show that Hilbert Edge Detection and Ranging is a competitive approach for nuclei counting and detection. HEDAR performs at the same or better level of error than commonly used solutions and recent state-of-the-art solutions and gives results that agree with ground truth data with a high level of accuracy.

In this paper we have highlighted both the benefits of this method and those scenarios where this method may falter. We believe this information to be essential before users should use this method on their own data. We expect that further development of this method, such as optimisation for larger nuclei and adaptations to the Hilbert-edge ranging stages, will further improve the performance and the applicability of this technique.

Acknowledgements. During this work, CJN was supported by an EPSRC (UK) Doctoral Scholarship (EP/K502832/1). PTGJ is supported by an EPSRC (UK) Doctoral Scholarship (EP/M507854/1). The work in this paper was supported by an academic grant from The Royal Society (UK; RF080232).

References

1. Bray, M.A., Fraser, A.N., Hasaka, T.P., Carpenter, A.E.: Workflow and metrics for image quality control in large-scale high-content screens. J. Biomol. Screen. **17**(2), 266–274 (2012)
2. Caicedo, J.C., et al.: Evaluation of deep learning strategies for nucleus segmentation in fluorescence images. bioRxiv (2018)
3. Carpenter, A.E., et al.: CellProfiler: image analysis software for identifying and quantifying cell phenotypes. Genome Biol. **7**(10), 1–11 (2006)
4. Fornaciari, M., Prati, A., Cucchiara, R.: A fast and effective ellipse detector for embedded vision applications. Pattern Recogn. **47**(11), 3693–3708 (2014)
5. Gurcan, M.N., Pan, T., Shimada, H., Saltz, J.: Image analysis for neuroblastoma classification: segmentation of cell nuclei. In: International Conference of the IEEE Engineering in Medicine and Biology Society, pp. 4844–4847 (2006)
6. Jaccard, P.: The distribution of flora in the Alpine Zone. New Phytol. **11**(2), 37–50 (1912)
7. Jackson, P.T.G., Obara, B.: Avoiding over-detection: towards combined object detection and counting. In: Rutkowski, L., Korytkowski, M., Scherer, R., Tadeusiewicz, R., Zadeh, L.A., Zurada, J.M. (eds.) ICAISC 2017. LNCS (LNAI), vol. 10245, pp. 75–85. Springer, Cham (2017). https://doi.org/10.1007/978-3-319-59063-9_7
8. Jia, Q., Fan, X., Luo, Z., Song, L., Qiu, T.: A fast ellipse detector using projective invariant pruning. IEEE Trans. Image Process. **26**(8), 3665–3679 (2017)
9. Jung, C., Kim, C.: Segmenting clustered nuclei using H-minima transform-based marker extraction and contour parameterization. IEEE Trans. Biomed. Eng. **57**(10), 2600–2604 (2010)
10. Kong, J., et al.: Automated cell segmentation with 3D fluorescence microscopy images. In: IEEE International Symposium on Biomedical Imaging, pp. 1212–1215 (2015)
11. Li, G., et al.: 3D cell nuclei segmentation based on gradient flow tracking. BMC Cell Biol. **8**(1), 1–10 (2007)
12. Ljosa, V., Sokolnicki, K.L., Carpenter, A.E.: Annotated high-throughput microscopy image sets for validation. Nat. Methods **9**(7), 637–637 (2012)
13. Nandy, K., Chellappa, R., Kumar, A., Lockett, S.J.: Segmentation of nuclei from 3D microscopy images of tissue via graphcut optimization. IEEE J. Sel. Topics Signal Process. **10**(1), 140–150 (2016)
14. Prasad, D.K., Leung, M.K.H., Quek, C.: ElliFit: an unconstrained, non-iterative, least squares based geometric ellipse fitting method. Pattern Recogn. **46**(5), 1449–1465 (2013)

15. Prasad, D.K., Leung, M.K., Cho, S.Y.: Edge curvature and convexity based ellipse detection method. Pattern Recogn. **45**(9), 3204–3221 (2012)
16. Ruusuvuori, P., Lehmussola, A., Selinummi, J., Rajala, T., Huttunen, H., Yli-Harja, O.: Benchmark set of synthetic images for validating cell image analysis algorithms. In: European Signal Processing Conference, pp. 1–5 (2008)
17. Xie, Y., Ji, Q.: A new efficient ellipse detection method. In: Proceedings of the 16th International Conference on Pattern Recognition, vol. 2, pp. 957–960 (2002)
18. Xu, H., Lu, C., Berendt, R., Jha, N., Mandal, M.: Automatic nuclei detection based on generalized Laplacian of Gaussian filters. IEEE J. Biomed. Health Inform. **21**(3), 826–837 (2016)
19. Xu, H., Lu, C., Mandal, M.: An efficient technique for nuclei segmentation based on ellipse descriptor analysis and improved seed detection algorithm. IEEE J. Biomed. Health Inform. **18**(5), 1729–1741 (2014)

Author Index

Printed in the United States
By Bookmasters